DEVELOPMENTS IN SEDIMENTOLOGY 18A

COMPACTION OF COARSE-GRAINED SEDIMENTS, I

FURTHER TITLES IN THE SERIES

1. L.M.J.U. VAN STRAATEN, Editor
DELTAIC AND SHALLOW MARINE DEPOSITS

2. G.C. AMSTUTZ, Editor
SEDIMENTOLOGY AND ORE GENESIS

3. A.H. BOUMA and A. BROUWER, Editors
TURBIDITES

4. F.G. TICKELL
THE TECHNIQUES OF SEDIMENTARY MINERALOGY

5. J.C. INGLE Jr.
THE MOVEMENT OF BEACH SAND

6. L. VAN DER PLAS Jr.
THE IDENTIFICATION OF DETRITAL FELDSPARS

7. S. DZULYNSKI and E.K. WALTON
SEDIMENTARY FEATURES OF FLYSCH AND GREYWACKES

8. G. LARSEN and G.V. CHILINGAR, Editors
DIAGENESIS IN SEDIMENTS

9. G.V. CHILINGAR, H.J. BISSELL and R.W. FAIRBRIDGE, Editors
CARBONATE ROCKS

10. P. McL. D. DUFF, A. HALLAM and E.K. WALTON
CYCLIC SEDIMENTATION

11. C.C. REEVES Jr.
INTRODUCTION TO PALEOLIMNOLOGY

12. R.G.C. BATHURST
CARBONATE SEDIMENTS AND THEIR DIAGENESIS

13. A.A. MANTEN
SILURIAN REEFS OF GOTLAND

14. K.W. GLENNIE
DESERT SEDIMENTARY ENVIRONMENTS

15. C.E. WEAVER and L.D. POLLARD
THE CHEMISTRY OF CLAY MINERALS

16. H.H. RIEKE III and G.V. CHILINGARIAN
COMPACTION OF ARGILLACEOUS SEDIMENTS

17. M.D. PICARD and L.R. HIGH Jr.
SEDIMENTARY STRUCTURES OF EPHEMERAL STREAMS

DEVELOPMENTS IN SEDIMENTOLOGY 18A

COMPACTION OF COARSE-GRAINED SEDIMENTS, I

EDITED BY

GEORGE V. CHILINGARIAN

Petroleum Engineering Department, University of Southern California, Los Angeles, Calif. (U.S.A.)

AND

KARL H. WOLF

Department of Geology, Laurentian University, Sudbury, Ont. (Canada)

ELSEVIER SCIENTIFIC PUBLISHING COMPANY
Amsterdam   Oxford   New York   1975

ELSEVIER SCIENTIFIC PUBLISHING COMPANY
335 JAN VAN GALENSTRAAT
P.O. BOX 211, AMSTERDAM, THE NETHERLANDS

AMERICAN ELSEVIER PUBLISHING COMPANY, INC.
52 VANDERBILT AVENUE
NEW YORK, NEW YORK 10017

LIBRARY OF CONGRESS CARD NUMBER: 73-85220

ISBN 0-444-41152-6

WITH 233 ILLUSTRATIONS AND 45 TABLES

COPYRIGHT © 1975 BY ELSEVIER SCIENTIFIC PUBLISHING COMPANY, AMSTERDAM

ALL RIGHTS RESERVED. NO PART OF THIS PUBLICATION MAY BE REPRODUCED, STORED IN A RETRIEVAL SYSTEM, OR TRANSMITTED IN ANY FORM OR BY ANY MEANS, ELECTRONIC, MECHANICAL, PHOTOCOPYING, RECORDING, OR OTHERWISE, WITHOUT THE PRIOR WRITTEN PERMISSION OF THE PUBLISHER,
ELSEVIER SCIENTIFIC PUBLISHING COMPANY, JAN VAN GALENSTRAAT 335, AMSTERDAM

PRINTED IN THE NETHERLANDS

This book is dedicated to

H.D. HEDBERG and H. FÜCHTBAUER
for their very important contributions
to our knowledge of compaction

and to our inspirers
MANOUCHEHR EGHBAL
N.A. EREMENKO

as well as to
our wifes and children
without whose patience, consideration
and understanding our research would
not have been possible

CONTRIBUTORS

D.R. ALLEN	3140 Lama Avenue, Long Beach, Calif., U.S.A.
H.J. BISSELL	Geology Department, Brigham Young University, Provo, Utah, U.S.A.
L.F. BROWN Jr.	Bureau of Economic Geology, University of Texas, Austin, Texas, U.S.A.
G.V. CHILINGARIAN	Petroleum Engineering Department, University of Southern California, Los Angeles, Calif., U.S.A.
A.H. COOGAN	Kent State University, Kent, Ohio, U.S.A.
K. GRANT	Division of Applied Geomechanics, C.S.I.R.O., Syndal, Vict., Australia
O.G. INGLES	Division of Applied Geomechanics, C.S.I.R.O., Syndal, Vict., Australia
R.W. MANUS	Kent State University, Kent, Ohio, U.S.A.
F.G. MILLER	Department of Petroleum Engineering, School of Earth Sciences, Stanford University, Stanford, Calif., U.S.A.
R. RAGHAVAN	Department of Petroleum Engineering, School of Earth Sciences, Stanford University, Stanford, Calif., U.S.A.
K.H. WOLF	Department of Geology, Laurentian University, Sudbury, Ont., Canada

CONTENTS

CHAPTER 1. INTRODUCTION
G.V. Chilingarian, K.H. Wolf and D.R. Allen 1

CHAPTER 2. MECHANICS OF SAND COMPACTION
D.R. Allen and G.V. Chilingarian 43

CHAPTER 3. COMPACTION AND DIAGENESIS OF CARBONATE SANDS
A.H. Coogan and R.W. Manus 79

CHAPTER 4. SUBSIDENCE
H.J. Bissell and G.V. Chilingarian 167

CHAPTER 5. ROLE OF SEDIMENT COMPACTION IN DETERMINING
GEOMETRY AND DISTRIBUTION OF FLUVIAL AND DELTAIC SANDSTONES
(Case study: Pennsylvanian and Permian rocks of North-Central Texas)
L.F. Brown Jr. 247

CHAPTER 6. THE EFFECT OF COMPACTION ON VARIOUS PROPERTIES OF
COARSE-GRAINED SEDIMENTS
O.G. Ingles and K. Grant 293

CHAPTER 7. IDENTIFICATION OF SEDIMENTS — THEIR DEPOSITIONAL
ENVIRONMENT AND DEGREE OF COMPACTION — FROM WELL LOGS
D.R. Allen 349

CHAPTER 8. MATHEMATICAL ANALYSIS OF SAND COMPACTION
R. Raghavan and F.G. Miller 403

APPENDIX A. CONVERSION TABLE 525

REFERENCE INDEX 527

SUBJECT INDEX 537

Chapter 1

INTRODUCTION

G. V. CHILINGARIAN, K. H. WOLF and D. R. ALLEN

The various aspects of compaction of coarse-grained sediments are considered in this book. Inasmuch as sandstones constitute the more important type of coarse-grained sediments, a brief introduction on the general properties of sandstones is also presented here. Some basic sedimentological concepts, often oversimplified, are presented in this chapter in order to increase the utility of this book also to civil and petroleum engineers who are concerned with compaction and subsidence problems.[1]

Compaction of sediments is the process of volume reduction which can be expressed as either a percentage of the original voids present, or of the original bulk volume. This process affects chiefly unconsolidated sediments and occurs mainly during the diagenetic stage. Diagenesis includes all physicochemical, biochemical, and physical processes modifying sediments between deposition and lithification (including cementation), at low temperatures and pressure characteristic of surface and near-surface environments. Epigenesis (or catagenesis, which is a term preferred by many geologists) includes all processes at low temperature and pressure that affect sedimentary rocks after diagenesis and up to metamorphism. Some compaction also occurs during the epigenetic stage.

In compaction studies, it is of utmost importance to consider the fact that the time interval for both diagenesis and epigenesis varies from one extreme to another, because it may determine the degree of chemical changes that accompany compaction.

The factors which usually contribute to the formation of clastic sedimentary rocks and influence compaction include: (a) fragmentation of source rocks, (b) mode and distance of transport, (c) dispersion and sorting, (d) chemistry and energy of depositional environment, and (e) chemical alterations and cementation. The latter diagenetic processes are extremely important and are virtually inseparable facets of compaction and lithification. The formation of a clastic rock is not an inexorable process that proceeds through a specific series of stages; instead, it consists of stages that are often interrelated so closely as to be inseparable. In some instances some of these stages might not occur at all.

Compaction is also a very important process in rocks that are easily soluble in groundwater, such as carbonate and evaporitic rocks, and can be of great significance for both petroleum and metalliferous exploration geologists.

---

[1] Supplemental information is presented in Introduction chapter of Vol. II.

Chapter 2 is devoted to mechanics of sand compaction. The reaction of sands or similar clastics to forces acting upon them during and after deposition, primarily the increasing overburden load which changes loose sediments into indurated rocks, is discussed in that chapter.

The mechanical compaction of sands is an important and, perhaps, underemphasized part of the history of sandstones, which comprise about 14—37% of all sediments (Pettijohn et al., 1972, p.4, variation in percentage according to different estimates). Compaction assumes great importance because of its effect on the rock voids that contain fluids and minerals vital to life itself.

Sand compaction studies in the field of man-induced subsidence, coupled with theories of the role of fluid-pressure support during compaction and consolidation, have enhanced our understanding of the mechanical processes of compaction.

DEFINITION OF SAND

The term "sand" is commonly used to denote texture and gives a broad idea of the size, arrangement and shape of the fragmental components. It is mainly used as a size term by most earth scientists. Even when restricting the term sand to a particular grain size, there are as many "standards" as there are disciplines interested in sands. One particular classification, however, has tended to rise above the others and to become accepted into a more or less common usage. With reference to clastic size grading, in the geological sciences the scale established by Wentworth in 1922 has achieved this status. His classification, which was subsequently modified somewhat, was based upon earlier work by J. A. Udden. Figure 1-1 compares several of the size scales commonly used today by various organizations in the U.S.A. The sand size ranges are often rearranged to fit the needs of a particular group. A comparison of the prevailing carbonate particle size scales are presented in Table 1-I.

A sand can be defined as an unconsolidated sedimentary deposit of clastic particles that vary in size between 1/16 mm and 2 mm and are derived from a source rock through mechanical means or as a result of chemical weathering, which is also active in breaking down of the host rock. This is one of the most commonly accepted definitions of sand. A sandstone is the consolidated or cemented rock equivalent of a sand.

From the geological point of view, the study of coarse-grained sediments has become a specialized field of investigation. Although the available information has been "snow-balling" during the past few years, several excellent books are available that can serve as a starting point, e.g., Folk (1968), Blatt et al. (1972) and Pettijohn et al. (1972).

INTRODUCTION

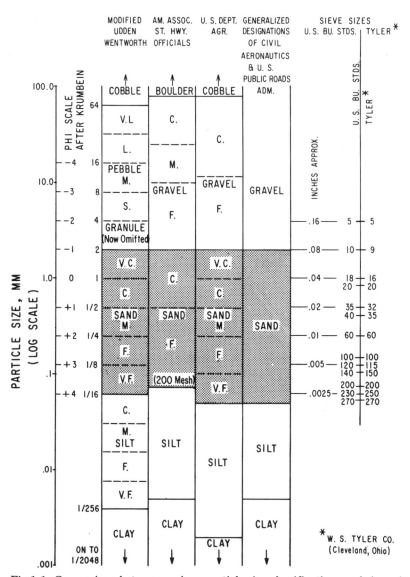

Fig.1-1. Comparison between various particle size classifications and sieve sizes.

TEXTURE

The texture represents the total characteristics of the particles of a sediment or a rock. These include grain size, shape, roundness, sphericity, distribution, packing, fabric, pore geometry, and type and degree of

TABLE 1-I

Particle size scales for carbonates (After Bissell and Chilingar, 1967, p.103)

Size in mm	Krynine (1948)	Howell (1922)	DeFord (1946)	Pettijohn (1949)	Mollazal (1961)	Folk (1959, 1962)	Bissell and Chilingar (1967)
—4.0—	Coarsely crystalline (grained)	Macro-crystalline	Mega-grained	Calci-rudite	Macro-crystalline — Coarse	Extremely coarsely crystalline	Macrocrystalline (macrograined) — Very coarse
—2.0—	Medium crystalline (grained)				Medium / Fine	Very coarsely crystalline	Coarse
—1.0—		Meso-crystalline	Meso-grained	Calc-arenite	Mesocrystalline — Coarse	Coarsely crystalline	Coarse
—0.5—					Medium	Medium crystalline	Mesocrystalline (mesograined) — Medium
—0.25—							Fine
—0.1—		Micro-crystalline	Pauro-grained	Calci-lutite	Fine	Fine	Very fine
—0.062—	Finely crystalline (grained)				Finely crystalline	Finely crystalline	Finely crystalline (grained)
—0.05—							
—0.025—						Very finely crystalline	
—0.016—			Micro-grained		Micro-crystalline		Aphanic — Microcrystalline (grained)
—0.01—		Crypto-crystalline				Aphano-crystalline	
—0.004—			Crypto-grained		Crypto-crystalline		Crypto-crystalline (grained)
—0.0025—							
—0.001—							

Phaneric / Aphanic divisions apply to DeFord (1946) and Pettijohn (1949) columns.

INTRODUCTION

cementation. Most of these characteristics are termed *index properties* by soils engineers. Many of the measurement techniques and practical uses of these properties were developed in this discipline. Although composition affects texture, it is not a textural property. Textural properties are very important in the study of sediment compaction because they are indicative of the degree of compaction and magnitude of forces which acted on that sediment in the past. Various properties also may indicate the future of the sediment if the magnitude and direction of forces acting upon it are altered.

Particle size

The particle size is perhaps the most commonly measured characteristic of a clastic, although not necessarily the most important. Size measurement is easily accomplished by sieving if the sand is unconsolidated or easily friable.

Numerous publications are available on sieving techniques, as exemplified by the excellent treatment by Folk (1968). Where a mixture of clay, silt, and sand are present, it is necessary to separate them. Size analysis techniques for clays, silts, and pebble-sized grains are also presented by Folk (1968). A large amount of sedimentological literature is available on (1) the reliabilities of the various techniques of size determination (sieving versus thin-section size analysis, sieving versus settling, etc.), (2) the means of graphical presentation of size analyses and selection of best approach, and (3) the interpretation of the size analyses results (e.g., as environmental indicators). (See, for example, Krumbein and Pettijohn, 1938; Pettijohn, 1957; Krumbein and Sloss, 1963; Griffiths, 1967; and Pettijohn et al., 1972.)

The Wentworth scale is a geometric scale with each division differing from the previous one by a ratio of 1/2. Other scales and sieve sizes that are often used are compared with the Wentworth scale in Fig.1-1. Many other scales exist with most of them agreeing fairly closely with Wentworth's, insofar as the gross limits of sand size are concerned. Screen size is a measure of the particle dimensions that will either pass through or be retained by an opening of a specified size. The two screen sizes compared on Fig.1-1 are quite similar and both are widely used. The screen sizes most important to a sedimentologist are those which correspond to divisions in the Wentworth scale. In soils engineering, the weight fractions retained by 20, 40, 60, 140, and 200 screen sizes are often required. Material passing through a particular screen opening is called the minus fraction, whereas the portion retained is termed the plus fraction.

The phi scale, first proposed by Krumbein in 1934, is now commonly used in sedimentological studies. It represents the negative logarithm (base 2) of the grain size in mm, e.g., 4 mm = -2ϕ and 2 mm = -1ϕ. The class sizes in Udden—Wentworth scale are all powers of 2 in this convenient log-scale expression. Inasmuch as phi intervals are equal, ordinary arithmetic paper can be used in plotting curves.

Wentworth (1933) and Lane (1938) examined the limits of grades in terms of the physical properties involved in grain transportation. They showed that given class limits agreed well with certain distinctions between suspension and traction loads. A natural separation in nature is not really distinct, however, until the colloidal sizes are reached at about 0.002 mm, which is the clay—silt boundary used by the United States Department of Agriculture.

Sizes of particles below 0.05 mm are generally measured in the laboratory by a hydrometric method based on Stoke's equation for the velocity of a freely-falling sphere (pipette analysis, see Fig.1-2).

Size distribution and classification

The graphical presentation of grain-size analysis is termed the size frequency distribution, which can be simply defined as the arrangement of the numerical data according to size. The two commonly used presentations are the histogram and the cumulative frequency curve. A typical histogram portrays the quantity of a size component on the y-axis and the particle size of the component on the x-axis (Fig.1-3A). The width for the class interval may be based on a grade size in the Wentworth scale or on a series of convenient sieve sizes. The data presented in Fig.1-3A are shown as a cumulative curve (percent finer) on Fig.1-3B. This sand can be classified as a well-

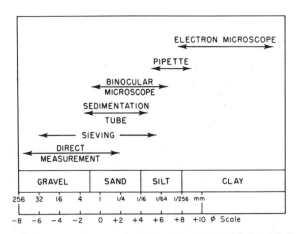

Fig.1-2. Range of applicability of different techniques of size analysis. (After Pettijohn et al., 1972, fig.3-1, p.70.)

INTRODUCTION 7

sorted sand even though it contains both a 5% fraction larger and a 10% fraction smaller than sand.

The smooth frequency curves, which are related mathematically with histograms, show the variation of the dependent variable as a continuous function of the independent variable. A common, but sometimes not very accurate, manner of drawing frequency curves is simply to superimpose a smoothed curve over the histogram bars. The frequency curve of a beach sand is presented in Fig.1-4. As shown in this figure, the arithmetic mean (the largest value) is not as good a measure of the central tendency as the other averages (Krumbein and Pettijohn, 1938, p.246; also see Table 1-IV, p.14).

Many variations in size distribution occur in nature. For example, Fig.1-5 compares a uniformly-sorted silty sand (curve *1*), a coarse- to medium-grained silt with one strongly predominant grain size (curve *2*), and a bimodal or skip-graded silty sand (curve *3*). This type of portrayal is most often used by soils engineers.

The classification of a mixture of particle sizes into a name term such as sandy silt and silty sand is difficult; many different classification schemes have been offered in the literature and are used for different purposes. Most of these use some type of component chart as illustrated in Fig.1-6—1-9. Names applied to the textural classes delineated in Fig.1-6 and 1-7 are presented in Tables 1-II and 1-III, respectively. Without such diagrams, the interpretation of names assigned to different mixtures by various authors is very difficult and uncertain.

Mode, kurtosis and skewness

A characteristic of natural sorting that appears when examining a histogram is the mode of the sand grain size, i.e., the predominance of one

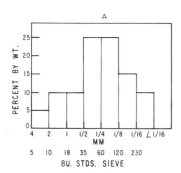
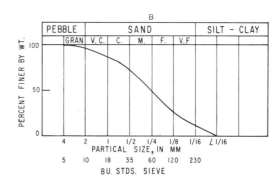

Fig.1-3. A. Conventional histogram (percentage of total sample weight for each class interval in mm and sieve sizes corresponding to Wentworth scale). B. Corresponding cumulative curve (% finer).

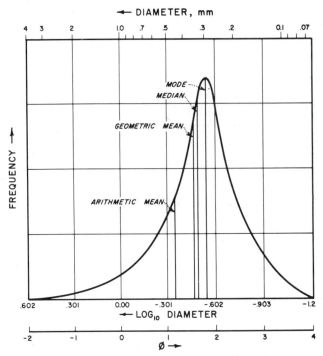

Fig.1-4. Frequency curve of a beach sand plotted with $\log_{10}$ diameter as an independent variable. The diameter scale and phi scale are added for comparison. Arithmetic mean = 0.440 mm; geometric mean = 0.352; logarithmic mean = 1.561 (the diameter equivalent of this value, 0.340 mm, is the geometric mean); median diameter = 0.320; modal diameter = 0.300 mm. (After Krumbein and Pettijohn, 1938, fig.114, p.246.)

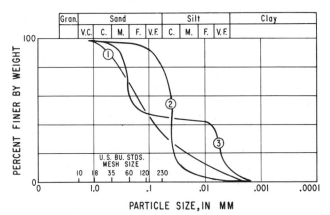

Fig.1-5. Cumulative curves (percent finer by weight) illustrating three common grain-size distributions: 1 = uniformly-sorted silty sand; 2 = coarse- to medium-grained silt with one strongly predominant grain size; 3 = bimodal or skip-graded silty sand.

INTRODUCTION

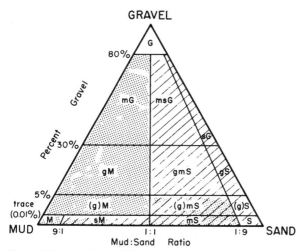

Fig.1-6. Textural classification of gravel-bearing sediments and sedimentary rocks. Classes are defined by the percentage of gravel (> 2 mm) and the ratio of sand (0.0625—2.0 mm) to mud (< 0.0625 mm). Modal size of gravel should be specified throughout, whereas the modal size of sand should be specified only in cross-hatched area. In stippled area, where practicable the terms "silty" or "clayey" should be substituted for "muddy". G = gravel; sG = sandy gravel; msG = muddy sandy gravel; mG = muddy gravel; gS = gravelly sand; gmS = gravelly muddy sand; gM = gravelly mud; (g)S = slightly gravelly sand; (g)mS = slightly gravelly muddy sand; (g)M = slightly gravelly mud; S = sand; mS = muddy sand; sM = sandy mud; and M = mud. (After Folk et al., 1970, fig.1, p.943.)

grain size over another. Figure 1-10 illustrates two common distribution modes: (A) well-sorted, unimodal grouping, and (B) bimodal grouping containing two distinct grain sizes. Bimodal distribution is commonly found in nature and is usually represented by one predominantly large grain size and a much smaller size which represents infilling between the large interstices. Curve 3 in Fig.1-5 represents a bimodal distribution because of the lack of material between medium sand and very fine silt, whereas curve 2 represents grain size distribution which is strongly unimodal and would exhibit a strong "peak" or kurtosis if plotted on a histogram.

Skewness is a measure of the symmetry of the grain-size distribution. Figures 1-10C and 1-10D illustrate positively and negatively skewed distributions.

The methods for quantifying both kurtosis and skewness are presented in Table 1-IV.

Particle shape, roundness and sphericity

Particle shape and roundness are two very significant factors in the mechanical compaction of sands. These two factors affect both bridging (the

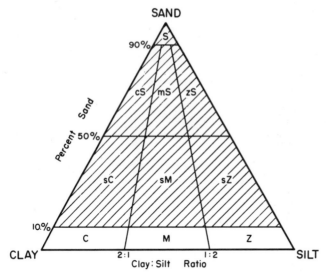

Fig.1-7. Textural classification of gravel-free sediments and sedimentary rocks. This diagram is an expansion of the bottom tier of Fig.1-6. Classes are defined by the percentage of sand (0.0625—2.0 mm) and the ratio of silt (0.0039—0.0625 mm) to clay (< 0.0039 mm). Modal size of sand must be specified for cross-hatched area. S = sand; zS = silty sand; mS = muddy sand; cS = clayey sand; sZ = sandy silt; sM = sandy mud; sC = sandy clay; Z = silt; M = mud; C = clay. (After Folk et al., 1970, fig.2, p.944.)

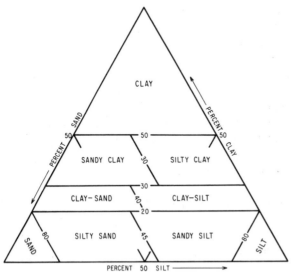

Fig.1-8. Three-component classification chart for clay—silt—sand mixtures. (After U.S. Department of Agriculture.)

INTRODUCTION

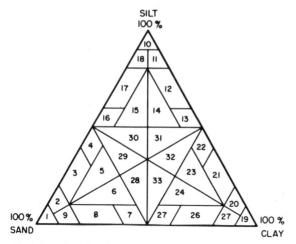

Fig.1-9. Classification scheme for clayey—silty—sandy rocks. (After Teodorovich, 1938, fig.1, p.70.) *1* = sand; *2* = slightly silty sand; *3* = silty sand; *4* = highly silty sand; *5* = clayey—silty sand; *6* = silty—clayey sand; *7* = highly clayey sand; *8* = clayey sand; *9* = slightly clayey sand; *10* = silt; *11* = slightly clayey silt; *12* = clayey silt; *13* = highly clayey silt; *14* = sandy—clayey silt; *15* = clayey—sandy silt; *16* = highly sandy silt; *17* = sandy silt; *18* = slightly sandy silt; *19* = clay; *20* = slightly silty clay; *21* = silty clay; *22* = highly silty clay; *23* = sandy—silty clay; *24* = silty—sandy clay; *25* = highly sandy clay; *26* = sandy clay; *27* = slightly sandy clay; *28* = clayey sandy loam; *29* = silty sandy loam; *30* = sandy "hypoaleurite"; *31* = clayey "hypoaleurite"; *32* = silty loam; and *33* = sandy loam. *1—9*: sand content > 50%; *10—18*: silt content > 50%; *19—27*: clay content > 50%.
In a compound term such as "clayey—silty" in clayey—silty sand (sand = 50—80%; silt = 10—40%; clay = 10—25%), the second term ("silty" in this case) designates the predominating component (silt content > clay content in this case). In sandy loam (sand = $33\frac{1}{3}$—50%; silt = 0—50%; clay = 0—50%), loam (sand = 0—50%; silt = 0—50%; clay = $33\frac{1}{3}$—50%), and hypoaleurite (sand = 0—50%; silt = $33\frac{1}{3}$—50%; clay = 0—50%), the content of none of the three components exceeds 50%.

ability of grains to resist deformation by sliding and rearrangement) and porosity. Particles have a certain geometric form or shape (flat, disk-shaped, tabular, bladelike, etc.) with varying degree of roundness. Roundness is an independent variable, not basically related to form or shape. A particle having a particular form or shape can be rounded to varying degrees.

Roundness is a measure of the sharpness of the particle edges, regardless of shape. One accepted method for determining roundness is to view the particles two dimensionally, and determine the ratio of the average radius of curvature of a particle's corners to the radius of the largest circle that can be inscribed in that particle. With the exception of the circle, this number is always less than one. Pettijohn (1957, pp.58—59) defined five roundness grades using a geometric progression based on the above definition: (1) angular, (2) subangular, (3) subrounded, (4) rounded, and (5) well-rounded.

TABLE 1-II

Names applied to the textural classes delineated in Fig.1-6 (After Folk et al., 1970, table 2, p.948)*

Symbol	Major textural class		Examples of usage — Consolidated equivalents only
	unconsolidated	consolidated	
G	gravel	conglomerate	granule conglomerate
sG	sandy gravel	sandy conglomerate	medium sandy boulder conglomerate
msG	muddy sandy gravel	muddy sandy conglomerate	muddy coarse sandy pebble conglomerate
mG	muddy gravel	muddy conglomerate	clayey pebble conglomerate
gS	gravelly sand	conglomeratic sandstone	granular very fine sandstone
gmS	gravelly muddy sand	conglomeratic muddy sandstone	bouldery muddy coarse sandstone
gM	gravelly mud	conglomeratic mudstone	pebbly siltstone
(g)S	slightly gravelly sand	slightly conglomeratic sandstone	slightly pebbly coarse sandstone
(g)mS	slightly gravelly muddy sand	slightly conglomeratic muddy sandstone	slightly cobbly silty fine sandstone
(g)M	slightly gravelly mud	slightly conglomeratic mudstone	slightly cobbly siltstone
S	sand	sandstone	well-sorted fine sandstone
mS	muddy sand	muddy sandstone	moderately well-sorted silty very fine sandstone
sM	sandy mud	sandy mudstone	fine sandy claystone
M	mud	mudstone	siltstone

*Modal size is specified for the italicized terms as shown in the list of examples. If pebbles, etc., are predominantly irregular, "breccia" should be substituted for "conglomerate" and "gravel".

Figures 1-11 and 1-12 show roundness images and classes. Current techniques for estimating roundness involve the use of a microscope examination of a number of grains and visual comparison with a standard chart. For obtaining reproducible results, the reader should refer to Griffiths (1967).

The degree of roundness commonly varies with size. Larger diameter sand or gravel particles are usually more rounded than the smaller ones. Maturity and degree of weathering affect this relationship. Freshly-broken fragments, which tend to be angular near the source, assume a greater degree of roundness as a result of weathering and abrasion during transportation.

Sphericity is sometimes confused with roundness. Although they are

INTRODUCTION

TABLE 1-III

Names applied to the textural classes delineated in Fig.1-7 (After Folk et al., 1970, table 3, p.949)*

Symbol	Major textural class		Examples of usage — consolidated equivalents only
	unconsolidated	consolidated	
S	*sand*	*sandstone*	well-sorted fine sandstone
zS	silty *sand*	silty *sandstone*	poorly sorted silty medium sandstone
mS	muddy *sand*	muddy *sandstone*	muddy bimodal fine and very fine sandstone
cS	clayey *sand*	clayey *sandstone*	clayey medium sandstone
sZ	*sandy* silt	*sandy* siltstone	fine sandy siltstone
sM	*sandy* mud	*sandy* mudstone	coarse sandy mudstone
sC	*sandy* clay	*sandy* claystone	fine sandy claystone
Z	silt	siltstone	siltstone
M	mud	mudstone	mudstone
C	clay	claystone	claystone

*Modal size is specified for the italicized terms as shown on the list of examples. Sorting category is specified, for textural classes S and zS only.

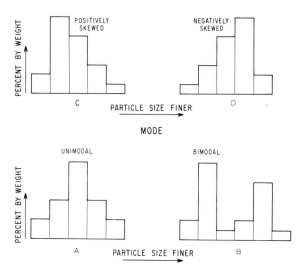

Fig.1-10. Histograms illustrating grain-size distributions. A. Unimodal. B. Bimodal. C. Positively-skewed. D. Negatively-skewed.

TABLE 1-IV

Summary of statistical measures used in grain-size analysis (Modified after Rieke et al., 1972, p.353. See Folk and Ward; 1957, and McCammon, 1962)

Measures of central tendency:
(1) Median (D_{50}) is the diameter which is larger than half of the grains in the distribution and smaller than the other half (i.e., the middlemost member of the distribution). It reflects the overall grain size as influenced by the origin of the rock and any subsequent alteration. It may be a very misleading value, however.
(2) Mean (D_M) is the measure of the overall average grain size:

$$D_M = \frac{D_5 + D_{15} + D_{25} \ldots + D_{85} + D_{95}}{10} \qquad (a)$$

or:

$$D_M = \frac{D_{16} + D_{50} + D_{84}}{3} \qquad (b)$$

(3) Mode (D_m) is the most frequently occurring grain diameter (peak of frequency curve). If two dominant grain sizes are present, then the frequency curve is bimodal.

Measure of dispersion:
Sorting (S_p) is a standard deviation measure of the grain sizes in a sample (Folk and Ward, 1957):

$$S_p = \frac{(D_{84} - D_{16})}{4} + \frac{(D_{95} - D_5)}{6.6} \qquad (c)$$

Measure of asymmetry:
Skewness (Sk_p) measures the non-normality of a grain-size distribution (Folk and Ward, 1957):

$$Sk_p = \frac{(D_{84} + D_{16} - 2D_{50})}{2(D_{84} - D_{16})} + \frac{(D_{95} + D_5 - 2D_{50})}{2(D_{95} - D_5)} \qquad (d)$$

A symmetrical curve has a Sk_p value of 0; Sk_p varies between the following limits: $-1 \leq Sk_p \leq 1$. Positive values indicate that the curve has a tail in the small grains, whereas negative values indicate that the curve is skewed toward the larger grains.

Measure of peakedness:
Kurtosis (K_p) is a measure of the degree of peakedness, that is, the ratio between the spread of the grain diameters in the tails and the spread of the grain diameters in the central portion of the distribution (Folk and Ward, 1957):

$$K_p = \frac{(D_{95} - D_5)}{2.44(D_{75} - D_{25})} \qquad (e)$$

INTRODUCTION

TABLE 1-IV — *continued*

Normal curves have a K_p of 1, whereas platykurtic (bimodal) distributions may have a K_p value as low as 0.6. A curve represented by a high narrow peak (very leptokurtic) may have K_p values ranging from 1.5 to 3.

D_n is the grain diameter in phi units at the n-th percentile.

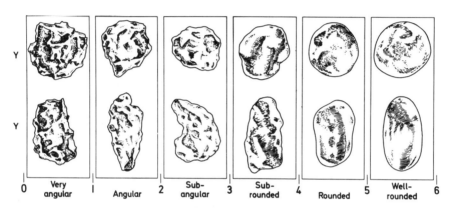

Fig.1-11. Roundness images and classes. Columns show grains of similar roundness but different sphericity. (After Powers, 1953, fig.1, as redrawn by Pettijohn et al., 1972, fig.A-2, p.586.)

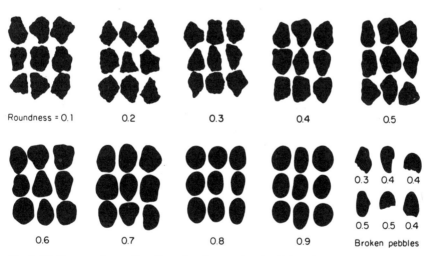

Fig.1-12. Images for estimating visual roundness. (After Krumbein, 1941, p.68; also in Griffiths, 1967, fig.6.3, p.113.)

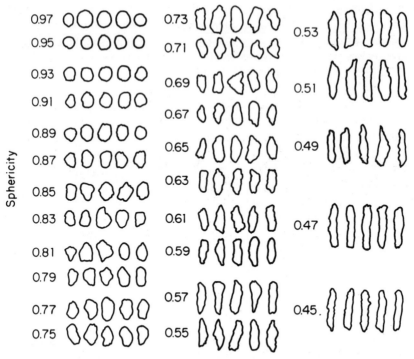

Fig.1-13. Images for estimating visual sphericity. (After Rittenhouse, 1943, p.80; also in Griffiths, 1967, fig.6.2, p.112.)

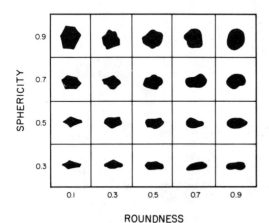

Fig.1-14. Visual chart for estimating roundness and sphericity of sand grains. (After Krumbein and Sloss, 1951, p.81; also in Krumbein and Sloss, 1963, fig.4-10, p.111.)

related to a certain degree, roundness is primarily a measurement of the angularity of a particle's corners, whereas sphericity is a measure of the degree the shape of the particle approaches that of a sphere. Images for estimating visual sphericity are given in Fig.1-13. True sphericity was defined by Wadell (1934) as s/S, where s is the surface area of a sphere of the same volume as the particle and S is the actual surface area of the solid. A capsule-shaped object could have a roundness factor of unity, whereas if its surface area were compared to that of a sphere of the same volume, the ratio would be far less than unity. A more practical formula for determining sphericity is d_n/D_s, where d_n is the true nominal diameter (i.e., the diameter of a sphere of the same volume as the particle) and D_s is the diameter of circumscribing sphere (Wadell, 1934).

Figure 1-14 is a combined sphericity—roundness chart for sands, in which sphericity is related to the ratio of breadth to length of the images, whereas the roundness is a measure of the curvatures of the image edges.

As pointed out by Krumbein and Pettijohn (1938, p.278), the factors that control shape and roundness include (a) original shape of the fragment, (b) durability of the material, (c) structure of the fragment (e.g., cleavage or bedding), (d) nature of geologic agent, (e) nature of the action to which the fragment is subjected and rigor of that action, and (f) time or distance through which the action is extended.

Sphericity factors, i.e., sphericity-mean, sphericity standard of deviation, sphericity-skewness, and sphericity-kurtosis, are more useful for distinguishing various depositional environments than are roundness factors. A real variation in the mean sphericity and mean roundness of a sediment is caused by sorting and abrasion, respectively.

SAND COMPOSITION AND SANDSTONE CLASSIFICATIONS

The mineralogic and chemical composition of terrigenous sediments vary from one extreme to another depending on numerous factors, which include source rock, climate, transportation mechanism, geologic time, depositional environment, diagenesis and epigenesis (for supplemental information see Wolf and Chilingarian, Vol. II, Ch. 3). The four main sources of sand are (1) plutonic rocks, which supply mostly quartz and feldspar; (2) terrigenous sediments, which supply mostly quartz and rock fragments; (3) volcanic rocks, which supply mostly rock fragments and glass; and (4) carbonate rocks, which supply carbonate debris (for more detailed discussions, see Folk et al., 1970). The depositional environments in which sediments form are very important in determining (a) regional distribution; size, shape, and thickness of the sandstone bodies; (b) stratigraphic relationships, (c) grain-size distribution, (d) mineralogic composition, and (e) depositional structures. The following are the most important environmental sandstone types: (1) fluvial and alluvial, (2) beach, (3) bar, (4) dune, (5) barrier island, (6) chenier plain, (7) desert, (8) intermontane basin, (9) lake, (10) glacial, (11) volcanic terrain, (12) marine shelf, (13) tidal, (14) continental slope (e.g., turbidites), and (15) bathyal.

Pure quartz sands are common and quartz grains make up a sizeable portion of many sands, especially in older, mature rocks that may have been reworked several times (multi-cycle). Quartz is a hard, durable mineral, that withstands weathering far better than most of the minerals with which it may have been associated in the original igneous source rock. The absence of heavy minerals in sands could be due to the severity of weathering and the length of exposure to an oxidizing or leaching environment regardless of age and/or intrastratal solution. Abundance of feldspars in many older sands could indicate that the sediments were exposed to rapid erosion, a short or rapid transport and deposition, and little or no reworking.

Another possible explanation is the predominance of mechanical processes over the chemical because the prevailing climate (e.g., hot and dry desert climates, cold and dry climates) was conducive mainly to mechanical weathering with little chemical weathering (see Folk, 1968).

Inasmuch as the softer, less durable rocks are often reduced to smaller sizes and constitute smaller fractions of the total sand composition, their presence is reflected in the textural features of sands and the configuration of cumulative curves and histograms.

Many igneous rocks are rich in feldspar (about 50%), whereas Holocene and Pleistocene sands average only about 19% feldspar (Pettijohn, 1957, p.123). Many studies show a general decrease in feldspar content with age, indicating a gradual loss of the softer or more easily destroyed minerals when compared with the more durable quartz components.

Arkoses and greywackes are rarely composed only of mineral grains with associated intergranular pore space. The pore space usually contains some silts, clays and various chemical cements. The interparticle material may be no more than a fine-grained expression of the larger mineral grains, which may or may not have been mechanically and/or chemically altered into other forms such as clay. Orthoquartzites, on the other hand, usually contain only cement.

If the interstitial material content exceeds about 35%, it is termed matrix and becomes a more visible portion of the rock. There are numerous sand classifications, most of which are basically methods of determining the relative proportions and composition of fragments and matrix.

The descriptive adjectives referring to composition include: arenaceous, argillaceous, calcareous, feldspathic, ferruginous, quartzose, and quartzitic.

These terms are then combined with a textural term into an expression such as "arenaceous siltstone".

The names of certain compositional mixtures have become firmly established in the literature and convey a certain meaning as to the make-up of sandstone. They can be generally grouped into five main divisions:

(1) Arkoses or arkosic sands. An arkose is predominantly a product of erosion and transportation of the weathering products of granites and gneisses; it contains feldspars (25—30%) in excess of rock fragments (up to 10%), e.g., chert, slate, and schist. Arkoses always contain some quartz and, perhaps, some dark minerals (mafics). A minor amount of fine detrital or clay material may be present in the pores of arkoses that may be either friable or well cemented.

(2) Greywackes (Phyllarenite of Folk, 1968; Folk et al., 1970). In sandstone classification, the term greywacke is considered by some geologists to be one of the most misused terms in sedimentary petrology. Virtually every writer seems to have a slightly different definition for greywacke (German "Grauewacke"). It was originally applied to the dark and tough Paleozoic Kulm sandstones, Harz Mountains, Germany, by G. S. Otto Lasius in 1789. In 1940, Krynine defined graywacke as a "clastic rock containing a substantial amount (20% or over) of dark rock fragments or dark-coloured ferromagnesian minerals". In 1945, Krynine stressed only the rock fragments: "chert, slate, schists, phyllites", etc. Folk (1954) followed this usage. In 1968, however, Folk replaced the term "graywacke" by "lithic arenite". The latter was defined by Williams et al. (1954, p.304) as a sandstone with less than 10% matrix, of which unstable, fine-grained rock particles constitute an important part. The term "phyllarenite", proposed by Folk (1968, p.131), is essentially equivalent to "lithic arenite", and is characterized by an abundance of detrital rock fragments (low-grade, metamorphic pelitic rocks, i.e., slate, phyllite, and mica schist; see Fig.1-15).

Williams et al. (1954, p.290) defined "wacke" as a sandstone containing 10% or more of argillaceous matrix materials. Pettijohn et al. (1972, p.159) retain the term graywacke for those sandstones containing significant matrix content (above 15%) and recognize two main classes, i.e., "lithic graywacke" and "feldspathic graywacke" depending on whether detrital rock particles or detrital feldspar dominate the rock fragment fraction.

(3) Lithic sands or sandstones. When the content of rock fragments (e.g., schists, slates, and cherts) exceeds that of feldspar, the sands are classified as lithics.

(4) Pure quartz sands or orthoquartzites. The orthoquartzites are composed mainly of quartz grains (90% or more), the remaining 10% being either chert or other rock fragments.

(5) Calclitharenites. Calclitharenites constitute the fifth common group of sands.

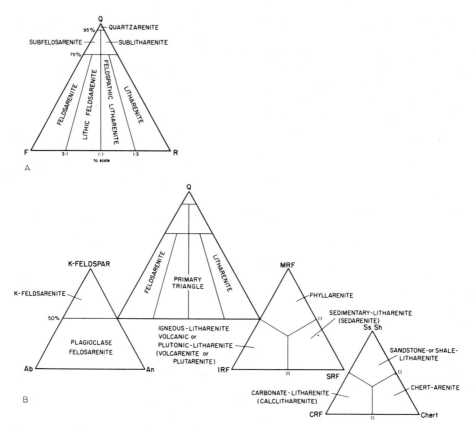

Fig.1-15. Sandstone classification of Folk et al. (1970, fig.8, p.955).
A. Primary arenite triangle. Q = monocrystalline and polycrystalline quartz (excluding chert); F = monocrystalline feldspar; R = rock fragments (igneous, metamorphic, and sedimentary, including chert).
B. Examples of second and third order arenite triangles that may be devised to refine the nomenclature given in the primary triangle. MRF = metamorphic rock fragments; IRF = igneous rock fragments; SRF = sedimentary rock fragments; CRF = carbonate rock fragments; Ab = albite; An = anorthite.

They are sands composed of over 50% carbonate detritus of sand grain size.

There are other various distinctive types of sands, such as volcanic tuffs, containing various vitreous fragments and shards, and gypsum sands, such as those found in New Mexico. Neither of these latter two groups are considered to be of importance because of limited distribution. The presence of some pyroclastic material, however, is pronounced in many sandstones (Teodorovich, 1965).

Many classification schemes have been proposed for sandstones and are in use at present. Each individual investigator must decide for himself as to which scheme and nomenclature suit his particular problem best. He may use one approach in one particular case and another in other instances. Three classification schemes are presented in Fig.1-15—1-17.

The term "litharenite" (a contraction of the term "lithic arenite") in Fig.1-16 was defined by McBride (1963) as a sandstone containing over 25% rock particles and less than 10% feldspar. "Feldspathic litharenite" was defined by McBride (1963, p.667) as a lithic arenite containing over 10% feldspar. He also defined a useful term "lithic subarkose" as a sandstone or arenite containing abundant, subequal amounts of rock fragments (> 10% but < 25%) and detrital feldspar (Fig.1-16). "Sublitharenite" of McBride (1963, p.667) contains 5—25% rock fragments, 0—10% feldspar, and 65—95% quartz.

Folk et al. (1970) classification includes not only mineralogic composition depicting source rocks but also textural "maturity concept", i.e., one refers to an immature, submature, mature, or supermature arkose, quartzite, or greywacke (Folk's phyllarenite, 1968, 1970), depending on the presence or absence of matrix, sorting, rounding, and mineralogic composition determined by the sedimentary environment. Textural maturity of a sediment is attained by the removal of clay-sized particles and the sorting and rounding of large grains (Fig.1-18). All three processes advance at different rates (Weller, 1960, p.91). The changes in relative mineralogic maturity in sandstones are presented in Fig.1-19. The relative mineralogic maturity of sediment is indicated by the amounts of feldspar and ferromagnesian minerals

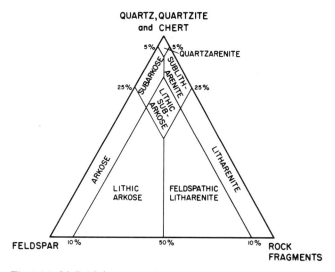

Fig.1-16. McBride's (1963, fig.1, p.667) classification of sandstones.

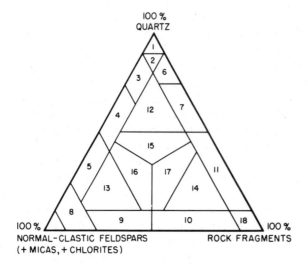

Fig.1-17. Classification of sandstones on the basis of mineralogic composition. (A) family of quartz sandstones (*1, 2, 3*); (B) arkoses (*4, 5*); (C) two-component mineral—petroclastic sandstones (*6, 7, 8, 9, 10, 11*) — the latter family can be subdivided into (a) predominantly quartz sandstones (*6, 7*), (b) predominantly feldspathic sandstones (*8, 9*), and (c) predominantly petroclastic sandstones (*10, 11*); (D) three-component mineral—petroclastic sandstones with absolute predominance of one of the components (*12, 13, 14*); (E) three-component mineral—petroclastic sandstones with relative predominance of one of the components (*15, 16, 17*); (F) ultrapetroclastic sandstones (*18*). (After Teodorovich, 1967, fig.1, p.76.)

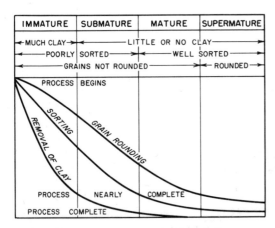

Fig.1-18. Diagram illustrating the concept of textural maturity of Folk (1951a, 1963). (After Weller, 1960, fig.26, p.91.)

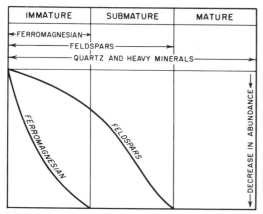

Fig.1-19. Diagram demonstrating the changes in relative mineralogic maturity in sandstones. (After Weller, 1960, fig.139, p.341.)

remaining in them. It is important to mention here that the compressibility of sands increases with increasing content of feldspars. The changes in textural and mineralogic maturity may be independent of each other depending on numerous environmental factors.

GENETIC CLASSIFICATION OF SANDS ON BASIS OF GRAIN-SIZE DISTRIBUTION

The whole topic of genetic classification of sands on the basis of grain-size distribution is complex, controversial, and in a state of flux, as rapid changes in ideas are occurring. Yet, this subject is briefly discussed here, using selected references, because sands of different depositional origin compact differently, probably largely because of differences in grain-size distribution and mineralogic composition.

Rukhin (1947, in: Rukhin, 1969, p.491) prepared a genetic classification diagram for recent sands on plotting average grain size versus sorting coefficient (Fig.1-20). The following depositional areas of sands can be distinguished on using his diagram, where I = area of river sands and sands deposited by other currents, II = area of near-shore sands (beach and shallow water) deposited in an environment of strong water agitation, III = area of bottom marine or lake sands, formed in weakly-agitated waters, and IV = area of aeolian sands. The hatched area represents the area of uncertainty with sandy—silty deposits containing considerable amount of grains less than 0.05 mm in size. Many river sediments probably would fall in this area. Preliminary experimental data obtained by the senior author show that certain ranges in compressibility values correspond to definite genetic areas in Fig.1-20.

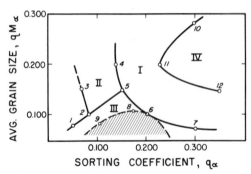

Fig.1-20. Genetic classification (energy index) of sands based upon grain size distributions. I = sands deposited by rivers and other currents; II = near-shore sands deposited in an environment of strong water agitation; III = bottom marine or lake sands, formed in weakly-agitated waters; and IV = aeolian sands. Dashed zone = area of uncertainty. (After Rukhin, 1969, fig.7-XV, p.491.)

Rukhin determines the average quantitative grain size (gM_α) and used the following two formulae, respectively: (1) $\log qM_\alpha = h + wm_1$, and (2) $q_\alpha = \sqrt{m_2 - m_1^2}$, where h = midpoint of logarithms of grain sizes (end values of a grade) of the fraction predominating in the sample, w = average value of differences between logarithms of end values of each grade, m_1 = first moment, and m_2 = second moment. The calculation method for medium-grained sands is presented in Table 1-V.

Grain-size distributions are commonly made up of more than one log-normal "population", and can provide information concerning the depositional processes. As pointed out by Visher (1967), three major populations are commonly developed (Fig.1-21A): A, transported by saltation; B, transported in suspension; and C, transported by rolling on the bed.

The junction point between the saltation population (A) and the traction or rolling population (C) often occurs near grain size of 2ϕ (phi) and its position is dependent upon the strength of the bottom current. The stronger the current, the coarser the size transported by saltation (Visher, 1967).

The lower the turbulent energy, the more suspension material is included in the distribution, and the finer is the grain size at the junction between suspension (B) and saltation (A) transported populations (Visher, 1967).

The strength of the current that produces saltation determines the grain-size range of saltation population. A strong current produces a population having a large size range (1¾ to 2½ϕ) (Visher, 1967).

Figure 1-21C shows a curve typical of modern streams and ancient sands of fluvial origin. There is an abrupt truncation at point "t". Wind-blown deposits show a preponderance of population transported by saltation and the curve exhibits excellent sorting of that population (Fig.1-21D).

TABLE 1-V

Method of calculating granulometric coefficients using moments (After Rukhin, 1969, p.489)*

Grade size		Interval, ω	h	Weight percentage frequency, x_1	Deviation, n	nx_1	n^2x_1
mm	$\log_{10}$						
1.4–1.0	0.146–0.000	0.146	0.073	0.1	+6	+0.6	3.6
1.0–0.71	0.000–$\bar{1}$.851	0.149	1.925	0.2	+5	+1.0	5.0
0.71–0.50	$\bar{1}$.851–$\bar{1}$.669	1.152	1.755	0.5	+4	+2.0	8.0
0.50–0.35	$\bar{1}$.699–$\bar{1}$.544	0.155	1.622	3.9	+3	+11.7	35.1
0.35–0.25	$\bar{1}$.544–$\bar{1}$.398	0.146	1.471	12.0	+2	+24.0	48.0
0.25–0.177	$\bar{1}$.398–$\bar{1}$.248	0.150	1.323	15.4	+1	+15.4	15.4
0.177–0.125	$\bar{1}$.248–$\bar{1}$.097	0.153	1.773	49.4	0	0.0	0.0
0.125–0.088	$\bar{1}$.097–$\bar{2}$.944	0.153	1.021	15.4	−1	−15.4	15.4
0.088–0.051	$\bar{2}$.944–$\bar{2}$.785	0.159	2.865	3.1	−2	−6.2	12.4
		0.151		100.0			

$h = \bar{1}.173$

$v_1\omega = +0.050$

$\log M_\alpha = \bar{1}.223$

$M_\alpha = 0.167$ mm

$\Sigma + = 54.7$
$\Sigma - = 21.6$

$\Sigma = + 33.1$
$v_1 = + 0.331$

$\Sigma = 142.9$
$v_2 = 1.429$
$v_1^2 = 0.110$

$v_2 - v_1^2 = 1.319$
$\sqrt{v_1 - v_1^2} = 1.148$
$\sigma = 0.173$

*For the first moment (m_1) and the second moment (m_2), Rukhin used symbols v_1 and v_2, respectively.

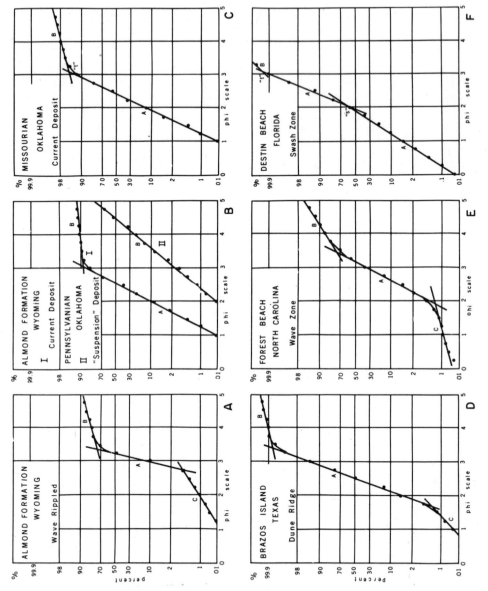

Fig.1-21. Selected grain-size distributions produced by differing depositional processes. (After Visher, 1967.)

Figure 1-21E represents a sediment sample collected off the coast of North Carolina from a depth of 11 ft (zone of wave action). There is a development of size population related to rolling (C), giving a similar appearance to that of Fig.1-21A. Effects of winnowing by wave action are also apparent.

The fourth class of distributions recognized by Visher (1967) is one related to beach processes (Fig.1-21F). The curve presented in this figure is typical of nearly 100 samples taken from the lower swash zone from beaches along the Gulf and East Coasts of the United States. The inflection at point "s" represents only a slight change in the slope of the saltation population (A). Visher (1967) pointed out that the position of inflection point "t" and percentage and sorting of population B (suspension) are similar to those found in wind-blown sand distributions, and may be in part controlled by provenance.

POROSITY, VOID RATIO, AND DENSITY

Virtually all detrital rocks are porous (i.e., contain void space) to some extent. These voids are extremely important in that they contain fluids that are, in many cases, beneficial to man. The voids in a sand are particularly important in the study of compaction, because compaction is associated with reduction in pore space. Under extremely high pressures, there is also a reduction in volume of solids; however, in most studies of compaction, reduction in solids volume has been ignored.

The relative volumes of voids and solids can be expressed in terms of (1) porosity and (2) void ratio. With few exceptions, porosity is preferred by geologists and petroleum engineers, whereas void ratio is used by soils and civil engineers. It should be pointed out here that the various disciplines in geology and engineering all have distinct sets of nomenclature and symbols for rock parameters.

Both porosity and void ratio are related to the bulk volume of a rock. Bulk volume V_b is defined as the sum of the volumes of the voids or pores, V_v or V_p, and the solids, V_s:

$$V_b = V_p + V_s \tag{1-1}$$

Porosity ϕ is the ratio of the void space to the bulk volume and is usually expressed in percent:

$$\phi = \frac{V_p}{V_b} \times 100 \tag{1-2}$$

When used in an equation, however, the decimal equivalent is usually used, i.e., 0.5 instead of 50%. Comparison charts to aid in visual estimation of porosity are presented in Fig.1-22.

Void ratio e is extremely important in compaction studies and is defined as the ratio of the voids volume to the solids volume:

$$e = V_p/V_s \tag{1-3}$$

also:

$$e = \phi/(1 - \phi) \tag{1-4}$$

or:

$$\phi = e/(1 + e) \tag{1-5}$$

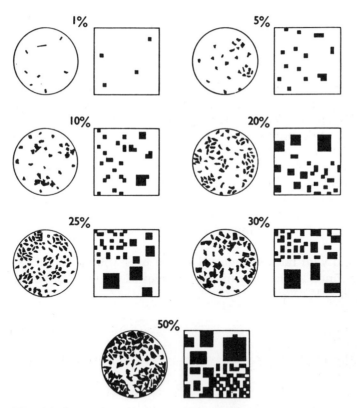

Fig.1-22. Comparison charts to aid in visual estimation of percentage. Circles after Terry and Chilingar (1955); squares after Folk (1951a).

Obviously, porosity can never exceed a value of 100% or 1.0 (fractional porosity), whereas void ratio often exceeds unity in fine-grained sediments and clays. Figure 1-23 shows the relationship between void ratio and porosity in the commonly occurring range.

Porosity is a measure of the volume of voids related to the bulk volume, which changes when compaction takes place. The volume of solids remains essentially constant under compression while the bulk volume decreases. The total volume change, therefore, cannot be represented by subtracting one porosity from another, because the porosities would be expressed in reference to different bulk volumes.

Equations 1-6 and 1-7 can be used in calculations because they relate the change in bulk and pore volumes to the volume of the solids, which is considered to be constant. A change in the bulk volume V_b can be expressed in terms of porosity:

$$\delta V_b = 1 - \frac{1 - \phi_1}{1 - \phi_2} \tag{1-6}$$

where δV_b = change in bulk volume, ϕ_1 = porosity at time 1, and ϕ_2 = porosity at time 2 (after compaction). Inasmuch as the volume of the solids remains constant, δV_b is also equal to δV_p and $\delta \phi$ in eq.1-7.

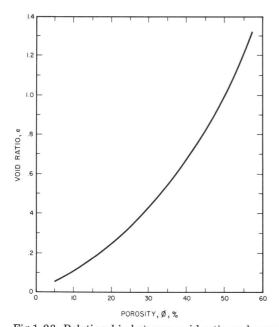

Fig.1-23. Relationship between void ratio and porosity.

The decrease in porosity per unit of the original bulk volume can be expressed as:

$$\delta\phi = \frac{\delta e}{1 + e_o} \qquad (1\text{-}7)$$

where $\delta\phi$ = change in porosity related to the original V_b, e_o = original void ratio, and δe = change in void ratio. Equation 1-7 is preferred when void ratios are used, because calculations are simplified.

Specific weight, γ, is often used in conjunction with porosity and void ratio. It is defined as the weight per unit volume, whereas density, ρ, is the mass per unit volume and is equal to γ/g where g is gravitational acceleration. The term "density", however, is often used to designate specific weight, which often results in erroneous calculations. Mass ρ is attracted by the earth with a force γ (= $\rho \times g$). For example, if the specific weight of water is equal to 62.4 lb/ft^3 and the gravitational acceleration, g, is equal to 32.174 ft/sec^2, then the density expressed in terms of slugs/ft^3 is equal to 1.94 (= 62.4/32.174).

Bulk specific weight can be either "dry" or "wet", depending upon the nature of the fluid in the pore spaces. The unit weight of a dry sand (only air is present in pore spaces) is equal to:

$$\gamma_{db} = (1 - \phi)\gamma_s \qquad (1\text{-}8)$$

where γ_{db} = dry weight per unit of bulk volume, and γ_s = specific weight of solids (grains).

The unit weight of a wet sand is expressed as:

$$\gamma_{wb} = (1 - \phi)\gamma_s + \phi\gamma_f \qquad (1\text{-}9)$$

or:

$$\gamma_{wb} = \gamma_s - \phi(\gamma_s - \gamma_f) \qquad (1\text{-}10)$$

where γ_{wb} = weight per unit of wet bulk volume, and γ_f = specific weight of fluid in the pores.

Quartz sands have an average specific gravity of 2.65 with reference to water (specific gravity = specific weight of a material at 60°F: specific weight of water at 60°F)*. If the weight of one cu ft of fresh water is assumed to be equal to 62.4 lb, then one cu ft of solid silica having a specific gravity of 2.65 would weigh 165.4 lb. If one cu ft of dry sand weighs 137.3 lb, then its dry bulk specific gravity would be equal to 2.2

---

*(°C × 1.8) + 32 = °F; [(°F + 40)/1.8] − 40 = °C.

INTRODUCTION

(= 137.3/62.4). The porosity of this dry sand can be calculated using Eq.1-8: $\gamma_{db} = (1-\phi)\gamma_s$ or $\phi = (\gamma_s - \gamma_{db})/\gamma_s = (2.65 - 2.2)/2.65 = 0.17$ or 17%. If the sand were saturated with water, its bulk specific weight would be equal to the weight of the solids (137.3 lb) plus the weight of the water (0.17 × 62.4 lb) or 147.91 lb/ft^3.

It can be easily illustrated on using idealized spheres that porosity is dependent upon the method of packing. If packed cubically, spheres of equal size would have a maximum possible void space of about 47.6% (Slichter, 1897—1898). If packed rhombohedrally, the porosity is reduced to a minimum of about 26% (Fig.1-24). It is obvious that sphere size does not change porosity when unit volumes with sides at least 2 radii in length are examined. In nature, owing to variation in size of grains and their angularity, usually the porosity of a sand or sandstone will be less than the values specified for spherical grains. It has also been demonstrated that in a mixture of spherical particles of different sizes, stacking arrangement does affect porosity. The introduction of a second set of spheres, small enough to fit in

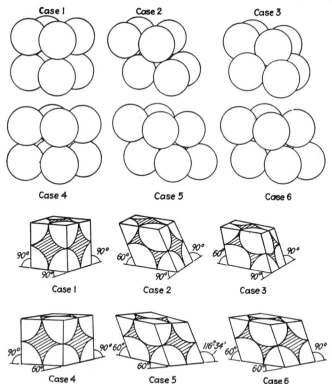

Fig.1-24. Several different ways of packing spheres. (After Graton and Frazer, 1935. See Table 3-I, p.98.)

the pore space between the larger set, can reduce porosity to about 13%. (See Wolf and Chilingarian, Vol. II, Chapter 3, for more details.)

Usually, finer-grained sediments exhibit greater porosity when deposited than coarse-grained ones. A well-sorted, well-rounded, loosely compacted medium- to coarse-grained sand may have a porosity of about 37%, whereas a poorly-sorted fine-grained sand with irregularly-shaped grains may have a porosity in excess of 50%. An admixture of irregular-shaped, tabular and bladed particles usually gives rise to a higher porosity because of particle bridging. The wide variability of porosity owing to depositional environment is best illustrated by the greywackes, which may have either a high or a very low porosity, depending on the amount of fine-grained material filling the pores. Clays and silts may have porosities as high as 50—80% when freshly deposited.

The terms "effective porosity" and "total porosity" are often used in petroleum geology and petroleum reservoir engineering studies. These terms differentiate between the interconnected pores through which fluids can move and the total pore space, regardless of its ability to transmit fluid.

PERMEABILITY

Permeability is a measure of the ability of a medium to transmit fluids and is related to effective porosity (Fig.1-25). A highly porous (total porosity) sandstone having low bulk specific weight could have low or no effective porosity and a low permeability, if the pores are sealed owing to cementation or to structural configuration. Relationship between effective porosity and permeability mainly depends on the grain-size distribution, pore sizes and their distribution, the number of pore interconnections and their widths, and pore path tortuosity.

The unit of permeability is the *darcy*, named after Henry Darcy, who in 1856 experimentally established the principles of water flow through porous media. He found that the velocity of flow through a homogeneous, porous medium is directly proportional to the hydraulic gradient and a constant of proportionality (called permeability) characteristic of a particular sand pack. Other workers extended his concept to include fluids having viscosities other than that of water. General formulas for both linear and radial flow of fluids were developed and are commonly used today (see Muskat, 1949; Pirson, 1958; Craft and Hawkins, 1959).

Permeability, which is usually determined empirically in the laboratory, refers to a specific rock property related to the viscous flow of fluids. The commonly-used formula for determining linear permeability is:

INTRODUCTION

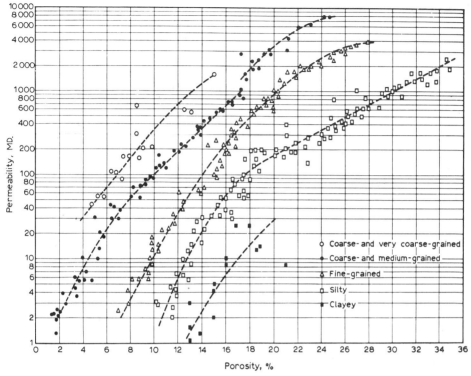

Fig.1-25. Relationship between porosity and permeability of very coarse-grained, coarse- and medium-grained, silty, and clayey sandstones. Very coarse-grained sandstones contain more than 50% of 1—2 mm fraction; coarse-grained: 0.5—1 mm; medium-grained: 0.25—0.5 mm; fine-grained: 0.1—0.25 mm. Silty—silt (< 0.1 mm) content in excess of 10%. Clayey—clay (< 0.004 mm) content in excess of 7%. Samples (cores) had irreducible water saturation. (After Chilingar, 1964, fig.2, p.73.)

$$k = \frac{q\mu L}{\delta p A} \qquad (1\text{-}11)$$

where k = permeability (constant of proportionality) in darcys, q = volumetric rate of flow in cc/sec, L = length of flow path in cm, μ = viscosity in centipoises, δp = pressure drop along distance L in atmospheres, and A = cross-sectional area in cm^2. Equation 1-11 can be rearranged for flow quantity:

$$q = \frac{k \delta p A}{\mu L} \qquad (1\text{-}12)$$

Permeability, k, is commonly expressed in millidarcys for convenience and is referred to flow of air unless otherwise specified. Various versions and derivations of Darcy's equation are used for both linear and radial flow measurements (see Muskat, 1949).

Permeabilities for a loose, unconsolidated sand commonly range from one to three darcys. Consolidated but slightly cemented sands may have a permeability range of 200 to 1,000 millidarcys, whereas permeabilities of well-consolidated, well-cemented ("tight") sandstones may range from 1 to 10 millidarcys. Consolidated shale permeabilities are normally much lower, perhaps 1/10,000 millidarcy or less (see Rieke and Chilingarian, 1974). It is sometimes convenient to express volumetric rate of flow, q, in barrels per day, cross-sectional area in square feet, length in feet, and pressure in pounds per square inch (psi). This can be done by multiplying the right hand side of eq.1-12 by 1.127 (conversion factor). Sometimes the term "perm" is used in order to eliminate the constant 1.127, i.e., 1 perm = 1.127 darcys.

Inasmuch as permeability is closely related to the pore path tortuosity, the permeability of a fracture or a vug can be virtually infinite.

For studies involving drainage into a wellbore, the formula for radial

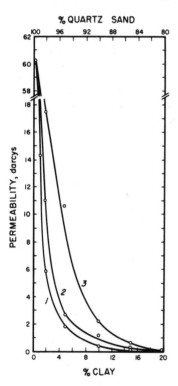

Fig.1-26. Effect of clay content on permeability of 0.35—0.50 mm sand fraction. *1* = Montmorillonite; *2* = polymictic clay; *3* = kaolinite. (After Tsvetkova, 1954.)

INTRODUCTION

flow is important. In addition, other variables such as turbulent flow, the number of layers having varying flow characteristics, and the relative permeabilities to oil, water and gas must be considered. In compaction studies covering geologic or relatively long periods of time and large vertical and horizontal distribution of sediments, an assumption of linear flow is usually adequate (see Rieke and Chilingarian, 1974).

Tsvetkova (1954) in her classical paper clearly showed that the presence of clays greatly reduces the permeability of sandstones (Fig.1-26). This is true especially in the case of montmorillonite clays. For example, the presence of 2% montmorillonite in coarse-grained sands lowers permeability 10-fold, whereas 5% of montmorillonite clay lowers permeability 30-fold. Sandstones are practically impermeable in the presence of 6—9% montmorillonite clay. On the other hand, sandstones containing 2—15% kaolinite clay can still remain quite permeable.

SURFACE AREAS OF SANDS AND SANDSTONES

Certain relationships have been established among surface area, permeability and porosity by several authors (see Langnes et al., 1972, p.243). Chilingar et al. (1963) developed the following formula for surface area per unit of pore volume, s_p, for sandstones in cm² /cm³ :

$$s_p = \frac{2.11 \cdot 10^5}{(F^{2.2} \phi^{1.2} k)^{1/2}} \tag{1-13}$$

where F = formation resistivity factor (= R_o/R_w; R_o is equal to the electrical resistivity of a formation 100% saturated with formation water and R_w is equal to the formation water resistivity), ϕ = fractional porosity, and k = permeability in millidarcys.

If values of 1.25 for tortuosity, τ, and 2.5 for shape factor, S_{hf}, are representative for unconsolidated sands, then the following formula can be used (Chilingar et al., 1963):

$$s_b = 5650 \left(\frac{\phi^3}{k}\right)^{1/2} \tag{1-14}$$

where s_b = surface area in cm² per unit of bulk volume in cm³, ϕ = fractional porosity, and k = permeability in darcys.

Tortuosity, τ, is equal to the square of the ratio of the effective length, L_e (tortuous path) to the length parallel to the overall direction of flow of the pore channels, L:

$$\tau = (L_e/L)^2 \tag{1-15}$$

Shirkovskiy (1971) presented several new formulas and discussed in detail the accuracy of different formulas proposed for determination of surface areas.

CAPILLARY PRESSURE CURVES

In addition to porosity and permeability determinations, capillary pressure curves (e.g., Fig.1-27) may aid in analyzing degree and nature of compaction.

The laboratory measurement of the interstitial water content of cleaned rock samples depends upon the principles of capillarity. As depicted in

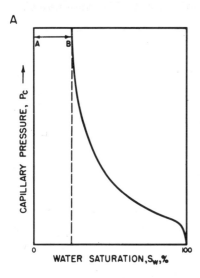

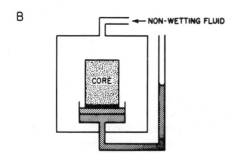

Fig.1-27. A. Capillary pressure curve. Irreducible interstitial water content = distance AB. B. Schematic diagram of apparatus for determining interstitial water content at various pressures.

INTRODUCTION

Fig.1-27 the sample is completely saturated with formation water and is placed in contact with a permeable diaphragm by means of a soft contact material. Both the diaphragm and the contact material must be completely saturated with water and have displacement pressures which exceed the maximum pressure to be used during the test. Thus, the nonwetting fluid cannot be transmitted through the diaphragm. The cell containing the sample is filled with oil and the pressure in the oil cell is set at the desired values. Water, displaced from the sample by the oil, raises the level in the U-tube which was previously filled to a definite level. Thus, the oil at a particular pressure may permeate all parts of the sample, while the water remains at essentially atmospheric pressure. The horizontal distance between the ordinate and the vertical portion of the curve (distance *AB*, Fig.1-27A) shows the amount of irreducible water content, which is present in minute interstices and cracks, and cannot be displaced even at very high pressures. These smaller pores usually do not affect the permeability of the rock.

Mercury injection technique, whereby mercury vapour acts as the wetting phase and the liquid mercury as the nonwetting phase, is also used to determine capillary pressure curves (see Chilingar et al., 1972).

COMPRESSIBILITY

In general, compressibility, c, can be defined as the rate of change of volume, V, with respect to the applied stress, σ, per unit of volume:

$$c = -\frac{1}{V}\frac{\delta V}{\delta \sigma} \qquad (1\text{-}16)$$

The total external stress σ is the sum of pore pressure, p_p, and intergranular stress σ'. The latter stress is also termed the effective pressure, p_e; thus:

$$\sigma' = p_e = \sigma - p_p \qquad (1\text{-}17)$$

Several different definitions of compressibility are used in the literature depending on the method of its determination (Gomaa, 1970; Rieke and Chilingarian, 1974): (1) bulk compressibility, (2) pore compressibility, (3) formation compressibility, (4) rock solids compressibility, and (5) pseudo-bulk compressibility.

(1) Bulk compressibility is commonly defined as the change in bulk volume, V_b, per unit of bulk volume, per unit change in external stress, while keeping the pore pressure, p_p, and temperature, T, constant:

$$c_b = -\frac{1}{V_b}\left(\frac{\delta V_b}{\delta \sigma}\right)_{p_p,T} \qquad (1\text{-}18)$$

Bulk compressibility can also be defined as the change in bulk volume, per unit of bulk volume, per unit change in effective pressure, p_e, while keeping the total external stress, σ, and temperature constant:

$$c_b = -\frac{1}{V_b}\left(\frac{\delta V_b}{\delta p_e}\right)_{\sigma,T} \tag{1-19}$$

The latter definition is preferred by the writers.
(2) *Pore compressibility* is the change in pore volume, V_p, per unit of pore volume, per unit change of external stress, σ, while keeping the pore pressure and temperature constant:

$$c_p = -\frac{1}{V_p}\left(\frac{\delta V_p}{\delta \sigma}\right)_{p_p,T} \tag{1-20}$$

Hall (1953, p.310) called pore compressibility "formation compaction"; whereas the present authors prefer the following definition for pore compressibility determined at a constant external stress, σ:

$$c_p = -\frac{1}{V_p}\left(\frac{\delta V_p}{\delta p_e}\right)_{\sigma,T} \tag{1-21}$$

where p_e is effective pressure ($= \sigma - p_p$).
(3) *Formation compressibility* can be defined as the change in pore volume, per unit of pore volume, per unit change of pore pressure, while keeping the external stress and temperature constant:

$$c_f = +\frac{1}{V_p}\left(\frac{\delta V_p}{\delta p_p}\right)_{\sigma,T} \tag{1-22}$$

Hall (1953, p.310) called this compressibility "effective" compressibility.
(4) *Rock solids compressibility* is equal to the change in rock solids volume, V_s, per unit of rock solids volume, per unit of external pressure at a constant temperature. If a rock sample is tested without a jacket or drainage ports, external stress will be equal to the pore pressure. Thus:

$$c_r = -\frac{1}{V_s}\left(\frac{\delta V_s}{\delta p_p}\right)_{\sigma = p_p;T} = -\frac{1}{V_s}\left(\frac{\delta V_s}{\delta \sigma}\right)_{\sigma = p_p;T} \tag{1-23}$$

or (Gomaa, 1970):

$$c_r = -\frac{1}{V_b}\left(\frac{\delta V_b}{\delta p_p}\right)_{\sigma = p_p;T} = -\frac{1}{V_p}\left(\frac{\delta V_p}{\delta p_p}\right)_{\sigma = p_p;T} \tag{1-24}$$

INTRODUCTION

(5) *Pseudo-bulk compressibility* as defined by Fatt (1958b) is the change in bulk volume, per unit change in pore pressure, per unit of bulk volume, at constant external stress and temperature:

$$c_{bt} = + \frac{1}{V_b}\left(\frac{\delta V_b}{\delta p_p}\right)_{\sigma, T} \tag{1-25}$$

In the field of Soil Mechanics, the term *coefficient of compressibility*, a_v, is widely used:

$$a_v = -\frac{\delta e}{\delta p} \tag{1-26}$$

where e is the void ratio, which is equal to the volume of voids, V_p, divided by the volume of solids, V_s ($e = \frac{V_p}{V_s} = \frac{\phi}{1-\phi}$; ϕ = fractional porosity), and p is the net overburden pressure. On assuming that volume of solids, V_s, does not change on compaction ($\Delta V_s = 0$), one can derive an equation relating void ratio, e, to bulk compressibility as follows:

$$c_b = -\frac{1}{V_b}\frac{\delta V_b}{\delta p} = -\frac{1}{(V_p + V_s)}\frac{\delta V_p}{\delta p} \tag{1-27}$$

Multiplying numerator and denominator by V_s and rearranging:

$$c_b = -\frac{1}{V_p/V_s + V_s/V_s}\frac{\delta V_p}{V_s \delta p} \tag{1-28}$$

and inasmuch as:

$$\delta e = \frac{\delta V_p}{V_s} \tag{1-29}$$

eq.1-28 becomes:

$$c_b = -\frac{1}{e+1}\frac{\delta e}{\delta p} \tag{1-30}$$

In terms of sample thickness, h, the compressibility can be expressed as:

$$c_b = -\frac{1}{h}\frac{\delta h}{\delta p} \tag{1-31}$$

In many compaction experiments, in which the pore pressure is kept atmospheric, the effective pressure and the applied external stress are equal.

REFERENCES

Arutyunova, N. M. and Sarkisyan, S. G., 1969. About some methods of analyzing data of granulometric analysis. *Lithol. Min. Resour., Akad. Nauk. S.S.S.R.*, (4): 95—103.
Blatt, H., Middleton, G. V. and Murray, R., 1972. *Origin of Sedimentary Rocks*. Prentice—Hall, Englewood Cliffs, N.J., 634 pp.
Bissell, H. J. and Chilingar, G. V., 1967. Classification of sedimentary carbonate rocks. In: G. V. Chilingar, H. J. Bissell and R. W. Fairbridge (Editors), *Carbonate Rocks, A*. Elsevier, Amsterdam, pp. 87—168.
Chilingar, G. V., 1964. Relationship between porosity, permeability and grain size distribution of sands and sandstones. In: L. M. J. U. Van Straaten (Editor), *Deltaic and Shallow Marine Deposits, I*. Elsevier, Amsterdam, pp. 71—75.
Chilingar, G. V., Main, R. and Sinnokrot, A., 1963. Relationship between porosity, permeability, and surface areas of sediments. *J. Sediment. Petrol.*, 33(3): 759—765.
Chilingar, G. V., Mannon, R. W. and Rieke III, H. H. (Editors), 1972. *Oil and Gas Production from Carbonate Rocks*. Am. Elsevier, New York, N.Y., 408 pp.
Chilingarian, G. V., Beeson, C. M. and Ershaghi, I., 1972. Porosity and permeability. In: R. W. Fairbridge (Editor), *Encyclopedia of Earth Sciences, IV A*. Reinhold, New York, N.Y., pp. 464—470.
Craft, B. C. and Hawkins, M. F., 1959. *Applied Petroleum Reservoir Engineering*. Prentice—Hall, Englewood Cliffs, N.J., 437 pp.
Fatt, I., 1958a. Pore volume compressibility of sandstone reservoir rocks. *J. Pet. Tech.*, 10(3): 64—66.
Fatt, I., 1958b. Compressibility of sandstones at low to moderate pressure. *Bull. Am. Assoc. Pet. Geol.*, 42: 1924—1957.
Folk, R. L., 1951a. A comparison chart for visual percentage estimation. *J. Sediment Petrol.*, 21: 32—33.
Folk, R. L., 1951b. Stages of textural maturity in sedimentary rocks. *J. Sediment. Petrol.*, 21: 127—130.
Folk, R. L., 1954. The distinction between grain size and mineral composition in sedimentary-rock nomenclature. *J. Geol.*, 62: 344—359.
Folk, R. L., 1959. Practical petrographic classification of limestones. *Bull. Am. Assoc. Pet. Geol.*, 43: 1—38.
Folk, R. L., 1962. Spectral subdivision of limestone types. In: W. E. Ham (Editor), *Classification of Carbonate Rocks. Am. Assoc. Pet. Geol., Mem.*, 1: 62—85.
Folk, R. L., 1968. *Petrology of Sedimentary Rocks*. Texas Hemphill's Book Store, Austin, 170 pp.
Folk, R. L. and Ward, W. C., 1957. Brazos River Bar, a study in the significance of grain-size parameters. *J. Sediment. Petrol.*, 27: 3—26.
Folk, R. L., Andrews, P. B. and Lewis, D. W., 1970. Detrital sedimentary rock classification and nomenclature for use in New Zealand. *N.Z. J. Geol. Geophys.*, 13: 937—968.
Gaither, A., 1953. A study of porosity and grain size relationships in experimental sands. *J. Sediment Petrol.*, 23: 180—191.
Gomaa, E. M., 1970. Compressibility of rocks and factors affecting them. *Am. Inst. Min.*

Metall. Pet. Eng., Annu. Stud. Contest Meet., April, Univ. Calif., Berkeley, Presented Pap.

Grabau, A. W., 1904. On the classification of sedimentary rocks. *Am. Geol.*, 33: 331—340.

Graton, L. C. and Frazer, H. J., 1935. Systematic packing of spheres — with particular relation to porosity and permeability. *J. Geol.*, 43: 785—909.

Griffiths, J. C., 1967. *Scientific Method in Analysis of Sediments.* McGraw—Hill, New York, N.Y., 508 pp.

Hall, H. N., 1953. Compressibility of reservoir rocks. *Trans. Am. Inst. Min. Metall. Eng.*, 198: 309—311.

Howell, J. B., 1922. Notes on pre-Permian Paleozoics of the Wichita Mountain area. *Bull. Am. Assoc. Pet. Geol.*, 6: 413—425.

Krumbein, W. C., 1941. Measurement and geologic significance of shape and roundness of sedimentary particles. *J. Sediment. Petrol.*, 11: 64—72.

Krumbein, W. C. and Pettijohn, F. J., 1938. *Manual of Sedimentary Petrography.* Appleton—Century—Crofts, New York, N.Y., 548 pp.

Krumbein, W. C. and Sloss, L. L., 1951. *Stratigraphy and Sedimentation.* Freeman, San Francisco, Calif., 497 pp.

Krumbein, W. C. and Sloss, L. L., 1963. *Stratigraphy and Sedimentation.* 2nd ed., Freeman, San Francisco, Calif., 660 pp.

Krynine, P. D., 1940. Petrology and genesis of the Third Bradford Sand. *Pa. State Coll. Bull.*, 29: 134 pp.

Krynine, P. D., 1945. Sediments and the search for oil. *Producers Mon.* 9: 12—22.

Krynine, P. D., 1948. The megascopic study and field classification of sedimentary rocks. *J. Geol.*, 56: 130—165.

Lane, E. W., 1938. Notes on the formation of sand. *Trans. Am. Geophys. Union*, 19: 505—508.

Langnes, G. L., Robertson, J. O. and Chilingar, G. V., 1972. *Secondary Recovery and Carbonate Reservoirs.* Am. Elsevier, New York, N.Y., 303 pp.

McBride, E. F., 1963. A classification of common sandstones. *J. Sediment. Petrol.*, 33: 664—669.

McCammon, R. B., 1962. Efficiencies of percentile measures for describing the mean size and sorting of sedimentary particles. *J. Geol.*, 70(4): 453—465.

Mollazal, Y., 1961. Petrology and petrography of Ely Limestone in part of the Eastern Great Basin. *Brigham Young Univ. Res. Stud., Geol. Ser.*, 8: 3—35.

Muskat, M., 1949. *Physical Principles of Oil Production.* McGraw—Hill, New York, N.Y., 922 pp.

Pettijohn, F. J., 1941. Persistence of heavy minerals and geologic age. *J. Geol.*, 49: 610—625.

Pettijohn, F. J., 1949. *Sedimentary Rocks.* Harper and Row, New York, N.Y., 1st ed., 526 pp.

Pettijohn, F. J., 1957. *Sedimentary Rocks*, 2nd ed. Harper, New York, N.Y., 718 pp.

Pettijohn, F. J., Potter, P. E. and Siever, R., 1972. *Sand and Sandstone.* Springer, New York, N.Y., 618 pp.

Pirson, S. J., 1958. *Oil Reservoir Engineering.* McGraw—Hill, New York, N.Y., 735 pp.

Powers, M. C., 1953. A new roundness scale for sedimentary particles. *J. Sediment. Petrol.*, 23: 117—119.

Rieke III, H. H. and Chilingarian, G. V., 1974. *Compaction of Argillaceous Sediments.* Elsevier, Amsterdam, 424 pp.

Rieke III, H. H., Chilingar, G. V. and Mannon, R. W., 1972. Application of petrography and statistics to the study of some petrophysical properties of carbonate reservoir rocks. In: G. V. Chilingar, R. W. Mannon and H. H. Rieke III, (Editors), *Oil and Gas Production from Carbonate Rocks.* Am. Elsevier, New York, N.Y., 408 pp.

Rittenhouse, G., 1943. A visual method of estimating two-dimensional sphericity. *J. Sediment. Petrol.,* 13: 79—81.

Rukhin, L. B., 1947. *Granulometric Method of Studying Sands.* Izd. L.G.U., Leningrad.

Rukhin, L. B., 1969. *Principles of Lithology (Study of Sedimentary Rocks).* Izd. Nedra, Leningrad, 703 pp.

Shirkovskiy, A. I., 1971. *Determination and Utilization of Physical Parameters of Porous Medium During Development of Gas Condensate Deposits.* VNIIZ—GAZPROM, Moscow, 48 pp.

Slichter, C. S., 1897—1898. Theoretical investigation of motion of ground waters. *U.S. Geol. Surv., 19th Annu. Rep.,* part 2.

Teodorovich, G. I., 1938. Towards question of classification of clayey—silty—sandy rocks. *Sov. Geol.,* 8—9: 67—74.

Teodorovich, G. I., 1958. *Study of Sedimentary Rocks.* Gostoptekhizdat, Leningrad, 572 pp.

Teodorovich, G. I., 1965. Expanded classification of sandstones based upon composition. *Izv. Akad. Nauk. S.S.S.R., Ser. Geol.,* 6: 75—95.

Teodorovich, G. I. and Tsvetkova, M. A., 1966. Lithological studies of oil reservoirs. *Izv. Akad. Nauk S.S.S.R., Ser. Geol.,* 7: 86—100.

Terry, R. D. and Chilingar, G. V., 1955. Summary of "Concerning some additional aids in studying sedimentary formations" by M. S. Shvetsov. *J. Sediment. Petrol.,* 25: 229—234.

Tsvetkova, M. A., 1954. Influence of mineralogic composition of sandy rocks on filtration capacity and oil production. *Tr. Inst. Nefti, Akad. Nauk S.S.S.R.,* 3.

Twenhofel, W. H., 1950. *Principles of Sedimentation.* McGraw—Hill, New York, N.Y., 673 pp.

Udden, J. A., 1898. The mechanical composition of wind deposits. *Augusta Libr. Publ.,* 1: 69 pp.

Udden, J. A., 1914. Mechanical composition of clastic sediments. *Bull. Geol. Soc. Am.,* 25: 655—744.

Uren, L. C., 1939. *Petroleum Production Engineering, Oil Field Exploitation,* 2nd ed. McGraw—Hill, New York, N.Y., 756 pp.

Visher, G. S., 1967. Grain size distributions and depositional processes. *Int. Sediment. Congr.,* 7th, 4 pp. (reprint).

Wadell, H., 1934. Shape determinations of large sedimental rock fragments. *Pan Am. Geol.,* 61: 187—220.

Wadell, H., 1935. Volume, shape and roundness of rock particles. *J. Geol.,* 40: 443—451.

Weller, J. M., 1960. *Stratigraphic Principles and Practices.* Harper and Row, New York, N.Y., 725 pp.

Wentworth, C. K., 1922. A scale of grade and class terms for clastic sediments. *J. Geol.,* 30: 377—392.

Wentworth, C. K., 1933. Fundamental limits to the size of clastic grains. *Sci.,* 77: 633—634.

Williams, H., Turner, F. J. and Gilbert, C. M., 1954. *Petrology.* Freeman, San Francisco, Calif., 406 pp.

Wolf, K. H. and Chilingarian, G. V., 1976. Diagenesis of sandstones and compaction. In: G. V. Chilingarian and K. H. Wolf (Editors), *Compaction of Coarse-Grained Sediments,* II. Elsevier, Amsterdam.

Chapter 2

MECHANICS OF SAND COMPACTION

D. R. ALLEN and G. V. CHILINGARIAN

INTRODUCTION

The mechanics of compaction has been approached from two different viewpoints for a long time, i.e., that of the geologist—sedimentologist and that of the soils engineer. The largest body of knowledge on properties affecting sediment compaction undoubtedly has been accumulated in the field of soil mechanics. Inasmuch as the terms compaction and consolidation are used by both disciplines, but with different connotations, it is necessary to present various definitions as used in each field. In the field of soil mechanics, compaction means a density increase caused by mechanical or hydraulic means such as tamping, vibrating, loading, or wetting; whereas in the geological sciences, compaction means the lessening of sedimentary volume owing to overburden loading, grain rearrangement, etc. In soils engineering, consolidation means the lessening of volume by the expulsion of pore water upon the application of a load, whereas in the field of geology it is commonly understood to mean the acquisition of structural competency by a sediment owing to reduction in volume, induration, cementation, etc. (see Table 2-I).

TABLE 2-I

Definitions of compaction and consolidation as used in the fields of geology and soil mechanics

Geology—Sedimentology usage	*Soils Engineering usage*
Compaction	
A lessening of sedimentary volume owing to overburden loading, grain rearrangement, etc.	An increase in bulk specific weight (or bulk specific gravity) caused by mechanical or hydraulic means such as vibrating, loading or wetting.
Consolidation	
The acquisition of structural competency by a sediment owing to reduction in volume, induration, cementation, etc.	The lessening of volume of a porous material (such as clay) by the expulsion of pore water upon the application of a load.

The writers of this chapter prefer the broader geological meanings. The soil mechanics definitions are much narrower in scope and can be easily fitted into the geological usages. Compaction includes consolidation as used by soils engineers.

For many years, the standards for the measurement of compaction were set by workers who used them for practical application, such as in the design of foundations and dams. These workers, however, were mainly concerned with relatively shallow depths and low overburden pressures. Geological compaction has been studied, discussed and theorized upon by numerous writers; however, it seems that it has not been investigated in sufficient detail relative to its significance in the formation of structures and fluid reservoirs.

Fuller (1908, p.33) theorized on the role of fluid pressure in compaction. Pratt and Johnson (1926) attributed subsidence above the Goose Creek Oil Field, Harris County, Texas, to fluid withdrawal. Poland and Davis (1969) also discussed the effect of fluid withdrawal on subsidence.

Terzaghi (1925), working in the field of soil mechanics, first developed the theory of "primary consolidation". His basic theory, which relates compaction in clays to loading and the expulsion of water owing to excess pore-water pressure, is discussed in detail by Terzaghi and Peck (1968). This theory establishes a relationship between dimensionless time and percent of total compaction in clays, based on applied load, excess pore-water pressure, length of the pore drainage paths, and permeability. Estimates of total settlement and the length of time required can be made for buildings and other structures built on clay and silt soils.

The application of these principles to processes occurring at greater depths ($> 1,000$ ft) remained unexamined for many years. In 1959, Hubbert and Rubey, in their classical paper on the role of fluid pressure in thrust faulting, made a significant contribution to the literature by summarizing the principles of fluid statics and effective pore-pressure support, and their application in geology.

Since about 1950, considerable research effort has been directed toward sand compaction under loading pressures greater than that usually used in the field of soil mechanics. Some of this work was spurred by several dramatic cases of surface subsidence over oil fields, notably those in Long Beach, California, and Lake Maracaibo, Venezuela. Additional interest in the mechanics of sand compaction has been created by numerous papers which have been written recently on the subject of "overpressured" zones in the Gulf Coast of the United States and in other areas around the world. The term "overpressured" usually means that the subsurface pore-fluid pressures are greater than those which would be developed at a particular depth by a static column of water, i.e., pressure greater than that normally expected to exist at that particular depth. The results of investigations using various approaches to compaction have proved to be of great value in better under-

STRESSES INVOLVED IN COMPACTION

The reduction of sedimentary volume by processes of compaction has been recognized for almost as long as sedimentary studies have been made. It is obvious to even the most casual observer that clays and muds decrease in volume upon dessication, and that a loosely-packed sand in a bucket can be reduced in volume by kicking and shaking the container. Sediment compaction is only a complex extension of these simple acts, compounded by (1) loading from each bed in a superjacent position to another and (2) support from the interstitial fluids.

The following terms, definitions of which are presented below, are used in discussing loads and stress changes during compaction:

Stress: total pressure (force per unit area, σ) acting at a point (see Fig.2-1D).

Effective stress: that part of the load (force per unit area) which is not counteracted by other forces and is available to cause compaction (p_e). Effective stress (p_e) = total stress (σ) — pore pressure (p_p) (see Fig.2-1E).

Hydrostatic gradient: pressure exerted by a column of water per unit of depth.

Hydrostatic pressure: pressure exerted at a certain depth by a column of water of that depth [(p in lb/ft^2) = (depth, D, in ft) × (specific weight, γ, in lb/ft^3)].

Geostatic gradient: total pressure exerted by the overburden and the interstitial fluids per unit of depth (see Fig.2-1D).

Geostatic pressure: pressure exerted by the weight of the overburden and the interstitial fluids at a particular depth; it is sometimes called the lithostatic pressure (see Fig.2-1D).

Skeletal loading: that portion of the geostatic load borne by the framework of solids in porous media; it is equivalent to effective pressure, p_e (see Fig.2-1E).

Compaction in any sediment having certain particle characteristics is controlled by the overburden placed upon it. This overburden load is usually caused by the continual deposition of new sediments. The stresses caused by the sediments and their fluid contents are modified by tectonic forces which cause folding, warping, and faulting.

Terzaghi defined the effective stress by the equation:

$$\sigma = p_e + p_p \tag{2-1}$$

where σ (or p_t) = total stress (overburden or geostatic load), p_e = effective overburden or intergranular loading, and p_p = the neutral stress, which is the pore or fluid pressure within the pore space. Terzaghi's neutral stress is an appropriate expression as long as the fluid gradient is hydrostatic. When the pore fluid pressure exceeds that which would be developed by a normal hydrostatic gradient, however, the term neutral stress may be misleading.

Commonly, the overburden weight creates the major stress which acts in a vertical direction. The direction of minor stress is perpendicular to the axis of major stress. The minor stress presumably exists in all directions in a horizontal plane as a lateral restraining force (see Fig.2-1A).

According to Hubbert and Willis (1957), in geologic regions where normal faulting is taking place, the greatest stress will be approximately vertical (= effective pressure of overburden), whereas the least stress will be horizontal, ranging from 1/2 to 1/3 of the effective pressure of the overburden (see Fig.2-1F). In the regions where there is folding and thrust faulting, on the other hand, the greatest stress can be horizontal and between 2 and 3 times the effective overburden pressure, which is the least stress (see Fig.2-1G).

Fig.2-1. Various stress conditions and concepts. A. Triaxial loading (two out of three principal stresses are equal: $p_x = p_y \neq p_z$). B. Hydrostatic loading (three principal stresses are equal: $p_x = p_y = p_z$). C. Uniaxial loading (force is perpendicular to one face of the cube, whereas the four faces perpendicular to this face are kept stationary). D. Total weight of sediment (sp. gr. = 2.6) saturated with pore fluid (sp. gr. = 1.0) is equal to (62.4 × 2.6) 0.80 + 62.4 × 0.2 = 142.27 lb; total pressure σ (force: area) is equal to 142.27 lb/ft². E. Weight of sediment cube (depicted in D) immersed in water to its top is equal to 79.87 lb, i.e., the total weight of saturated porous sediment (142.27 lb) acting downwards minus the buoyant force (62.4 lb), which acts upwards and is equal to the weight of displaced fluid. Effective stress p_e (79.87 lb/ft²) = total stress σ (142.27 lb/ft²) —pore pressure p_p (62.4 lb/ft²). Scale was adjusted to read zero after filling the tank with water, but before placing saturated sediment block on top. F. Relative stress conditions in the case of normal faulting. G. Relative stress conditions in the case of thrust faulting.

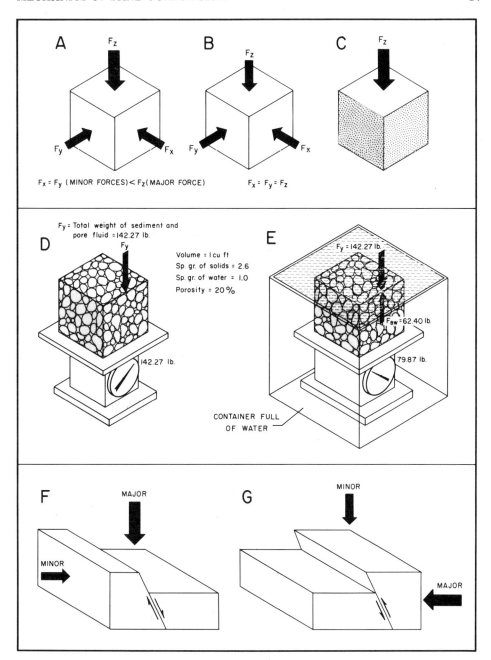

According to Hooke's law, the horizontal strain can be expressed as follows:

$$\epsilon_x = \frac{\sigma_x}{E} - \nu \frac{\sigma_y}{E} - \nu \frac{\sigma_z}{E} \qquad (2\text{-}2)$$

where ϵ_x = horizontal strain, σ_x, σ_y, σ_z = effective stresses along x and y (horizontal) and z (vertical) axes, E = Young's modulus, and ν = Poisson's ratio. Inasmuch as ϵ_x is essentially equal to zero and the lateral stress σ_x is equal to lateral stress σ_y for rocks in compression, then:

$$\sigma_x = \sigma_y = \sigma_h = \frac{\nu}{1-\nu} \sigma_z \qquad (2\text{-}3)$$

where σ_h = horizontal stress in general. On assuming a Poisson's ratio, ν, of 0.18 to 0.27 for consolidated sedimentary rocks, the horizontal compressive stress would range from 0.22 to 0.37 psi/ft of depth. According to Harrison et al. (1954), for soft shales and unconsolidated sands found in the Gulf Coast of Texas and Louisiana, which can be considered to be in a plastic state of stress, the horizontal stresses are in excess of 0.37 psi/ft of depth. Faulting will occur in cemented rocks at stresses that will only cause plastic deformation in uncemented rocks.

On assuming no effective fluid flow, the total vertical stress at a point in a porous media containing fluid can be expressed as:

$$p = [\gamma_s(1-\phi) + \gamma_w \phi]D \qquad (2\text{-}4)$$

where p = pressure at a certain depth, lb/ft² *, γ_s = specific weight of the solids, lb/ft³, γ_w = specific weight of pore fluid, lb/ft³, ϕ = porosity, fraction, and D = depth, ft. The pressure, p, is the geostatic pressure (weight per unit area) that would be exerted on an impermeable membrane if it were completely supporting all overlying material.

The effective stress can be calculated on using the equation:

$$p_e = [\gamma_s(1-\phi) + \gamma_w \phi - \gamma_w]D \qquad (2\text{-}5)$$

Equations 2-4 and 2-5 differ by the quantity $(\gamma_w D)$, which reflects the buoyant effect of the pore fluid (Archimedes Principle) and is equal to Terzaghi's neutral stress.

The question has been raised, however, as to the validity of application of these simple equations to a porous rock in which constituent particles are

---

*In order to convert to lb/inch² (psi), divide by 144.

never completely surrounded by the pore fluid owing to grain contact areas that may be quite large (see Chapter 3, Vol. II of this book). These objections were thoroughly discussed by Hubbert and Rubey (1959, 1960). On the basis of thorough mathematical analysis, they proved that the fluid pressure acts over the whole of any surface passed through the porous solid. As long as there is a continuous interconnected fluid phase, which at some places either acts directly upon the point in question or indirectly owing to pressure transmitted through the matrix or cement, then the neutral stress is fully effective.

In the presence of vertical dynamic flow and resulting fluid drag pressure on the grains, the effective pressure is either increased or decreased, depending upon the flow direction. This effect gives rise to "quicksand", a case where the intergranular loading has been reduced to nearly zero by upward water seepage and the skeletal structure has no bearing strength. Although this effect is taken into account in many ground-water calculations, it is not too important in compaction studies.

If loads were always equal and pore fluid pressures always hydrostatic, pressure calculations would be indeed simple. Inasmuch as this is usually not true, however, a thorough understanding of the loading processes and load transfer from the pore fluids to the skeletal framework of the sands is necessary. Compaction of sand or other clastics occurs as a result of loading or a change in loading; and an imbalance must exist between the applied load and the ability of the sediment structure to resist this load. In any depositional environment, once a particle has reached the end of its transport phase, a continuing deposition of sediments increases the load on those beneath. The magnitude of this loading and various changes in loading, which either cause compaction or prevent it from occurring, are presented in Fig.2-2. On assuming that the strata in the illustration consist of sands and shales, and that all pores contain fluid, curve *1* represents the hydrostatic gradient, i.e., the pressure, owing to fluid column, exerted per unit depth. Curve *3* shows the overburden (geostatic) pressure gradient. The effective unit load on the sand grains, i.e., the intergranular pressure, is shown on curve *2*. The latter pressure gradient is equal to the difference between the gradient of curve *3* and that of curve *1*. Figure 2-2 is a graphical representation of eq.2-1, 2-4, and 2-5.

Over geologic time, it would be extremely rare for any deposit not to undergo many load changes. Two such cases are illustrated in Fig.2-2.

Case 1: If the fluid level is lowered in an unconfined aquifer (i.e., no cap rock, and fluid is present as a continuous phase to the surface) to a depth of 500 ft below surface (assuming no residual capillary water), the hydrostatic pressure shifts to zero at that point (curve *1a*), and the geostatic and intergranular pressure gradients become identical down to a depth of

500 ft (curves *2a* and *3a*). The intergranular pressure would increase owing to the loss of the supporting hydrostatic pressure (from *2* to *2a*), whereas the total overburden weight is reduced because of loss of the weight of water fraction (from *3* to *3a*). At a depth of 500 ft, each one would assume a normal pressure gradient (curves *2b* and *3b*) owing to the presence of pore fluid from that point downward.

Case 2: It is assumed that the entire section is water saturated, but that the hydrostatic pressure in the confined aquifer (i.e., impermeable cap rock is present above aquifer) at a depth of 1,500 ft has been reduced to zero by

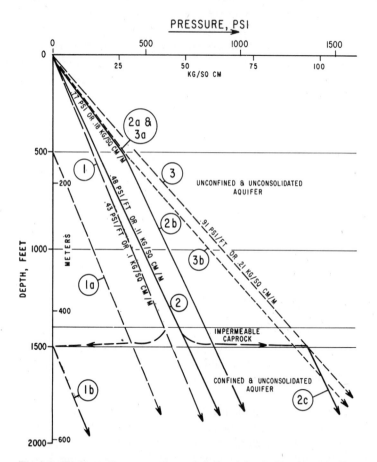

Fig. 2-2. Hydrostatic, geostatic and skeletal load changes as influenced by changes in fluid levels and pressure in unconfined and confined aquifers. *1, 1a, 1b* = hydrostatic pressure gradient; *2, 2a, 2b, 2c* = intergranular pressure gradient; *3, 3a, 3b* = total overburden pressure (geostatic) gradient. Sp. gr. solids = 2.7; sp. gr. water = 1; porosity, ϕ = 35%.

pumping (curve *1b*). Inasmuch as the pores are all saturated with water, the hydrostatic pressure can be zero only at that point and will increase with depth, at a normal hydrostatic gradient; therefore, curve 1b is parallel to curve *1*. The skeletal structure at a depth of 1,500 ft assumes the full overburden load, and the intergranular gradient shifts from curve *2* to curve *2c*, because of the loss of neutral (hydrostatic) pressure support.

In many areas of the world, overpressured zones exist at depth, i.e., the pore pressures are higher than hydrostatic and, in many cases, approach geostatic pressures. Figure 2-3 illustrates changes in the skeletal loading if the

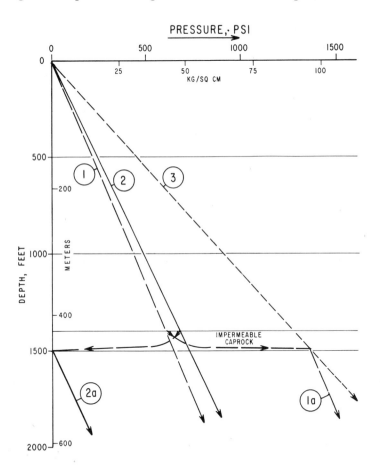

Fig. 2-3. Hydrostatic, geostatic and skeletal load changes under overpressured reservoir conditions. *1, 1a* = hydrostatic pressure gradient; *2, 2a* = intergranular pressure gradient; *3* = total overburden pressure (geostatic) gradient. Sp. gr. solids = 2.7; sp. gr. water = 1; and porosity, ϕ = 35%.

pore pressure at the top of confined zone at 1,500 ft is raised to become equal to the geostatic pressure. The hydrostatic pressure gradient (curve *1*) would become equal to the geostatic gradient (curve *3*) at 1,500 ft (curve *1* shifts to curve *1a*) and the intergranular pressure (curve *2*) is reduced to zero (curve *2* shifts to curve *2a*). The entire weight of the overburden is then resting on the pore fluid at the top of confined zone. With increasing depth below 1,500 ft, some of the load will be transferred from the pore fluids to the skeletal structure. This is illustrated by the fact that the slope of curve *1a* is steeper than that of curve *3* and that curve *2a* is parallel to curve *2*.

The bulk specific gravities of these undercompacted, high-pressure formations are lower than those of well-compacted rocks having similar lithologies. This has been demonstrated to be true by actual oil well drilling (examination of cores) and by logging techniques. An example of an overpressured area off the coast of California, where the pressure gradient at a depth of about 9,000 ft approaches geostatic gradient, is presented in Fig.2-4. Various well log parameters, reflecting rock bulk specific gravity (commonly called bulk density) and fluid content, show the presence of overpressured zones. Most logs vary in some pattern with depth, especially those sensitive to compaction or density. Figure 2-5 is an example of two

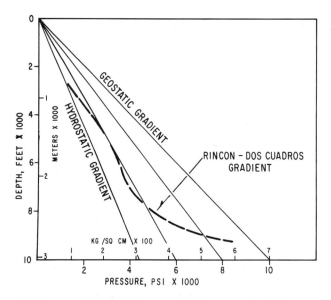

Fig.2-4. Variation in pressure gradient with depth in Rincon-Dos Cuadros overpressured reservoir in California. (After McCulloh, 1969, fig.11, p.38.)

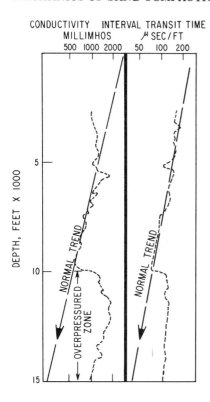

Fig.2-5. Change in electrical conductivity and sound velocity in an overpressured area. (Modified after Mathews and Kelly, 1967, fig.5, p.100. Courtesy of *Oil Gas J.*)

common log measurements (conductivity and sonic transit time) that clearly show the presence of an overpressured zone below a depth of 10,000 ft. A normal compaction trend that increases with depth will continuously decrease both rock conductivity and sonic travel time. The deviation from the normal trend observed on both curves is indicative of either a change in degree of compaction, fluid salinity, or both. A zone having relatively high pore fluid pressure can be expected to have a porosity greater than that normally found at an equivalent depth. This phenomenon is found to exist in most subsiding basins. It should be kept in mind, however, that the forces creating the overpressured zone might have generated the high pressures after both compaction and cementation had taken place. If so, with the exception of a small pore volume expansion owing to elasticity, the pore volumes could be virtually identical to those of similar sands having lower pore pressures.

EXPERIMENTAL LABORATORY SAND COMPACTION

Soil engineers and scientists, working primarily with near-surface clays and soils, have long recognized compaction and subsidence owing to loading. They made estimates of compaction magnitude and allowed for this compaction in foundation design. The concentration of research work in this field has been directed towards clays and other fine-grained sediments and soils because of their high initial porosity and susceptibility to compaction at low overburden pressures, whereas sands and coarser clastics are essentially stable at pressures of around 200 psi or less. Inasmuch as high loading pressures are not found in most construction foundations, there was little need for investigation of the compressibility of sands at high pressures.

Loose sands have void ratios (see p.28 for definition) ranging from about 0.4 to 1.0. Although typical foundation loads cause little or no compaction in sands, vibration or tamping can drastically alter grain arrangement and bulk specific gravity (Fig.2-6). As early as 1925, Terzaghi compacted sands and found that porosity decreases from 40 to 30% on applying a load of 1,250 psi. Tickell et al. (1933) compacted loose quartz sands at pressures of up to 4,150 psi and found a reduction of only 1 or 2 porosity percent. Gilboy (1928, in: Terzaghi and Peck, 1968) showed that unconsolidated sands exhibit greater compressibility when scale-shaped particles, such as micas, were included in the test sample.

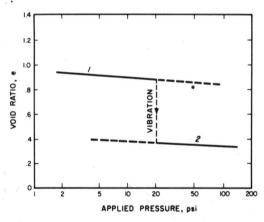

Fig.2-6. Relationship between void ratio and applied load before and after application of vibration. 1 = very loose sand (ϕ = 47—48%), and 2 = very compact sand (ϕ = 26—27%). (See Chapter 3, Vol. II, on diagenesis regarding terms "loose", "compact", etc.)

Carpenter and Spencer (1940) compacted consolidated cores from oil wells using triaxial loading and found only negligible compaction up to pressures of 8,000 psi. These authors, however, referred to work by other investigators on unconsolidated sands, which showed a large compaction potential. The above authors did not consider this to be applicable to deeply buried sediments. They also investigated the effect of temperatures up to 140° F on compaction and found no apparent relationship.

These early investigations of the compressibility of sands were either largely ignored or overlooked for many years. Most of the work done in the 1940s and early 1950s concentrated on the compressibility or compaction potential of consolidated sandstones, carbonate rocks, or well-rounded, uniformly graded quartz sands. Problems associated with compaction and surface subsidence owing to ground water and oil and gas production stimulated research work in this field in the 1950s and 1960s.

Hall (1953, p.310) utilized the reduction in pore space, V_v, at a constant overburden pressure, to obtain the "effective" compressibility values $[= -(1/V_v)(\delta V_v/\delta p_i)_{p_t}]$ of sandstones and limestones. In his equation, p_i, is the internal fluid pressure in the voids of the rock and p_t is the constant overburden pressure. He introduced another compressibility term, "formation compaction" $[= (1/V_v)(\delta V_v/\delta p_t)_{p_i}]$, which is part of the total rock compressibility. Hall's values for "effective" reservoir rock compressibilities ranged from $1 \cdot 10^{-7}$ to $3.4 \cdot 10^{-6}$ psi^{-1} by gradually reducing static fluid pressure in the core (0—26% porosity range) at a constant external hydrostatic pressure p_t of 3,000 psi. According to Hall, as porosity increases, the "effective" rock compressibility decreases. Obviously, this correlation is not applicable in the case of unconsolidated sands. Hall (1953, p.310) applied a constant external hydrostatic pressure to his lucite-jacketed cores and allowed the pore pressure to be reduced in steps.

Geertsma (1957) demonstrated that the bulk and pore volume compressibilities of cemented sandstones are functions of the elastic and viscous deformation of the cemented rock matrix and rock bulk material and the porosity. On the basis of mathematical analysis, he also showed that for many oil sands, which show isotropic elastic behaviour, pore volume compressibility measured in the laboratory under uniform and constant pressure is about twice the compressibility in the reservoir.

Handin and Hager (1957) tested a variety of rock types under high confining pressures in order to determine their compressive and shear strengths and ductility. They examined sandstone samples having various degrees of cementation. Their results showed that under high confining pressures, rock strengths were quite similar for the poorly and moderately-well cemented samples, whereas the behaviour of well-cemented samples was similar to that of quartzite. The only significant differences in properties were found to exist in the case of ductility. The lesser cemented sands showed a much higher ductility than did the well-cemented and quartzitic sands.

Fatt (1958) compacted a number of consolidated sandstone cores having relatively low porosity in the pressure range 0—15,000 psi in a triaxial (hydrostatic) compaction apparatus and found them to be only slightly compressible. His experiments probably constitute the first elaborate research work using actual core samples and approximating underground conditions. He classified the samples according to composition and texture and found that the compressibility of poorly-sorted sands was much higher than that of well-sorted ones.

Roberts and De Souza (1958) tested several samples of clay, shale, and unconsolidated quartz sand in a uniaxial compaction apparatus. It should be pointed out, however, that in the field of soil mechanics, a laterally restrained, uniaxial compression test is often referred to as triaxial loading. They found the sands to be either as, or more compressible than, clays in the pressure range 5,000—10,000 psi. These authors also described the sounds of fracturing sand grains (popping) with increasing pressure, photographed the results, and attributed most of the compaction at high loading pressures to this phenomenon.

Handin et al. (1963) conducted a comprehensive series of experiments on the deformation of cemented sandstones under triaxial pressures as high as 50,000 psi. They found that beyond the realm of elastic compressibility, some samples showed grain fracturing along small shear planes, whereas in the case of others, fractures were distributed throughout the sample. Handin et al. stated that, generally, frictional resistance was high and deformation was cataclastic (breakage across grains and cement) at high confining pressures of about 1.2 kbars. Owing to the frictional resistance to slippage, shearing and fracturing occurred across the grains because the major minerals, of which most sands are composed (quartz, etc.), preferentially fracture rather than flow.

Van der Knaap and Van der Vlis (1967) uniaxially compacted both sand and shale core samples from Lake Maracaibo, Venezuela, oil wells at pressures up to 6,000 psi. These authors found that the sands and shales were both compressible (Fig.2-7) and estimated that their contribution to compaction and subsequent subsidence should be about equal. They also showed that the rates of shale compaction were very slow compared to unconsolidated sands under similar loading conditions. This difference in compaction rate is due to the low permeability of shales and their relatively strong internal structure.

Roberts (1969) uniaxially tested a large number of samples of oil sands, beach sands, and mixtures of quartz sand and other materials prepared in the laboratory. The comparison of these tests showed that sands were either as compressible as, or more compressible than, clays in the pressure range 1,000—10,000 psi. Both Roberts and De Souza (1958) and Roberts (1969)

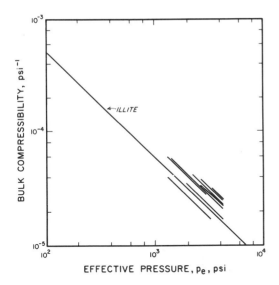

Fig.2-7. Compressibility of a number of sand samples from the post-Eocene formations of the Bolivar Coast. Illite curve is after Chilingar and Knight, 1960, fig.2, p.104. (After Van der Knaap and Van der Vlis, 1967, fig.5, p.90.)

concluded that in deep sedimentary deposits the compressibility of sands can be either equal or greater than that of clays. Roberts (1969) interpreted the sharp "break" present on some of his sand compression curves (plotted as void ratio versus log of pressure) as being the point at which sand grains commenced shattering (Fig.2-8). The shattering-point pressure was considered to be related to the initial density, angularity and grain size in these unconsolidated sands. Densely-compacted, well-rounded, smaller-sized grains have a high break-point pressure, whereas lesser compacted sands with angular, large grains tend to have a lower break-point pressure. According to Roberts (1969), clays undergo most of the compaction at low pressures and high void ratios, whereas at pressures above 5,000 psi unconsolidated sands may be more compressible.

A similar break in slope of the void ratio-versus-log pressure curve is often present in the case of clays and soils. This break is considered to be an indication of the magnitude of previous overburden load, called the preconsolidation pressure by soils engineers. The portion of the compression curve at pressures above this point is termed the "virgin" range, meaning that the specimen has not been previously subjected to loads of that magnitude (Fig.2-9). The resemblance of this change of slope on the void ratio-versus-log pressure curves of both clay and unconsolidated sand samples may be coincidental and not related in origin if fracturing of sand grains occurs.

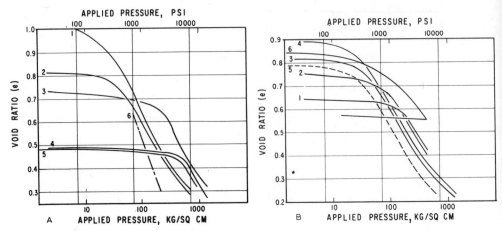

Fig.2-8. Relationship between void ratio and applied pressure for ground quartz, ground feldspar, ground dolomite, and various natural sands. A. Curve *1* = 20—40 ground quartz (loose); *2* = 20—40 ground quartz (dense); *3* = 20—40 Ottawa Sand (loose); *4* = 20—40 Ottawa Sand (dense); *5* = graded Ottawa Sand; *6* = core sample of oil sand from California. (After Roberts, 1969, fig.2, p.370.) B. Curve *1* = Rhode Island Sand, Sandy Point; *2* = Plum Island Sand; *3* = 20—40 ground quartz; *4* = 20—40 ground feldspar; *5* = 20—40 ground dolomite; *6* = 100—325 ground quartz. (After Roberts, 1969, fig.3, p.371.) Numbers, e.g., 20—40, refer to sieve sizes, U.S. Bureau of Standards.

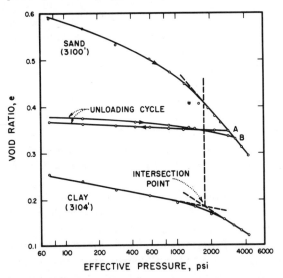

Fig.2-9. Relationship between the effective pressure (difference between total pressure and pore pressure) and void ratio for adjacent clay and sand samples from a post-Eocene formation of the Bolivar Coast, Venezuela. (After Van der Knaap and Van der Vlis, 1967, p.89.)

On discussing the cause of oil field subsidence, Allen and Mayuga (1969) compared uniaxial compaction data from oil well cores of arkosic sands and siltstones with that for shallow clays and silts. They concluded that surface subsidence was caused by compaction and that the sands constitute the major compacting material in Wilmington Oil Field, California (Fig.2-10). The shales (and siltstones) appeared to have already achieved an indurated state relative to the shallow-zone sands. With sufficient time, however, the shales may also have contributed about 1/3 of the total volumetric compaction. It is interesting to note that at comparable depths in this field, the shale and sand porosities are often similar, even though the apparent structural strength of the shales is far greater than that of the sands.

Chilingar et al. (1969) compared the results of compaction tests of sands, shales and clays performed by several investigators. This comparison was expanded by Sawabini in 1971 with the inclusion of his work and that of Kohlhaas and Miller (1969) who used unconsolidated Wilmington Oil Field cores (Fig.2-11). Sawabini used Wilmington Oil Field sand cores and simulated reservoir loading (triaxial, approaching hydrostatic), pore pressures, and temperature (Fig.2-12). He reduced the pore pressure incrementally as might occur in an oil field during producing operations. Before mounting in the compaction apparatus, his samples were imbibed with brine under vacuum, to ensure that the pores were saturated with water.

The samples tested were fine- to coarse-grained arkosic sands; most of them were well sorted and loose, and contained 10 to 15% infill of silt and clay material. Sawabini's tests showed high compressibilities while using a simulated overburden pressure of 2,500—3,000 psi and incrementally reducing the pore pressure to atmospheric. His results, presented as void ratio-versus-pressure curves, average slightly higher than those obtained from (1) one-dimensional tests in which the pore pressure was atmospheric, and (2) tests using triaxial loading and atmospheric pore pressure (Colazas, 1971). Pore and bulk volume compressibilities determined by Sawabini (1971) also are somewhat higher than those obtained by other investigators using unconsolidated sands and both triaxial and uniaxial loadings, but atmospheric pore pressure. Inasmuch as the data spread is wide, however, it is not certain that appreciable difference exists between the uniaxial and triaxial (or hydrostatic) test methods. It appears that in the case of triaxial or polyaxial loading, stress is applied to a larger number of grain contacts at random in different directions. This causes grains to slide past each other and rearrange themselves to a closer packing, at least during the initial stages of compaction. Figure 2-13 illustrates the appearance of Sawabini's samples before and after compaction and the fluids collected.

Colazas (1971) also tested unconsolidated Wilmington oil sands, some from the same approximate depth interval as Sawabini's (1971). His samples

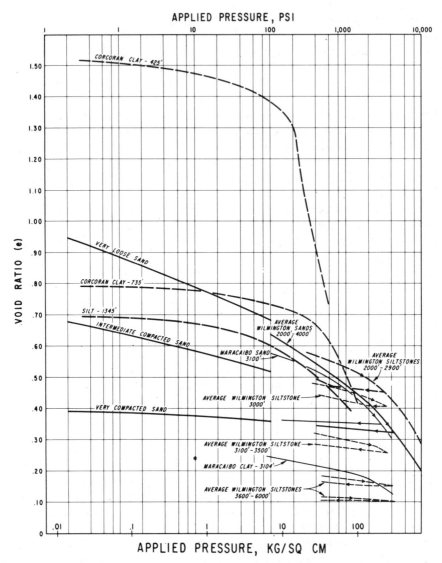

Fig.2-10. Comparison of compression tests on Wilmington Oil Field sands and shales, shallow sediments from water-bearing strata, and oil-zone shale and sand from Lake Maracaibo, Venezuela. Void ratio, e is plotted versus applied pressure. (After Allen and Mayuga, 1969, fig.7, p.415.)

were cleaned, mounted in a foil sheath and compacted in a triaxial compaction apparatus (see Fig.2-14). The samples, tested at room temperature, contained no pore fluids and the pore pressure was atmospheric. The average

MECHANICS OF SAND COMPACTION

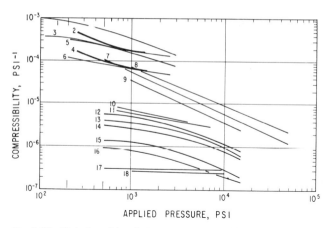

Fig. 2-11. Relationships between compressibility and applied pressure for different rock types obtained by various investigators using different loading methods. (Modified after Sawabini, 1971.)

Curve No.	Investigator	Sand or rock type	Load type	Compressibility [*5]	
1	Sawabini (1971)	unconsolidated arkosic sands, wet[*1]	triaxial (?) (hydrostatic)	pore	$\left[-\dfrac{1}{V_p}\left(\dfrac{\delta V_p}{\delta p_e}\right)_\sigma\right]$
2	Kohlhaas and Miller (1969)	unconsolidated arkosic sands, dry	uniaxial	pore	
3	Sawabini (1971)	unconsolidated arkosic sands, wet[*1]	triaxial (?) (hydrostatic)	bulk	$\left[-\dfrac{1}{V_b}\left(\dfrac{\delta V_b}{\delta p_e}\right)_\sigma\right]$
4	Kohlhaas and Miller (1969)	unconsolidated arkosic sands, dry	uniaxial	bulk	
5	Colazas (1971)	unconsolidated arkosic sands, dry	triaxial	pore	
6	Colazas (1971)		triaxial	bulk	
7	Chilingar, Rieke, and Sawabini (1969)	illite clay, wet[*2]	uniaxial	bulk	$\left[-\dfrac{1}{V_b}\left(\dfrac{\delta V_b}{\delta p_e}\right)_{p_p}\right]$
8	Chilingar, Rieke, and Sawabini (1969)	illite clay, dry	uniaxial	bulk	$\left[-\dfrac{1}{V_b}\left(\dfrac{\delta V_b}{\delta p_e}\right)_{p_p}\right]$
9	Knutson and Bohor (1963)	Repetto Formation, wet[*1]	net confining	"pore"	$\left[-\dfrac{1}{V_p}\left(\dfrac{\delta V_p}{\delta \sigma}\right)_{p_p}\right]$

Fig.2-11 — *continued*

Curve No.	Investigator	Sand or rock type	Load type	Compressibility
10	Knutson and Bohor (1963)	Lansing—Kansas City Limestone, wet*[1]	net confining	"pore"
11	Carpenter and Spencer (1940)	Woodbine Sandstone, wet	net confining	pseudo-bulk $\left[-\frac{1}{V_b}\left(\frac{dV_v}{dp}\right)\right]$
12	Fatt (1958b)	feldspathic greywacke, wet*[3]	net confining*[4]	bulk $\left[-\frac{1}{V_b}\left(\frac{\delta V_b}{\delta p_e}\right)_{p_p}\right]$
13	Fatt (1958b)	greywacke, wet*[3]	net confining	bulk
14	Fatt (1958b)	feldspathic greywacke, wet*[3]	net confining	bulk
15	Fatt (1958b)	lithic greywacke, wet*[3]	net confining	bulk
16	Fatt (1958b)	feldspathic quartzite, wet*[3]	net confining	bulk
17	Podio et al., (1968)	Green River Shale, dry	net confining	bulk
18	Podio et al., (1968)	Green River Shale, wet*[3]	net confining	bulk

*[1] Saturated with formation water.
*[2] Saturated with distilled water.
*[3] Saturated with kerosene.
*[4] Net confining pressure, p_e = external hydrostatic pressure on a jacketed specimen, σ, minus $0.85p_p$; where p_p is the pore pressure.
*[5] p_e = effective pressure, σ = total overburden pressure, V_b = bulk volume, $V_p = V_v$ = void or pore volume.

test results of Colazas were very similar to those of Sawabini, as shown in Fig.2-11. Colazas also compared the void ratio-versus-pressure curve of Ottawa Sand (well-rounded, uniform, clean quartz sand) with that of the same sand having clay admixtures (Fig.2-15). A frequency distribution graph of a typical Wilmington Oil Field sand, similar to those tested by Sawabini (1971) and Colazas (1971), is presented in Fig.2-16.

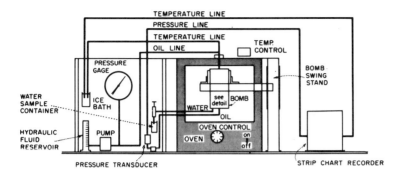

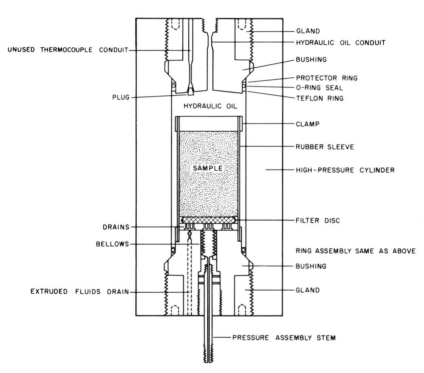

Fig. 2-12. Schematic diagram of compaction equipment used by Sawabini et al. (1971, fig.1 and 2.) Top: assembled compaction apparatus; bottom: assembled pressure vessel. (Courtesy of *J. Sediment. Petrol.*)

The following formulas were used by Sawabini (1971) in plotting his data and for the conversion of Kohlhaas and Miller's (1969) data (Fig.2-11). Colazas's data were converted by the writers of this chapter using the same formulas:

Pore volume compressibility, $c_p = \left[\dfrac{1}{V_p} \dfrac{(\delta V_p)}{(\delta p_e)} \right]_{\sigma, T}$ (2-6)

Bulk volume compressibility, $c_b = \left[\dfrac{1}{V_b} \dfrac{(\delta V_b)}{(\delta p_e)} \right]_{\sigma, T}$ (2-7)

and in terms of void ratio, $c_b = \left[\dfrac{1}{1 + e_o} \dfrac{(\delta e)}{(\delta p_e)} \right]_{\sigma, T}$ (2-8)

where V_p = original pore volume, V_b = original bulk volume, δV_p and δV_b = changes in pore and bulk volume, respectively, with respect to the original volumes, e_o = original void ratio, δe = change from the original void ratio at a particular p_e, and p_e = effective applied pressure. In Sawabini's (1971) tests, p_e was the external pressure, σ, minus the measured pore pressure, p_p. The p_e was considered to be equal to the external applied pressure for the test data of both Colazas (1971) and Kohlhaas and Miller (1969), because the pore pressure was atmospheric. Tests are conducted at a constant overburden load, σ, and constant temperature, T.

Pore volume compressibilities of unconsolidated arkosic sands range from about 10^{-3} to 10^{-4} psi at effective pressure range of 0 to about

Fig.2-13. Core sample before and after compaction; and the expelled fluids. Round filter disk before and after compaction is also shown. Overburden pressure = 3,000 psi. (After C. T. Sawabini, personal communication, 1971.)

Fig.2-14. Jacketed core samples at different stages of loading. *1* = 200 psi (sleeve setting pressure); *2, 3* = 2,500 psi pressure. (After Colazas, 1971, part IV, p.135.)

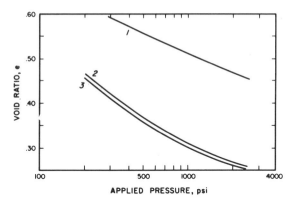

Fig.2-15. Relationship between void ratio and applied pressure for pure Ottawa Sand and Ottawa sand mixed with clays. *1* = Ottawa Sand; *2* = mixture of 50% by weight of Ottawa Sand and 50% illite clay; *3* = mixture of 34% Ottawa Sand, 33% illite, and 33% kaolinite. (After Colazas, 1971, fig.81—85, pp.184—186.)

3,000 psi. Bulk volume compressibilities are lower and range from about $4 \cdot 10^{-4}$ to $8 \cdot 10^{-5}$. Corresponding void ratio range is from 0.6 to 0.35 (or 0.40). These ranges appear to hold true for both uniaxial and triaxial (or hydrostatic) loadings, although triaxial loading tests appear to give slightly

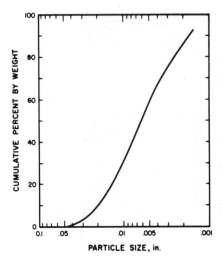

Fig. 2-16. Grain size distribution curve of typical Wilmington Oil Field arkosic sand used in compression tests of Sawabini (1971) and Colazas (1971).

higher compressibilities. Owing to the wide range (spread) of individual test data, however, the differences seen on the average curves may not be significant. Both Colazas's (1971) and Kohlhaas and Miller's (1969) samples were oil-zone cores that had been cleaned, dried and mounted in metal sleeves by the application of pressure. Sawabini's (1971) samples were obtained from the same intervals, but contained original pore fluids. Instead of cleaning his samples, he imbibed them with formation brine under vacuum. Thus, the differences in compressibility obtained by different authors could be due to the method of preparation of samples and of jacketing.

Newman (1972) presented the compressibilities and porosities of 256 rock samples from 40 reservoirs (197 sandstone samples from 29 reservoirs and 59 samples of carbonate rocks from 11 reservoirs). Porosities of these samples ranged from less than 1 to 35%. He used the following formula in determining pore volume compressibility, c_p:

$$c_p = \frac{1}{V_p} \frac{dV_p}{dp_e} \qquad (2\text{-}9)$$

where V_p = pore volume of the sample at a given net (effective) pressure, dV_p = incremental change in pore volume resulting from an incremental change in net pressure, and dp_e = incremental change in net pressure. The above equation assumes that most of the pore volume changes results from the net pressure difference. A common pressure base of 75% of the litho-

MECHANICS OF SAND COMPACTION 67

static pressure was used for comparison. The lithostatic pressure gradient was assumed to be equal to 1 psi/ft of depth.

Newman (1972) showed that pore volume compressibilities for a given initial porosity can vary widely, depending on the rock type. He found major differences between consolidated sands (Fig.2-17), friable sands (Fig.2-18), and unconsolidated sands (Fig.2-19). Relationship between pore volume compressibility and initial porosity for carbonate rocks is presented in Fig.2-20. In the opinion of the writers, more useful correlation can be obtained if samples of similar lithologies, exposed to similar overburden pressures and similar initial porosities, are used. In Newman's experiments,

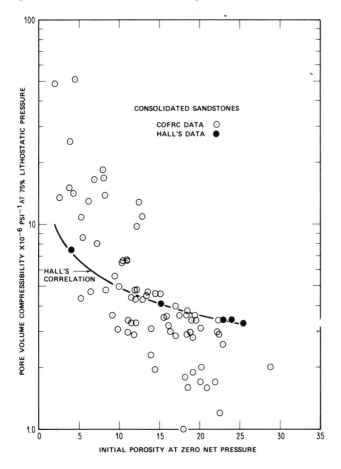

Fig.2-17. Relationship between pore volume compressibility at 75% lithostatic pressure and initial sample porosity at zero net pressure for consolidated sandstones. (After Newman, 1973, fig.4, p.133. Courtesy of the Society of Petroleum Engineers of AIME.)

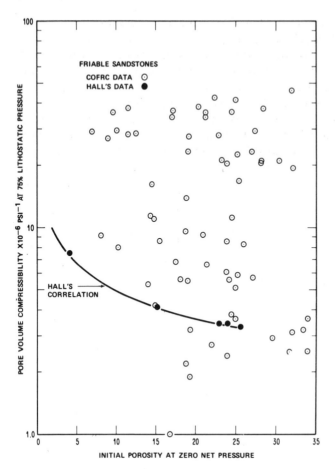

Fig.2-18. Relationship between pore volume compressibility at 75% lithostatic pressure and initial sample porosity at zero net pressure for friable sandstones. (After Newman, 1972, fig.5, p.133. Courtesy of the Society of Petroleum Engineers of AIME.)

81% of the samples were measured at 74°F, whereas the remaining 19% were tested at temperatures ranging from 130 to 275°F. He did not observe any temperature effect on compressibility. The compaction apparatus used by Newman is shown in Fig.2-21. Newman's experiments undoubtedly represent a major effort towards the understanding of compaction processes and it would be rewarding to see them presented as graphs of compressibility versus pressure or void ratio versus pressure.

The following conclusions can be reached on examining the laboratory experimental results obtained by several investigators: Unconsolidated sands are readily compactable on applying overburden loads of 200 psi or higher.

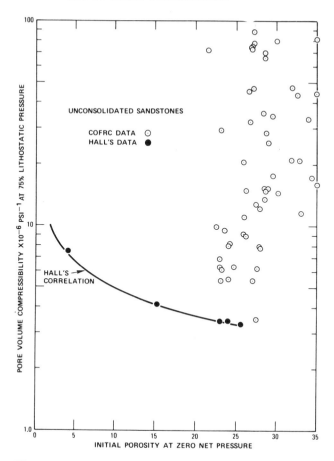

Fig. 2-19. Relationship between pore volume compressibility at 75% lithostatic pressure and initial sample porosity at zero net pressure for unconsolidated sandstones. (After Newman, 1972, fig.6, p.133. Courtesy of the Society of Petroleum Engineers of AIME.)

The degree to which they can be compacted is dependent upon the (a) original packing and/or bridging, (b) original void ratio, (c) shape of the grains, (d) roundness of grains, (e) sphericity of grains, (f) composition of the sand, and (g) size grading. At least up to 140°F and an overburden load of about 3,000 psi, compressibility does not seem to be dependent to any large degree upon temperature or pore fluid type.

Well-sorted, rounded, well-packed, clean quartz sands do not compact readily except when the applied loads are sufficient to break the sand grains. The load necessary to crush or fracture grains is not as high as one might expect, because pressures upon contact points may be considerably

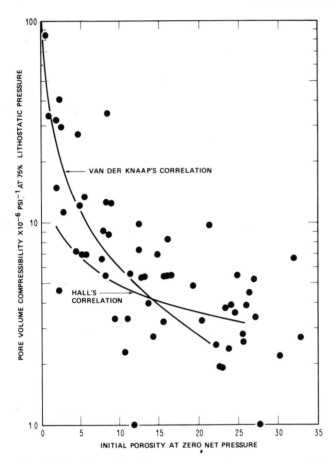

Fig.2-20. Relationship between pore volume compressibility at 75% lithostatic pressure and initial sample porosity at zero net pressure for carbonate rocks. (After Newman, 1973, fig.3, p.132. Courtesy of the Society of Petroleum Engineers of AIME.)

amplified above the average load pressure by mechanical advantage. Loosely-packed, irregularly-shaped grains, particularly those composed of comparatively brittle minerals such as feldspar, compact readily by grain rearrangement and shattering of sharp grain points. When present in the pore spaces, micas, clays, and other fine-grained materials appear to act somewhat like a dry lubricant and increase sand compressibility. From available test data of the four major types of sands, arkosic sands exhibit the greatest compactibility, because of composition and, possibly, grain angularity.

Volcanic tuffs, which are commonly composed largely of glass-like shards with a high bridging tendency, might be highly susceptible to

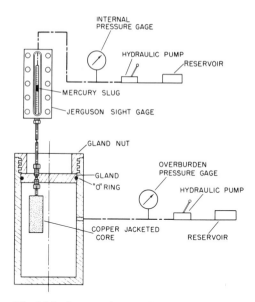

Fig. 2-21. Compaction equipment for measuring pore volume compaction and compressibility used by Newman (personal communication, 1972). (After Fatt, 1958a, fig.1, p.362. Courtesy of the Society of Petroleum Engineers of AIME.)

compaction if the applied load is sufficient to crush the skeletal framework (see Chapter 3, Vol. II, on diagenesis).

Various compaction experiments on unconsolidated sands have shown an elastic rebound or bulk volume expansion of only about 1/2%, after release of applied load. Thus, compaction (as much as 1/3—1/2 of the original pore space at loadings of 3,000 psi or less) is largely irreversible.

Consolidated and cemented sands, when compressed at pressures less than that required to fracture the samples, showed a bulk volume decrease of about 1—2% (Wilhelmi and Somerton, 1967). This can be largely attributed to elastic deformation of the skeletal structure.

MAN-INDUCED COMPACTION AND SURFACE SUBSIDENCE

The best examples of the role of fluid pressure and load transfer in compaction have been inadvertently created by man's use of his natural resources, primarily water and oil. Extraction of these fluids has caused widespread subsidence in many areas of the world. The role of reduction in pore fluid pressure in causing the transfer of load to the sand skeleton was recognized in 1925 by Minor and by Pratt and Johnson in 1926. These authors attributed the surface subsidence of about 3 ft over the Goose Creek

Oil Field, Harris County, Texas, to fluid withdrawal and subsequent subsurface compaction. Meinzer and Hard (1925) recognized that with lowering of the hydrostatic head, artesian aquifers compacted owing to the transfer of a portion of the overburden load from the pore fluids to the sandstone framework.

Since about 1925, the problems of surface subsidence caused by the extraction of fluids have become world wide. Owing to reservoir compaction, surface subsidence as great as 29 ft has occurred over Wilmington Oil Field, Los Angeles County, California (Allen and Mayuga, 1969). (See Fig. 4-20, p. 237.)

Poland and Davis (1969) reported that an area of about 3,000 sq miles in the San Joaquin Valley, California, was subsiding to some extent as a result of compaction caused by ground water extraction. They reported a surface subsidence of 23 ft near Mendota, California. This subsidence was attributed to compaction of silts, silty sands, and clays, and is considered to be largely irreversible. According to these authors, the differences in subsidence from one area to another, when compared to the amount of hydrostatic head decline, were chiefly due to variations in mineral composition, cementation, grain sorting, mode of packing, degree of previous consolidation, and lateral confinement.

Examples of subsidence caused by man's use of natural resources have been documented in the U.S.A., U.S.S.R., Mexico, Japan, England, Union of South Africa, China, Germany, and Italy. (See Chapter 4 for detailed discussion.)

CHRONOLOGY OF SAND COMPACTION

The initial porosities of freshly deposited sands range from about 25 to 50% and higher, dependent upon particle shape, size grading, roundness, packing arrangement and composition.

Initial packing arrangement and void ratio strongly influence the compressibility of a sand. Bridging of particles can form a strong skeletal framework that resists deformation, but can be abruptly disarranged by vibration.

Compaction takes place when an overburden load is applied or transferred to the skeletal framework. Under normal conditions of deposition and geostatic support, this load at any point is equal to the weight of the overburden less the neutral or pore pressure.

At the time of deposition, shales that are normally found interbedded with sands have a much higher porosity than the sands. As deposition proceeds and the overburden load increases, initially the shale intervals compact at a greater rate than do the sands. Inasmuch as the total load is supported jointly by the skeletal framework and the pore water, water is slowly extruded from the shales into the more porous and permeable sands. "Plate-like" or "blade-like" clay components of the shales, containing adsorbed water, initially have very little internal self-supporting structure. The shale pore water, which carries a great portion of the geostatic load (Terzaghi's excess pore pressure), bleeds into the nearby sands where the pore water is relatively less pressured. Inasmuch as the sands are generally more permeable than the shales and are, possibly, hydraulically connected at great distances to areas of pressure relief, the intrusion of water from shales causes flow or migration. Eventually, the shales cease to compact at a rapid rate. During the period of relatively rapid shale compaction, the sands also support a portion of the overburden load and compact to some extent. As the overburden load becomes greater, the rate of sand compaction increases and, eventually, may exceed the rate of shale compaction. At shallow depths, perhaps in the order of 1,000 ft or less, clays and shales may have a greater tendency to compact upon pore fluid pressure reduction than sands, whereas at greater depth sand compaction may be greater.

Initial sand compaction occurs as a result of repacking and rearrangement of grains into a closer-fitting system. Grain rotation, sliding along flat faces, and compression and plastic flow of micas, clays and other relatively soft minerals contribute to compaction during this stage. Taylor (1950) suggested that presence of three or more face contacts per grain in a thin section indicate that mechanical compaction has taken place. She classified grain contacts caused by pressure into three classes: concavo-convex, tangential-long, and sutured. Taylor also photographed the development of pressure cracks in individual grains that had not severed, apparently due to confinement. The end result of pressure solution is formation of the sutured (wavey) grain contacts.

At high pressure and/or temperature ranges, grains undoubtedly deform by sliding along twinning planes or by true plastic deformation of even the hardest minerals. These phenomena, however, are not necessary to reduce mechanically the porosity of sands to 20% or lower.

Generally, sands and shales of similar composition and with a similar geologic history will attain similar porosities and densities, but the time-rate of reaching the final porosity may be different. If, along the geological path of burial and compaction, the sands are cemented, compaction may virtually cease for a time except for that portion related to elastic deformation. If pressures and temperatures are continually increased, plastic flow, sliding and breaking of cementing material will occur eventually. Cementation of grains is considered to be overwhelmingly the most important single factor limiting mechanical sand compaction.

Excessive pore-fluid pressures form in case of rapid sediment deposition if the pore fluids cannot escape due to the presence of cementation (permeability) barriers, fault barriers, pinch-outs, etc. The Gulf Coast area of the United States is a prime example of rapid deposition and geosynclinal basin subsidence. Overpressured zones are common in this area and must be carefully watched for and predicted while drilling deep wells. These overpressured zones are detectable by shale natural bulk specific gravity (density) measurements and examination of various formation well logs. A great portion of the overburden load is borne by the fluids in such a case and, consequently, relatively porous, undercompacted sands can exist at greater depths than thought possible only a few years ago.

Excessive pore-fluid pressures can exist for many other reasons, e.g., (1) tectonic folding and compression in an area where pore fluids cannot migrate and (2) uplift of sediments with pore pressures generated by greater overburden loads. Excessive pore-fluid pressures have also been attributed to a natural build-up in large anticlines caused by the presence of a two-fluid (oil—water) pressure system (differential fluid densities — Jodry, 1972, pp.47—49).

During diagenesis and metamorphism, large volumes of water could be released and thus create the overpressured formations. This is considered by some authors to be the mechanism involved in the formation of high-pressured geothermal reservoirs.

It is unlikely that any previously deeply buried sand unpacks to a lesser porosity as a result of uplift, although elastic expansion probably would increase porosity by a small amount. Loose sand cores, cut at depth, often swell when brought to the surface owing to elastic expansion and extrusion of pressured pore fluids. This would not take place, however, as long as the lateral restraint of the formations is present.

One known area of artificially induced formation expansion has been documented by Allen and Mayuga (1969). High-pressure water injection into deep oil reservoirs caused surface uplift exceeding 1 ft. The total area of uplift covers about 10 sq miles with an average uplift of about 1/2 ft or less. This effect is believed to be due to elastic expansion of the sand grains when the pore-fluid pressure is raised and a larger part of the overburden load is transferred to the interstitial fluid.

NEEDED AREAS OF RESEARCH ON COMPRESSIBILITY OF SANDS

Certain aspects of sand compaction do not appear to have been adequately covered by researchers. A quantitative evaluation of the effects of composition, size grading and particle shape on compaction is needed.

Inasmuch as all of these factors are interrelated in nature, a thorough, separate evaluation of each one may be difficult to achieve from collected samples. Representative laboratory samples, however, could be prepared for evaluation purposes.

Comparison tests are needed on the greywackes and lithic sandstones in particular. The majority of the unconsolidated sand samples tested and described were arkosic. Many other samples tested by various investigators were fabricated from quartz sands.

Comparison of the compressibility of cemented sandstones having different composition might also be useful in studies of rock fracturing and other oil field operations. In the case of studying cemented sandstones, other parameters such as the modulus of elasticity and Poisson's ratio become important. Attempts have been made to apply these concepts to unconsolidated sand but with little success. The nature of unconsolidated sand compaction (grain rearrangement, etc.) precludes the use of these variables.

The writers suggest that on testing any particular sand type of certain mineralogic composition, the sample should be sieve sized and comparisons be made to others of similar composition but with different size gradings and particle shapes. Experiments should be performed not only on *homogeneous* sand samples of different compositions but also on *layered* samples such as sand—silt, sand—clay, sand—silt—clay, etc. Layer thickness should be varied and both vertical and horizontal compressibilities measured. Further study in these areas should contribute to the understanding of sediment compaction.

REFERENCES

Allen, D. R. and Mayuga, M. N., 1969. The mechanics of compaction and rebound, Wilmington oil field, Long Beach, California. In: *Land Subsidence. IASH/UNESCO, Tokyo*, 89(2): 410—413.

Athy, L. F., 1930. Density, porosity, and compaction of sedimentary rocks. *Bull. Am. Assoc. Pet. Geol.*, 14: 1—24.

Carpenter, C. B. and Spencer, G. B., 1940. Measurements of compressibility of consolidated oil-bearing sandstones. *U.S. Bur. Mines, R.I.*, 3540: 20 pp.

Chilingar, G. V. and Knight, L., 1960. Relationship between pressure and moisture content of kaolinite, illite, and montmorillonite clays. *Bull. Am. Assoc. Pet. Geol.*, 44: 101—106.

Chilingar, G. V., Rieke III, H. H. and Sawabini, C. T., 1969. Compressibilities of clays and some means of predicting and preventing subsidence. In: *Symp. Land Subsidence. IASH/UNESCO, Tokyo*, 89(2): 377—393.

Colazas, X. C., 1971. *Subsidence, Compaction of Sediments and Effects of Water Injection, Wilmington and Long Beach Offshore Oil Fields*. Thesis, Univ. South. Calif., 198 pp. (unpubl.).

Fatt, I., 1958a. Pore volume compressibility of sandstone reservoir rocks. *J. Pet. Tech.*, 10(3): 64—66.

Fatt, I., 1958b. Compressibility of sandstones at low to moderate pressure. *Bull. Am. Assoc. Pet. Geol.*, 42: 1924—1957.
Fuller, M. L., 1908. Summary of the controlling factors of artesian flows. *U.S. Geol. Surv. Bull.*, 319: 44 pp.
Geertsma, J., 1957. The effect of fluid pressure decline on volumetric changes of porous rocks. *Trans. Am. Inst. Min., Metall. Pet. Eng.*, 210: 331—340.
Hall, H. N., 1953. Compressibility of reservoir rocks. *Trans. Am. Inst. Min. Metall. Pet. Eng.*, 198: 309—311.
Handin, J. and Hager, R., 1957. Experimental deformation of sedimentary rocks under confining pressure: Tests at room temperature on dry samples. *Bull. Am. Assoc. Pet. Geol.*, 41: 1—50.
Handin, J., Hager, Jr., R. V., Friedman, M. and Feather, J. N., 1963. Experimental deformation of sedimentary rocks under confining pressure; pore pressure tests. *Bull. Am. Assoc. Pet. Geol.*, 47: 717—755.
Harrison, E., Kiesnick, Jr., W. F. and McGuire, W. J., 1954. The mechanics of fracture induction and extension. *Trans. Am. Inst. Min., Metall. Pet. Eng.*, 201: 254—255.
Hedberg, H. D., 1936. Gravitational compaction of clays and shales. *Am. J. Sci., Ser. 5*, 31: 241—287.
Hough, B. K., 1957. *Basic Soils Engineering.* Ronald, New York, 513 pp.
Hubbert, M. K. and Rubey, W. W., 1959. Role of fluid pressure in mechanics of overthrust faulting, I. Mechanics of fluid-filled porous solids and its application to overthrust faulting. *Bull. Geol. Soc. Am.*, 70: 115—166.
Hubbert, M. K. and Rubey, W. W., 1960. Role of fluid pressure in mechanics of overthrust faulting, a reply to discussion by H. P. Laubscher. *Bull. Geol. Soc. Am.*, 71: 617—628.
Hubbert, M. K. and Willis, D. G., 1957. Mechanics of hydraulic fracturing. *Trans. Am. Inst. Min., Metall. Pet. Eng.*, 210: 153—166.
Jodry, R. L., 1972. Pore geometry of carbonate rocks (basic geologic concepts). In: G. V. Chilingar, R. W. Mannon and H. H. Rieke III (Editors), *Oil and Gas Production from Carbonate Rocks.* Am. Elsevier, New York, N.Y., 408 pp.
Knutson, C. F. and Bohor, B. F., 1963. Reservoir rock behavior under moderate confining pressure. In: C. Fairhurst (Editor), *Rock Mechanics.* Pergamon, New York, N.Y., pp.627—658.
Kohlhaas, C. A. and Miller, F. G., 1969. Rock compaction and pressure transient analysis with pressure-dependent rock properties. *Soc. Pet. Eng., 44th Annu. Fall Meet., Denver, Colo., SPE Pap.*, 2563: 7 pp.
Matthews, W. R. and Kelly, J., 1967. How to predict formation pressure and fracture gradient. *Oil Gas J.*, 65, 8: 62—106.
Meinzer, O. E. and Hard, H. H., 1925. The artesian water supply of the Dakota Sandstone in North Dakota, with special reference to the Edgeley quadrangle. *U.S. Geol. Surv., Prof. Pap.*, 520E: 73—95.
Minor, H. E., 1925. Goose Creek oil field, Harris County, Texas. *Bull. Am. Assoc. Pet. Geol.*, 9: 286—297.
McCulloh, T. H., 1969. Geologic characteristics of the Dos Cuadros offshore oil field. In: *U.S. Geol. Surv., Prof. Pap.*, 679: 29—46.
Newman, G. H., 1973. Pore-volume compressibility of consolidated, friable, and unconsolidated reservoir rocks under hydrostatic loading. *J. Pet. Tech.*, 25(2): 129—134.
Podio, A. L., Gregory, A. R. and Gray, K. E., 1968. Dynamic properties of dry and water-saturated Green River Shale under stress. *J. Soc. Pet. Eng.*, 8(4): 389—404.

Poland, J. F. and Davis, G. H., 1969. Land subsidence due to the withdrawal of fluids. In: D. J. Varnes and G. Kiersch (Editors), *Reviews in Engineering Geology, II.* Geol. Soc. Am., Boulder, Colo., pp.187—269.

Pratt, W. E. and Johnson, D. W., 1926. Local subsidence of the Goose Creek oil field. *J. Geol.*, 34: 577—590.

Rieke III, H. H., Ghose, S., Fahhad, S. and Chilingar, G. V., 1969. Some data on compressibility of various clays. *Proc. Int. Clay Conf.*, 1: 817—828.

Rittenhouse, G., 1971. Mechanical compaction of sands containing different percentages of ductile grains: a theoretical approach. *Bull. Am. Assoc. Pet. Geol.*, 55: 92—96.

Roberts, J. E., 1969. Sand compression as a factor in oil field subsidence. In: *Symp. Land Subsidence, IASH/UNESCO, Tokyo*, 89(2): 368—376.

Roberts, J. E. and De Souza, J. M., 1958. The compressibility of sands. *Proc. ASTM*, 58: 1269—1277.

Sawabini, C. T., 1971. *Triaxial Compaction of Unconsolidated Sand Core Samples Under Producing Conditions at Constant Overburden Pressure of 3,000 psig and a Constant Temperature of 140°F*. Thesis, Univ. South. Calif. 178 pp. (unpubl.).

Sawabini, C. T., Chilingar, G. V. and Allen, D. R., 1971. Design and operation of a triaxial, high-temperature, high-pressure compaction apparatus. *J. Sediment. Petrol.*, 41(3): 871—881.

Taylor, J. M., 1950. Pore space reduction in sandstones. *Bull. Am. Assoc. Pet. Geol.*, 34: 701—716.

Terzaghi, K., 1925. Principles of soil mechanics. IV. Settlement and consolidation of clay. *Eng. News-Rec.*, 95: 874—878.

Terzaghi, K. and Peck, R. P., 1968. *Soil Mechanics in Engineering Practice.* Wiley, New York, N.Y., 729 pp.

Tickell, F. G., Mechem, O. E. and McCurdy, R. C., 1933. Some studies on the porosity and permeability of rocks. *Trans. Am. Inst. Min. Metall. Eng.*, 103: 250—260.

Van der Knaap, W. and Van der Vlis, A. C., 1967. On the cause of subsidence in oil producing areas. *Trans. World Pet. Congr.*, 7th, 3: 85—95.

Whalen, H. E., 1968. Understanding and using frac pressures in well planning. *Pet. Eng.*, 40, 10: 75—82.

Wilhelmi, B. and Somerton, W. H., 1967. Simultaneous measurement of pore and elastic properties of rocks under triaxial stress conditions. *J. Soc. Pet. Eng.*, 7(3): 283—294.

Chapter 3

COMPACTION AND DIAGENESIS OF CARBONATE SANDS

A. H. COOGAN and R. W. MANUS

INTRODUCTION

Coarse-grained carbonate sediments compact under the pressure of overlying sediment, with increasing subsurface temperatures, through the mechanism of solution at the points of grain contact and diffusion of the solute into the pore fluid. Three main factors affect the compaction of carbonate sands: (1) the inherited factors such as packing, grain size and shape, and sorting; (2) the inhibitory factors which retard compaction, principally synsedimentary cementation; and (3) the dynamic factors which are connected with the subsurface environment, including the rate of loading, the geothermal gradient, the overburden pressure, the nature of the formation fluid, and others.

The carbonate, mainly calcium carbonate, sands considered here are the grain-supported sands, the grain diameters of which are greater than silt size. Carbonate sands have been classified by a variety of schemes (Folk, 1959; Ham and Pray, 1962; Bissell and Chilingar, 1967; Bathurst, 1971b) based on mineralogy, grain size (e.g., calcarenites), grain-to-matrix ratios (e.g., sparites), grain types (e.g., skeletal limestones) and framework or packing (e.g., grainstones). The terminology used for carbonate sediments and rocks in this chapter is not a complex one. The simplest terminology which conveys the relationships under discussion is used, so that the terms gravel, sand and silt are used as size terms and are used for unconsolidated and uncemented aggregates of those size classes regardless of mineralogical composition. Mineralogy is designated with an appropriate modifier — quartzose, calcareous, aragonitic, etc. The limestones (the cemented and consolidated sediments) are named according to Dunham's (1962) classification which stressed the grain-to-grain, grain-to-mud relationships more than other classifications. The limestones under discussion here are mainly the grainstones, that is the mud-free, grain-supported sediments and their lithified equivalents. In fact, most of those discussed are oolitic grainstones simply because more is known about them than about other compacted limestones of sand size.

Until recently it was maintained that carbonate sands compact very little, if at all, owing to the widespread phenomenon of early cementation (Weller, 1959). This conclusion is still somewhat true because many carbonate rocks are little compacted; however, some are very much compacted. The details of the compactive process and the exact parameters of the environment in which compaction proceeds are unknown. For the present, one must be satisfied with simple numerical approximations to the conditions of compaction; a comprehensive mathematical model is still lacking as are the data to formulate one (see Chapter 3, Vol. II, for discussion). Many carbonate sands begin as uncemented aggregates of organically-derived particles with mineralogies unstable for most of the postsedimentary environments which they will occupy. Such sands can be altered and completely cemented in a geologic instant to stable assemblages of particles, with altered or preserved sedimentary fabrics, or can be preserved for tens of millions of years essentially unaltered. These quick-change artists, these houdinis of mineralogical composition contrast sharply with the conservative, little-changing clastic sands, mainly quartzose, discussed elsewhere in this book.

The most fruitful approach to understanding the compaction of carbonate sands so far has been (1) to compare and contrast them with clastic or artificial sands, (2) to compare the processes and environments of compaction, and (3) to examine compacted rocks for evidence and characteristics of compaction. In this chapter the writers are concerned mainly with the grain-to-grain relationships. If carbonate rocks and sediments compact in the subsurface by a process of massive dissolution and transfer of the material away from the site, and the process leaves no trace of its existence or cannot be interpreted, then these types of rocks are disregarded here. Experimental studies of compaction have added certain limits and parameters which serve as tools for studying the process of compaction of carbonate sediments.

It is important to know the natural grain fabric resulting from sedimentation, and before the compactive stress is applied during compaction, to understand the changes in grain-to-grain relationships. This requires a consideration of the packing of particles, spherical and nonspherical, in regular and randomly ordered packs of single and multicomponent mixtures, because these relationships serve as the basis for quantitative estimates of the amount of cement derived from local-source solution and reprecipitation and the amount of compaction in selected carbonate grainstones. Once having established a practical starting point, i.e., a quantitative estimate of the grain fabric in the uncompacted state, approximations of the degree of compaction for limestones consisting of several kinds of particles can be made from measurements of grain volume and from grain-to-grain relationships.

FACTORS AFFECTING THE COMPACTION OF CARBONATE SANDS

General statement

A survey of the literature and analysis of the evidence for compaction of carbonate sands and of the environment of compaction lead to the conclusion that there are three groups of factors which affect the compaction of carbonate sands. The first group of factors (Fig.3-1), termed *inherited*, reflects the original composition of the sand and includes particle size and shape, size and shape sorting, particle mineralogy, and sedimentary fabric (packing). The second group of factors, which inhibit compaction and thus can be termed *inhibitory*, are related to physically and biologically induced chemical changes during alteration. The alteration can be penecontemporaneous with sedimentation or subaerial exposure or can occur during other diagenetic stages. The main factors in this group are early cementation and dolomitization. The third group of factors, which are related mainly to the burial of the carbonate sand, are termed *dynamic* and include two main factors, changes in overburden pressure and in temperature. Other dynamic factors are the rate of loading, length of burial time, composition and amounts of connate water and other fluids in the pore space, and pore fluid pressure.

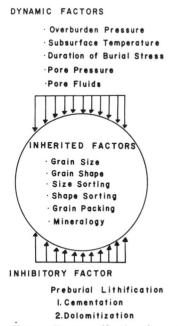

Fig.3-1. Factors affecting the compaction of carbonate sands.

In general, the inherited factors are mainly passive and affect the degree of compaction for a given compactive stress during burial. The inhibitory factors generally reduce grain-to-grain compaction, or if they are not present, their absence simply allows compaction to proceed more efficiently. The dynamic factors cause compaction of the sediment independently or in combination with other factors.

Inherited factors

The principal factors in the original carbonate sand which usually affect subsequent compaction are particle shape and size, size and shape sorting, fabric or packing, the homogeneity of packing, and particle mineralogy.

Grain size

The size of grains is important in compaction independent of its relationship to sorting, original porosity, or packing. In carbonate sands the grain size is clearly related to the grain type which, in turn, is related to the biological or physical source of the particles. For example, coarse carbonate gravels cannot consist of planktonic foraminiferal tests which do not grow to gravel-size proportions. As a result, the environments of origin of carbonate particles and of their deposition strongly control the sediment type and size. Grain size alone is important in compaction owing to the differences in surface area of particles. In addition, smaller particles give rise to larger surface areas per unit volume. As compaction proceeds, the thickness of the calcium carbonate dissolved from the particle at the point of contact, or by a through-flowing acidic fluid or fluid undersaturated in calcium carbonate, is relatively greater for small particles than for larger ones. If in the course of compaction, a 0.005-mm thick slice of aragonite is removed from a grain, this amounts to a 2% change in radius for a medium size grain (diameter = 0.5 mm), whereas it amounts to a 10% change in the radius of a very fine sand (diameter = 0.1 mm). As a result, given other constant conditions, it would be expected that finer sized sands should compact more than coarser ones. There is no statistical evidence known to the authors that this is true for carbonate sands, probably because so many other factors are involved that the effect of grain size is combined with them in a still undecipherable mix. Grain size also plays a role in cementation which in turn affects compaction.

Grain shape

Particle shape is important in compaction and is critical in any consideration of the original packing configuration of the sediment as a result of sedimentation. Almost all analyses of the packing of materials have been done with spherical particles (Graton and Fraser, 1935; Brown and Hawksley, 1945; Bernal and Finney, 1967; Ridgway and Tarbuck, 1967; Coogan, 1970; Morrow, 1971; Rittenhouse, 1971; and others). These experiments and analyses are generally based on the ideal geometric arrangements of spheres which correspond to one of the regular packing configurations or to some unordered natural arrangement (Fig.3-2). It is the natural precompaction packing configuration of the sediment which must be quantified in order to evaluate the extent of subsequent compaction (Coogan, 1970). This information has been lacking for most carbonate sands. With the exception of a small contribution by Graton and Fraser (1935) on the packing and porosity of mica and minor consideration by Dunham (1962) on the packing of leaf-like and other irregular algal particles, almost no information has been available on the packing of irregular-shaped particles either in single- or varied-shaped particulate sands. This applies to Recent and ancient carbonate rocks, with the exception, of course, of nearly spherical oolitic sands.

Importance of the problem of varied shapes in carbonate sands was recognized by Maiklem (1968) who studied the hydraulic properties of bioclastic carbonate grains. He recognized four basic shapes: blocks, rods, plates, and spheres. Dendroid and irregular particles can be added to the list. While he does not consider the effect of these shapes on subsequent compaction of the sediment, his paper is useful in recognizing the complexity of the particle shapes in carbonate sands.

The shape of carbonate particles is a reflection of the (1) abundance, availability and type of contributing organisms; (2) growth-size distributions of the organisms; and (3) broken size distributions of the contributing skeletal materials. The latter depend on the mechanisms of size and shape reduction (crushing, abrasion, chipping, solution or decay-disintegration) and on the skeletal structure of the material (internal cavities, anisotropism of hardness, and crystal bundles). These inherited parameters of the shape distribution change with the biological province, the local environmental setting of the carbonate environment, and geologic age. To this list of parameters can be added the later history of hydraulic modification of the grain

 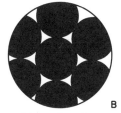 

Fig.3-2. Regular and random packing of uniform spheres, displayed as a monolayer. A. Regular open cubic packing. B. Regular close hexagonal (rhombohedral) packing. C. Unordered packing.

shape distribution owing to transport of the particles as sediment to its site of deposition, method of transport (bottom traction, saltation, suspension), length of time, distance and selectivity of transport of the original shape distribution due to the competency and capacity of the agent, and the rate of supply of the material. The subsequent mode of deposition, whether by dumping or gradual (i.e., rate of sediment accumulation), partial loss of the load, and post-depositional reworking by benthic animals and plants are additional processes which may affect the final shape distribution of a carbonate sediment.

In general, carbonate sands are mixtures of two or more of the shapes discussed by Maiklem (1968). There are exceptions, however, such as nearly single-component algal sands and gravels and crinoidal ossicle sands, both of which may form from nearly single-constituent, preservable biotic populations that are barely transported. It was shown that the shape directly affects the settling velocity. When compared to calculated velocities based on mineral density, this velocity may be as much as four times slower. Block-shaped grains such as coral and coralline algae show the least differences between calculated and actual settling rates. Plate-shaped grains show the greatest differences and these differences in settling velocity, which certainly must be reflected in the sedimentary structures of the undisturbed carbonate sand, directly contribute to variations in the sedimentary packing arrangement (Fig.3-3).

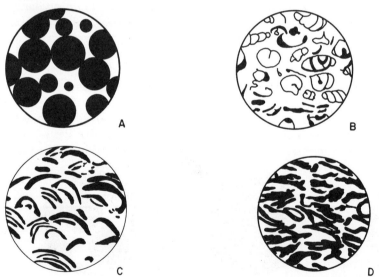

Fig.3-3. Diagrammatic illustrations of random packing of varied carbonate grain shapes, randomly cut. A. Spheres. B. Rods (whole gastropod shells). C. Plates (pelecypod valves). D. Irregular plates (plate algae). Not to scale.

In comparison to quartzose sand particles, carbonate particles are extremely diverse in shape and the parameters of shape, i.e., sphericity and angularity. Many grains in quartzose sands have sphericities of 0.7—0.8 (Rittenhouse, 1971) or can be described as spheroidal in contrast to the disks, blocks, plates and rods of common carbonate grains. Carbonate particles tend to approach sphericity less rapidly than quartzose grains in their respective environments, but tend to round faster owing to their lower hardness. Although carbonate grains before abrasion are generally smooth, occasionally they have angular ribs or spikes. They are rapidly rounded in environments with any appreciable wave motion or abrasion. In general, the diverse shape of carbonate particles is more important in packing, and hence in compaction, than the angularity of the grains.

Finally, the biological processes that produce carbonate particles with abundant internal cavities (e.g., gastropods, foraminifers, cephalopods) produce a sediment which differs considerably from that made of clastic particles or carbonate particles with a more solid construction (e.g., corals, crinoid ossicles). The former sediment may have a very high internal (intraparticulate) porosity which, when filled with sediment or fluid, could modify the effect of overburden pressure, at least at shallow and moderate depths.

Size sorting

Sorting in sedimentary rocks is an important character of the fabric of the sediment that modifies the effects of compactive forces. In addition to variations in grain size usually described by the terms *well to poorly sorted*, the distribution of sorting patterns in the sediment, or the arrangement of clusters having varied sorting, is important in understanding the effect of compaction on a grain sediment. This is especially true for the phenomenon known as bridging, which occurs where large grains receive the bulk of the vertically applied stress and small grains between them are less affected. Other kinds of sorting, i.e., sorting by shape, sorting of clusters of grain sizes, sorting of fabrics (Morrow, 1971), and sorting of intraparticulate pore space, are also important factors affecting compaction. Familiarity with the concepts of regular and irregular packing in sediments, however, is required before discussing them.

Packing

The term *packing* refers to the spacing or density pattern of grains or pores in a sediment or rock. It is in part synonymous with the term *fabric*. Fabric, however, also refers to larger-scale sedimentary features such as crossbeds and burrowed structures. The difference can be illustrated with an example: A cross-bedded sediment composed of grains with homogeneous, irregular close packing may have an average grain volume of 63% and pore volume of 37% calculated over the observed surface of the sample. An intensely burrowed sand may have the same packing and average grain volume, but obviously different gross fabric.

Original packing is important in compaction because it represents the natural arrangement of the grains and pores on which the compactive stress will be applied. Regularly ordered, grain-supported sands of spherical particles may have grain volumes as high as 74.1% for regular rhombohedral packs and as low as 52.4% for regular cubic packs. Naturally sedimented sands appear to have unordered average grain volumes between 50% and 70%. Grain-supported carbonate sands of predominantly nonspherical particles with poor size sorting, such as branching algae, may have grain volumes considerably below the loosest spherical cubic packing of 52.4%, approaching 40, 30, or even 20%. It should be emphasized that the lower the initial grain volume, the more the sediment must be compacted; therefore, chemical compaction should proceed more readily in sediments with initial close-packing configurations. A detailed discussion of packing configurations, measurements of packing, and variations in packing is presented in the following sections.

Mineralogy

Chemical and mineralogical studies of Modern and Pleistocene shallow- and deep-water carbonate sediments (Chave, 1962; Chave et al., 1962; Friedman, 1964, 1965; Milliman, 1966; Matthews, 1967, 1968; Land and Goreau, 1970; Benson and Matthews, 1971; Huang and Pierce, 1971) show that these sediments are composed predominantly of metastable carbonate minerals (high-magnesium calcite and aragonite) with lesser amounts of the more stable low-magnesium calcite. The original mineralogy is largely a direct product of the metabolic processes of the plants and animals which contribute the skeletal grains to the sediment or, and to a much smaller extent, of physicochemical processes. Most of the magnesium present substitutes for calcium by solid solution in the calcite lattice in such common marine animals as foraminifers, sponges, coelenterates, holothuridians, crinoids, brachiopods, molluscs, and ostracodes. In contrast, the skeletal parts of madreporarian corals, many pelecypods, gastropods and cephalopods are aragonitic, as is the principal physicochemical grain carbonate sediment, i.e., oolitic sand. Thus carbonate sand differs greatly from clastic sand in being more varied in particle size and particle shape. It is less varied in particle mineralogy, but has high differential particle solubility owing to different proportions of skeletal contributions to the sediment.

Major element composition varies with carbonate grain size. Analyses by Taft and Harbaugh (1964) of southern Florida sediment show that there is a progressive decrease of the calcium/magnesium ratio with decreasing grain size. Samples from the Bahamas (Taft, 1967) show that sand-size carbonate particles tend to have the highest proportion of aragonite. In contrast, the Recent lime muds of southern British Honduras (Matthews, 1966) contain as much as 44% high-magnesium calcite derived from skeletal disintegration. In summary, the mineralogy of carbonate particles is controlled by biogenic productivity and subsequent alteration. The striking mineralogical difference between modern sediments and ancient carbonate rocks is that the latter commonly have little of their original mineralogy. Thus the lithification* process usually involves changes in particle mineralogy. Minor element composition, however, is probably not a major factor in lithification during compaction.

Inhibitory factors

Certain factors can inhibit the compaction of carbonate sands so that at moderate depths of burial, e.g., between 1,500 and 5,000 m, there may be little physical evidence of compaction even though the rock is lithified. The cause for the lack of compaction in carbonate sands is generally considered to be the result of early cementation (Weller, 1959), which may preserve loose grain packing and uncrushed fragile shells even at great depths. Early lithification* may occur: (1) penecontemporaneously with sedimentation, even as part of the sedimentary process; (2) during postdepositional periods of subaerial exposure of the carbonate terrain owing to slight regional uplift or low stands of sea level; (3) to various submarine and shallow subsurface changes in pore water composition related to a fresh-water lens or artesian upwelling; or (4) to biological activity. The extensive phenomenon of early lithification of carbonate sediments (Bathurst, 1971b) led many (Weller, 1959) to conclude that most limestones show no evidence of compaction. Lithification before burial tends to add strength to the sedimentary unit, fills the pores with cement, and removes fluids which might aid in dissolution. The principal processes of early lithification include cementation and dolomitization.

Cementation

An extensive literature on diagenetic cementation of carbonate sands (LeBlanc and Breeding, 1957; Bathurst, 1958, 1959, 1971a, b; Friedman, 1964; Pray and Murray, 1965; Larsen and Chilingar, 1967; Matthews, 1967; Bricker, 1971; and many others) reveals the source and characteristics of the

---

*Lithification: that complex of processes that converts a newly deposited sediment into an indurated rock.

cement, the driving force and conditions of cementation, and the chemical reactions which illustrate that many sedimentary and diagenetic environments favour dissolution, reprecipitation, and cementation. Purdy's (1968) summary emphasizes the diagenetic environment. He showed that carbonate sediments are transformed into limestone in subaerial, submarine, and subsurface environments. While not extensively distributed or massively pervasive, cementation, principally as minor pore filling, occurs in submarine shallow tropical waters, for example, in the oolite and grapestone bars of the Bahamas (Purdy, 1968; Taft, 1968), and in the deep sea as reported by Milliman (1966), Fischer and Garrison (1967) and Thompson et al. (1968). In contrast, subaerial processes are of considerable importance. They involve solution, replacement and cementation, which affect major portions of the carbonate terrain of a sedimentary basin, principally through the efficacy of percolating meteoric water in the vadose zone where solution and replacement occur together (Matthews, 1967; Benson and Matthews, 1971; Bricker, 1971).

It is difficult to explain the source of large amounts of cement which are present in carbonate rocks. As Bathurst (1971b) has pointed out, petrographic evidence commonly suggests that cementation of limestones begins early, i.e., before compaction, and that about half of the volume of calcium carbonate in the rock has come from a source outside that of the immediate sediment. Redistribution of calcium carbonate by dissolution and reprecipitation is inadequate to fill the pores of most carbonate sediments in the uncompacted state. The possible sources of cement are the local dissolution of aragonite, influx of cement in pore-filling sea water, and pressure solution in subsurface environments.

Cementation in the subsurface below the depth of a few meters is poorly understood. The factors which affect the chemical equilibrium are many, although temperature and pressure appear to dominate. From the geological standpoint, it appears that a given carbonate sediment may reach a state of complete cementation at any depth of burial between the surface and that at which metamorphism begins. Once the sediment is completely cemented, compaction through grain-to-grain adjustments is unlikely.

Dolomitization

Recent dolomitization at or near the surface of sediments has been described by many investigators since the early 1960's from South Australia, the Persian Gulf, Florida, and the Bahamas (Alderman and Skinner, 1957; Shinn, 1964; Deffeyes et al., 1965; Shinn et al., 1965; Atwood and Bubb, 1970; Butler, 1971). Penecontemporaneous dolomitization is part of the process of sedimentation on carbonate tidal flats. In the most general way, this type of dolomitization occurs as the result of influx of magnesium-bearing tidal waters into the carbonate sediment, concentration of the pore water by evaporation, and precipitation of calcium carbonate or calcium sulfate. The process results in a magnesium-rich brine with the consequent substitution of the magnesium for calcium to form dolomite in the sediment over thousands of years. The process may produce near stoichiometric dolomite or may partly dolomitize limestones of any texture to depths of a meter or so and over areas of several thousands of square kilometers.

The second major process of extensive dolomitization is held to be a result of movements of magnesium-rich brines by gravity flow from an area of brine concentration, such as a tidal flat or restricted pond, into contemporaneous or older carbonate rocks of any texture. This process of seepage dolomitization has been proposed and partially documented in West Texas, the Persian Gulf, Bonaire in the Antilles, and Great Inagua Island, Bahamas (Adams and Rhodes, 1960; Chilingar and Bissell, 1961; Deffeyes et al., 1965; Bubb and Atwood, 1968; Murray, 1969; Butler, 1971).

Both of these processes of dolomitization can produce a lithified dolomite or dolomitic limestone extensively cemented by calcite, with a substantial stratigraphic thickness and areal extent. The rigid, cement-bound framework of the resultant rock should resist compactive stress equivalent to many hundreds of meters of overburden.

Dynamic factors

The dynamic factors, those causing or contributing to compaction of the carbonate sand, are the most difficult to assess independently in the field, because there are almost no data on natural compactive processes below depths of about a meter. Experimental data on compaction of coarse-grained sediments is available for some carbonate sands (Fruth et al., 1966; Ayer, 1971) and are discussed later. In this section, the various factors directly affecting compaction are outlined together with their independent roles.

The two principal factors are: (1) the increase in overburden pressure with depth (also referred to as increase in depth of burial, maximum effective stress or applied load, depending on the locale and the author), and (2) the increase in temperature with burial, generally discussed in terms of geothermal gradients or in the laboratory as measured temperatures. Other factors include the rate of loading, the length of geological time or measured time, the composition and movement of interstitial fluids, and the static or dynamic pore pressure.

Maximum effective stress or overburden pressure

The maximum effective stress to which a stratigraphic unit or individual grains are subjected in a sediment or rock with pore space can be expressed according to the formula presented by Terzaghi (1936) as: $a = S - p$, where S is the total load, p is the pore-fluid pressure and a is the effective or matrix stress. The formula draws immediate attention to the close relationship between the effective load (grain-to-grain stress of the sediment) and the buoyancy provided by the pore fluids. For compaction to proceed, the pore fluid must migrate as a gas or fluid out of the sediment in response to the applied pressure. The fluid usually can migrate until the sediment is matrix- or grain-supported. Most carbonate sands reach this stage at the end of sedimentation stage. Further fluid migration is necessary as the grains are crushed or welded together and as cementation occurs in the diminished pore space.

Very little data are available on the pore pressure in rock units or in sediments undergoing compaction, after the grain-supported state is reached. Generalized numerical estimates of the total load (Weller, 1959) show that there is increasingly greater total load with depth. Below a depth of about 175 m and porosity of less than 37%, the weight of pore water must be considered. Above a depth of about 175 m, the effect of hydraulic lift is assumed to be completely effective. (See Chapter 2 for detailed discussion.)

In carbonate rocks the situation is more complex than in sandstones owing to early cementation. There are, however, a few instances, for example, in the Bikini and Eniwetok drill holes (Schlanger, 1963), where some intervals of carbonate grain sediments, which are buried as deeply as 900 m, exhibit little mineralogical alteration to low-magnesium calcite and little cementation, such that the sediment recovered is friable and not much compacted. The JOIDES program of drilling on the Blake Plateau off north-eastern Florida included recovery of cores of unconsolidated, essentially uncemented and uncompacted, planktonic carbonate sands and muds as old as Eocene and Paleocene (Schlee and Gerard, 1965; Bunce et al., 1966). In the more common case, for example, in the Bahamas and Florida where carbonate rocks of Pleistocene and Late Tertiary age crop out, the sedimentary column is cemented from the surface downward, owing to early diagenetic cementation in the submarine, vadose and phreatic zones. In such cases, consideration of the effects of present-day overburden load on grain relationships is made substantially more difficult. Thus a further study of the effect of compaction progressively deeper in the stratigraphic column by drilling and use of better tools for evaluating compaction is needed (see Chapter 7).

Extrapolations of data on the pressure due to overburden load have been made (Weller, 1959; Maxwell, 1960) for depths far in excess of those penetrated by current drilling methods. Maxwell (1960) estimated that pressures at a depth of 14,000 m should approximate 3,100 kg/cm^2, an extrapolation slightly beyond Weller's greatest depth. Although the reported data and extrapolations are for sand—shale sequences, the overburden load for wholly carbonate sections should be of similar magnitude.

Temperature changes

From the surface downward to a depth where all porosity has been eliminated by compaction, the temperature may be expected to increase by several hundred degrees centigrade (Weller, 1959). Experimental work by Sippel and Glover (1964) has shown that the effect of temperature on the solubility of calcium carbonate is much greater than the effect of pressure at moderate depths. For example, at a depth of 1,000 m and a concentration of 220 mg/l calcium carbonate, a 10% increase in temperature at constant pressure would produce a decrease of about 30 mg/l in the concentration. As depth increases, the effect of change in pressure on calcium carbonate solubility becomes more pronounced. It is not clear how these changes affect the micro-environment of grain-to-grain, grain-to-pore and grain-to-cement relationships in a buried sediment, because the temperature probably affects the whole of the sediment equally, whereas the pressure is concentrated at the points of contact. In an extrapolation of Sippel and Glover's (1964) and Sharp and Kennedy's (1965) data, Bathurst (1971b) projects the solubility of calcite with increasing temperature between zero and 600°C for pressures of 412 kg/cm^2 (400 bars) to 1,712 kg/cm^2 (1,400 bars). The graph (Bathurst, 1971b) shows that a calcium carbonate solubility of 11 mg/kg of solution would be reached at about 260°C for a pressure of 200 kg/cm^2, whereas a temperature of over 400°C would be required to reduce the solubility to the same level at a pressure of 1,712 kg/cm^2. At a depth of 6,000 m, the solubility should be reduced a whole order of magnitude, hence reducing the rate of dissolution and retarding compaction. On the other hand, even at depths of 8,000 m and temperatures of about 400°C, calcium carbonate is still soluble and in the presence of favorable conditions (fluid movement, unstable mineral species) it would be expected that compaction probably would continue. In general, one might expect that in areas of low geothermal gradient compaction should proceed more rapidly.

The most readily obtainable data on geothermal gradients and their effect on compaction are provided by Maxwell (1964) for clastic sedimentary basins. His data show that temperature is an important variable, but one about which there is little reliable information, especially at great depths. Bottom-hole temperature readings probably give reliably consistent results, at least on the average, to depths of 5,000—7,000 m. For greater depths the published temperature data are scarce. It is useful to consider the range of geothermal gradients encountered by Maxwell (1964) in his study of quartzose sandstone porosity at depth. The range is between 7° and

12°C/350 m. He found that clastic sands had lower porosities in areas of high geothermal gradients, probably reflecting the change in solubility of quartz with temperature and increased cementation.

Rate of loading

The rate of application of stress to the compacting sediment is one of the two time factors important in compaction. As discussed previously, Terzaghi's (1936) formula ($a = S - p$) shows that compaction cannot proceed much unless the pore fluid is able to migrate. Only when the rate of loading is slow in terms of fluid migration will there be a compaction equilibrium at all times. The rate of fluid movement is directly related to permeability. In the case of carbonate sediments and rocks with inter- and intraparticulate pore space, the degree of communication between the pores is important. Low permeability may result in long periods of time for attainment of equilibrium process, because of its influence on the expulsion of large quantities of water where adjacent pressure differentials are small and friction is relatively great. In the case of rapid loading, longer time is needed for the compactive system to equilibrate.

In carbonate stratigraphic sections, the likelihood of occurrence of extremely rapid loading is not great. The average rate of deposition in carbonate sedimentary basins may be as rapid as in clastic basins; for example, more than 5,000 m of shallow-water sedimentary rocks in the Bahamas have accumulated since the Middle Cretaceous at a rate of deposition at least equal to the rate of subsidence. It probably is true, however, that locally, for example off the present Mississippi River mouth, sediment loading in this clastic depocenter is much more rapid than in any carbonate sedimentary province. Scarcity of overpressured, shallow-depth sedimentary sequences in carbonate areas tends to suggest that loading was either sufficiently slow or that communication and egress of fluids was great enough, so that equilibrium was maintained in carbonate sand bodies under compaction.

Length of time of burial

The duration of burial is the second time factor affecting compaction. Like the rate of loading, duration of burial is related to the ability of the sediment or rock to expel pore fluids and provide space for the collapse of the grains or movement of grains closer together and for cementation. It is important to note that duration of burial is not the same as the geological age of the carbonate sand. Instead, it is the duration of burial as the sediment is subjected to a given stress. For ancient rocks this is difficult to evaluate even if the history of the sedimentary basin is known in some detail. Experimental work in one petroleum company laboratory has suggested that for quartzose sandstones with the same packing, pore fluid pressure, geothermal gradient, salinity, pH of formation water, and average grain size, an increase in duration of burial, for example from 5 to 500 million years, should reduce the porosity from 35% to 5%, with a corresponding increase in the amount of compaction.

Formation fluids

The existence of varied interstitial fluids in the pores of incompletely compacted sediments, especially fluids which wet the surfaces of the grains, probably plays a role in the process of compaction. There exists a potential for retardation of compaction through pressure solution in carbonate sands which have hydrocarbon-wet surfaces, because the relative permeability to water would be small and the consequent movement and uniform distribution of cement would be difficult. Oil-wet particle surfaces could substantially reduce the potential for solution of carbonate minerals and the redeposition of calcium carbonate as cement in the immediately adjacent pore space.

The actual effect of hydrocarbons on cementation in the pore space of carbonate rocks is unknown. Oil and gas have been suspected of controlling or maintaining open pore space in clastic sandstones in many fields and in the Smackover Jurassic oolite, but the documentation usually has been questionable. Close association among presence of oil, higher than expected porosity, and slight evidence of compaction does not necessarily mean that the hydrocarbons have preserved the porosity; in fact, it may mean only that the high porosity has served to localize the oil. It has been suggested that the presence of gas does not help preserve porosity in the clastic sediments in south Louisiana (Atwater and Miller, 1965). There, an average decrease in porosity of 1.2% per total volume for each 350 m of burial, observed in 3,000 gas reservoirs, indicates little effect of gas presence on porosity. This rate of porosity decrease was essentially that observed for the water-filled sands, too. From the general knowledge on pressure solution and reprecipitation now available, it does seem unlikely that oil or gas would greatly inhibit pressure solution as long as a thin film of water surrounds each grain, which is the case in the typical water-wet oil sand, whether quartzose or carbonate.

In a study of the differences in compaction of sands in petroliferous and water-bearing strata of Early Cretaceous reservoirs of the Ust' Balyk and West Surgut Fields, Western Siberia, Zaripov and Prozorovich (1967)* compared the total number of impressed contacts between grains of quartz, feldspar, and rock fragments. Their findings show markedly higher numbers of impressed contacts for quartz grains in the water-wet beds over those in

---

*For other references see Chapter 3, Vol. II, on sandstone diagenesis.

the oil-wet beds at the same depths (2,100—2,300 m). They found the differences to be statistically significant. Although they concluded that in the oil-saturated sandstone the dissolution of quartz ceased or slowed down after the reservoir had been filled, compaction of the reservoir sand continued chiefly through plastic deformation of rock fragments so that both oil- and water-wet beds continued to compact. In addition, they noted that coarser sands showed more effects of pressure solution owing to greater pressure per grain per contact and the greater permeability of the sediment.

Formation waters may increase in salinity dramatically with increase in depth of burial. In general, the diagenesis of connate sea water may involve a reduction in the pH locally owing to the decomposition of organic matter not previously oxidized near the surface, resulting in a slight increase in magnesium and calcium ion concentration of the solution. With further burial and compaction, some of the salty water should be expelled. Where porous and permeable carbonate units are interbedded with other sediments, this could mean expulsion of brines into or through the carbonate sediment. Review of the literature indicates there are no data on the relative amounts of compaction of carbonate sands interbedded with clastic shales and those which are part of massive, thick carbonate units.

Influx of substantial amounts of low-pH water, as the result of (1) exposure to a fresh-water supply at the surface, (2) upwelling artesian flow into a carbonate sand body, or (3) expulsion from a clastic sediment into a carbonate body, would greatly affect the mineralogy of the original carbonate sediment tending to convert the carbonate particles to low-magnesium calcite. Maintenance of a static, low-pH fluid in the pores of a carbonate sediment is unlikely. Lack of experimental work on compaction of carbonate sands in the presence of moving acidic waters hinders further analysis, but one might speculate that a through-flowing acidic fluid would tremendously enhance compaction by accelerating massive dissolution.

Summary of geological factors favoring compaction of carbonate sands

Considering the dynamic, inhibitory and inherited factors affecting compaction, it appears that the compaction of a carbonate sand is more likely in geological settings where the following conditions prevail, recognizing the fact that if compactive stresses are high enough, deformation will occur regardless of other factors:

Dynamic:
    (1) Burial is deep and effective stress is high.
    (2) Geothermal gradient is low.
    (3) Rate of loading is low.

(4) Pore pressure is low in a static fluid, or there is sufficient permeability to allow fluid expulsion.
(5) Duration of burial under stress is long.
(6) Through-flowing fluids are acidic and undersaturated with respect to calcium carbonate.
(7) Grain surfaces are water-wet.

Inhibitory:
(1) Early cementation is minimal.
(2) Dolomitization is minimal.
(3) Diagenesis and alteration to low-magnesium calcite is minimal.

Inherited:
(1) Average grain size is small.
(2) Grains are moderately to well sorted.
(3) Packing, as a result of deposition and penecontemporaneous processes such as burrowing, is the most close packing possible for the particular sediment.
(4) The mineralogy of the carbonate sediment is of the least stable kind for the subsurface environment.

PRECOMPACTION GEOMETRY — PACKING

General pertinency of packing to compaction

The critical problem facing the geologist who desires to measure the amount of compaction, which has taken place in a grain carbonate rock, is to determine the point of zero compaction. He must choose the numerical value which represents the starting point or the packing arrangement of the loose, uncemented sediment before any adjustment, rotation, slippage, peeling, fracture, or pressure solution has occurred.

Two principal sedimentary characteristics which affect the packing are the shape of grains and their geometric arrangement. Almost all theoretical and experimental studies of packing have used spherical particles or points in space (Kelvin, 1887; Graton and Fraser, 1935; Marvin, 1939; Matzke, 1939; Brown and Hawksley, 1945; Bernal and Finney, 1967; Ridgway and Tarbuck, 1967; Coogan, 1970; Morrow, 1971; Rittenhouse, 1971). The principles are well understood and, for geologists interested in sediments, have been presented at length by Graton and Fraser (1935), who dealt with the regular packing of spheres and progressed to more irregular, random packing arrangements. Naturally, there is criticism of using spheres as model grains (see Chapter 2). Morrow (1971) has extended the analysis to systems of heterogeneous and homogeneous regular and irregular packing. Rittenhouse (1971) has related the effects of compaction on certain packing arrangements to the maximum degree of cementation to be expected from local-source cement during the compactive process. In this section, the writers present an outline of the regular packing configurations and then the unordered ones, which can be derived from them, as a basis for choosing methods of measuring packing and hence compaction in carbonate sands.

Regular packing of spheres

By far the most complete analysis of regular packing of spheres is that of Graton and Fraser (1935), who summarize the configurations which are stable against the force of gravity acting alone. There are two types of layers, the square and simple rhombic (Fig.3-4A, B), and there are three simple ways of stacking these layers one on top of another (Fig.3-5A—F). Because two of the three ways of stacking square layers are identical to, but differently oriented from, two of the three ways of stacking simple rhombic layers, there are six fundamental regular arrangements. Two of these six arrangements repeat as to form (grain volume, porosity), but differ in symmetry and hence tortuosity and permeability.

The six regular geometric arrangements of four spheres are named according to crystallographic terminology the cubic, orthorhombic and rhombohedral packing of square layers (Fig.3-5A—C) and the orthorhombic,

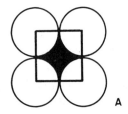

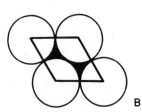

Fig.3-4. Plan view of two types of regular packing layers. A. Simple square layer. B. Simple rhombic layer.

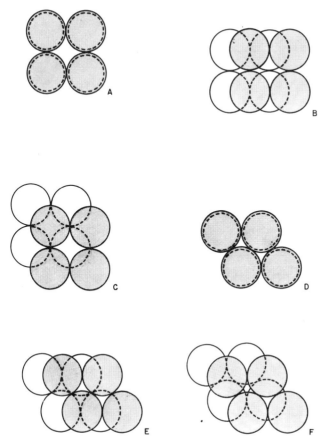

Fig.3-5. Six regular packing configurations of uniform spheres; view from above. Cases 1 to 3 are arrangements of square layers (Fig.3-4A). Cases 4 to 6 represent similar offsets for stacks of rhombic spheres (Fig.3-4B). A. Cubic, case 1; four spheres sit directly above four other spheres. B. Orthorhombic, case 2; four spheres offset one-half sphere in one direction with regard to underlying spheres. C. Rhombohedral, case 3; four spheres offset one-half sphere distance in two directions with regard to underlying spheres. D. Orthorhombic, case 4. E. Tetragonal, case 5. F. Rhombohedral, case 6.

tetragonal and rhombohedral packing of simple rhombic layers (Fig.3-5D—F). For each of these six cases of regular packing, there is a fixed ratio of grain to pore space and each has a definite stability related to the number of grain contacts below a given grain (Table 3-I). In addition, the six packing types can be characterized in terms of coordination numbers and the number of other grains touched by an arbitrarily chosen central grain. For example, in case 1 (cubic packing), where uniform spheres sit precariously atop other spheres (Fig.3-5A), each central sphere touches six others

TABLE 3-I

Grain and pore volume, stability, and number of grain contacts of the six regular packing types

Packing type	Grain volume (%)	Pore volume (%)	Stability	Number of contacts by each grain	
				below	all sides
Case 1, cubic	52.36	47.64	low	1	6
Case 2, orthorhombic	60.46	39.54	medium	2	8
Case 3, rhombohedral	74.05	25.95	high	4	12
Case 4, orthorhombic	60.46	39.54	low	1	8
Case 5, tetragonal	69.81	30.19	medium	2	10
Case 6, rhombohedral	74.05	25.95	high	4	12

(Fig.3-6A). When displayed as a monolayer, four of the six are in the same plane and two in another plane.

For case 6 (rhombohedral packing), where spheres lie in the "holes" between underlying spheres, each central sphere touches eight others (Fig.3-6B), but only six are shown when displayed as a monolayer. In the two respective cases, the grains may be said to have six- or eight-fold coordination. Coordination numbers, which are common in crystallographic terminology, are almost impossible to derive for naturally compacted sediments and rocks. Their packing must be described in other less precise terms, partly because the packing of natural grains is seldom regular to any appreciable extent and partly because of cementation.

The porosity of regular packing configurations is fixed and directly related to the geometric arrangement of the pack. Because the sorting of grains of the six regular cases is regular, the sorting of their pores is also regular. In cubic packs the pores appear as diamond shaped in the display of spheres as a monolayer. Graton and Fraser (1935) made extensive determinations of the geometry of the pore space, including serial sections of the pores in the regular packs (Fig.3-7). By way of contrast, two different monolayers are shown (Fig.3-7), which emphasize pores and grains at different positions for the case 1 and case 6 packing. It is clear that the space in rhombohedral packs is not only less (25.9% porosity versus 47.6% for cubic packs), but the geometry of the pore space is markedly different. In rhombohedral packs, there is less grain surface area exposed to wetting fluids and the capillary pressure and surface tension for the whole pack are higher. All of these characteristics affect the subsequent compactive forces as they relate to solution of the grains under pressure and the maintenance of pore pressure.

COMPACTION AND DIAGENESIS OF CARBONATE SANDS

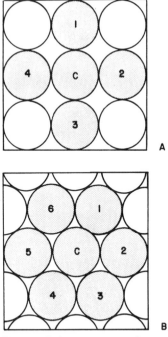

Fig.3-6. Sphere contacts in regularly-ordered packing of uniform spheres, shown as a monolayer. A. Cubic packing; each sphere touches 6 others, 2 are not in the plane of the monolayer. B. Rhombohedral packing; each sphere contacts 8 others, 2 are not in the plane of the monolayer.

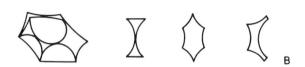

Fig.3-7. Geometry of pore space in cubic and rhombohedral packing seen as though the solid spheres were removed. A. Cubic pores. B. Rhombohedral (case 6, Fig.3-5F) pores. Left-hand figure of each set represents the three-dimensional pore space in the unit cell. The right-hand figures represent different cross sections of this three-dimensional pore.

Stability of the pack is directly related to the number of grain contacts vertically and laterally made by each grain in the pack. Any sphere acted upon by gravity and supported from below requires at least three points of support in order to attain equilibrium and a fixed and stable position. Of the six packing types, only the two rhombohedral ones (cases 3 and 6, Fig.3-5C and F) fulfill this requirement because only in these two cases do spheres of any layer have three or more points of support from those of the underlying layer. In the orthorhombic and tetragonal (cases 2 and 5, Fig.3-5B and F) packs, each sphere receives support from below at only two points and is balanced in the cusp between the two underlying spheres except for the lateral support afforded by neighbors in its own layer. In the least stable configuration, the cubic and orthorhombic packs (case 1 and 4, Fig.3-5A and D), each sphere is perched tenuously on the pinnacle of a single underlying sphere and requires lateral support from neighboring spheres. In general, of any two uncemented packs, the one that is more stable has the lower porosity and the higher grain volume. In the more stable pack each sphere touches the largest number of adjacent spheres and the vertical spacing between layers is smaller. Such stable packs should undergo the least amount of consolidation by their own weight or any minor disturbance such as slight sedimentary loading near the depositional surface. In addition, stable packs should have to adjust the least amount in attaining complete compaction — that is, the configuration with 100% grains and no pore space — through the application of stress.

Random packing of spheres

Definition and origin

There are only two explicitly definable states of a packed bed. The first is when the bed is completely ordered and the grain volume is at a maximum for the configuration. The second is when the grain volume is at a minimum so that any decrease in grain volume would result in a cloud of separate particles, that is, a quick sand. Intermediate states are forms of random packing. The random packing of spheres or other grains occurs under natural conditions of sedimentation. No examples of the six ordered packing types have been reported for naturally sedimented carbonate grain sediments to the knowledge of the authors, even for nearly spherical oolitic sediment, with the exceptions of possibly small clusters of ordered grains of cubic or rhombohedral packing occupying small regions of an otherwise randomly-packed sand. These random packs may exhibit either less or more complicated kinds of orderliness than characterize the six simple cases or may contain combinations of ordered and unordered packing. They cannot, however, be characterized as lacking in order entirely, although disorderly packing appears generally in random packing. The stability of random packs is intermediate, and their variable grain volume and porosity are commonly described in terms of averages for the whole pack of sample, mainly for convenience and because averaging measurement techniques are used.

It has been shown (Graton and Fraser, 1935) that random packing will occur in simple cases where spheres are poured freely and fortuitously into a container. Vigorous jarring might cause slight reordering, especially where the initial packing is loose; however, random packing, which is fairly typical of natural sands, is usually stable. On the average, random packing is statistically reproducible. It is important to note that random packing, once established in the midst of a large sand body, cannot be completely eliminated or translated into one of the more systematic types of packing by any amount of mere jarring. This is because the spheres are locked into position and openings, which are more than large enough to accommodate a whole sphere, may be left unfilled by the manner of sphere accumulation. These openings are protected by arching of the surrounding spheres and cannot be filled without the lifting of some stable individual grains. Hence, random packing, which is not the most stable theoretical configuration, has a great natural stability such that carbonate sands can become rapidly and permanently fixed in a randomly-packed state by early cementation around grain boundaries near the depositional surface.

Chance packing, the natural fortuitous combination of systematic regions of ordered packing, mainly of the case 6 and case 1 (Fig.3-5F and A) configurations commonly occurs with intervening random, truly orderless zones in carbonate sands of nearly uniform grain shapes and sizes, such as oolitic sands. No direct quantitative data are available on the relative amounts of such chance packs, but indirect evidence supplied by Graton and Fraser (1935) and by Morrow (1971) indicates that chance packs do occur in natural sands.

Density of random packs

The grain volume of varied types of random packs is important to determine as a basis for establishing the zero point of compaction. Samples of spherical grains prepared in the laboratory in relatively large containers with more than ten million grains, or collected from various sedimentary environments, have measured porosities of about 37% which corresponds to grain volume of about 63%. According to Graton and Fraser (1935), more dense, minute rhombohedral colonies with local grain volumes of 74.1% constitute substantial portions, perhaps as much as one-half of the volume of generally less dense random packs, as indicated by inspection. The remainder of the pack is represented by random colonies with grain volumes as low as those present in cubic packing (52%).

There are two reproducible states of random packing which have been called "loose" and "dense" by Scott (1960); the difference being about 3% grain volume. Scott filled rigid cylinders of known volume and nonrigid balloons with steel ball bearings 3.1 mm (very coarse sand) in diameter and by shaking cascaded them into containers. Graphical extrapolation provided a method of obtaining the grain volume of the packs in an infinite array, free from wall effects. He recommended the values of 63.7% grain volume for dense random packing and 60.1% grain volume for loose random packing of spheres. Following his work, Bernal and Mason (1960) investigated the coordination of close randomly-packed spheres by packing the same size steel balls and soaking them in black paint. After separation, the balls show points of contact not covered by paint. The mean number of contacts was 10—11 for a grain volume of 64%. This compares to the theoretical value of 8—10 contacts per grain for regular spherical packs (cases 2, 4, 5; Fig.3-5B, D, E, and Table 3-I) of comparable grain volume. They concluded that the volume of close random packing is repeatable, that it must be mathematically determinable, although so far undetermined, and that close random packing is one of minimum energy. Furthermore, any lowering of the stability of a packed bed must necessarily increase the distances between its spheres and increase the energy. Subsequently, Bernal and Finney (1967) further refined the estimates of volumes of random sphere packs. Using measured coordinates of a random model, which included a complete analysis of local density variations as well as determining an overall average in the absence of wall effects, they confirmed the values shown in Table 3-II for spherical packs of loose and dense random packing.

Effect of grain size

Although theoretically grain size has no influence on the porosity or grain volume percentage of sediments, this does not prove to be true for assemblages of natural sands. As the sand size decreases, friction, adhesion

TABLE 3-II

Loose and dense random packing of spheres

Type of packing	Average grain volume (%)	Method of determination
Random, close	63.4	measurement of coordinates of sphere centers after removing outer layers
Random, loose	60.1	porosity measurements and extrapolation to infinite volume without wall effects

and bridging become important because of the increasing ratio of surface area to volume and mass. Therefore, the smaller the grain size, the less the percentage of grain volume in randomly-packed sediments of small grain sizes. Determinations by Ellis and Lee (1919) on 36 samples ranging from coarse sand to silt show grain volumes (recalculated from porosity) varying from 46% to 61% (Table 3-III). The average grain volume of the 36 samples was 55.91%. As the grain size decreases, the tendency toward bridging and consequent looser packing is undoubtedly augmented by increased variation in the shape and differences in the method of deposition. No equivalent data are available for carbonate sediments.

Mixtures of sizes of grains

The two main effects which occur when smaller spheres are added to a bed of larger spheres are (1) decrease in the grain volume percentage owing to forcing apart of the larger spheres, and (2) increase in the grain volume percentage because of filling of the voids, if they fit readily into the voids between the large spheres. The situation is considered here for binary, tertiary and more complex mixtures of grains.

Binary mixtures

If an assemblage of uniform spheres is compared with an otherwise similar binary-size mixture, two effects are noticeable which appear to work in opposite senses but which are not of equal value. The result of size mixing is to increase the grain volume percentage by filling voids. As the large and small spheres become increasingly different in diameter, the increase in grain volume percentage is more pronounced. This applies both to the case of a single sphere in an assemblage of smaller ones and to a number of large spheres in such an assemblage up to the point where the number of large spheres dominate the pack. Decreasing the grain volume percentage due to

TABLE 3-III

Grain volume of sands sorted by size

Size grade*	Grain volume (%)
Coarse sand	61—59
Medium sand	59—52
Fine sand	56—51
Fine sandy loam	50—46

*Fine sandy loam is a mixture of fine sand (>.08 mm) and finer sediment where the fine sand content equals 50% or more.

mixing spheres is caused by looser packing of small spheres about the larger ones, resulting in an increase of the "wall effect" throughout the pack.

The ratio of the diameter of a small sphere, which can just pass through the pore throat between larger spheres into the interstitial void, to the diameter of the larger sphere, is known as the critical ratio of entrance. This ratio is 0.154 for rhombohedral regular packing and is 0.414 for cubic packing (Fraser, 1935).

When spheres of two sizes are mixed, the smaller size dominates the general structure of the pack so long as the proportion of these spheres is sufficiently great to keep most of the large spheres separated from one another. When the proportion of large spheres increases beyond this limit (small sphere control of the pack), two alternative situations arise depending on the relative sizes of the small and large spheres. When the small spheres have a diameter less than the critical ratio of entrance and the number of large spheres is sufficient so that they touch one another, are self-supporting, and bridge the smaller spheres, then the large spheres take control of the pack. From this point on, any further increase in the number of large spheres will decrease the grain volume percentage of the pack. On the other hand, when the small spheres have diameters exceeding the critical ratio of entrance, and the number of large spheres increases beyond the limit of domination by the small spheres, the two sizes will interfere with one another as more large spheres are added. This mutual interference will lower the grain volume percentage over what it would otherwise be. Nevertheless, grain volume percentage would still increase as the proportion of large spheres continues to increase.

In the case of binary pack of sand and pebble-size grains, the same conclusions should be evident. Fraser (1935) tested the degree of disturbance of the packing of sand caused by a pebble contained within the sand. He experimented with four different sizes of pebbles collected from Revere Beach, Massachusetts, where the sand was fairly uniform and contained occasional pebbles. At places where a pebble just protruded above the top of the sand, a pebble—sand sample was carefully removed without disturbing the packing of the moist sand. The porosity of the entire sample was determined and recorded as total porosity. The volume of the pebble was determined separately and subtracted from the total volume (Table 3-IV). As expected, a large pebble has a strong effect on the grain volume percentage.

Multicomponent packs (Tertiary, Quaternary, Quinary)

Measurements of the grain volume of various three-component mixtures of spheres have been made by Fraser (1935). He mixed sands in which the diameters of the spheres were approximately 1.3, 2.3, and 8.1 mm, so that each size was greater than the critical ratio for the next larger size in chance packing. The lower grain volume percentages were invariably obtained in mixtures in which one size strongly predominated. With increasing complexity of the mixture, the grain volume percentage tended to increase. It is remarkable that average grain volume of 67% (33% porosity) can be secured consistently from very diverse mixtures of grain sizes (Table 3-V). One conclusion to be drawn from these data is that it apparently is impossible to predict grain volume percentage of a tertiary mixture of spheres on the basis of percentage of components. It is equally, if not more, unlikely that a useful prediction can be made for more complex mixtures. Much work remains to be done on the study of random sphere packings because there is no satisfactory geometrical probability analysis capable of explaining the value of the voidage of a randomly-packed bed (Ridgway and Tarbuck, 1967), and why it should be so very reproducible. We are even further from an explanation of the packing of spheres of different sizes or the packing of nonspherical particles.

Experiments by Ridgway and Tarbuck (1967) indicate that the highest grain volume for multicomponent packs is achieved where the size ratio between the pairs of grain sizes is at least seven. Given a spherical grain with the radius r, the sizes of the secondary, tertiary and higher order sizes which will just fit in the void of a bed of rhombohedrally-packed primary spheres has been calculated (Table 3-VI).

TABLE 3-IV

Grain volume of beach sand around a pebble (From Frazer, 1935)

Sample	Portion occupied solely by pebble (%)	Grain volume of sample (%)	Grain volume without pebble (%)
1	3.09	59.45	58.17
2	69.88	86.82	56.22
3	63.56	83.45	54.83
4	65.54	83.48	51.94

TABLE 3-V

Grain volume of tertiary mixtures (calculated from measured porosity)

Sample	Percentage of grain sizes by sphere diameters			Grain volume (%)
	1.3 mm	2.3 mm	8.1 mm	
1	10.51	9.63	79.86	66.68
2	29.57	41.39	29.04	66.16
3	41.43	29.84	28.55	66.87
4	10.17	44.48	45.34	66.61

TABLE 3-VI

Radii of spheres fitting in the void of rhombohedrally-packed primary spheres of a given radius r

Sphere	Radius
Primary	r
Secondary	$0.414\,r$
Tertiary	$0.225\,r$
Quaternary	$0.177\,r$
Quinary	$0.116\,r$

Studies of natural sands of mixed grain sizes confirm some of the data obtained for spherical packs. Data from Stearns (1927), strikingly portrayed in Fig.3-8, show that a wide range of mixtures of sand sizes may result in the same grain volume percentage. Similarly Gaither (1953) experimented with natural sands and showed that adding 8 g of coarse (0.5—1.0 mm) and medium (0.62—0.25 mm) sand to 84 g of medium (0.25—0.5 mm) sand caused an increase of only 1.7% in grain volume. Earlier, King (1898, reported in Gaither, 1953) had determined that on decreasing the size sorting, by mixing sands having average diameters of 0.48 mm (medium sand) and 0.09 mm (fine sand), the grain volume of well-sorted, wholly fine sands increased from 60—65% to 76%.

Packing heterogeneities

Small-scale packing heterogeneities have been investigated by Morrow (1971) in terms of the distribution of pore sizes which is indicative of grain size heterogeneities in sediments without clay or cement. Packing heterogeneity occurs where there are variations in particle size or packing in different areas of a rock. For example, heterogeneities may be characterized by distinct local variations of particle sizes (Fig.3-9) and pore sizes. This could occur without variation in the average porosity or grain volume. In sands, such heterogeneities would be found where local variations in grain sorting exist. Another example of packing heterogeneity could be an essentially homogeneous random pack of spheres of equal size (Fig.3-10), but in which there are alternate regions of rhombohedral close packing and simple cubic packing. Interestingly, data provided by Brown and Hawksley (1945) can be plotted as a map (Fig.3-11) which shows that in a pack of uniform spheres the distribution of regions of tight, intermediate and loose packing is a random one.

Capillary pressure drainage curves can be used to characterize packing heterogeneities in rock, because the slope of the curve reflects the pore size

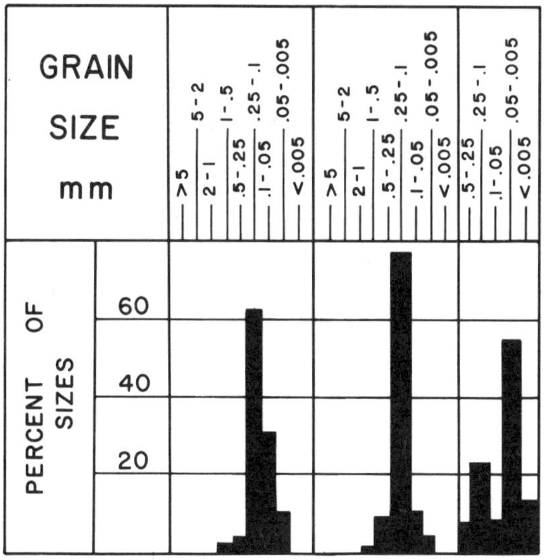

Fig. 3-8. Variations in grain size sorting of three sands with grain volumes between 67 and 68%. (Data from Stearns, 1927.)

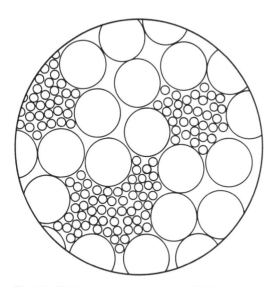

Fig. 3-9. Heterogeneous random distribution of primary and secondary size spheres, shown as a monolayer. Radius of secondary spheres is less than 0.4 r and greater than 0.1 r of primary sphere (radius = r), or smaller than the critical value for a cubic pack. This fabric could result from burrowing by organisms in lagoonal sands.

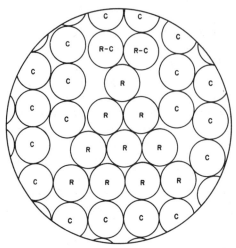

Fig.3-10. Heterogeneous packing of uniform spheres containing central portion of rhombohedral packs R and marginal cubic packs C, shown as a monolayer. Large pores are present at transitions from one packing to the other.

Fig.3-11. Distribution of density of packing in tightly-packed groups of uniform spheres. Percentage for (1) tight packing (grain volume > 79%) is 44% — light dot pattern; (2) intermediate grain volume (79%–71%) is 17% — dark dot pattern; and (3) loose grain packing (grain volume < 71%) is 40% — white. (Based on data from Brown and Hawksley, 1945.)

distribution of porous sand. In addition, larger interconnected pores tend to drain faster than smaller ones. The capillary pressure curve, according to Morrow (1971), proved to be useful in determining the apparent pore size (hence, grain size) distributions of sands. Based on experiments with artificial packs in which clusters of particles are surrounded by a matrix of coarser particles, Morrow concluded that the irreducible wetting phase saturation of such mixtures is much higher than the saturations of corresponding well-mixed, more homogeneous aggregates of the same particles. Because the heterogeneity of packing largely determines the magnitude of the irreducible wetting phase saturation (Fig.3-12) in completely wetted system, the capillary pressure curve in turn provides an index of packing heterogeneity. Accordingly, a parallel bundle of capillaries, or a rock with any other pore configuration that allows complete drainage, has zero heterogeneity. Theoretically, rhombohedral and cubic packs having spheres of equal size will have irreducible saturations of about 5.5% and 4.2%, respectively. The liquid will be held as pendular rings at points of contact between particles (Harris and Morrow, 1964). In random packs of equal spheres, irreducible saturation falls in the range of 6—8%, representing only a slight loss of homogeneity from the regular packs. Other well-mixed packs, including artificial sands (Morrow, 1971) give values in the range of 6—10%. The irreducible saturation values of 15—40% are commonly observed for sedimentary rocks and indicate a significant variation in packing heterogeneity and possibly cementation. The effects of porosity and permeability on irreducible wetting phase saturation are shown by Chilingar et al. (1972) for a variety of carbonate rocks.

Obviously a variety of packing types can give the same irreducible saturation. The reasons for a given region remaining more saturated than other areas might range from gross differences in particle size to subtle differences in packing arrangement. Also, even though packing of coarse

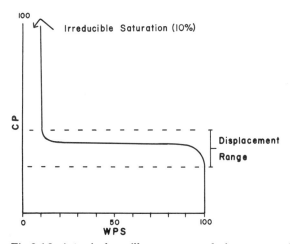

Fig.3-12. A typical capillary pressure drainage curve for a well-sorted, randomly-packed, coarse sand without clay matrix or cement. Irreducible wetting phase saturation is dependent on pore geometry. CP = capillary pressure, WPS = wetting phase saturation (%).

particle clusters surrounded by a finer matrix are clearly heterogeneous, no indication of this may be given by the capillary pressure curve, because the packing will appear to be homogeneous and the sample will drain to a uniform saturation throughout on a gross scale. Nevertheless, natural sands will likely contain mixtures of various types of heterogeneity (Graton and Fraser, 1935) so that, on the average, values for irreducible wetting phase saturation should provide a reasonable measure of rock heterogeneity where cementation is insignificant. Where capillary pressure curves, porosity, and permeability are obtained for rocks which have also been examined in thin section for patterns of grain distribution, pore types and cement, these physical measurements will enhance understanding of packing in rocks and hence the state of compaction (see Chilingar et al., 1972).

Recent and Pleistocene oolitic grainstones from the Bahamas and Florida were characterized by Robinson (1967) in terms of porosity, permeability, and capillary pressure curves. Slightly cemented, slightly altered, well-sorted oolitic grainstone with 41.5% porosity and 8,895 md permeability had a low threshold entry pressure of 0.2 kg/cm^2 and almost no displacement zone. The L-shaped capillary pressure curve was typical of well-sorted, well-connected pores. In contrast, altered, leached and cemented oolitic grainstone, with equally high porosity (45.7%), lower permeability (935 md) and less well-sorted pore types (mixture of intra- and interparticulate pores), had lower threshold entry pressures, but a wide displacement zone of nearly 7.0 kg/cm^2. The irreducible wetting phase saturation varied as much as 30% for these oolitic grainstones which had little difference in porosity.

Random packing of irregular grains

The shape of sedimentary grains is probably never spherical, not even in oolitic sands, and these irregularities in shape should result in a larger possible range of grain volume percentages in sediments, because irregular grains theoretically may be packed either more tightly or more loosely than regularly-packed spheres (see Chapter 2). It is difficult to determine the effect of grain shape on grain volume percentage in natural sediments because of the difficulty of obtaining odd-shaped particles of the same size by screening. Determination of the effect of grain shape on porosity and packing is commonly hindered by the second independent variable, grain size. As a result, a study of the average packing conditions by determining the porosity in a sand, which has grains of different sizes and shapes, generally supplies little information concerning the effect of shape on grain volume. Studies of carefully-sized materials, which range in shape from spheres to flat plates, by Fraser (1935), however, showed expectable decreases in grain volume percentage with departure from spherical shape (Table 3-VII). For lead and sulfur shot, the values at loosest packing are not reliably comparable with those for natural sands because the greater specific gravity of these materials affects the compactness of the unjarred assemblages. The shot was screened and the portion used had an average diameter of 1.5 mm (coarse sand). The porosity was measured before and after the shot was consolidated by jarring when dry and when saturated with water. If the averages of the lead and sulfur shot tamped down are taken as representative of uniformly-sized spherical grains, then packs of dry spheres have a grain volume of 53.28% and wet ones 66.68% (Table 3-VII). Comparing the other samples with the shot, two have higher grain volume percentages and two lower. The shape of the marine sand grains is reported to be fairly uniformly disk-shaped and may be packed more tightly to the third dimension, with flat sides together, which would allow for the high grain volume percentage displayed. The grain volume percentage of beach sand does not differ widely from that of the spherical shot. Assuming that variations in grain volume of compacted dry sands are due to the influence of the grain shape, there exists a range of 9.63% grain volume variation (from 66.22 to 56.59%) for a suite of rather typical samples.

Wet meterials pack more loosely than dry materials owing to the bouyancy of water. The difference between packing of rounded and angular grains (wet and dry) in compacted samples, however, proves that wetting the sand increases the effect of angularity on the packing. Of the samples tested, flat and needle-like grains have the greatest effect on packing. For example,

TABLE 3-VII

Influence of grain shape on grain volume of wet and dry, loose and tamped sands (After Fraser, 1935)

Material	Specific gravity	Grain volume (%)			
		dry		wet	
		loose	compacted	loose	compacted
Lead shot	11.21	59.94	66.72	57.60	61.11
Sulfur shot	2.02	56.62	66.65	55.86	61.76
Marine sand	2.68	61.48	63.22	57.04	64.96
Beach sand	2.65	58.83	63.45	53.45	61.54
Dune sand	2.68	58.83	66.40	55.07	60.66
Crushed calcite	2.66	49.50	59.24	55.50	57.26
Crushed halite	2.18	47.95	56.49	—	—
Crushed mica	2.83	6.47	13.38	7.62	12.72
Crushed quartz	2.65	58.80	58.80	56.12	56.04

crushed mica had a grain volume of less than 10%. This amazingly low grain volume was not increased above 14% by prolonged jarring. Pressures sufficient to burst a strong glass container were required to increase the grain volume percentage of the wet mica to 32.6% (Fraser, 1935).

Angularity, or departure from spherical shape, may increase or decrease the grain volume of packs. Most commonly it decreases the percentage grain volume and, according to Fraser (1935), the only type of angularity found to increase grain volume is that in which the grains are mildly and uniformly disk-shaped.

Changes in grain packing with jarring

It has been shown that any aggregation of grains has a fairly definite range of grain volume which represents varying approaches to perfection in packing. Moderately well-sorted, medium-grained, randomly-packed natural sands have grain volumes between 53.5% and 62.2% depending on the looseness of packing, angularity of the grains, wet or dry state when measured, and other factors. Fraser (1935) took the analysis one step further by tamping a beach sand with a grain volume of 54.0% (wet) and 53.70% (dry) for 12 min. The grain volume increased nearly 10%. Further jarring produced no additional settling, however. In fact, further jarring loosened the packing (Table 3-VIII).

Pebble and sand grains are deposited in a more stable state than are smaller particles, because the effect of buoyancy by the water during sedimentation is smaller for coarser grains. Accordingly, the grain volume percentage and packing configuration of sands and gravels as initially deposited cannot be greatly increased until stress is sufficient to crush the grains, or solution removes portions of the grains to provide space for closer packing. When artificially deposited under water, sand derived from the St. Peter Sandstone (Athy, 1930) could be made to settle about 11% of initial

TABLE 3-VIII

Grain volume variations with duration of tamping in beach sand (After Frazer, 1935)

Tamping duration (min)	Grain volume (%)	
	wet	dry
0	54.00	53.70
1	57.50	60.47
2	59.23	62.00
4	61.53	61.80
8	61.74	62.18
12	61.74	62.18

volume by continued jarring at normal atmospheric pressure. When a pressure of 280 kg/cm^2 was applied, the additional increase in compaction was only 2%.

Summary of packing of natural sands

Natural, randomly-packed, nonspherical, moderately well-sorted and rounded sands have average grain volumes between 50 and 70%; the majority of values falls between 54 and 67%. Based on these data, Rittenhouse (1971) concluded that regular orthorhombic packs (grain volume of 60.46%) of spherical particles approximate natural randomly-packed sands on the average, but that the analogy is semi-quantitative. In measuring oolitic grain volume in thin section, Coogan (1970) found that undisturbed, in-place, oolitic sand has an average grain volume of 64.8% and subaerially weakly cemented oolitic grainstone of Pleistocene age has a grain volume of 64.4%. Experiments have repeatedly shown (Gaither, 1953) that well-sorted, well-packed, wet sands have grain volumes about 67%.

There seems to be overwhelming evidence that the values cited are typical of a wide range of natural sands of varying composition and sorting. Somewhat lower values, however, were obtained by Pryor (1971) from porosity data on modern sands collected from varied environments. Expressed as grain volume percentage, Pryor found values between 50% and 60% (average 55%) for modern river sands, 44—61% (average 54%) for beach sands, and 49—56% (average 52%) for dune sands. His values appear to be for angular sands and there may be some disturbance from side-wall effects when the samples were pushed into small-diameter core barrels.

At present, the point of zero compaction for common, randomly-oriented, mixed sands of moderate sorting, reasonably high sphericity, and moderately close packing can be taken at about 60—65% grain volume. For further refinement in specific cases, artificial mixtures must be prepared and tests made to determine additional reference points. This is especially true for Recent carbonate sediments and for ancient carbonate rocks.

EVIDENCE FOR COMPACTION OF CARBONATE SANDS

General statement

Aside from the nearly universal recognition of stylolites in limestones, most references to compaction of carbonate sands refer in a general way to apparent "overclose" packing, to fracturing of grains, or to grain contacts which appear to reflect solution or impression. All of these effects also are observed in barely compacted carbonate sands (Carozzi, 1961; Bishop, 1968; Purdy, 1968). Of course, many do result from compaction but these features do not in themselves provide any direct lead to the amount of compaction which has occurred in the rock since deposition. Few attempts have been made to quantify compaction, that is, to measure the amount of compaction which has occurred in carbonate grainstone (Kahle, 1966; Coogan, 1970), owing to the difficulty in determining a value for zero compaction. The present state of the art recognizes two useful approaches to measuring compaction in lime grainstones and, together with well-established methods of carbonate petrography, these can provide considerable information on the degree of compaction.

Measuring packing and compaction, petrographic techniques — packing density and packing index

The two principal measurable effects of the compaction of carbonate sand are: (1) increase in grain volume percentage as compared to the original uncompacted sediment, and (2) increase in closeness of the grains as reflected in the increased number of grains which touch each other in a compacted sand. The first effect, increase in grain volume percentage, can be directly measured in thin sections of lime grainstones using one or more of the point counting methods. In 1956, Kahn proposed a new parameter, *Packing Density* and defined it as:

$$PD = \left(m \sum_{i=1}^{n} gi/t \right) \times 100 \qquad (3\text{-}1)$$

where n is the total number of grains in a given traverse across the thin section, gi is the grain intercept of the ith grain, m is the magnification constant, and t is the length of the traverse. Packing Density is essentially the equivalent of grain volume percentage as measured in two dimensions of a thin section and is closely comparable with three-dimensional grain volume if sufficient area of the thin section is traversed. A comparison of grain volume percentage determined by weight and volume measurements and by point counting (Gaither, 1953) show that the average of about 7% greater grain Packing Density was obtained from thin section measurements. In measuring oolitic sand volume of Bahamian samples, a grain volume calculated from volumetric measurements was about 2% lower than that determined using thin section measurements. Most of the difference is attributed by Gaither (1953) to shortcomings of preparation techniques.

A simpler form of the Packing Density expression can be used in grainstones where no consideration is given to recording cement and pore space separately. A simplified expression used in measuring compacted oolitic grainstone is (Coogan, 1970):

$$\bar{x}PD\% = (gi/TL) \times 100 \tag{3-2}$$

where $\bar{x}PD$ is the arithmetic mean of the separate measurements made on all traverses of the thin section and TL is the total of all traverses. This method of determining Packing Density is shown in Fig.3-13, where a single traverse line crosses six ooliths, pores, cement and some shaly matrix. The six ooliths occupy 73% of the linear distance of the traverse. The summation of successive traverses yields the mean Packing Density of the sample. In this example, the volumes of cement, pores, and matrix are 7, 12, and 8%, respectively. As compaction proceeds, grains are pushed closer together resulting in a higher grain volume percentage, cement, and matrix volume, and lower pore volume. The minimum value for Packing Density corresponds to the minimum value of the grain volume percentage for the uncompacted sediment; the maximum value is 100%.

The Packing Density measurement is itself affected by compaction. Calculation of the standard deviation for several samples of uncompacted oolitic grainstones shows values as high as 10% for one standard deviation ($PD = 65\% \pm 10\%$). Furthermore, the short traverse data (Fig.3-13) show a nearly normal distribution of Packing Density values, with a mean at about 65% (Fig.3-14) and a small value for kurtosis. In comparison, data from deeply-buried Jurassic Smackover oolitic grainstones (Fig.3-14) show that

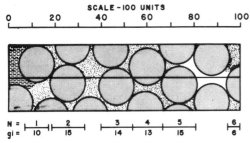

Fig.3-13. Illustration of the measurement of Packing Density and Packing Index using a thin section of spheres, shown as a monolayer. (a) Number of ooliths crossed by traverse, $N = 6$; (b) sum of grain intercepts, gi: $10 + 15 + 14 + 13 + 15 + 6 = 73$; (c) total length of traverse, $TL = 100$ line units; (d) packing density calculation for single traverse, $PD\% = (gi/TL) \times 100 = 73\%$; (e) grain-to-grain contacts, $GC = 2$; (f) packing index calculation for single traverse, $PI\% = (GC/N) \times 100 = 2/6 \times 100 = 33.3\%$; (g) other data: grains touching = 3; grain-to-matrix contacts = 1; grain-to-cement contacts = 3; grain-to-pore contacts = 2; length of matrix intercepts = 8 units; length of cement intercepts = 7 units; length of pores intercepts = 12 units. Stippled pattern is cement; white is pore space; dashed pattern is shaly matrix.

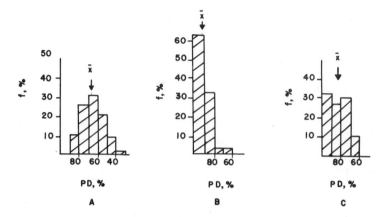

Fig.3-14. Histograms of Packing Density sorting for three oolitic grainstones. A. Uncompacted oolitic sand from the Bahamas ($\bar{x}PD$ = 64.4%; SD = 10.1%). B. Jurassic Smackover grainstone from a depth of 3,150 m ($\bar{x}PD$ = 90.5%, SD = 5.8%). C. Jurassic Smackover oolitic grainstone from a depth of 3,687 m ($\bar{x}PD$ = 83.1%, SD = 7.5%). Width of histogram columns is equal to 10% Packing Density interval. PD = Packing Density, f = frequency, $\bar{x}$ = mean.

the mean shifts toward the 100% limit with strong asymmetry and a forced reduction in the value of the standard deviation. Comparing the three samples illustrated in Fig.3-14, it can be seen that with increasing compaction, the mean of the Packing Density shifts, the standard deviation is somewhat reduced, and a strong kurtosis may develop. The reason for the variation in kurtosis is not clarified solely by the histograms. Inspection of the samples, however, shows that grain-size sorting is a crucial factor. In the case of the Smackover oolite buried to 3,687 m the grains are poorly sorted. The tendency of large grains to bridge across the smaller ones would cause the smaller ones to absorb less of the compactive stress. This could result in certain areas of the rock exhibiting a lower degree of compaction than others. In the case of the sample from a depth of 3,150 m, the grain size sorting is good and, hence, the stress should be more evenly distributed.

The second feature, increased number of grains in contact, is represented in the *Packing Index* proposed by Masson (1951) and used with modification by Kahle (1966) and Coogan (1970) in measuring the compaction of oolitic and pelletoidal lime grainstones. As originally defined, the Packing Index is based on grain boundary measurements. Use of Masson's index requires the counting of intersections (one per grain) between traverse lines and grain boundaries (either grain-to-grain, grain-to-cement, or grain-to-pore space). Contacts are counted as the traverse leaves the grain. Masson's formula is:

$$PI = 100 \, (Ng/Ng + Nn) \qquad (3\text{-}3)$$

where Ng is the number of grain-to-grain contacts and Nn is the number of grain-to-cement contacts.

In 1956, Kahn proposed a similar measure of closeness of grains called the *Packing Proximity* for which the formula is:

$$Pp = q/n \times 100 \tag{3-4}$$

where q is the number of grain-to-grain contacts, n is the total number of contacts (as well as the total number of grains), and $0 < q < n$. The maximum value for this expression is 100%. For oolitic grainstones, the expression may be written as:

$$\overline{x}PI\% = GC/N \times 100 \tag{3-5}$$

where GC is the number of grain-to-grain contacts along the line of traverse, and N is the number of grains along the same traverse. Values for repeated traverses are summed to derive the mean Packing Index. An illustration (Fig.3-13) of the method of measurement of Packing Index and Packing Proximity shows that Masson's (1951) Packing Index (in the example, 40%) is not the same as Kahn's (1956) Packing Proximity (in the example, 53.3%). The latter index is the same as the simplified Packing Index used by Coogan (1970). The rationale behind the use of the Packing Index is the idea that lime grainstones which are compacted should have more grains touching each other than those which are uncompacted.

All packing indices measure similar relationships. The Packing Index is more quickly determined than the Packing Density. It is less useful in an evaluation of compaction in rock, however, because the Packing Index is affected more by the total number of grains counted, the shape of the grains, their sorting, and the effects of cementation and diagenesis which change grain contact configurations. The dependency of the Packing Index on the total number of grains counted may cause difficulties as can be seen in Fig.3-15 and Table 3-IX. In Fig.3-15, 10 spherical grains are drawn to occupy a traverse length of 100 linear units through the centers of the spheres in row *1* for which $N = 10$, $TL = 100$, $PD = 100\%$, $GC = 9$, and $PI = 90\%$ (Table 3-IX). In this situation the maximum Packing Index for the line is 90% and not 100% as stated by Kahn (1956), because there are only 9 grain-to-grain contacts possible (contacts are counted as the traverse leaves the grain). In rows *2—6* (Fig.3-15), 5 spheres occupy 50% of the traverse length but are arranged so that all the grains touch each other in row *2*, whereas none touch in row *6*. Rows *3—5* have intermediate numbers of grains touching each other. As shown in Table 3-IX, the Packing Density for all the spheres in rows *2—6* is the same (50%), but the Packing Index ranges from zero to 80% depending on the spacing of the grains. Approximately

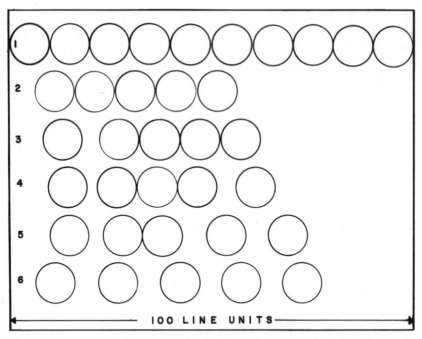

Fig.3-15. Diagram used in measuring Packing Density and Packing Index for single-line traverses of 5 and 10 uniform spheres in a row, as presented in Table 3-IX.

TABLE 3-IX

Relationship among the Number of grains (N), Grain Contacts (GC), Grains Touching (GT), Packing Density (PD) and Packing Index (PI) for idealized packed spheres (Fig.3-15); number of grains and grain contacts were measured in thin section

N	TL	Σgi*	$PD\%$	GC	GT	$PI\%$	Fig.3-15, row no.
10	100	100	100	9	10	90	1
9	100	90	90	8	9	89	—
9	100	90	90	7	8	78	—
8	100	80	80	7	8	87	—
8	100	80	80	6	7	75	—
8	100	80	80	5	6	62	—
5	100	50	50	4	5	80	2
5	100	50	50	3	4	60	3
5	100	50	50	2	3	40	4
5	100	50	50	1	2	20	5
5	100	50	50	0	0	0	6

*Σgi = sum of measurements of grain intercepts of the ith grain.

COMPACTION AND DIAGENESIS OF CARBONATE SANDS 119

200 or more grains must be counted before a maximum value for the Packing Index is approached for any given Packing Density (Fig.3-16). Wide ranges in grain size, as well as differences in apparent size owing to the random intersection of the thin section plane with the grain, also reduce the chance of intersecting grain-to-grain contacts and affect the Packing Index measurement.

In spite of the difficulties of directly relating Packing Density for short traverse data to Packing Index (Table 3-IX), it does appear possible to relate the mean Packing Density for a number of samples to the mean Packing Index. In Fig.3-17, the mean values of the Packing Index are plotted versus mean Packing Density for short line traverse measurements on 24 thin sections of oolitic grainstones. A linear regression was calculated using the expression:

$$y\ (PI\%) = a + b\ x\ (PD\%) \tag{3-6}$$

where (Σ over $i = 1 \ldots n$) $a = \Sigma y_i - b\Sigma x_i/n$ and $b = [n\Sigma x_i y_i - \Sigma x_i y_i]/[n\Sigma x_i^2 - (\Sigma x_i)^2]$ (Freund and Williams, 1958). Thus, $\bar{x}(PI\%) =$

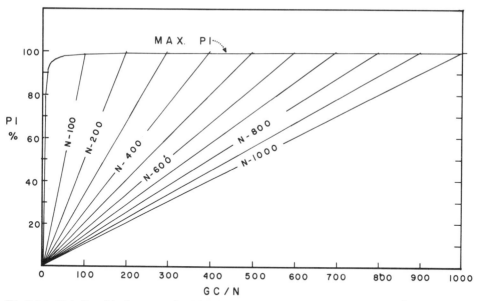

Fig.3-16. Relationship between Packing Index (*PI*) and *GC/N* ratio for different numbers of grains. The maximum *PI* increases rapidly from zero with increasing numbers of grain contacts per number of grains. Diagonal lines show plot of *GC/N* for samples of differing total number of grains (e.g., *N* = 200). As *N* and *GC/N* increase above 100, the maximum *PI* flattens. *PI* = Packing Index, *N* = number of grains, *GC* = number of grain contacts, Max. = maximum.

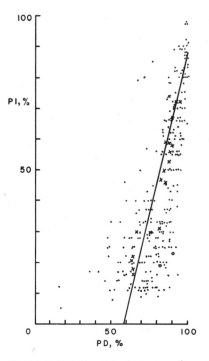

Fig.3-17. Relationship between the mean Packing Index (*PI*) and mean Packing Density (*PD*) based on short-line traverse measurements on thin sections of compacted and uncompacted oolitic grainstones. Solid points are individual line measurements. Small x's and open circles are mean values for 40 traverses each.

$-125 + 2.13\bar{x}(PD\%)$. This result is considered tentative because few samples counted were completely uncompacted grainstones. Nevertheless, the regression is based on traverses of over 8,000 grains in oolitic grainstones from surface and subsurface localities ranging in age from Mississippian to Recent.

The minimum value for the mean Packing Index varies with the shape of the grains and the grain volume. Randomly and loosely arranged packs of spherical grains have values as low as 2.5%, but the normal range for Recent oolitic sands is about 18—20%; the maximum is 100%. The minimum Packing Density projected for zero Packing Index is 54% (Fig.3-17), a value close to that of the grain volume percentage of the ideal cubic (case 1) packing (Fig.3-5A).

In summary, these two measurements of packing are important in establishing the extent of compaction. The Packing Index is less useful but more rapid to calculate. It may disclose one anomalous compaction feature, the one indicated by high Packing Density and low Packing Index, involving a situation in which the grain volume percentage has been reduced but the grains seldom touch each other. In general, use of the Packing Density is recommended for routine thin-section analysis of compaction.

Other measurements and evidence of compaction

Many petrographic features may indicate that carbonate particles have been subjected to compactive stress. Special notice must be taken of the work on grain-contact relationships in clastic sands. The change from the apparent "floating" grains and grains with tangential contacts, to grains having concavo-convex and sutured microstylolitic contacts (Fig.3-18) is evidence of compaction. In her study of pore space reduction, Taylor (1950) considered the number of impressed and stylolitic contacts which were an index of compaction. This procedure is risky, however, because the original shape of the grain (spherical, elongate, discoidal, or platy) will determine to a great extent the shape of the grain contact. For example, longitudinal contacts (Fig.3-18) occur readily between elongate grains as a result of sedimentary packing without any influence of compactive stress.

The composition, hardness, ductility, and susceptibility to solution of grains are important in determining the final grain-contact shape as a result of compaction. Hard quartz grains tend to penetrate soft calcite ooliths without producing microstylolitization, whereas adjacent ooliths in the same thin section show mutual solution features. In a study by Hays (1951, cited by Gaither, 1953) it was noted that quartzose sands with a large amount of carbonate grains had an abnormally high percentage of sutured contacts compared with other quartzose sands buried to equivalent depths, but devoid of carbonate grains.

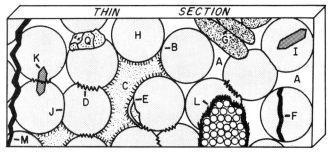

Fig.3-18. Petrographic effects in randomly-packed, compacted oolitic grainstone, shown as a monolayer. A = pore space, B = tangential contact between grains, C = cement, D = microstylolitic contact between grains, E = ruptured cortex of oolith, F = calcite-filled, prelithification fracture, G = longitudinal contact, H = oolith, I = authigenic quartz, J = indented grain contact, K = authigenic quartz incorporating microstylolite, L = macrostylolite, M = postlithification fracture, and Q = polycrystalline quartz grain indenting ooliths.

The data from Taylor (1950) for sandstones show a definite trend toward increased number of contacts per grain and more pressure solution with increasing depth (Table 3-X). There is a corresponding increase in values of Packing Density and Packing Index, because the counting of the total number and/or the number of different kinds of grain contacts is a form of packing index measurement. She reported between 0.63 and 1.6 contacts per grain for randomly-packed artificial sands and Gaither (1953) estimated a value of 0.85 for the number of contacts per grain for freshly-deposited clastic sand. Naturally, the size, shape and sorting of the grains are important factors in determining the number of grain contacts. The similarity between the determination of Packing Index and Taylor's method of counting types of grain contacts suggests that the latter has little to add to the measurement of compaction experienced by a sand. The study of the kinds of grain contacts, the mineralogical composition of the grains, and their relationships to cement, however, may yield useful information concerning compaction of sands.

In a comparison of artificially-compacted, coarse-grained carbonate sediments with naturally-compacted grainstones, Ayer (1971) recognized 12 structural and petrographic deformation features related to individual grains. Somewhat modified, these are:

TABLE 3-X

Number and types of grain contacts in synthetic and natural, buried, Wyoming sands (From Taylor, 1950; Gaither, 1953)

Grain contact data	Synthetic sands		Formation name and depth			
	Taylor	Gaither	Mesa Verde, 865 m	Shannon, 1,370 m	Lower First Wall Creek, 2,075 m	Morrison, 2,540 m
Average number of grain contacts	1.6	0.85	2.5	3.5	4.4	5.2
Floating grains (%)	16.6	46.0	0.3	0	0	0
Tangential contacts (%)	59.4	77.0	51.9	21.4	0.9	0
Long contacts (%)	40.8	17.0	38.1	59.8	51.6	45.0
Concavo-convex (%)	0	6.0	9.6	19.1	28.5	23.1
Micro-stylolitic (%)	0	0	0	0	18.5	31.8

(1) Point contacts with no grain interpenetration (also called tangential contacts).

(2) Concavo-convex contacts, which result from the yielding of one grain.

(3) Linear contacts which result from mutual yielding of grains; these are not to be confused with point contacts between elongate grains.

(4) Sutured contacts (microstylolitic), which show small grain interpenetration.

(5) Plastic flow or distortion of grains.

(6) Crushing, which results from repeated microfaulting or shattering.

(7) Radial fractures which are confined to the radius of the grain and do not pass through the center.

(8) Diagonal fractures of the grain from the surface through the center.

(9) Faulting or rupture, which breaks the grain in two or more parts that remain nearly contiguous.

(10) Splitting, which breaks and separates the grain into two more-or-less equal parts; the displacement is perpendicular to the fracture trace.

(11) Buckling (crinkling) of the outermost portion of the grain, which results in a small sharp crest. This feature is common to some oolitic grainstones.

(12) Spalling, which flakes off the outermost portion of the grain. This feature is common to some deformed oolitic grainstones.

It is not clear whether or not concavo—convex contacts are distinctly separate from flow features. It appears that several of the 12 deformational types are related not only to the degree of compaction but also to the spherical shape of the oolith grain. Such deformational features should be rare or indiscernible in non-oolitic carbonate particle grainstones.

Another approach to measuring compaction in thin section using grain-contact relationships was proposed by Allen (1962). He measured the degree of indentation of grains by determining the ratio of the length of the grain margin touching another grain (fixed margin) to the length of the remaining boundary (free margin), which is in contact with pore space or cement. An indented grain is considered as fixed when the fixed margin exceeds the free margin in length and as free in the opposite case. His measurement of indentation is called the Condensation Index:

$$\text{Condensation Index} = \% \text{ Fixed Grains}/\% \text{ Free Grains} \qquad (3\text{-}7)$$

In part, this index relates the number of "floating" and tangential grains to the number of impressed and sutured grains and is similar to Taylor's approach. A further attempt to use grain contacts in the form of the free and fixed margin ratio develops an index of compaction defined as:

Fixed-Grain Compaction Index = (Number of Fixed Grains/Total Grains) × 100   (3-8)

In the latter case, the Fixed Grain is one with more than 50% fixed margin (Allen, 1962). The maximum possible value for the Condensation Index is infinity, whereas the maximum value for the Fixed-Grain Compaction Index is 100%; the minimum value for both is zero. A comparison of these indices with Packing Density, Packing Index, average number of grain contacts per grain, and Compaction Index is shown in Table 3-XI for four oolitic grainstones (Fig.3-19—3-22). For example, the Condensation Index of the Bahamian uncemented oolitic sand is zero and that of the strongly compacted Smackover oolite is 0.62. The latter is an inconveniently low-sounding value for a rock with a Packing Density of 95.9%, considering the fact that the maximum value for the Condensation Index is 99% to infinity. The Condensation Index scale is not an arithmetic scale and the values do not progress uniformly with increasing compaction. Values for the Fixed-Grain Compaction Index are zero for the uncompacted Bahamian oolite and 38% for the Smackover oolite. On the Compaction Index scale derived from the Packing Density (Coogan, 1970), to be discussed later, Smackover oolite is 87% compacted.

Unfortunately, the indices used by Taylor and Allen have not been related to the uncompacted state of carbonate sands so as to allow an estimate of the zero point of compaction on their scales. In addition, it is

TABLE 3-XI

Comparison of various measurements of compaction for four oolitic grainstones (Fig.3-19—3-22, Table 3-XV)

Measurement	Specimen*				Maximum possible value
	1	2	13	3	
Mean Packing Density (%)	64.8	70.3	82.9	95.9	100
Mean Packing Index (%)	18.6	22.5	18.6	66.5	100
Compaction Index (%) (Coogan, 1970)	zero	15	51	87	100
Condensation Index (%) (Allen, 1962)	zero	0.13	0.11	0.62	99 to infinity
Fixed-grain Compaction Index (%)	zero	10.8	8.1	38	100
Grain contacts per grain	1.2	1.4	0.8	3.7	14.1**
Figure no.	(3-19)	(3-20)	(3-21)	(3-22)	

*For sample descriptions, see Table 3-XV.
**From Marvin (1939).

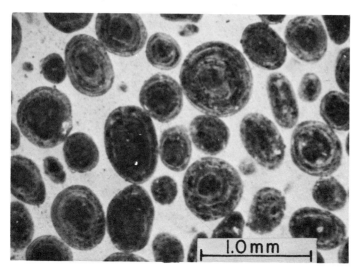

Fig.3-19. Loose, uncompacted, oolitic sand, impregnated with plastic in situ on an oolite bar top to preserve fabric. Browns Cay, Bahamas.

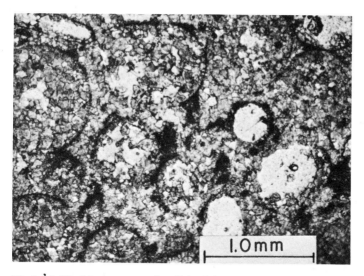

Fig.3-20. Weakly-compacted oolitic doloarenite, a dolomitized oolitic grainstone. Jurassic Smackover Formation, Pan American Petroleum Corp., No. 1, Parker, Van Zandt Co., Texas, depth = 3,917 m.

doubtful that their scales could be used with much precision for irregularly shaped grains of bryozoans, crinoid columnals, plate algae or other biologically derived carbonate particles. For example, the Condensation

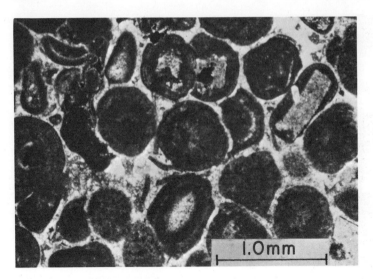

Fig.3-21. Compacted oolitic grainstone, Mississippian Ste. Genevieve Formation, Roane Co., Tennessee, outcrop.

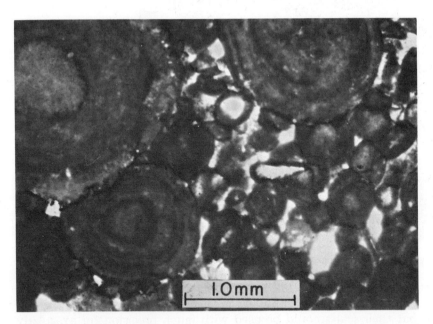

Fig.3-22. Compacted oolitic grainstone showing interpenetrating grains, microstylolitic contacts and cement. Jurassic Smackover Formation, Tenneco Oil Co., No. 1, Lowe, Clairborne Parish, La., depth = 3,149 m.

Index is zero for Bahamian oolite (Fig.3-3A) and gastropod sand (Fig.3-3B), yet the Packing Index of the two sands is quite different. Inasmuch as both of the indices using the fixed-grain concept are based on grain-contact relationships, both suffer from the same deficiencies in use that one encounters on using the Packing Index. These indices, therefore, cannot be recommended as measurements of compaction.

Compaction index

A Compaction Index based on Packing Density was formulated by Coogan (1970) for oolitic grainstones. An increase in Packing Density reflects the movement of grains closer together, presumably as the result of increased overburden pressure. Thus, any increase in Packing Density may be thought of as an increase in the amount of compaction from some Packing Density value, which represents zero compaction for the particular carbonate sand (depending on the particle shapes, sorting, and sedimentary packing configurations), to 100% Packing Density, representing a rock composed entirely of grains or one that is 100% compacted (no porosity). It is recognized that subsequent compaction through gross dissolution of calcium carbonate along stylolitic surfaces might occur (Park and Schot, 1968). Based on the concept of increasing Packing Density, a revised Compaction Index is presented (Fig.3-23), which arbitrarily relates Packing Density or grain volume percentage in thin section to the amount of compaction. New data have been added for special carbonate grain shapes.

An initial zero compaction index value was calculated for several different kinds of carbonate sands (curves $A-F$, Fig.3-23). For sands composed principally of ooliths, the value of 65% Packing Density or grain volume percentage was taken as zero compaction based on determinations of the average Packing Density of unburied, naturally-packed, well-sorted oolitic sand from the Bahamas (Coogan, 1970). Two different kinds of samples were examined. One, a Recent oolitic sand from an oolite bar top, 1.5 km east of Browns Cay was sampled by impregnation with epoxy resin, preserving the undisturbed cross-bedding fabric of the sand. Thin sections of this sand had a Packing Density of 64.8%, a sample standard deviation of 10%, and a Packing Index of 18.6%. A check on these values was made by measuring the Packing Density of oolitic grainstone from subaerially weakly-cemented oolitic rock of Pleistocene age on Joulters Cay, northeastern edge of Andros Island, Bahamas. The Joulter's Cay oolitic grainstone had a Packing Density of 64.4% (SD = 9.6%) and a Packing Index of 21.6%. As a point of reference, the mean grain volume of 9 cemented Pleistocene oolitic grainstones from Florida and the Bahamas, converted from Robinson's (1967) bulk porosity data, is 60.6% (39.4% porosity; 6,777 md permeability). Six very lightly cemented oolitic grainstones had a mean grain

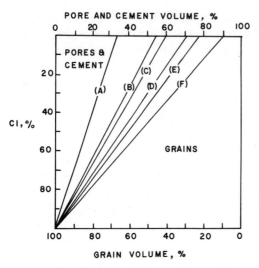

Fig.3-23. Relationship between Compaction Index (*CI*) and grain volume percentage (hence also pore plus cement volume) for selected mud-free carbonate sands. The range of grain volume percentage is from 10 to 100%; the range of pores plus cement is from 90 to zero % depending on the original packing (curves $A-F$) and amount of compaction. A = oolitic grainstone, B = grapestone, C = angular, mixed skeletal sand, D = valves of the pelecypod *Noetia ponderosa* Say, E = valves of *Anomia simplex* Orbigny, and F = red algal particulate sand (data from Dunham, 1962).

volume of 58.7% (41.3% porosity; 47,000 md permeability). Based on earlier cited values of grain volume of randomly-packed spheres and the Packing Density measurements from the oolitic sands, a Packing Density value or grain volume percentage of 65% was assumed to be an average value for the state of zero compaction for oolitic sands on the Compaction Index scale.

Attempts to relate other types of carbonate grain sands to this scheme are just beginning, because of the lack of knowledge on the average packing of most other typical carbonate sands. Nevertheless, new determinations by the writers on shell materials collected from the west coast of Florida and from the Bahamas, backed with data from the literature, provide useful approximations of the grain volume of grapestone and single-component and mixed shell sands. These data provide zero compaction reference points for the Compaction Index chart (Fig.3-23).

The data for grapestone provided by Fruth et al. (1966) were checked by determining the porosity of grapestone sand collected in the Berry Islands, Bahamas. The most densely packed samples with the least wall effect have a grain volume of 55.3%, about 3% above the value determined by Fruth et al. (1966). The 55.3% value is used as the zero compaction point on the Compaction Index line for grapestone (curve B, Fig.3-23).

Estimates for skeletal lime grainstones (curves C—E, Fig.3-23) are derived from three sets of data for different-shaped particles. A shell "hash" was collected from a 10-cm high beach ridge on the causeway between Fort Myers Beach and Sanibel Island, Florida. The mean grain volume based on porosity determinations for this coarse sand is 39.5%. The grains are very angular and consist of broken particles of numerous bay or inlet dwelling pelecypods and gastropods. The sample has a mineralogical composition of about half calcite, half aragonite. The zero value for mixed skeletal grainstone Compaction Index line (curve C, Fig.3-23) is based on analysis of this collection.

In determining the porosity for more loosely packed skeletal sand, a collection of valves of the taxodont pelecypod *Noetia ponderosa* Say, the ponderous ark shell, were used. The sample was scooped from piles of shells washed up on a narrow beach facing the Gulf of Mexico, about 1 km north of the bridge between Sanibel and Captiva Islands, Florida. The single valves were hand-sorted by size until a collection of several hundred valves, each about 3 cm wide, was available. The greatest grain volume measured in several containers of different shapes and sizes was 29.0%. This value is used as the zero compaction point for curve D (Fig.3-23). *Noetia ponderosa* is an aragonitic, nearly equivalve species with a deep body cavity, suboval outline, and strong umbos.

The loosest packed single-component skeletal sand for which the grain volume was measured is a pebble sand consisting of oval, curved to flat, thin valves of *Anomia simplex* Orbigny, the Common Jingle Shell. The shells were hand-sorted by type and size from the Sanibel—Captiva collection described above. The curved to flat shape of the valve makes it analogous to many types of fossil grain shapes, for example, some of the plate algae. The size of the shells selected for porosity determinations ranged from 1 to 3 cm across the valve. Porosity was measured for packs of the Jingle shells in different sized containers and the values were extrapolated to a container of infinite dimensions. Calculated as grain volume, the value entered as the zero compaction point on the Compaction Index chart (curve E, Fig.3-23) is 22.1%.

Even looser packing has been reported by Dunham (1962) for red-algal particulate sand (curve F, Fig.3-23) which had a grain volume of 10%. This value is close to that reported by Graton and Fraser (1935) for packed mica and may represent the loosest packing fabric for carbonate particulate sands.

Using the Compaction Index chart (Fig.3-23), it can be seen that an oolitic grainstone with a Packing Density of 90% (compressed from 65%) has a Compaction Index of 68%. In other words, this oolitic grainstone has been compacted to a point where it is 68% of the way from a completely uncompacted oolitic sand to a completely compacted rock composed of 100% grains. Oolitic grainstone with a Packing Density of 95% is about 85% compacted (Fig.3-23). On the other hand, with a Packing Density of 65% a grapestone is 35% compacted, a skeletal sand is 42% compacted, a sand of *Noetia ponderosa*-shaped shells is 51% compacted, a sand of *Anomia*-shaped particles is 55% compacted and a sand of red algae is 62% compacted (Fig.3-23). More refinements based on shape factors and more determinations of grain volumes are needed to develop widespread applicability of the Compaction Index, but the scheme has the potential of producing comparable values for compaction of limestones regardless of the original differences in packing of the grains. As the Compaction Index is refined and more closely linked to depth of burial, it may be possible to use it to determine the previous depth of burial of now outcropping compacted carbonate grainstones.

EXPERIMENTAL STUDIES OF COMPACTION

Summary of experimental work

Only a small amount of experimental work has been performed on the compaction of carbonate sands and gravels which could serve as a basis for understanding subsurface compaction of these sediments. In 1966, Fruth et al. reported on the compaction of five Bahamian carbonate sediments subjected to pressures as high as 1,000 kg/cm^2 and their work stands as a solid reference point for compaction of carbonate sands. Subsequently, Fruth subjected similar sediments to three types of hydrostatic compaction tests and Ayer (1971) compared the results with ancient compacted grainstones. Earlier, Terzaghi (1940), Hathaway and Robertson (1961), Robertson et al. (1962) and Robertson (1967) reported on the consolidation of calcium carbonate mud. There also has been some interest in the loss of water from high-porosity carbonate sediments on the sea floor under slight consolidation stress from the standpoint of engineering geology (Miller and Richards, 1969). Others have experimented with the compaction of clastic sands (Maxwell, 1960), which have some bearing on understanding the compaction of lime grainstones. Finally, other work on the compaction of spheres (Matzke, 1939; Marvin, 1939) is pertinent to understanding one of the extreme limits of compaction of grain-supported sediment. In this section, the writers first review the experimental work on the compression of spheres, then the pertinent work from papers on the compression of sands in general, and, finally, the experimental compaction of carbonate sands.

Experimental compaction of spheres

Studies on the compaction of spheres are pertinent to an understanding of the compaction of lime grain sediments, because they provide a limiting condition for the maximum compaction of grains without cement. In an experiment on the compression of randomly-packed, small, spherical lead shot, Marvin (1939) eliminated all void space between particles in a cylinder under pressure of 16,000 kg/cm^2. The particles of shot, sized at 2.54 mm (granule size), were compressed at 0.25 cm/min and the shape of the compressed shot and average number of contacts per grain were determined visually under a microscope after discrete stages of compression. The number of grain contacts increased (Table 3-XII) until under a total pressure of 16,000 kg/cm^2, when all space was eliminated, solidly-packed lead polyhedra had an average of 14.16 faces per grain. In a second experiment uniform spherical shot were packed as rhombic regular layers (case 6, Fig.3-5F) and compressed. Each particle, which originally had 6 contact points and 6 free sides facing pores, ended with 12 contacts and no new faces, in contrast with the earlier experiment in which new faces were produced in irregular packs. In subsequent experiments, smaller shot (1.27 mm — very coarse sand) were compressed and, except for particles on the periphery of the cylinder, the average contacts per grain for 624 grains of shot were 14.7. Compressed contact faces formed at the original packing contacts were larger than those produced solely from pressure. Neither tamping before compaction nor changing the shot size changed the average of 14 contacts per grain after compression. Thus, it can be assumed, that for nearly spherical grains of about uniform size, compaction without concurrent cementation will produce tetrakaidecahedral grains.

Mixing proportions of small and large shot (1.27 and 2.54 mm in diameter) in ratios (small to large) of 1:4, 1:1 and 4:1 and compressing the mixture under a total load of 18,000 kg/cm^2 resulted in different average

TABLE 3-XII

Average number of contacts for 100 observations on spherical, uniform, randomly-packed lead shot compressed under different pressures (After Marvin, 1939)

Pressure (kg/cm^2)	Average number of grain contacts per grain
450	8.41
2,160	10.97
4,530	12.91
11,000	13.62
16,000	14.16

numbers of polyhedral faces. The number of faces per grain for the 1:4 mixture was 9.5 for small shot and 20 for large shot. Equal mixtures had an average of 12 faces for small shot and 19 to 30 for large shot. In the 4:1 mixture, the small shot averaged 10 to 16 faces per grain, whereas the large shot exhibited 25 to 36 faces per grain with the mode of 31 faces. The large shot in the 4:1 mix had more contacts, presumably because the large shot was surrounded mostly by small shot. In the equal mix of sizes, some of the large shot, by touching each other, bridge over small shot and protect the latter until compression becomes severe. The number of contacts varied with the size and shape of the particles under compression.

The Packing Density for all samples of lead shot at the end of the compression experiments (Marvin, 1939; Matzke, 1939) was 100%. The Packing Index, based on calculations of fitted outlines of irregular dodecahedra, is close to 100%, and the Compaction Index (Coogan, 1970) is 100%.

Experimental compaction of clastic sands — results related to carbonate sands

In a series of compaction experiments on quartzose sandstones, Maxwell (1960) attempted to isolate and study independently the following variables: (1) pressure, (2) time, (3) temperature, (4) composition of the sediment, (5) composition of the fluid, and (6) fluid dynamics, i.e., static versus moving fluids. The general impact of the various factors has been considered. In his experiments, the consolidated sands were greatly fractured, interpenetrated and rotated. The larger grains appeared stronger. Most coherent grains showed strain and fractures which are related to points of contact. Solution and redeposition of silica did occur and was enhanced by increased temperature and through-flowing solutions. At pressures equivalent to a depth of about 9,000 m of overburden, the grain volume was increased to about 70%. Chemical processes were limited to solution, transportation, and precipitation of silica, similar to the results for carbonate sands. Great emphasis was placed on the importance of fluid movement. For example, Maxwell cited calculations which show that 32,000 cc of water derived from compacting strata have passed through each square centimeter of sediment now found at a depth of 1,750 m in the Cenozoic Ventura Basin of California. Deeply-buried sediments release large amounts of waters saturated with silica (or calcium carbonate, in the case of limestones). This upward moving fluid should lead to upward transportation and deposition of considerable volumes of cement in overlying sediments. It is important to note, however, that cores recovered from the Blake Plateau as part of the JOIDES program (Schlee and Gerard, 1965; Bunce et al., 1966) contain

Early Tertiary carbonate sediments, which are buried 150 m (at a depth of 1,180 m below sea level). These sediments are barely consolidated or lithified. Similar uncompacted and very slightly cemented carbonate sediments, with almost normal sea water as connate fluid, were found on the West Florida shelf in the Early and Late Tertiary part of the section. Unlithified and mineralogically unaltered sediments also were found in the Bikini and Eniwetok cores (Schlanger, 1963) recovered from beneath an overburden of over a thousand meters. It seems, therefore, that in these instances there was no significant upward flushing of the fluids. As discussed in the section on experimental compaction of carbonate sediments, fracturing owing to rapid loading is far more typical of experimentally compacted clastic sands than naturally compacted ones. Pressure solution of grains, development of supersaturated solutions, and precipitation are common where fluids can move. This results in microstylolitic contacts between grains and cemented pore space.

In other experiments on clastic sands, Fatt (1958) showed that the compressibility (reciprocal of the bulk modulus) is a linear function of composition for a given grain shape and sorting under pressures up to 1,055 kg/cm^2. He chose an idealized model of carbonate composition and texture as a pack of spherical solids with holes in it. He concluded that the effect of the internal fluid pressure on compressibility varies from rock to rock and estimated it to be only 85% effective in counteracting the overburden pressure. According to him, the effectiveness of pore pressure may range from 50 to 100%. For example, the net overburden pressure at 8,700 m is 1,100 kg/cm^2 (based on rock and water density of 2.3 g/cc and 1.0 g/cc, respectively). If the 0.85 effectiveness factor is neglected, then the pressure is 985 kg/cm^2. Based on experimental work and theoretical analysis, however, Rieke and Chilingarian (1974) concluded that pore fluid pressure is 100% effective in counteracting overburden pressure.

Compression of sands to 246 kg/cm^2 may cause a sharp drop in volume indicating crushing of the grains at that pressure. Plots of grain content versus the bulk volume compressibility (Fatt, 1958) show that intercepts on the y-axis (compressibility) tend to decrease with increasing pressure. This would indicate that for well-sorted silica sands the compressibility decreases with increasing pressure to 844 kg/cm^2 and is independent of pressure above that value. This indicates that at very high pressures the grains in all sandstones are in such close contact and extreme state of strain that the intergranular material does not contribute to the compressibility. Calcite and quartz have about the same compressibility, but calcite has a very low compressive strength. Compressibility is commonly obtained by a triaxial measurement, or the sample is hydrostatically stressed. Under these conditions, behavior of calcite and quartz is close. When subjected to a uniaxial stress, however, the low compressive strength of 2,600 kg/cm² (compared to 25,300 kg/cm² for quartz) results in crushing of the calcite.

Experimental compaction of carbonate sands

Using a triaxial apparatus designed to permit systematic evaluation of pressure, temperature and strain rate, Fruth et al. (1966) compacted carbonate sediments from the Great Bahama Bank which had been selected to reflect the major sedimentary facies of the Banks. The materials studied by them included sediments from the oolite, oolitic, grapestone, and skeletal facies (Purdy, 1963), all of which are carbonate or muddy carbonate sands. The sediments were subjected to pressures of 1,025 kg/cm² after determining the initial pore volume. The change in grain volume was calculated from the loss of porosity which was plotted versus increasing confining pressure. In addition, the study included petrographic examination of the compacted sediment for evidence of fracturing, cortical rupture, crushing, and interpenetration of grains. The results of the experiment show that all the sediments became compacted, including the carbonate muds which are not discussed here.

The sediments from the oolite facies (Purdy, 1963) consist of polished subspherical ooids and other well-rounded grains, most of which have oolitic coatings (Fig.3-24). Slightly cemented grain aggregates and organic material are common. Size analyses showed a median diameter of 0.35 mm, a Trask sorting coefficient of 1.31, and that less than 0.5% of the sediment was less

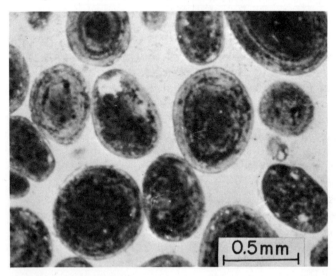

Fig.3-24. Oolite facies sand from top of the oolite bar at Browns Cay, Bahamas.

than 0.125 mm in size. The grain volume of the sediment was 60%. This oolite grain sand, with the predominance of solid grains and small amount of mud, was highly fractured by compression. There was negligible initial consolidation because the grains were self-supporting and the fine-grained matrix was very small. Fractures tended to radiate from points of initial contact, but in some cases a single fracture extended completely across the grain (Fig.3-25). The compacted grain volume is 78%. Only a few "floating" grains and rare tangential contacts remained after compression; instead, long contacts predominate. Considerable spalling and buckling of the borders of the ooliths occurred as a result of compaction (Fig.3-25). In spite of the high stress, shells in the oolite sediment are barely affected. Ooliths surrounding a gastropod show spalling and microfaulting of the cortex, whereas a lining of acicular aragonite crystals within the central cavity of the gastropod are hardly affected. Similar spalling may be seen in a compacted oolitic

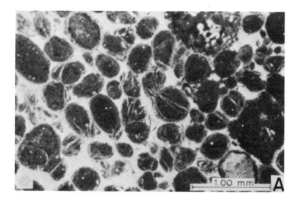

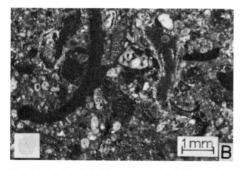

Fig.3-25. Artificially compacted Recent carbonate sediments. A. Fractured oolitic grain sand showing marginal peeling and grain penetration after stress of approximately 1,000 kg/cm^2. B. Skeletal facies sediment compacted under approximately 1,000 kg/cm^2. (After Fruth et al., 1966, published with permission of the Soc. Econ. Paleontol. and Mineral., *J. Sediment. Petrol.*)

grainstone (Fig.3-18 and 3-33), but the intense fracturing of grains occurring under laboratory conditions is seldom duplicated in oolitic rocks compacted in the subsurface.

The oolitic facies sediment consists of (1) numerous oolitically coated grains, which are rounded and subspherical; (2) friable and coherent grain aggregates; and (3) skeletal grains with and without oolitic coating (Fig.3-26). The median diameter of grains is 0.35 mm (medium sand), sorting coefficient is 1.4, and about 4% of the sediment is less than 0.125 mm (fine sand) in size. The initial grain volume was approximately equal to 60%. Under stress of 1,025 kg/cm^2, the grain volume increased to 84% (CI = 52%). Although fine matrix (4%) tended to cushion the effects of the stress on grains, the oolitic facies sediment showed abundant grain fracturing. Penetration of grains and buckling of the grain edges are also common.

The grapestone sediment (Fig.3-27) consisting of friable and coherent aggregates of skeletal and non-skeletal grains with intraparticulate pore space, had an initial grain volume of about 52%. The grains larger than 0.25 mm in diameter were polished and included well-rounded pellets, shell fragments, and subspherical ooids; the finer grains were more angular. The median diameter is 0.41 mm (medium sand), the sorting coefficient is 1.86, and about 8% of the sediment is below 0.125 mm in size. Compaction

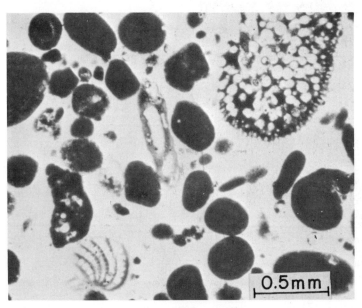

Fig.3-26. Oolitic facies sand from the tidal channel between oolite bars at Sandy Cay, Bahamas.

increased the grain volume to 84% grains (CI = 65%), as in the case of the oolitic facies sediment. Among the samples tested, penetration of grains is most pronounced in the grapestone sand. Fracturing at grain contacts was not common, probably because of flowage of soft grains and the abundance of small grains. After stress, all originally "floating" grains were moved into contact. Most of the grains show evidence of penetration, many with minute fractures.

The skeletal facies sediment contains abundant skeletal particles of whole foraminiferal tests, mollusc shells and fragments, and non-skeletal grains of mud and organic aggregates (pellets and pelletoidal grains) (Fig.3-28). The skeletal grains are rounded to angular and are either smooth or rough, whereas the inorganic grains are ovoid. The median diameter is 0.32 mm (medium sand), the sorting coefficient is 2.67, and 22.5% of the sediment is less than 0.125 mm in size. In Dunham's (1962) classification, this may be called a skeletal packstone sediment as opposed to the grainstone sediments mainly under discussion here. Similar packing of deep-sea sediment is shown in Fig.3-29. The original grain volume approximated 38%, which could be accounted for by separation of some of the grains by mud matrix. In addition, the effect of poor shape sorting on the packing also might result in low initial grain volume percentage. After compaction, the grain volume was close to 88% (CI = 80%) and the principal visual effect was the deformation of the skeletal grains. For example, a moderate-sized shell fractured under stress, whereas the grains contained in the intraparticulate space were not affected.

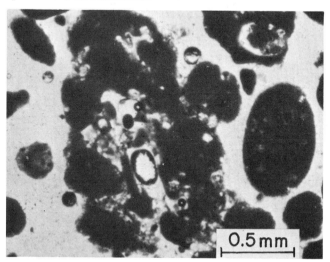

Fig.3-27. Grapestone facies sand from a grapestone bar top, middle Fish Cay, Berry Islands, Bahamas.

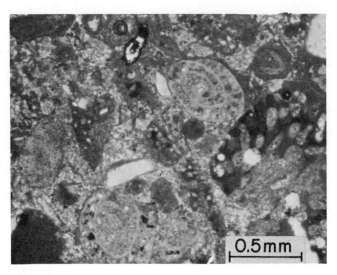

Fig.3-28. Skeletal facies sand from the burrowed tidal flat near Pigeon Cay, Bimini Lagoon, Bahamas.

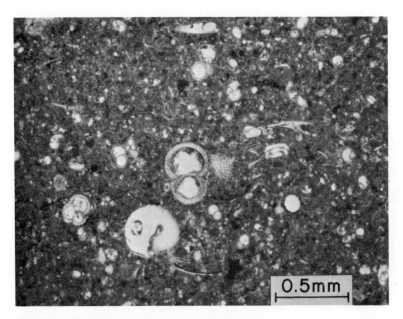

Fig.3-29. Deep-sea, planktonic-foraminiferal, packstone-textured sediment from the Tongue of the Ocean, Bahamas.

Fruth et al. (1966) showed that artificial compaction of these carbonate sediments produced many features present in naturally-compacted limestones. Their experiments, however, generally did not produce sediments with textures and fabrics which in composite are similar to those common in compacted lime grainstones. It should be remembered that the experiments were designed to obtain continuous porosity versus pressure curves and the thin sections were made only of the sediment carried to the maximum pressure. It is true that loss of porosity and increase in grain volume accompany both artificial and natural compaction of carbonate sands and that fracturing and grain interpenetration occur in both; however, the mechanism of grain interpenetration during subsurface compaction is through the dual process of pressure solution and mechanical deformation, which results in microstylolitic contacts between the grains. Such contacts were not produced much in these experiments. In addition, grain fracturing is far less common in naturally-compacted grainstones.

Carbonate oolitic and grapestone sediments were subjected by Fruth (Ayer, 1971) to three types of hydrostatic compaction* tests. In the first type, the confining pressure was increased in a series of increments and the fluid consisting of reconstituted sea water was allowed to drain freely. This type of test produced structural deformation of the sediment characterized by more sutured contacts, more longitudinal contacts, and less buckling. In the second type of test, the confining pressure was increased at a constant rate and the sea water allowed to drain freely. The result was a compacted sediment with a higher number of concavo-convex contacts, spalling, and diagonal fractures, but an intermediate number of cases of buckling. In the third type of test, the confining pressure and pore pressure were increased simultaneously at the same rate. After maximum pressures were attained, the pore pressure was decreased at a selected rate. In the three types of experiments, the average maximum confining pressure ranged from about 1,000 to 1,500 kg/cm^2. In the third type of test, the compacted sediment had the smallest number of point contacts, fewer concavo-convex contacts than in the type-two test, and the greatest amount of buckling.

Comparison of the test materials with naturally-compacted oolitic grainstones showed some differences and similarities. Ayer (1971) concluded, however, that to some extent the predominance of the type of deformation in a naturally-compacted rock was indicative of the depth of burial and might be related to the type of loading. In geological terms, the type-one tests may be similar to the compaction effects of periodic sedimentation with the concurrent release of pore fluids. The type-two tests might be compared with continuous compaction owing to nearly constant rate of sedimentation and concurrent pore fluid release. The type-three tests

---

*Method described by Fruth et al. (1966).

might be compared with continuous sedimentation of a sediment confined by impermeable seals which are later broken by faulting. The latter case is probably rare for carbonate sedimentary realms.

Petrographic examination shows that the radial and diagonal fractures and the crushing features are restricted to artificial tests and are not found in the compacted oolitic grainstones. The fractures are found in desiccated grainstone sediment, however. In contrast, almost no spalling or buckling features are found in deeply-buried samples and the degree of suturing increases with depth, as reported by Taylor (1950) for clastic sands.

In conclusion, it appears that laboratory experiments on the compaction of carbonate sediments have revealed important and useful data, and, while not always directly comparable with naturally-compacted rocks, should be encouraged. Further experimentation involving longer periods of time, more reactive pore fluids, and varied temperature conditions is desirable.

COMPACTION AND LITHIFICATION OF CARBONATE SANDS

General problem

Experimental works on the compaction of lead shot as well as clastic and carbonate sediments and the consideration of regular packing in sediments provide a necessary theoretical base. They, however, do not elucidate the perplexing facets of compaction of carbonate sands under natural conditions, because of concurrent changes in carbonate mineral species, solution, and cementation of grains.

In the simplest case, the lithification of a wholly grain carbonate sediment consisting of nearly spherical particles may be thought of as proceeding in one of three ways (Fig.3-30). Assuming the sediment has an initial grain volume of 65%, the simplest form of lithification would consist of the introduction of cement between the particles of the sediment in an amount equal to that of the initial porosity of 35%, without compaction or solution of the particles. The resultant rock consists of 65% grains, as did the sediment, and the 35% cement is obtained from a non-local source, presumably from solution of another carbonate sediment, or as a precipitate from the pore water (Bathurst, 1971b). A completely lithified beach rock on a modern tropical beach cemented by aragonite (Friedman, 1968) would serve as one example of this type of lithification. Many ancient lime grainstones appear to have been cemented without compaction or noticeable solution (Fig.3-31—3-33). This widespread and, locally, pervasive cementation doubtless prevents further compaction of the carbonate sands in many otherwise compactive-prone settings.

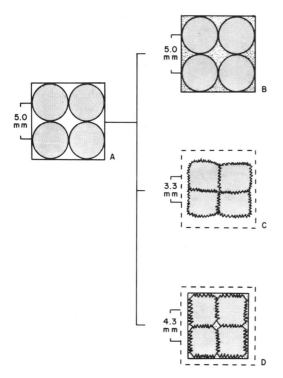

Fig.3-30. Three lithification processes of spherical grains, shown as a monolayer. A. Original sediment and grains; grain volume = 65%, pore volume = 35%, 6—8 contacts per grain. B. Simple cementation without compaction; grain volume = 65%, cement volume = 35%, 6—8 contacts per grain. C. Complete compaction without cementation; grain volume = 100%, pore plus cement volume = 0%, 8—14 contacts per grain; vertical center-to-center (of spheres) distance reduction = 33%. D. Compaction by pressure solution and cementation; grain volume = 75%, cement volume = 25%, 8 and more contacts per grain; vertical reduction in center-to-center distance = 14%.

A second type of lithification process, starting with the same sediment having a 65% grain volume, would consist of compressing, squeezing and dissolving the grains until they are so compacted that the rock is composed of 100% grains and no cement. Carbonate grainstones lithified in this manner are extremely rare, if they exist at all, except in an intense stage of metamorphism. The experimental compaction of lead shot by Marvin (1939) and of carbonate sediments by Fruth et al. (1966) are partial illustrations of this type of compaction.

Finally, there is the intermediate case wherein the sediment is compacted and at the same time cemented, resulting in a rock with some grain volume above the initial uncompacted value of 65% and below the

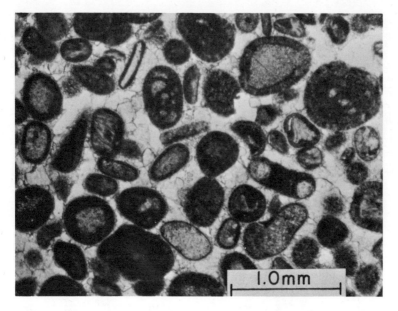

Fig.3-31. Weakly-compacted skeletal oolitic grainstone from the Mississippian Fredonia Limestone Member, Ste. Genevieve Formation, Harrison Co., Ind.

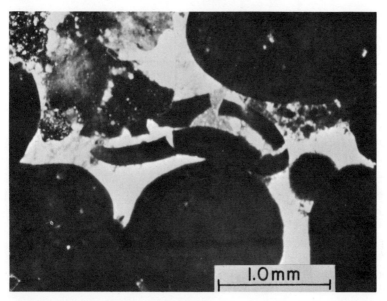

Fig.3-32. Compacted oolitic grainstone from the Jurassic Smackover Formation, showing quartz grain penetrating and breaking carbonate grain. From the North Central Oil Co., J. F. Roper No. 1 well, Freestone Co., Texas, depth = 3,696 m.

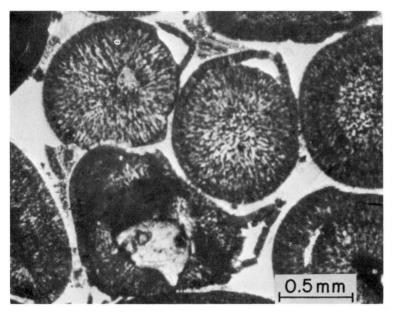

Fig.3-33. Compacted oolitic grainstone showing spalling of oolith margins. Collected from the outcropping Union Member, Greenbriar Formation (Mississippian), Monroe Co., W. Va.

theoretical maximum of 100% grains. This appears to be the usual case for compacted lime grainstones.

Sources of cement during compaction of carbonate sands

One of the principal unknowns in our understanding of carbonate sand compaction is the source of various cements found in compacted grainstones (Bathurst, 1971b). It is widely observed, for example, that Paleozoic rocks which have not been buried to great depths, such as the Pennsylvanian quartzose sandstones of Kentucky and Indiana, are not much compacted or cemented. Nevertheless, the carbonate rocks interbedded with them or immediately underlying them, for example, the Ste. Genevieve Group oolitic grainstones (Fig.3-31), are cemented. If the cement was precipitated from sea water during the syngenetic stage or from interstitial fluids during the diagenetic stage, the source of such a tremendous volume of cement, representing about 35% of the volume of several cubic miles of rock, can be accounted for with varying degrees of ease. If the cement, on the other hand, must be attributed solely to the solution of grains at the points of contact during compaction, the problem arises as to whether there is sufficient material to completely cement the rock without causing large, obvious solution cavities to form (Matthews, 1967).

In a general discussion of the cementation of sediments by carbonate minerals, Mackenzie and Bricker (1971) focused on the composition, chemical behavior and mass transport of some common carbonate cements. From a theoretical standpoint, cement generated through pressure solution and precipitation might have the following characteristics:

(1) Environment of cementation
    (a) At depth in the sedimentary column.
    (b) In combination with other previously precipitated cements.

(2) Timing
    (a) Long after deposition of the sediment.

(3) Cementation process (single or multiple event?)
    (a) Either as a single process over a long period of time or more commonly as multiple events.

(4) Mineralogy
    (a) Either monomineralic or polymineralic.

(5) Source
    (a) Either locally derived or imported.

The mass of materials dissolved or precipitated in a given carbonate sediment volume has been difficult to predict when the rock is buried, owing to the unpredictable conditions of temperature, pressure, and solution chemistry. Inasmuch as cements derived directly from stationary pore water can only occupy less than about 1% of the pore volume of a sediment after compaction (Mackenzie and Bricker, 1971) and only 8% can come from the transformation of aragonite to calcite, major cementation requires that cement be transported into the sediment from an outside source. The mechanisms of mass transport of sediment are bulk fluid flow and diffusion, with associated dissolution of sediment particles and pore cementation. The solubility of carbonate minerals in water containing CO_2 decreases with temperature and the solubility trends probably account for one or more phases of mutual replacement of silica and carbonate cements during compaction. The mass transfer of cement involved in burial of a carbonate sediment with its contained water has been calculated by Mackenzie and Bricker (1971) to a depth of 3.5 km for a region with a geothermal gradient of $0.3°C/1,000$ m. It was found to be $3.65 \cdot 10^{-3}$ moles $CaCO_3/1,000$ g of water or about 0.4 g $CaCO_3/1,000$ g H_2O, assuming a constant P_{CO_2} of $10^{-2.5}$ atmospheres, a pH of 7.4, and equilibrium with calcite. Thus, for every 1,000 cm^3 of pore space, about 0.15 cm^3 of calcium carbonate will be precipitated, resulting in a decrease of pore space of 0.02%. Under these conditions, porosity reduction in a sediment from an initial 30 to 1% would require the flowage of about 1,500 l of sea water through a 1,000 cm^3 volume of sediment. At a flow rate of 700 cm/year, less than $1.5 \cdot 10^6$ years would be required to transmit this volume of fluid. Even at much lower flow rates the time required for nearly complete cementation would be small. In regions of higher geothermal gradient, the mass transfer and time required for cementation could be reduced greatly.

In an extensive discussion of cementation, Bathurst (1971b) called on three main mechanisms which produce substantial amounts of carbonate cement in rocks: (1) the local dissolution of aragonite, (2) influx of sea water, and (3) pressure solution. He maintains, however, that none of these processes can produce the large amount of cement that is known to be present in limestones. While the authors agree with Bathurst (1971b, p.422) that the exact process may not be clear in specific cases, there is no agreement that the process need be a single one or that it is entirely elusive.

Effect of grain size and shape on cementation

The effect of grain size and shape on cementation has been investigated experimentally for quartzose sands (Heald and Renton, 1966) and some of their conclusions apply to carbonate sands. Grain size was found to have an important effect on rates of cementation by freely circulating solutions. Well-sorted coarse sands were cemented faster than fine sands because of their greater permeability. Finer sands were cemented much more rapidly than coarser ones (by a factor of 2.5), however, where the influx of cementing solutions was the same for both cases. On the basis of area differences, very fine sand theoretically would be cemented 16 times faster than very coarse sand and over 90% of the original porosity in the very coarse sand would be present after all the porosity was eliminated in the very fine sand. In addition, if a bed or lens of coarse sand were surrounded by fine sand, porosity reduction due to influx of cementing solutions would eventually cease in the coarse sand after entry of solutions had been prevented by more advanced cementation of the surrounding fine sand.

Rates of cementation were also found to vary with the angularity of the grains. Where influx of cementing solutions is the same, cementation proceeds considerably faster in highly angular sands than in rounded sands of the same grain size, mainly as a result of the greater surface area. Because initial consolidation after sedimentation is greater in angular sands, the combined effect of cementation and compaction would result in much more rapid reduction of porosity in highly angular sands. Since fine sands (for the same particle shapes) tend to be more angular, both the factors of size and shape favor the more rapid cementation of fine sands for an equal influx of cement.

The more rapid cementation of fine sands would result in more rapid increase in strength of their sand-rock body. Hence, upon burial, compaction, and cementation, fine sand bodies are more likely to become sufficiently indurated to resist increasing compaction than coarse sand bodies if supplied with the same amount of cement. In many instances one should be able to observe that either the fine sand is more cemented and the coarser sand is more porous, or that both are equally cemented but that the coarse sand has a greater Packing Density.

Varying cement composition

There is a commonly observed variation in mineralogical composition of carbonate cement in compacted grainstones. These differences may be attributed to relative proportions of aragonite, calcite, high-magnesium calcite, and dolomite which formed during penecontemporaneous diagenesis. Most cement in ancient limestones (Bricker, 1971), however, is low-magnesium calcite. Several investigators (Bathurst, 1971a; Choquette, 1971; Oldershaw, 1971) have emphasized the differences in form and composition of what appear to be late and early cements. Commonly, interparticulate pore spaces and intraparticulate skeletal pores are lined with a narrow zone of small prismatic nonferroan calcite crystals which constitute less than 20% of the total cement (Oldershaw, 1971). The remaining space is filled with large, randomly-oriented crystals of ferroan calcite. Petrographic evidence suggests that small nonferroan calcite material was precipitated before compaction, which caused fracturing of shells, whereas the large ferroan calcite was deposited after compaction. The origin of the early cement is postulated to be penecontemporaneous with sedimentation. The late cement, on the other hand, owes its origin to pressure solution, or mass transfer and migration to the pore site from other dissolving sites. Evidence of very late, partly postinduration cementation was offered by Choquette (1971) for some Ste. Genevieve grainstones in Illinois where the cement is epigenetic ferroan calcite formed in the presence of strongly reducing subsurface waters.

Geometrical and packing considerations

Evaluation from a geometrical standpoint of the amount of imported versus locally-derived cement produced by pressure solution has been made by Rittenhouse (1971) for clastic sands and part of his study is pertinent to the cementation of carbonate sands. Because the relative amounts of porosity loss due to solution and cementation vary greatly depending on grain shape, angularity, packing, the direction of applied stress, and grain

TABLE 3-XIII

Loss of porosity by solution versus that by precipitation of cement for four packing models (Data from Rittenhouse, 1971)

	Packing type			
	cubic	cubic, rotated 45°	orthorhombic	orthorhombic, rotated 30°
Porosity loss from solution (%)	13.4	18.4	10.3	10.0
Porosity loss from cementation (%)	9.2	4.2	4.2	4.5

mineralogy, the amount of cement derived from solution of the grains versus the amount of cement imported from other sources must also vary. Calculations for ideal packs of cubic and orthorhombic configurations by Rittenhouse (1971) showed that if the percent of sphere radii in segments of spheres removed by solution was equal to 28%, the solution alone could cause a reduction in porosity of 13.3% (from 47.6 to 34.3%) for cubic packs, if dissolved matter were removed from the system. If the material were entirely precipitated in adjacent pores there would be an additional loss of porosity of 9.1% (from 34 3 to 25.2%). The amounts and ratios of porosity loss owing to solution and to precipitation vary with the packing (Table 3-XIII).

Estimation of the maximum amount of cement that can be derived from solution at points of grain contacts for any given amount of pore space reduction can be made for moderately well-sorted to poorly-sorted, rounded to very angular sands (but not for sands having extremely irregularly shaped carbonate particles) by using the case of regular orthorhombic packing model rotated 30°. For example, if the porosity of sand is reduced from 35 to 25%, a reduction of 10%, the maximum amount of cement derived from the solution of grains at their points of contact would be 1.7% according to Rittenhouse (1971) (see Fig.3-34). The application of the same analysis to compacted oolitic grainstones (Fig.3-34—3-39) provides the very interesting results that for deeply-buried samples, there is sufficient cement generated through solution and reprecipitation not only to fill all available pore space but also to supply cement for export (Table 3-XIV; see also Fig.3-31—3-33). These data tend to confirm Coogan's (1970) conclusions that some of the cement in the Smackover (Texas, Louisiana) and Greenbriar (West Virginia) oolitic grainstones is late. The calculations show that the source of the cement can be locally derived by compaction of the oolitic sand to a Packing Density value of 82 to 88%. In the case of the Ste. Genevieve oolitic grainstone (Fig.3-31), however, it is clear that a minimum of 13.6% of the cement present would have to come from sources other than those resulting from

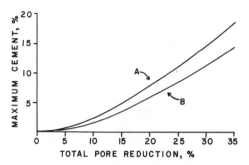

Fig.3-34. Graph showing the method of determining the maximum amount of cement, which can be derived from solution of grains and subsequent reprecipitation in adjacent pore space for a given reduction of total pore volume. Curve A represents the maxima for any sand and should be used for well-rounded, well-sorted sands. Curve B represents the estimated maxima for sands that are very poorly sorted, extremely well-sorted, or very angular. To use the diagram, enter the amount of pore reduction on the abscissa and read the maximum amount of cement derived from solution and reprecipitation for any particular type of sand on the ordinate (e.g., if original pore space = 35% and the present pore space = 15%, then the total reduction in pore space = 35—15 = 20%; using curve A, the maximum amount of cement = 7.6%). (After Rittenhouse 1971, published with permission of the Am. Assoc. Pet. Geol., Tulsa, Okla.)

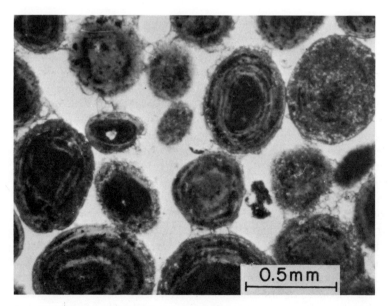

Fig.3-35. Uncompacted Pleistocene eolian oolitic grainstone from the outcrop at Joulters Cay, Bahamas.

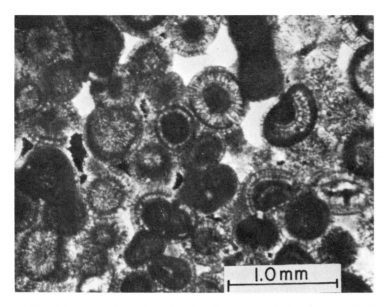

Fig.3-36. Interpenetrating grains in the compacted oolitic grainstone of the Jurassic Smackover Formation, from Trahan No. 1 Brown well of Tenneco Oil Co., Clairborne Parish, La., depth = 3,034 m.

TABLE 3-XIV

Proportions of solution-derived cement versus total cement volume for selected oolitic grainstones (Fig.3-31—3-33)

	Sample formation name		
	Ste. Genevieve	*Smackover*	*Greenbriar*
Location	Illinois	Texas	West Virginia
Fig. no.	(3-31)	(3-32)	(3-33)
Original porosity (estimated)	35.0	35.0	35.0
Packing density (%)	67.4	88.7	82.7
Total cement volume (%)	32.6	11.3	17.3
Maximum cement from solution and reprecipitation (%)	19.0	19.0	19.0
Difference between total cement volume and solution-derived cement (excess or deficit) (%)	−13.6	+7.7	+1.7

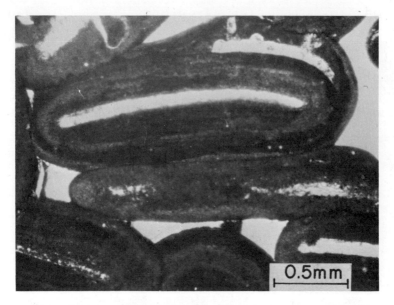

Fig.3-37. Straight contacts, bridging, and grain interpenetration in a compacted oolitic grainstone from J. F. Roper No. 1 well of the North Central Oil Co., Jurassic Smackover Formation, Freestone Co., Texas, depth = 3,696 m.

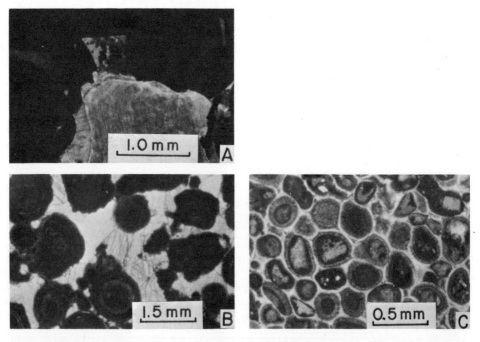

Fig.3-38. Three compacted Jurassic, Cretaceous, and Mississippian oolitic grainstones. A. Illustration shows penetration of harder clastic grain into oolith and the authigenic growth of corners into another oolith; from the Jurassic Smackover Formation, North Central Oil Co., J. F. Roper No. 1 well, depth = 3,696 m. B. Illustration shows fractured cement in the Cretaceous El Abra Limestone from the Pemex Cazones No. 2 well, San Luis Potosi, Mexico, depth = 2,421 m. C. Closely-packed, but generally not touching, grains in the Mississippian Greenbriar Formation outcrop in Monroe Co., W. Va.

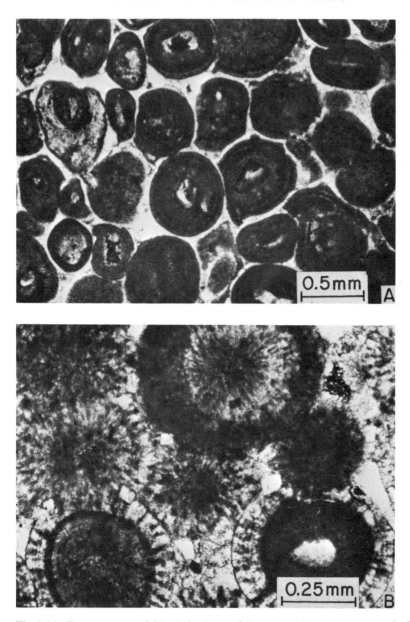

Fig.3-39. Two compacted Mississippian and Jurassic oolitic grainstones. A. Closely-packed grains which are not touching in the Mississippian Ste. Genevieve Formation outcrop, Roane Co., Tennessee. B. Authigenic crystal growth between compacted ooliths in the Jurassic Smackover Formation, Tenneco Oil Co., No. 1 Waller well, Clairborne Parish, La., depth = 4,022 m.

local pressure solution. It should be noted, however, that the Packing Density for this rock is close to the value expected for a completely uncompacted oolitic grainstone. This in itself suggests that no compaction has taken place and that all the cement is either of syngenetic or early diagenetic origin, owing to solution and alteration of near-surface sediments.

PRESSURE SOLUTION AND STYLOLITIZATION

General statement

Pressure solution is important in the compaction of a carbonate sediment because it is an effective agent of bulk volume reduction through the gross dissolution of calcium carbonate, as well as the agent of pore volume reduction by the physical reduction of the void space. Pressure solution also acts as an agent supplying cement to void space adjacent to the site of solution. In addition, it has been suggested by many authors that the calcium carbonate released in the process of pressure solution may serve as a major source of exportable pore-filling cement.

Stylolitization process

Physical features develop in response to dissolution at points of contact between mineral grains subject to pressure. Dissolution occurs as a result of local increase in chemical potential and thus of solubility of calcium carbonate where grains are subjected to increased elastic strain at the contact points. Where grains are not in contact and the elastic strain is consequently less, the relative solubility of the calcium carbonate is lower and ions diffusing from the more saturated contact areas will tend to precipitate near the unstrained surfaces (Bathurst, 1958, 1971b). The distribution of strain at the surface of contacting grains is a function of their relative sizes, shapes and orientations. Solubility also varies with the mineralogy of the grains and hence the geometry of the contact surface between carbonate grains can be expected to be highly variable. Attempts to classify the geometry of pressure solution features, especially stylolites, have been made by Amstutz and Park (1967) and Trurnit (1968).

For the process of pressure solution to continue, stress must be passed across the contact from grain to grain, while a solution film is simultaneously maintained between the grains so that dissolved ions may diffuse through it to sites of lower stress. Two mechanisms have been offered to explain the pressure solution:

(1) The undercutting mechanism (Bathurst, 1958) calls for the estab-

lishment of an irregular surface between the grains at a microlevel. Solution is visualized as taking place over only a small part of the surface at one time, undercutting higher relief portions of the surface.

(2) The solution film mechanism (Weyl, 1959) calls for the maintenance of a continuous film of solution between the grains, which is able to support shear stress and allow the diffusion of ions. The maintenance of the solution, it is argued, is implied in the force of crystallization noted in experiments (Becker and Day, 1916; Taber, 1916).

A stylolite is a complex interface between two bodies of rock, commonly limestone, which exhibits mutual column and socket interdigitation. The long axes of the columns and sockets are perpendicular to the interface and are generally parallel to one another. The interspace between the two bodies of rock separated by a stylolite commonly is characterized by a thin seam of insoluble residue which became concentrated as the rock was dissolved. The difference between microstylolitic grain contacts and those which cut across many grains is simply one of scale.

Microstylolitic solution can occur after the precipitation of first generation cement, because a thin fringe of cement does not completely prevent grain-to-grain movement. When the second generation cementation is sufficiently advanced, however, relative movement between grains must end. Consequently, grain-to-grain pressure solution must occur before the complete precipitation of pore-filling cement. Grain-to-grain movement and, hence, pressure solution obviously will be inhibited seriously in cases where appreciable cementation precedes the development of stress conditions sufficient to produce stylolitization, for example, in cases of vadose zone cementation.

Because pressure solution appears to be a response to overburden pressure, it would seem desirable to correlate some function of pressure solution with depth of burial. Unqualified generalizations regarding such relationships, however, appear to be tenuous now. For example, Dunnington (1967), in discussing stylolitic carbonate petroleum reservoirs, indicates that depths of greater than 600 m have been required generally for stylolitization, whereas Schlanger (1964) has described numerous microstylolites from limestones on Guam which were probably never buried more deeply than 90 m.

Bulk volume loss through stylolitization

Stylolitization can cause compaction of a sequence of carbonate rocks owing to substantial losses of rock volume through solution. Reductions in thickness of the order of 30% are common and Dunnington (1967) reported that reductions of up to 40% were recorded. The loss of bulk volume through stylolitization may be diagrammatically illustrated (Fig.3-40);

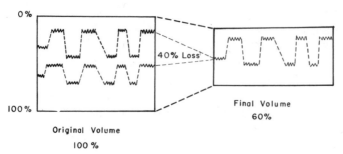

Fig.3-40. Diagrammatic illustration of the bulk volume loss of a limestone through stylolitization.

however, it is very hard to reconstruct lost bulk volume by matching or restoring stylolitic surfaces. If the interpenetration lengths of stylolite columns on each seam are added, a minimum magnitude of unit thinning may be obtained. This must be a minimum value because the degree of solution occurring at the column tips cannot be determined. Another approach to estimating the bulk volume loss is to calculate the volume of the insoluble residue from the thickness of the seam and then convert it to bulk volume by using the average insoluble material content of the undissolved rock. The tendency of stylolites to develop in more argillaceous parts of limestones, however, makes this approach inaccurate because of inherent sample bias (Dunnington, 1967). A measure of the volume of rock dissolved in stylolitization may also be based upon the amount of apparent offset of an oblique vein by a cross-cutting stylolite (Pettijohn, 1957).

Pore volume loss through pressure solution

Pore volume is also reduced by pressure solution as a result of closer packing of grains, grain corrosion, and filling of pores with cement generated by pressure solution. The degree of pore space reduction to be expected by the solution of grains at points of contacts for several ordered packing arrangements of spheres has been demonstrated by Rittenhouse (1971), who also estimated the volume of cement which could be generated by pressure solution. Inasmuch as pore volume equals bulk volume minus grain volume and the original porosities for ordered packing arrangements of spheres are known (Table 3-I), the degree of pore volume reduction can be shown as a function of decrease in bulk volume (Fig.3-41). The data for this model were obtained following the method devised by Rittenhouse (1971) and do not reflect the added influence of void-filling cement, which was locally generated and precipitated through pressure solution. In Fig.3-41, the degree of pore volume reduction (as a percent of original porosity lost) is plotted

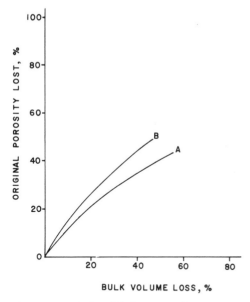

Fig.3-41. Relationship between the percentage of original porosity loss (% ϕ Loss) and the percentage of bulk volume loss (% BV Loss) for regular cubic and orthorhombic packs of spheres. A = cubic, B = orthorhombic.

versus the bulk volume reduction owing to pressure solution for cubic and orthorhombic regular packs of spheres. A reduction of 40% in bulk volume, causes a 35 to 44% reduction in pore space without cementation, depending on the packing arrangement. Noticeable decrease in porosity has been observed in the immediate vicinity of stylolites (Harms and Choquette, 1965; Dunnington, 1967). Inasmuch as the pore fluids in a region undergoing pressure solution are highly saturated, there is a temptation to explain this pore reduction as being largely due to cementation. Examination of Fig.3-41, however, reveals that the decrease in porosity is due at least as much to tighter packing as it is to cementation. An example of tighter packing accompanying a stylolite occurs in an oolitic grainstone from the Silurian Noix Oolite Member of the Edgewood Formation, Louisiana, Missouri (Fig.3-42).

Pressure solution as a source of cement

It is frequently suggested that pressure solution may play a significant role in providing a source of allochthonous calcium carbonate which is subsequently precipitated as cement, commonly as second generation cement. Because of the high degree of calcium carbonate saturation of the

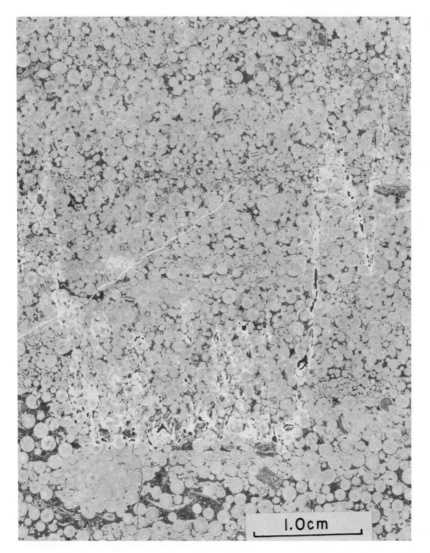

Fig.3-42. Oolitic grainstone from the Silurian Noix Oolite Member of the Edgewood Formation, La., Mo., showing loose and tight packing of ooliths in an area of stylolitization.

pore solutions, it is not likely that pressure-solution generated cement could travel far before being precipitated in the available pore space. Accordingly, the amount of dissolved calcium carbonate which is being produced must exceed the remaining pore space volume in order for the cement to be exported from the immediate vicinity. Examination of Fig.3-43, however, which shows the relationship between the bulk volume reduction and the ratio of cement generated to the remaining pore volume sheds doubt upon the efficacy of the process. In Fig.3-43 a comparison was made between the data obtained for regular cubic and orthorhombic packs of spheres according to the method used by Rittenhouse (1971). In addition, data was generated (Fig.3-43) for a regular pack of cylinders arrayed so that the ends of the cylinders appear stacked in a cubic regular order. As shown in Fig.3-43, the available pore space exceeds the volume of produced cement until a point is reached where the cement generated (CG) divided by the pore volume remaining equals unity. There the bulk volume is reduced by an amount equal to the original porosity. Indeed, in the case of cubic and orthorhombic packs of spheres (Fig.3-43, curves A, B), the bulk volume reduction would have to be in excess of that commonly supposed to occur before allochthonous cement could be generated. Of course, this assumes local precipitation of cement so long as pore space is locally available. In the case of cubic packs of cylinders (Fig.3-43, curve C), the original porosity (21.5%) is lower and the bulk volume loss required is more in keeping with most estimates of volume reduction.

Because it is often considered that ions derived from pressure solution

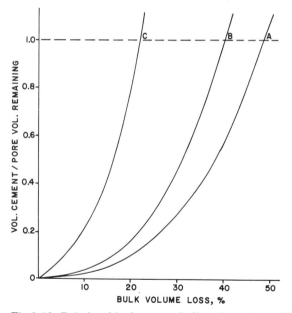

Fig.3-43. Relationship between bulk volume loss (% BV Loss) and the ratio of cement generated (CG) to remaining pore volume (ϕ) for regular packs of spheres and cylinders. A = cubic packs of spheres, 47.6% original porosity; B = orthorhombic packs of spheres, 39.5% original porosity; C = cubic packs of cylinders, 21.5% original porosity.

travel in fluids parallel to the direction of stress (Bathurst, 1971b), it is tempting to suggest that pressure solution is a self-limiting process. There is a reduction in bulk volume until the precipitated cement fills the remaining local pore space so that the permeability is reduced to zero and calcium carbonate can no longer be transported. At this point pressure solution ceases. The condition may be modified somewhat in the case where cement is transported outside the immediate pore vicinity early in the pressure solution process by rapidly flowing solutions. In this case the curve may rise above the CG/ϕ ratio of 1.0 (Fig.3-43). In addition, if permeability is increased by the presence of transparticulate pores, such as fractures, this process may be modified.

TABLE 3-XV

Comparison of Packing Density, Packing Index and Compaction Index of uncompacted and compacted oolitic grainstones

Description	Packing Density (%)	Packing Index (%)	Compaction Index (%)	Fig. no.
(1) Uncompacted, uncemented oolitic sand, Browns Cay, Bahamas.	64.8	18.6	0	3-19
(2) Weakly-compacted doloarenite; a dolomitized, weakly-compacted oolitic grainstone showing the relic oolitic texture of grain margins from a depth of 3,917—3,919 m in the Pan American Petroleum Corp., No. 1 Parker, Van Zandt Co., Texas, Jurassic, Smackover Formation.	70.3	22.5	15	3-20
(3) Compacted oolitic grainstone, Jurassic Smackover Formation from a depth of 3,149 m in the Tenneco Oil Co. No. 1 Lowe, North Haynesville Field, Clairborne Parish, La.	95.9	66.5	87	3-22
(4) Weakly-compacted, skeletal oolitic grainstone from the Mississipian Ste. Genevieve Formation, Fredonia Limestone Member, Harrison County, Ind.	67.4	30.0	7	3-31
(5) Compacted Smackover oolitic grainstone from a depth of 3,696 m in the J. F. Roper No. 1.	88.7	59.5	69	3-32
(6) Compacted oolitic grainstone, Mississippian Greenbriar	82.7	31.2	50	3-33

TABLE 3-XV — *continued*

Description		Packing Density (%)	Packing Index (%)	Compaction Index (%)	Fig. no.
	Formation, Union Member, Monroe Co., W. Va. outcrop.				
(7)	Pleistocene eolian oolitic grainstone, uncompacted, Joulters Cay outcrop, Bahamas.	64.4	21.6	0	3-35
(8)	Compacted oolitic grainstone, Jurassic Smackover Formation from a depth of 3,034 m in the Trahan No. 1 Brown, Clairborne Parish, La.	88.3	74.0	66	3-36
(9)	Compacted oolitic grainstone, Jurassic Smackover Formation from a depth of 3,696 m in the J. F. Roper No. 1.	84.3	43.7	55	3-37
(10)	Compacted El Abra Formation; oolitic lump grainstone from a depth of 2,421 m, Pemex Cazones No. 2.	78.1	34.6	37	3-38b
(11)	Compacted oolitic grainstone, Jurassic Smackover Formation from a depth of 3,696 m in the North Central Oil Co., J. F. Roper No. 1, Freestone Co., Texas.	86.2	46.4	60	3-38a
(12)	Compacted oolitic grainstone from the Greenbriar Formation, Union Member, Monroe Co., W. Va.; outcrop.	92.4	23.0	78	3-38c
(13)	Compacted oolitic grainstone, Mississippian Ste. Genevieve Formation, Roane Co., Tennessee; outcrop.	82.9	18.6	51	3-39a
(14)	Compacted oolitic grainstone, Jurassic Smackover Formation from a depth of 4,022 m in the Tenneco Oil Co., No. 1 Waller, Clairborne Parish, La.	93.3	74.0	82	3-39b

GLOSSARY

Chance packing: The natural and fortuitous combination of systematic colonies of particles (chiefly case 6, rhombohedral, Fig.3-5) with intervening or surrounding haphazard or unordered zones of particles.

Compaction: The decrease in bulk volume of sediments as a result of stress accompanied by a loss of pore volume.

Compaction Index: A measure of the degree of compaction of a grain sediment or rock, scaled in percent, derived from the grain volume measurement and adjusted for the original grain volume percentage of the uncompacted sediment, according to grain type.

Condensation Index: The index of compaction based on the determination of the ratio of fixed to free grains, as determined in thin section.

Consolidation: The adjustment of a saturated sediment in response to increased load and consequent loss of water.

Fixed Grain: A grain which has its fixed margin (length of the margin of the grain in contact with another grain) longer than its free margin. The measurement is made in thin section.

Fixed-Grain Compaction Index: An index of compaction based on the number of fixed grains compared to total grains in the rock.

Grain volume: The volume of solid grains in the rock divided by the total bulk volume. Grain volume can be expressed as a fraction or, more commonly, as a percentage of the total rock or sediment bulk volume.

Haphazard packing: All packings of spheres which do not correspond to one of the six systematic packing configurations. In haphazard packing, more or less order may prevail, but commonly a disorderly arrangement predominates. Haphazard packing is a special case of random packing.

Irreducible wetting-phase saturation: A minimum wetting-phase saturation. Fluid is immobile even at very high pressure and occupies minute interstices and cracks in the sediment or rock and is present as thin films around grains. In most natural systems, the wetting phase is either fresh or salt water. Some carbonate rocks, however, are oil wet.

JOIDES: The consortium of oceanographic institutions in the United States of America banded together to investigate the strata of the ocean floors (*J*oint *O*ceanographic *I*nstitutions for *D*eep *E*arth *S*ampling).

Packing: The spacing or density pattern of grains or pores in a sediment or rock.

Packing Density: The grain volume percentage of a sediment or rock as measured in thin section.

Packing heterogeneity: Variations of packing in a sediment or rock owing to differences in particle size and shape or the presence of alternate regions of variously homogeneously packed particles (see chance packing, haphazard packing, and random packing).

Packing Index: The index of closeness of grains to each other in a sediment or rock on the average. Packing Index is determined from the ratio of number of grains touching each other to the total number of grains as counted in thin section.

Packing Proximity: One of the forms of the Packing Index, specifically that proposed by Kahn (1956).

Random packing: A general term to encompass all cases of packing of particles which are not one of the six systematic, regular, ordered packings of spheres.

Regular packing: The ordered packing of spheres, one of the six systematic cases of packing (Fig.3-5).

Unordered packing: The arrangement of grains in a sediment or rock which does not correspond to any ordered arrangement or chance packing.

REFERENCES

Adams, J. E. and Rhodes, M. L., 1960. Dolomitization by seepage refluxion. *Bull. Am. Assoc. Petrol. Geol.*, 44: 1912—1920.

Alderman, A. R. and Skinner, H. C. W., 1957. Dolomite sedimentation in the South-East of South Australia. *Am. J. Sci.*, 255: 561—567.

Allen, J. R. L., 1962. Petrology, origin and deposition of the highest lower Old Red Sandstone of Shropshire, England. *J. Sediment. Petrol.*, 32: 657—697.

Amstutz, G. C. and Park, W. C., 1967. Stylolites of diagenetic age and their role in the interpretation of the Southern Illinois fluorspar deposits. *Mineralium Deposita*, 2: 44—53.

Athy, L. F., 1930. Density, porosity and compaction of sedimentary rocks. *Bull. Am. Assoc. Pet. Geol.*, 14: 1—24.

Atwater, G. I. and Miller, E. E., 1965. The effect of decrease in porosity with depth on future development of oil and gas reservoirs in southern Louisiana. *Bull. Am. Assoc. Pet. Geol.*, 49: 344.

Atwood, D. K. and Bubb, J. N., 1970. Distribution of dolomite in a tidal flat environment, Sugarloaf Key, Florida. *J. Geol.*, 78: 499—505.

Ayer, N. J., 1971. *Statistical and Petrographic Comparison of Artificially and Naturally Compacted Carbonate Sediments*. Thesis, Univ. Ill., Urbana, Ill., 92 pp.

Bathurst, R. G. C., 1958. Diagenetic fabrics in some British Dinantian limestones. *Liverp. Manch. Geol. J.*, 2: 11—36.

Bathurst, R. G. C., 1959. Diagenesis in Mississippian calcilutites and pseudobreccias. *J. Sediment. Petrol.*, 29: 365—376.

Bathurst, R. G. C., 1971a. Two generations of cement. In: O. P. Bricker (Editor), *Carbonate Cements*. John Hopkins Univ. Press, Baltimore, Md., p.296.

Bathurst, R. G. C., 1971b. *Carbonate Sediments and Their Diagenesis*. Elsevier, Amsterdam, 620 pp.

Beall, A. O. and Fischer, A. G., 1969. Sedimentology. In: *Initial Reports of the Deep Sea Drilling Project*, 1. U.S. Gov. Print. Off., Washington, D.C., pp.521—593.

Becker, G. F. and Day, A. L., 1916. Note on the linear force of growing crystals. *J. Geol.*, 24: 313—333.

Bernal, J. D. and Finney, J. L., 1967. Random packing of spheres in non-rigid containers. *Nature*, 214: 265—266.
Benson, L. V. and Matthews, R. K., 1971. Electron microprobe studies of magnesium distribution in carbonate cements and recrystallized skeletal grainstones from the Pleistocene of Barbados, West Indies. *J. Sediment. Petrol.*, 41: 1018—1025.
Bernal, J. D. and Mason, J., 1960. Co-ordination of randomly packed spheres. *Nature*, 188: 910—911.
Bissell, H. J. and Chilingar, G. V., 1967. Classification of sedimentary carbonate rocks. In: G. V. Chilingar, H. J. Bissell and R. W. Fairbridge (Editors), *Carbonate rocks. Developments in Sedimentology*, 9A. Elsevier, Amsterdam, 471 pp.
Bishop, W. F., 1968. Petrology of Upper Smackover Limestone in North Haynesville Field, Clairborne Parish, Louisiana. *Bull. Am. Assoc. Pet. Geol.*, 52: 92—128.
Bricker, O. P. (Editor), 1971. *Carbonate Cements*. John Hopkins Univ. Press, Baltimore, Md., 376 pp.
Brown, R. L. and Hawksley, P. G., 1945. Packing of regular (spherical) and irregular particles. *Nature*, 156: 421—422.
Bubb, J. N. and Atwood, D. K., 1968. Recent dolomitization of Pleistocene limestone by hypersaline brines, Great Inagua Island, Bahamas. *Bull. Am. Assoc. Pet. Geol.*, 52: 522.
Bunce, E. I., Emery, K. O., Gerard, R. D., Knox, S. T., Lidz, L., Saito, T. and Schlee, J., 1966. Ocean drilling on the continental margin. *Science*, 150: 709—716.
Butler, G. P., 1971. Origin and controls on distribution of arid supratidal (sabkha) dolomite, Abu Dhabi, Trucial Coast. *Bull. Am. Assoc. Pet. Geol.*, 55: 332—333.
Carozzi, A. V., 1961. Distorted oolites and pseudoolites. *J. Sediment. Petrol.*, 31: 262—274.
Chave, K. E., 1962. Factors influencing the mineralogy of carbonate sediments. *Limnol. Oceanogr.*, 7: 218—233.
Chave, K. E., Deffeyes, K. S., Weyl, P. K., Garrels, R. M. and Thompson, M. E., 1962. Observations on the solubility of skeletal carbonates in aqueous solutions. *Science*, 137: 33—34.
Chilingar, G. V. and Bissell, H. J., 1961. Dolomitization by seepage refluxion (discussion). *Bull. Am. Assoc. Pet. Geol.*, 45: 679—683.
Chilingar, G. V., Mannon, R. W. and Rieke III, H. H., (Editors), 1972. *Oil and Gas Production from Carbonate Rocks*. Am. Elsevier, New York, N.Y., 409 pp.
Choquette, P. W., 1971. Late ferroan dolomite cement, Mississippian carbonates, Illinois Basin, U.S.A. In: O. P. Bricker, (Editor), *Carbonate Cements*. John Hopkins Univ. Press, Baltimore, Md., 376 pp.
Coogan, A. H., 1970. Measurements of compaction in oolitic grainstone. *J. Sediment. Petrol.*, 40: 921—929.
Coogan, A. H., Bebout, D. G. and Maggio, C., 1972. Depositional environments and geologic history of the Golden Lane and Poza Rica Trend, Mexico, an alternative view. *Bull. Am. Assoc. Pet. Geol.*, 56: 1419—1447.
Deffeyes, K. S., Lucia, F. J. and Weyl, P. K., 1965. Dolomitization of Recent and Plio-Pleistocene sediments by marine evaporite waters on Bonaire, Netherlands Antilles. In: L. C. Pray and R. C. Murray (Editors), *Dolomitization and Limestone Diagenesis — A Symposium. Soc. Econ. Paleontol. Mineral., Spec. Pap.*, 13: 71—88.
Dunham, R. J., 1962. Classification of carbonate rocks according to depositional texture. In: W. E. Ham (Editor), *Classification of Carbonate Rocks — A Symposium. Am. Assoc. Pet. Geol., Spec. Publ.*, 14: 108—121.
Dunnington, H. V., 1967. Aspects of diagenesis and shape change in stylolitic limestone reservoirs. *Proc. World Pet. Congr., 7th., Mexico, 1967*, 2: 339—352.

Ellis, A. L. and Lee, C. H., 1919. Ground waters of western San Diego County, California. *U.S.G.S. Water Supply Pap.*, 446: 121—123.
Fatt, I., 1958. Compressibility of sandstones at low to moderate pressures. *Bull. Am. Assoc. Pet. Geol.*, 42: 1924—1957.
Fischer, A. G. and Garrison, R. E., 1967. Carbonate lithification on the sea floor. *J. Geol.*, 75: 488—496.
Folk, R. L., 1959. Practical petrographic classification of limestones. *Bull. Am. Assoc. Pet. Geol.*, 43: 1—38.
Fraser, H. J., 1935. Experimental study of porosity and permeability of clastic sediments. *J. Geol.*, 43: 910—1010.
Freund, J. E. and Williams, F. J., 1958. *Modern Business Statistics*. Prentice Hall, Englewood, N.J., 539 pp.
Friedman, G. M., 1964. Early diagenesis and lithification in carbonate sediments. *J. Sediment. Petrol.*, 34: 777—813.
Friedman, G. M., 1965. Occurrence and stability relationships of aragonite, high-magnesian calcite, and low-magnesian calcite under deep-sea conditions. *Bull. Geol. Soc. Am.*, 76: 1191—1196.
Friedman, G. M., 1968. The fabric of carbonate cement and matrix and its dependence on the salinity of water. In: G. Müller and G. M. Friedman (Editors), *Recent Developments in Carbonate Sedimentology in Central Europe*. Springer, Berlin, pp.11—20.
Fruth, L. S., Jr., Orme, G. R. and Donath, F. A., 1966. Experimental compaction effects in carbonate sediments. *J. Sediment. Petrol.*, 36: 747—754.
Füchtbauer, H., 1961. Zur Quarzneubildung in Erdoellagerstaetten. *Erdöl. Kohle*, 14: 169—173.
Gaither, A., 1953. A study of porosity and grain relationships in experimental sands. *J. Sediment. Petrol.*, 23: 180—195.
Graton, L. C. and Fraser, H. J., 1935. Systematic packing of spheres with particular relation to porosity and permeability. *J. Geol.*, 35: 785—909.
Ham, W. E. (Editor), 1962. *Classification of Carbonate Rocks — A Symposium*. Am. Assoc. Pet. Geol., Tulsa, Okla., 279 pp.
Harms, J. C. and Choquette, P. W., 1965. Geologic evaluation of gamma-ray porosity device. *Trans. Soc. Prof. Well Log Analysts, Annu. Logging Symp. 6th, Dallas, Texas, 1965*, C, pp.1—37.
Harris, C. C. and Morrow, N. R., 1964. Pendular moisture in packing of equal spheres. *Nature*, 207: 706—708.
Hathaway, J. C. and Robertson, E. C., 1961. Microtexture of artificially consolidated aragonitic mud. *U.S. Geol. Surv., Prof. Pap.*, 424C: 301—304.
Hays, F. R., 1951. *Petrographic Analysis of Deep Well Cores*. Thesis, Dep. Geol., Univ. Cincinnati, Cincinnati, Ohio (not seen).
Heald, M. T. and Renton, J. J., 1966. Experimental study of sandstone cementation. *J. Sediment. Petrol.*, 36: 977—991.
Huang, T. and Pierce, J. W., 1971. The carbonate minerals of deep-sea bioclastic turbidites, southern Blake Basin. *J. Sediment. Petrol.*, 41: 251—260.
Kahle, C. F., 1966. Some observations on compaction and consolidation in ancient oolites. *Compass*, 44: 19—29.
Kahn, J. S., 1956. Analysis and distribution of the properties of packing in sand-size sediments, I. On the measurement of packing in sandstones. *J. Geol.*, 64: 385—395.
Kelvin, Lord, 1887. On the division of space with minimum partitional area. *Phil. Mag.*, 24: 503—514.

King, F. H., 1898. Principles and conditions of the movements of ground water. *U.S. Geol. Surv., 19th Annu. Rep.*, 3: 208—218.
Land, L. S. and Goreau, T. F., 1970. Submarine lithification of Jamaican reefs. *J. Sediment. Petrol.*, 40: 457—462.
Larsen, G. and Chilingar, G. V., 1967. *Diagenesis in Sediments. Developments in Sedimentology*, 8. Elsevier, Amsterdam, 551 pp.
LeBlanc, R. J. and Breeding, J. G. (Editors), 1957. *Regional Aspects of Carbonate Deposition. Soc. Econ. Paleontol. Mineral., Spec. Pap.*, 5: 1—178.
Mackenzie, F. T. and Bricker, O. P., 1971. Cementation of sediments by carbonate minerals. In: O. P. Bricker (Editor), *Carbonate Cements*. John Hopkins Univ. Press, Baltimore, Md., 376 pp.
Maiklem, W. R., 1968. Some hydraulic properties of bioclastic carbonate grains. *Sedimentology*, 10: 101—109.
Marvin, J. W., 1939. The shape of compressed lead shot and its relation to cell shape. *Am. J. Bot.*, 26: 280—288.
Masson, P. H., 1951. Measurement of grain packing in sandstone. *Am. Assoc. Pet. Geol., Annu. Meet., St. Louis*, not seen (abstract).
Matthews, R. K., 1966. Genesis of Recent lime mud in southern British Honduras. *J. Sediment. Petrol.*, 36: 428—454.
Matthews, R. K., 1967. Diagenetic fabrics in biosparites from the Pleistocene of Barbados, West Indies. *J. Sediment. Petrol.*, 37: 1147—1153.
Matthews, R. K., 1968. Carbonate diagenesis: equilibration of sedimentary mineralogy to the subaerial environment; coral cap of Barbados, West Indies. *J. Sediment. Petrol.*, 38: 1110—1119.
Matzke, E. B., 1939. Volume—shape relationships in lead shot and their bearing on cell shape. *Am. J. Bot.*, 26: 288—295.
Maxwell, J. C., 1960. Experiments on compaction and cementation of sand. *Geol. Soc. Am., Mem.*, 79: 105—132.
Maxwell, J. C., 1964. Influence of depth, temperature and geologic age on porosity of quartzose sandstone. *Bull. Am. Assoc. Pet. Geol.*, 48: 697—709.
Milliman, J. D., 1966. Submarine lithification of carbonate sediments. *Science*, 153: 994—992.
Miller, D. G., Jr. and Richards, A. F., 1969. Consolidation and sedimentation — compression studies of a calcareous core, Exuma Sound, Bahamas. *Sedimentology*, 12: 301—316.
Morrow, N. R., 1971. Small-scale packing heterogeneities in porous sedimentary rocks. *Bull. Am. Assoc. Pet. Geol.*, 55: 514—522.
Murray, R. C., 1969. Hydrology of south Bonaire, Netherlands Antilles — A rock selective dolomitization model. *J. Sediment. Petrol.*, 39: 1007—1013.
Oldershaw, A. E., 1971. The significance of ferroan and non-ferroan calcite cements in the Halkin and Wenlock Limestones (Great Britain). In: O. P. Bricker (Editor), *Carbonate Cements*. John Hopkins Univ. Press, Baltimore, Md., pp. 225—229.
Park, Won-choon and Schot, E. H., 1968. Stylolitization in carbonate rocks. In: G. Müller and G. M. Friedman (Editors), *Recent Developments in Carbonate Sedimentology in Central Europe*. Springer, Berlin, pp. 66—74.
Pettijohn, F. J., 1957. *Sedimentary Rocks*. Harper and Row, New York, N.Y., 2nd ed., 718 pp.
Pray, L. C. and Murray, R. C. (Editors), 1965. *Dolomitization and Limestone Diagenesis — A Symposium. Soc. Econ. Paleontol. Mineral., Spec. Publ.*, 13: 1—180.

Pryor, W. A., 1971. Reservoir inhomogeneities of some Recent sand bodies. *Soc. Pet. Eng., Am. Inst. Min., Metall. Petrol. Eng.*, Preprint Paper SPE 3607, presented at Annu. Meet., 1971, 12 pp.

Purdy, E. G., 1963. Recent calcium carbonate facies of the Great Bahama Bank. 2. Sedimentary facies. *J. Geol.*, 71: 472—497.

Purdy, E. G., 1968. Carbonate diagenesis: an environmental survey. *Geol. Romana*, 7: 183—228.

Ridgway, K. and Tarbuck, K. J., 1967. Random packing of spheres. *Br. Chem. Eng.*, 12: 384—388.

Rieke III, H. H. and Chilingarian, G. V., 1974. *Compaction of Argillaceous Sediments.* Elsevier, Amsterdam, 424 pp.

Rittenhouse, G., 1971. Pore-space reduction by solution and cementation. *Bull. Am. Assoc. Pet. Geol.*, 55: 80—91.

Robertson, E. C., 1967. Laboratory consolidation of carbonate sediment. In: A. F. Richards (Editor), *Marine Geotechnique*. Univ. Ill. Press, Urbana, Ill., pp.118—127.

Robertson, E. C., Sykes, L. R. and Newell, M., 1962. Experimental consolidation of calcium carbonate sediment. *U.S. Geol. Surv., Prof. Pap.*, 350: 82—131.

Robinson, R. B., 1967. Diagenesis and porosity development in Recent and Pleistocene oolites from southern Florida and the Bahamas. *J. Sediment. Petrol.*, 37: 355—364.

Scott, G. D., 1960. Packing of equal spheres. *Nature*, 188: 908—909.

Schlanger, S. O., 1963. Subsurface geology of Eniwetok Atoll. *U.S. Geol. Surv., Prof. Pap.*, 260 BB: 991—1066.

Schlanger, S. O., 1964. Petrology of the limestones of Guam. *U.S. Geol. Surv., Prof. Pap.*, 403 D: 1—52.

Schlee, J. and Gerard, R., 1965. Cruise report and preliminary core log M/V Caldrill I — 17 April to 17 May, 1965. *J.O.I.D.E.S. Blake Panel Rep.*, 64 pp., unpublished.

Scholle, P. A., 1971. Diagenesis of deep-water carbonate turbidites, Upper Cretaceous Monte Antola flysch, northern Apennines, Italy. *J. Sediment. Petrol.*, 41: 233—250.

Sharp, W. E. and Kennedy, G. C., 1965. The system $CaO-CO_2-H_2O$ in the two-phase region calcite + aqueous solution. *J. Geol.*, 73: 391—403.

Shinn, E. A., 1964. Recent dolomite, Sugarloaf Key. In: *Guidebook for Field Trip No. 1, Geol. Soc. Am., Conv. 1964*, pp.62—67.

Shinn, E. A., Ginsburg, R. N. and Lloyd, R. M., 1965. Recent supratidal dolomite from Andros Island, Bahamas. In: L. C. Pray and R. C. Murray (Editors), *Dolomitization and Limestone Diagenesis — A Symposium. Soc. Econ. Paleontol. Mineral., Spec. Publ.*, 13: 112—123.

Sippel, R. F. and Glover, E. D., 1964. The solution alteration of carbonate rocks, the effects of temperature and pressure. *Geochim. Cosmochim. Acta*, 28: 1401—1417.

Stearns, N. D., 1927. Laboratory tests on physical properties of water-bearing materials. *U.S. Geol. Surv., Water Supply Pap.*, 596 F: 163-169.

Taber, S., 1916. The growth of crystals under external pressure. *Am. J. Sci.*, 4: 532—556.

Taft, W. H., 1968. Yellow Bank, Bahamas. A study of modern marine carbonate lithification. *Bull. Am. Assoc. Pet. Geol.*, 52: 551.

Taft, W. H. and Harbaugh, J. W., 1964. Modern carbonate sediments of South Florida, Bahamas and Espiritu Santo Island, Baja California: A comparison of their mineralogy and chemistry. *Stanford Univ. Publ., Univ. Ser., Geol. Sci.*, 8(2): 1—133.

Taylor, J. M., 1950. Pore-space reduction in sandstone. *Bull. Am. Assoc. Pet. Geol.*, 34: 701—716.

Terzaghi, K., 1936. Simple tests determine hydrostatic uplift. *Eng. News-Record*, 116: 872—875.
Terzaghi, R., 1940. Compaction of lime mud as a cause of secondary structure. *J. Sediment. Petrol.*, 10: 78—90.
Thompson, G., Bowen, V. T., Melson, W. G. and Cifelli, R., 1968. Lithified carbonates from the deep sea of the equatorial Atlantic. *J. Sediment. Petrol.*, 38: 1305—1312.
Trurnit, P., 1968. Pressure-solution phenomena in detrital rocks. *Sediment. Geol.*, 2: 89—114.
Weller, J. M., 1959. Compaction of sediments. *Bull. Am. Assoc. Pet. Geol.*, 43: 273—310.
Weyl, P. K., 1959. Pressure solution and the force of crystallization — a phenomenological theory. *J. Geophys. Res.*, 64: 2001—2025.
Zaripov, O. G. and Prozorovich, G. E., 1967. O razlichiyakh epigenetischeskogo uplotneniya vodonasyshchennykh i neftenasyshchennykh porod-kollektorov (na primere produktivnykh gorizontov Surgutskogo svoda, Zapadnaya Sibir'). (Concerning the differences between the epigenetic consolidation of water-saturated and oil-saturated reservoir rocks, using as an example the productive horizons of the Surgut anticline, Western Siberia.) *Dokl. Akad. Nauk S.S.S.R.*, 176: 1131—1133.

Chapter 4

SUBSIDENCE

H. J. BISSELL and G. V. CHILINGARIAN

INTRODUCTION

Reference to a dictionary reveals that the term *subsidence* refers to a "sinking to a lower level". Sinking — or subsidence as it is known in the language of geologists and engineers — is a commonplace phenomenon from the standpoint of soil compaction and settling, but it is to the geologist that subsidence of the earth's crust takes on a real meaning. This chapter consists of two parts: (1) Geologic Subsidence and (2) Present-day Subsidence. In the first part, the discussion centers around the general theme of subsidence of the earth's crust (not just the soil) to form the various depocenters (that is, centers of sedimentation and deposits of sediments), both in the geologic past and at the present time. In essence, therefore, the first part relates to various sedimentary basins and other areas of sediment accumulation; more emphasis is placed upon the formation of these depocenters in time and space than in theoretical appraisal of causal relationships. Enough information is provided so that students can be asked to estimate the amount of compaction that had occurred in some of the basins described. The second part consists of a discussion of aspects and causes of present-day subsidence and compaction, and techniques of measuring them. Details of adjustments within sediments during and since accumulations of sediments in the basins are presented in Chapter 2.

Oil and gas are obtained, for the most part, from sedimentary rocks that are now buried in depocenters which experienced subsidence in the past. To produce these natural "fossil" fuels requires the expertise of geologists and engineers alike. Therefore, it is of utmost concern to scientists and laymen, as well, to investigate some of the numerous ramifications of subsidence to determine what effect, if any, subsidence had (and may have) upon the generation, migration, and accumulation of hydrocarbons.

GEOLOGIC SUBSIDENCE

General considerations

The Earth, as a mere orb in space, has the form of a simple ellipsoid; its circumference at the equator is 24,902.4 miles. The diameter varies slightly in different directions, but its average length is 7917.78 miles. The surface area of the Earth is 196,950,000 sq miles. It is not necessary that an extended discussion be given regarding the methods that have been utilized by geologists, geodesists, and other scientists to show that the Earth seemingly consists of a series of spheres of different composition and densities around a high-density core. For the sake of simplicity, the Earth may be compared to an avocado that has been cut in two: the inner large seed is the core, the fleshy meat is the mantle, whereas the green rind is the crust. The rind is not necessarily smooth, nor is it of uniform thickness; the crust of the earth similarly is not a smooth sphere, nor is it of uniform thickness. Contrarily, its average thickness in continental areas is approximately 35 km, whereas beneath oceanic areas it has been computed to be as little as 5 km. If the concept that the earth's crust is the outermost shell of lithosphere, and as such is distinguished by low earthquake-wave velocities, is a valid one, then it would follow that its base is bounded by the so-called Mohorovičić discontinuity (commonly referred to as the Moho). The abrupt increase in seismic-wave velocities below the Moho suggests to geodesists that rocks having greater density occur beneath this separation, or discontinuity. It is the observable part of the earth's crust on which attention is focused, because it is the outermost "rind" that has experienced subsidence and uplift, and it is that part of the earth in which geologists can study the layered rocks.

It is through several methods of isotopic dating of ancient sedimentary rocks now exposed in the earth's crust that geologists point out that sediments were accumulating in depressed parts of this crust well in excess of three billions of years ago. It is not the purpose in this chapter to discuss the pros and cons of "permanency of oceans and continents", or of the relationship of geologic subsidence to the plate-tectonic model, but rather to focus attention to the continents and the record of Precambrian to Holocene depocenters. It is germane in this discussion to recall that "what went down has come up". The fact that subsidence does occur is proved in stratigraphy — some of the highest mountains on continents display layered rocks and they contain fossils (corals, brachiopods, algae, etc.), proving that in the geologic past some elevated areas were once at much lower levels, namely, beneath sea level. It is also important to retain at least a modicum of the spirit of Uniformitarianism: if the Present is not entirely the *Key* to the Past, then possibly it is a *Clue* to the Past!

Diastrophism

In his Monograph on Lake Bonneville, Gilbert (1890, p.3, 340) stated: "I find it advantageous to follow J. W. Powell in the use of *diastrophism* as a general term for the processes of deformation of the earth's crust... It is convenient also to divide diastrophism into orogeny (mountain-making) and epeirogeny (continent-making)." Gilbert thus coined the terms *orogenesis* and *epeirogenesis*; the former relates to the episodic processes that operate along mobile belts, thereby forming great linear crustal disturbances of mountain chains, whereas the latter pertains to the more general broad vertical upwarps and downwarps. Respective single movements are termed an orogeny and an epeirogeny. A third term (not coined by Gilbert) has been added by geologists, namely, taphrogeny. *Taphrogenesis* is a general term that refers to normal (and some normal upthrust) faults that rupture the earth's surface and form graben-like and horst-like areas of negative and positive mobility, respectively. Thus, orogeny, epeirogeny, and taphrogeny are to be considered under the term diastrophism. Furthermore, a credo of this chapter is that diastrophism is the cause of sedimentation. It is contended that subsidence of various depocenters, both past and present, was not initially caused by sedimentation. Contrarily, it is believed that crustal and subcrustal orogenic, epeirogenic, and taphrogenic mobility was caused by forces other than mere weight of wet sediment on a stable crust. Forces that produce mobility of the crust can hardly be considered to be due only to the weight of static columns of sediment and rock; rather, it is contended that dynamics within the crust and upper mantle are causative forces of diastrophism. The reasons for this contention are as follows:

(1) Aggregate thickness of sedimentary rocks in some depocenters is as much as 75,000 ft; this represents compacted and indurated sediment that accumulated in shallow, mostly marine realms. The average density of the sedimentary rocks in these depocenters is only about four-fifths of that of the crustal material that had to be displaced to permit subsidence. Thus, the weight of sediment alone could not have caused the heavier rocks to be pushed aside.

(2) Commonly, sedimentation and subsidence rates tend to keep pace with each other, indicating that subsidence does not depend upon weight of sediment.

(3) There are huge linear troughs on ocean floors where sedimentation is nil or is negligible.

(4) There is a systematic arrangement of mobile belts on the earth and not a helter-skelter chaos.

(5) In very many cases, orogeny, epeirogeny, and taphrogeny have proceeded almost independently of sedimentation, or in the presence of very little sedimentation.

Scheidegger and O'Keefe (1967), however, maintain the thesis that mechanisms are possible whereby (ortho-)geosynclines are caused by the deposition of sediments in shallow water. According to these authors, this can be done in conformity with the principle of isostasy. The writers agree that once a depocenter starts to subside, it can behave according to principles of isostasy (also see Rieke and Chilingarian, 1974).

Forty years ago the book entitled *"The Deformation of the Earth's Crust"* by Bucher (1933) was published. This thought-provoking book relating to diastrophism seemingly has escaped the attention of some of today's writers on sedimentation and tectonics. Bucher attempted to assemble pertinent geologic facts that bear on the problem of crustal deformation and to derive inductively an hypothetical picture of the mechanics of diastrophism. He stated the facts in carefully worded generalizations, which he suggested should be termed "laws" of nature and not merely theses of men. He considered the hypothetical picture as merely tentative, good only as long as nothing in sight seemed better. Bucher expected these "laws" to be scrutinized and, if found wanting, rejected. In his words: "The writer expects his logic and his hypothetic ideas to be scrutinized as severely as and, if found wanting, rejected as determinately as he has done with the ideas of others. We can hope to make headway in our search for a road to understanding only by taking the task of critical reasoning as seriously as we take that of accurate observation." Perhaps as he penned those words, Bucher recalled a somewhat similar statement by that great geologist Grove Karl Gilbert who, in his classic work entitled *"Studies of basin-range structure"* (1928), said: "Men are prone to take the familiar as a matter of course and to ask the meaning of the strange only." It is germane to the discussion at this point to repeat Bucher's first seven "laws" in order that the reader may realize that for more than a half-century geologists have written at great length concerning subsidence and sedimentation. Many geologists believe that the former geologic process is the cause (not the result) of the latter.

Law 1. Aside from the effects of erosion and deposition, the earth's surface deviates from the simple form of an ellipsoid, to which it may be referred, through outward (upward) and inward (downward) deflections.

Law 2. In the progress of crustal deformation, the direction of radial displacement is reversible.

Law 3. In ground plan, the forms of crustal elevations and depressions are of two types. First, elevations and depressions which are essentially equidimensional (swells and basins); and second, others that show a distinct linear development with one horizontal dimension decidedly greater than the other (welts and furrows).

Law 4. On the present face of the earth, excessive heights and excessive depths of crustal deformation are limited to "welts" and "furrows", that is, to elevations and depressions of distinctly linear outline.

Law 5. On the present face of the earth, "welts" and "furrows" do not occur independently, but are closely associated and lie side by side in relatively long and narrow belts.

Law 6. "Welts" and "furrows" similar in form and dimensions exist at the level of the ocean bottoms as well as on the continental platforms.

Law 7. In modern "welts" and "furrows", the relative elevations of points varies greatly along the strike, giving rise to deeper hollows in the "furrows" and to higher groups of mountain peaks separated by sags in the "welts".

By now, the reader undoubtedly realizes that as a downward fold of the crust of the earth, a geosyncline is a synonym for the neutral word "furrow" as used by Bucher. The word "furrow", however, focuses attention on the sedimentary contents of the depocenter. The review of the literature shows that it was Dana who introduced the term "geosyncline" for a depression of the earth's crust which has received sediments of excessive thickness (Dana, 1873).

At this point in the discussion, it seems appropriate to cite an example or two that bear out the topic under consideration, namely, subsidence and concomitant sedimentation. Both examples are from publications that are more than a half-century old; they are given here to point out the fact that subsidence of depocenters was a subject of serious consideration by petroleum geologists and researchers. The first dates to the year 1916 when Johnson and Huntley published a tectonic sketch map showing the generalized structure of the Dakota Sandstone (Lower Cretaceous) in front of the Rocky Mountains, from Colorado to Alberta. They termed the major area of downwarping (of subsidence) the "Moosejaw Synclinorium", aligned northwesterly across northwestern North Dakota into southern Saskatchewan. Relief from high mountains to the base of the Cretaceous is known to be at least 4 miles.

The second example is provided by Schuchert (1923). Fifty years after Dana coined the term geosyncline, Schuchert published the first systematic representation of geosynclines on an impartial basis of actual thickness of sediments. His paper is a classic and should not be overlooked by anyone interested in the subject of subsidence of the crust. On his map 5 (p.217), Schuchert reconstructed from the available information a paleogeographic map of North America for Upper Cambrian and Lower Ordovician times. Sediments of the portions of these two systems under consideration total 15,000 ft in northeastern Tennessee and southwestern Virginia, are almost 23,000 ft in northeastern Alabama, but are only 7,000 ft in the Arbuckle Mountains of eastern Oklahoma.

These two examples were provided, largely to emphasize the significance of Bucher's Law 8 and some of its corollaries, which states: "Laws 1 to 7 have been valid throughout the geological past as far as can be judged from available records." (a) Intensely folded sediments are several times thicker than the formations of the same age in undisturbed regions. (b) Excessively thick sediments form relatively narrow, elongated belts. (c) The largest part of the terrigenous sediments that fill the geosynclines was derived from highlands closely adjoining them and not from the large continental lowlands of the "swells". (d) Coarse waterlaid sediments (e.g., conglomerates, breccias, and arkoses) that indicate rapid erosion of rapidly risen highlands, are practically limited to geosynclinal belts (tillites of continental glacial origin excluded).

As a prelude to the section that follows, it is apropos to the discussion to recall that Gilbert introduced the term "orogenesis" for the processes that create what Bucher termed "welts" and "furrows". The liability, Bucher stated, to greater vertical movement may be spoken of as greater "mobility". He felt justified in terming "orogenic belts" also as the "mobile belts" of the earth's surface. This led him to give his opinion 2 which states: "The zones of crustal folds, of 'welts' and 'furrows', are the outcrops at the surface of relatively narrow sections of the earth's crust which display greater mobility than the remainder of the crust." Temptation to attempt to relate this mobility to the plate-tectonic theory is avoided here, because such a discussion is far beyond the scope of this chapter.

Subsidence and depocenters

In this chapter, the non-genetic term *depocenter* (Murray, 1952, p.224) identifies centers or areas of sediment accumulation; the term has long been utilized in this sense of a definition for sediments of Late Paleozoic age in the Cordilleran area of the western United States (Bissell, 1962a). It is true that many depocenters are at the same time geosynclines, whereas some others are intracratonic basins formed by epeirogenesis or by taphrogenesis. Most geosynclines seemingly are related in one manner or another with orogenesis. This does not necessarily mean that orogenic activity typifies the "welts" continuously during development and filling of adjacent "furrows". For example, in northern Alabama the Appalachian geosyncline accumulated about 30,000 ft of sediments during Cambrian to Early Pennsylvanian times, but orogeny did not occur until later in the Paleozoic. Also, about 70,000 ft of sediments formed in geosynclines (Cordilleran and the Rocky Mountain) in southeastern Idaho and adjacent areas from Cambrian through Cretaceous times without strong orogenic pulses being characteristic until the Cretaceous. During the past few decades several attempts have been made to classify geosynclines; names such as deltageosyncline, zeugogeosyncline, autogeosyncline, paraliageosyncline, taphrogeosyncline, miogeosyncline, eugeosyncline, and others have been proposed. Justification can be seen to retain the latter two names and, possibly, the name taphrogeosyncline is applicable also; however, of the host of other names little can be said favoring their perpetual use.

The term *basin* has been variously applied by geologists to identify areas of sediment accumulation. As a result, certain confusion has arisen as to just what characterizes a basin. The same, of course, could be said about the term geosyncline. Perhaps much, if not all, of the confusion could be avoided if geologists identified a particular realm (that is, a depocenter) of sediment accumulation in time and space as a *sedimentary basin*. Its definition is that of Landes (1951, p.328) and is quoted in the 1957 American Geological Institute Glossary of Geology: "A segment of the earth's crust which has been downwarped, usually for a considerable time, but with intermittent risings, and sinkings. The sediments in such basins increase in thickness toward the center of the basin." It is in this sense of definition that the writers of this chapter have made reference to basins of the Cordilleran geosyncline (Paleozoic to Early Mesozoic), that is, downwarped segments of the crust. So interpreted, this ancient geosyncline, which began to subside in the Late Precambrian and ceased to subside by Mid-Triassic time, was characterized by more than one sedimentary basin in time and geographic realm. These basins were most actively filled with sediments during times of negative tectonism, that is, when subsidence was greatest. Thus, the sediment accumulation was also greatest. Subsidence, which was initiated first, was followed by sediment accumulation that kept pace with subsidence in large measure.

Dallmus (1958) stated the crux of the whole problem of study of basins: "Because oil accumulations occur in sedimentary basins, it is logical to reason that they are related to each other, and therefore the forces that played a part in the origin and growth of a basin must have affected all of its constituents, solid, liquid, and gaseous. This approach then leads directly to the mechanics of basin evolution, and so to the problem of diastrophism in general." Dallmus pursued the definition of sedimentary basin further; he believed that in an attempt to solve the problem of basin evolution, the clue is in the word "evolution", because it implies a unit in the process of growth, that is, a dynamic unit. So interpreted, he regarded a dynamic basin as any portion of the earth's crust which is actively sinking as a unit with respect to the center of the earth. In his words (Dallmus, 1958, p.884): "As distinct from a dynamic basin, a sedimentary basin is a topographic depression receiving sediments, and its size and shape are controlled only by the existing topography. Although a sedimentary basin may coincide with a dynamic basin, this is not necessarily so. A sedimentary basin may comprise several dynamic basins or parts of such basins. Also, a dynamic basin may be divided into separate sedimentary basins by pre-existing topography at the time of subsidence." Dallmus classified dynamic basins into two fundamental classes: (1) primary and (2) secondary. His primary dynamic basins formed because of concentric downbending of the earth's crust; the Williston Basin, Michigan Basin, and the Eastern Interior Basin were cited as examples for the United States. Secondary basins, he argued, are formed by the local downbreaking of the crust on the top and flanks of regional uplifts; they are secondary features on a much larger tectonic unit. Such terms as grabens, troughs, half grabens, asymmetric basins, and others identify secondary basins.

Dallmus did a great service in calling attention to the improper method of depicting restored cross sections across dynamic and other basins; as he noted, we are so accustomed to looking at flat maps and cross sections constructed with sea level as a horizontal line that we seem to have forgotten that the Earth is a sphere. Map-makers seem to have overlooked the curvature of the earth completely, or have ignored it in preparing cross sections. Figure 4-1 is a NASA photograph of a part of the Earth showing this curvature; north is at the top of the picture, and The Great Salt Lake of Utah is located by an arrow. The Williston Basin, for the most part, is situated below the cloud cover that is northeast of The Great Salt Lake. Dallmus used the Williston Basin as his prototype to illustrate a dynamic basin and its geometry, as reconstructed with the curvature of the earth shown to true scale and also in exaggerated projection (1958, fig.1, p.885). As he pointed out, a 5° (degrees of arc) primary dynamic basin is 375 miles in diameter; the difference in length between the arc and subtended chord is 19,913 ft. For a 6° dynamic basin, the median height is 28,372 ft. Thus, at the center of a nearly circular former dynamic basin (such as the Williston Basin), a thickness of sediments equal to the median height for the corresponding diameter can accumulate before the floor of the basin becomes concave with respect to a horizontal line. This relationship cannot be shown if one makes the conventional restored cross section with sea level drawn as a horizontal line and with the vertical scale exaggerated. Yet, the literature is full of just such restored cross sections, each one claiming to depict an ancient geosyncline or other depocenter.

Sedimentary basins have been defined as geologically depressed areas, with thick sediments in the interior and thinner sediments at the edges (Glossary of Geology and Related Sciences, 1960). In discussing sedimentary basins, Kamen-Kaye (1967) followed this definition, but with the following qualifications: "With respect to epeirogeny it will be noted that 'depressed' is the operative adjective in the above definition. However, for convenience

SUBSIDENCE 175

Fig.4-1. NASA photograph of part of the Northern Hemisphere of the Earth; geographic north is at the top of the photograph, and an arrow locates The Great Salt Lake of Utah. Dark sinuous bands to the left of The Great Salt Lake identify north—south trending mountain ranges that contain the sedimentary record of the ancient Cordilleran geosyncline. Light bands adjacent to the ranges are alluvial-filled valleys.

in dealing with the systematics of basin, the term 'depression' may be replaced by the more comprehensive term 'subsidence', and subsidence may be qualified further as basement subsidence." It was his contention that basin form continues in existence as long as basin strength is able to balance the stresses and strains that accumulate in the depositional prisms. A concept which Kamen-Kaye (1967, p.1833) introduced, with which the writers heartily concur and which they endorse, is quoted here: "If the balance of strength fails, a new phase (orogeny) of the basin tectonic cycle occurs, and the cycle commonly ends as it began, in epeirogeny. The part of the basin which remains uncompressed during the new phase may be called the *epeirogen*, in antithesis to the *orogen*. If the balance of strength does not fail, the continuing initial phase of epeirogeny proceeds through a subsidiary cycle of algebraic summation or net subsidence of basement." The crux of it all appears to lie within the last four words: *net subsidence of basement.* Dallmus (1958, p.888) provided factual data relating to the amount of sediment (thickness) that can accumulate in a dynamic basin before the floor of the basin becomes concave with respect to a horizontal line. It is essentially axiomatic that epeirogenic forces initiate and perpetuate subsidence of a basin or an epeirogen (largely geosynclinal) and that sedimentation is the result (not the cause) of this type of diastrophism. Aggregate thickness of sediments in a dynamic basin, epeirogen, geosyncline, taphrogenic basin, call it what one will, depends upon the net subsidence of the basement. This, possibly, is a function of causative forces in the crust and upper mantle; isostasy cannot be ruled out entirely.

Kamen-Kaye (1967) reviewed some important sedimentary basins throughout various continents, relating their areas, sediment volume and thickness, and epeirogenic behavior. He observed that in some of these basins (the Gulf of Mexico was taken as a classic example) basement subsidence was as much as 40,000 ft. It was his contention that for some of these special cases of excessive basement subsidence, a special term is needed to describe them. Accordingly, he proposed the new term *hypersubsidence*, which he defined (p.1838) as follows: "A phase of subsidence which is attained when the algebraic sum of terms in the epeirogenic cycle is equivalent to a final basement depression of 15,000 ft, an amount which may be definitive in the quantitative classification of major epeirogenic sedimentary basins." He noted that numerous wells have been drilled well below a depth of 15,000 ft and that rocks brought up from these depths display no unusual modification of their physicochemical properties. His suggestion, that a downbending of the geoisotherms may be necessary for these hypersubsident sediments to remain virtually unchanged, demands serious consideration. Thus, Kamen-Kaye has indeed provided geologists with some new and fundamental concepts; furthermore, he argued for more aggressive search for oil and gas in the hypersubsident parts of basins. The area of hypersubsidence in the Gulf of Mexico basin, he pointed out, may be as great as 65% of the total basin. Four major areas of hypersubsidence discussed by Kamen-Kaye are illustrated in Fig.4-2.

Some sedimentary basins

In this chapter no attempt is made to make a comprehensive survey of the depocenters around the world. Rather, a few examples have been

SUBSIDENCE 177

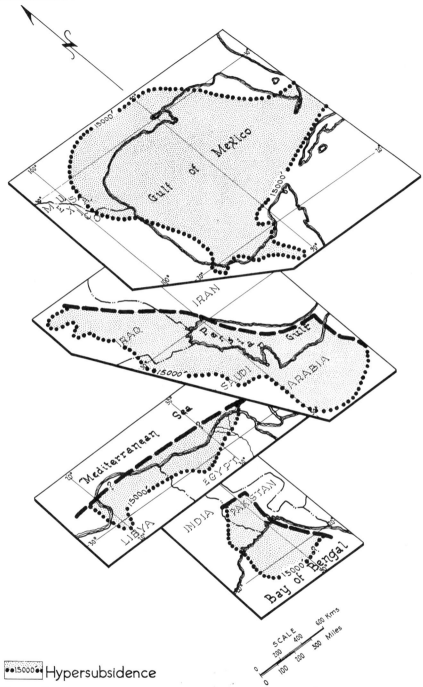

Fig.4-2. Four major areas of hypersubsidence. (After Kamen-Kaye, 1967, fig.3, p.1839. Courtesy of the American Association of Petroleum Geologists.)

selected illustrating magnitude of subsidence and hypersubsidence of some realms of sedimentation in time and space. Some of these are eugeosynclines and miogeosynclines, whereas others are intracratonic, taphrogenic, or other types of basins. In addition, an example of the Pleistocene depocenter in Utah is provided.

Beltian Trough

Eardley (1951, fig.163, p.287) illustrated the Beltian Trough as a generally north-south trending depocenter in the interior of western North America of Late Precambrian sediments. During the past decade geologists have added much information concerning thicknesses and lithologic types of sediments which accumulated in that ancient geosyncline. For example, available data indicate that this Precambrian trough was wider and thus extended farther west into eastern Nevada. Seemingly, it extended southerly into what is now the Death Valley region. This trough-like epeirogen of subsidence was the forerunner of the Late Precambrian and Paleozoic Cordilleran miogeosyncline. Rocks of the Belt Supergroup and correlative units in Canada and adjacent Montana consist largely of clastic sediments (orthoquartzites, siltstones and shales), with considerable thicknesses of carbonates. From 15,000 to 20,000 ft of such strata are present in the Lewis and Clark Range and other areas of Montana. Various changes in thickness and lithology typify the sedimentary and metasedimentary rocks of Late Precambrian age as they are traced in a southerly direction in the ancient geosyncline. For example, the name Big Cottonwood Series is applied in northern Utah where conglomerates, quartzites, phyllites, some carbonates, along with mafic sills and flows aggregate as much as 12,000 ft of rocks. East and southeast of Salt Lake City, about 20,000 ft of Late Precambrian quartzites, slates, conglomerates, alleged tillite, argillites, and shales are present. Farther east, in the Uinta Mountains, about 25,000 ft of similar rocks occur (this area was an embayment that extended easterly from the Beltian geosyncline). In the eastern part of the Uinta Mountains, rocks of Early Precambrian age (dated isotopically at about 2 billion years BP) are referred to as the Red Creek Metaquartzite (that includes schists, amphibolites, marble and pegmatite) and total 20,000 ft in thickness. They are overlain by 24,000 ft of the Uinta Mountain Group, which consists mostly of maroon orthoquartzite with minor amounts of shale.

Hypersubsidence, i.e., subsidence resulting in accumulation of sedimentary rocks having a thickness in excess of 15,000 ft, characterized most of the westerly part of the Beltian geosyncline. Much of the sediment that accumulated in this rapidly subsiding depocenter consisted of *diamictite*. The latter is defined by Crittenden et al. (1971) as "a nonsorted sedimentary rock consisting of sand and/or larger particles in a muddy matrix". In the area around Pocatello, Idaho, from 20,000 to 23,000 ft of Late Precambrian (and possibly some Early Cambrian) diamictite, quartzite, conglomerate, argillite, siltstone, shale, and minor volumes of limestones comprise the geosynclinal section. Farther northwest in the Metaline District, Washington, rocks referred to the Belt Supergroup occur as the base of the Late Precambrian pile, being overlain by thick diamictites, volcanics, shales, quartzites, argillites, and some conglomerates. Hypersubsidence in this part of the Beltian geosyncline accounted for the accumulation of more than 25,000 ft of sediments (Park and Cannon, 1943). Diamictites, volcanics, conglomerates, sandstones, siltstones, and shale comprise the Late Precambrian—Early Cambrian section in the Nelson area in British Columbia, Canada. A rather large spectrum of similarly-dated sedimentary and metasedimentary rocks typifies the section in the Nahanni River area, Northwest Territory, Canada (Ziegler, 1959; Gabrielse et al., 1965); aggregate thickness of rocks is in excess of 30,000 ft. Units equivalent to the Purcell Supergroup typify the basal 10,000 ft, consisting there of sandstone, limestone, dolostone, siltstone, and shale. The remaining 20,000 ft, up to and including some lower Cambrian rocks, consist of diamictite, siltstone, shale, dolostone, much sandstone, and limestone. Again, this is a classic example of hypersubsidence in a westerly part of the Beltian geosyncline, one in which about 30,000 ft of gravity-driven clastics (diamictites), volcanics, and some carbonates accumulated in varying sedimentary realms that displayed differences in bathymetry. That is, much sediment likely accumulated in water depth in excess of a few hundred fathoms (possibly as "turbidites"), whereas carbonates formed in shallower marine depocenters. Volcanics accumulated partly in submarine and in part in subaerial environments. Thus, during the Late Precambrian—Early Cambrian times a transition from the shelf and craton on the east into a miogeosyncline west of it and a eugeosyncline farther west (the then-existing western North American border) typified the depocenters of subsidence and hypersubsidence. It would appear that this great pile of sediment aggregating tens of thousands of feet in thickness, that accumulated on the continental border, represents a critical thickness with respect to crust—mantle equilibrium.

Rocks of Late Precambrian age do not form areally-extensive outcrops across the area of Oregon, northern Nevada and, generally, south towards Death Valley of California—Nevada. There are, however, some sections of these rocks in scattered outcrops in eastern Nevada and western Utah. These rocks have been the subject of investigation by Misch and Hazzard (1962) who stated (p.292): "Precambrian outcrops in the region are scattered and isolated in different ranges, and display various kinds and degrees of metamorphism. Hence, correlation commonly is difficult. The stratigraphically best defined major unit is a thick sequence of largely quartzose and argillaceous metasediments which is concordantly succeeded by Lower Cambrian Prospect Mountain Quartzite." These geologists studied sediments and metasediments of Late Precambrian age in no fewer than 13 localities in east-central to northeastern Nevada, northwestern Utah, and south-central Idaho (1962, fig.1, p.291). They assigned the entire section of from 5,000 to almost 9,000 ft of metasediments to the McCoy Creek Group. Young (1960, p.158) had previously referred about 10,000 ft of similar metasediments to his "Pierpont group" that crops out in the Schell Creek Mountains of eastern Nevada. These rocks (regardless of which nomenclature is used) accumulated in a part of an ancient geosyncline that displayed subsidence.

The discussion which relates to the Beltian trough or geosyncline to this point has stressed the fact that sediments and metasediments thicken from east to west across the ancient linear depocenter. Stewart (1972) described this area as representing a narrow and slightly sinuous belt of Upper Precambrian and Lower Cambrian strata in North America that extends from Alaska and northern Canada to northern Mexico, a distance of 2,500 miles. His generalized isopach map of Upper Precambrian and Lower Cambrian strata in this belt (fig.2, p.1348) shows a thickening from 0 ft on the east to 15,000—25,000 ft in areas 100—300 miles to the west. His 20,000-ft isopach, for example, trends in a general southwesterly direction across about central Nevada, whereas the 5,000-ft line is drawn near Las Vegas. So interpreted, approximately 12,000 ft of quartzite, shale, siltstone, and carbonates document the record of Late Precambrian sedimentation for the Death Valley segment of the ancient epeirogen. From what has been mentioned in previous pages, it would seem that a tectonic hinge line that trended southwesterly across southwestern Utah into southern Nevada separated shelf and cratonic belts of sedimentation from the geosyncline farther west. The inception of this Wasatch—Las Vegas Hinge Line in Late Precambrian time indicates that the Cordilleran geosyncline was a negative mobile belt well back into the Late Precambrian. Just when the Beltian geosyncline was replaced by the Cordilleran geosyncline is not an event to easily pin-point in time and space. This would be like attempting to date the dawn of the Paleozoic Era.

The Upper Precambrian and Lower Cambrian rocks are absent or are very thin in the shelf—craton area that existed along a sinuous belt east of the ancient Beltian—Early Cordilleran geosyncline. The sediments thicken progressively westward in the depocenter and diamictites and volcanics are present in this thick prism. Algal dolostones (the Noonday Dolomite) occur low in the sedimentary prism of Late Precambrian rocks in the Death Valley region of California and Nevada. The Pahrump Group identifies lower units of Lower Precambrian sediments that rest on older Precambrian gneiss and schist. Limestone, siltstone and shale, and dolostone typify the lower half of the Pahrump, whereas diamictite of the Kingston Peak Formation occurs above. Stewart (1972, fig.1, p.1346) showed that the Late Precambrian—Early Cambrian totals more than 16,000 ft in this area. In all, therefore, what is now the Death Valley region of California experienced hypersubsidence during Late Precambrian and Early Cambrian times in this epeirogen. The writers concur with Stewart (1972, p.1345) that "... a change occurred in the tectonic framework of western North America after deposition of the Belt Supergroup, and that this change marked the beginning of the Cordilleran geosyncline". Subsidence and hypersubsidence typified the Beltian geosyncline, and geologic hypersubsidence on an unparalleled scale was perpetuated into Paleozoic and Early Mesozoic times in this major downwarped sector of the earth's crust. Diamictites and post-diamictite rocks were the initial deposits in the Cordilleran geosyncline (Stewart, 1972, p.1345).

The reader may rightfully ask how, if at all, this discussion of geologic subsidence and hypersubsidence of dynamic basins, geosynclines, and related depocenters, along with accumulation of overthickened prisms of sediments, relates to the plate tectonic theory. Stewart (1972) has presented an interesting explanation, one with which the writers concur. Stewart (p.1345) stated: "The distribution pattern and lithologic characteristics of the Upper Precambrian and Lower Cambrian sequence fit the recently developed concept that thick sedimentary sequences accumulate along stable continental margins subsequent to a time of continental separation. The depositional pattern of the sequence is unlike that of underlying rocks, a relation consistent with the idea of a continental separation cutting across the grain of previous structures. The pattern is, on the other hand, similar to that of overlying Lower and Middle Paleozoic rocks, suggesting that the diamictite and post-diamictite rocks were the initial deposits in the Cordilleran geosyncline. The units of volcanic rock near the bottom of the sedimentary sequence indicate volcanic activity related to thinning and rifting of the crust during the continental separation." So interpreted by Stewart, the assemblage that he referred to the Windermere Group and equivalent rocks (>850 million years old) represent supracrustal rocks that consist of relatively unmetamorphosed sedimentary and volcanic rocks resting upon metamorphic and plutonic rocks. A major change took place about 850 million years ago and, in terms of plate tectonics theory, this change marks the time of continental separation (Stewart, 1972, p.1345). His sedimentary model is shown in Fig.4-3. It was pointed out by Stewart (p.1356) that: "... the occurrence of significant amounts of tholeiitic basalt near the bottom of the sedimentary sequence suggests volcanic activity related to the thinning and rifting of the crust during the separation". He added this note (p.1356): "The linear belt of westward-thickening detrital rocks deposited

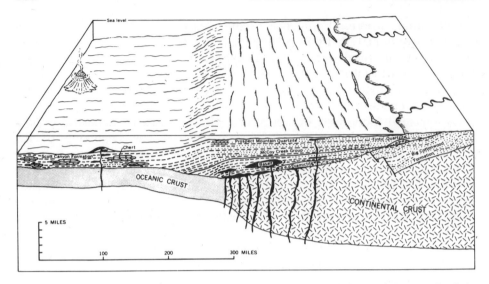

Fig.4-3. Diagrammatic cross section showing Upper Precambrian and Lower Cambrian rocks in the northern Great Basin, Nevada and Utah. (After Stewart, 1972, fig.3, p.1350. Courtesy of the Geological Society of America.)

subsequent to the volcanic activity indicates an accumulation along a stable continental margin ... that is perhaps similar in setting and character to the thick sedimentary accumulation ... off the east coast of the North American continent." This interesting subject is discussed later in this chapter.

Cordilleran geosyncline

Reference is again made to Fig.4-1, the NASA photograph, in which The Great Salt Lake is located by an arrow. A dark band just to the east (right) of the lake identifies the approximate eastern boundary of the earlier (Beltian) and also the later (Cordilleran) geosynclines. Accordingly, another dark band (mountain range) west of it and in the lower-left defines the western boundary. Any restored cross section showing the most complete and maximum record of subsidence and hypersubsidence with concomitant sedimentation should be constructed from left to right passing just south of the lake. Available geologic data would provide a wealth of objective geologic information for such a restored cross section. As pointed out in preceding paragraphs, the Cordilleran geosyncline, in which a thick prism of essentially complete Paleozoic and Early Mesozoic sediments accumulated, differed in tectonic (and plate tectonic) setting from its Late Precambrian predecessor, the Beltian geosyncline. The writers are, therefore, in agreement with many other geologists in their interpretation that the inception of the

Cordilleran geosyncline occurred in Late Precambrian time. This geosyncline was a broadly expanded belt of negative mobility, thereby contrasting with the sinuous and narrower Beltian trough-like geosyncline. It lasted through the Paleozoic, terminating in about Mid-Triassic time. Sometime in the Early Paleozoic (Ordovician?), a well-defined eugeosyncline typified its western half, the miogeosyncline was farther east, and a shelf and craton lay still farther east. The Wasatch—Las Vegas hinge line, as during the duration of the Beltian geosyncline, separated the craton from the miogeosyncline.

A wealth of information has been accumulated by many geologists concerning the sedimentary rocks of the miogeosynclinal part of the depocenter. The Cordilleran geosyncline, which is a classic, is divided into a couplet: two adjacent and parallel structures consisting of a true geosyncline (eugeosyncline) on the west and a miogeosyncline (lesser geosyncline) east of it. The two terms are commonly shortened to *eugeocline* and *miogeocline*. Reference to Fig.4-3 will enable the reader to visualize how the eugeocline (at the left) is related to the miogeocline east of it. This figure should be compared with the NASA photograph (Fig.4-1) for correct scale and curvature of the Earth.

Any attempt at constructing an isopach map for total Paleozoic and Early Triassic sedimentary rocks for the entire duration of the Cordilleran geosyncline is unrealistic, because its diastrophic behavior during Cambrian through Devonian periods differed somewhat from that during the remainder of the Paleozoic through Early Triassic times of greater negative mobility. Furthermore, various episodes of orogeny have translated some plates easterly, thus stacking up some stratigraphic sections a few tens of miles to the east of their original depocenters. With these facts in mind, Fig.4-4 was assembled showing subsidence and sediment accumulation in the Late Precambrian—Early Cambrian Beltian miogeocline and the Cambrian through Devonian Cordilleran miogeocline. The belt of negative mobility, which established the Beltian downwarp of the earth's crust, was perpetuated more or less along the same lineaments during development of the Cordilleran geosyncline. Record of Cambrian sedimentation is best preserved in the miogeocline, where thicknesses of 7,000—8,000 ft of total Cambrian rocks are common. Locally, thicknesses of from 12,000 to 13,000 ft have been measured. Commonly, a sandstone formation (now an orthoquartzite to incipient metaquartzite, such as Prospect Mountain, Tintic, Brigham, etc.) forms the base of the Cambrian System, with shale, limestone and dolostone making up the remainder of the prism.

As subsidence continued throughout the Ordovician, Silurian, and Devonian periods, from 5,000 to 12,000 ft of additional carbonates, shales, and some sandstones accumulated in the miogeocline. Geologists interpret these rocks to have originated from sediments that accumulated in "knee-

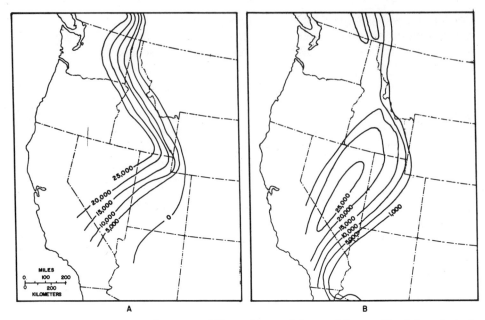

Fig.4-4. Generalized isopach maps of Upper Precambrian and Lower Cambrian strata in western North America. A. For the Beltian miogeocline. B. For the Cordilleran miogeocline (Cambrian through Devonian time).

deep" water upon a slowly subsiding crust; seemingly, sedimentation kept pace with the rate of subsidence. Greater thicknesses of sediments accumulated in the eugeocline, giving rise to thick shales, argillites, cherts, graywackes, and spilites that formed in deeper waters.

With the inception of the Antler orogeny in Devonian time, a great belt of diastrophism extended from southern Nevada and across the central and west-central part of the state into western Idaho (possibly, it extended farther north). Effects of this orogeny tapered off in Mississippian—Pennsylvanian time, after some 6,000 ft or more of synorogenic conglomerates, orthoquartzites, shales and some limestones formed in a subsiding trough near Eureka, Nevada. Some geologists interpret the Late Devonian and Mississippian sediments (some of which were termed flysch-like mudstone) as having accumulated on the continental shelf that was then the Cordilleran miogeocline. Sediments that gave rise to similar lithic units formed in another rapidly sinking trough that extended from southern to central Idaho. From 5,000 to 6,000 ft of interbedded carbonates, shales, orthoquartzites, and sandstones also accumulated in various depocenters (sedimentary basins of the miogeocline). Figure 4-5 illustrates various sedimentary basins that experienced maximum subsidence during

Fig.4-5. Isopach map of total Mississippian sedimentary rocks in western Utah, eastern Nevada, and part of southern Idaho.

SUBSIDENCE

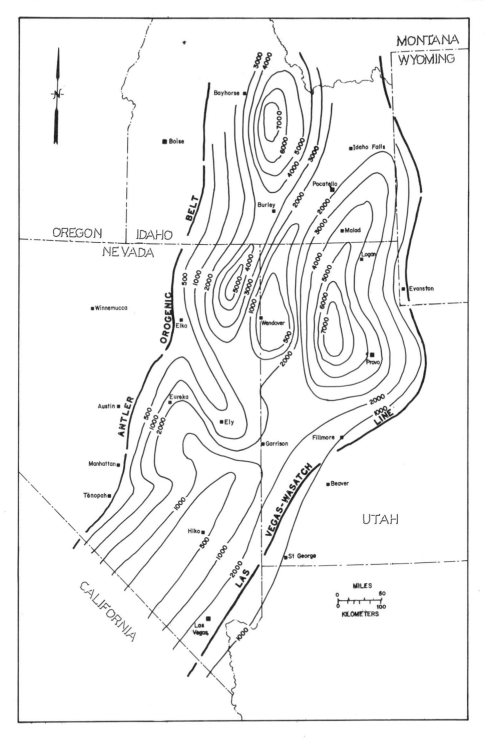

Mississippian time in part of the Cordilleran miogeosyncline in eastern Nevada, western Utah, and southern Idaho. The great Oquirrh Basin of northwestern Utah, the Sublett Basin of southern Idaho, the Elko and Ely Basins of eastern Nevada, and the Bird Spring Basin of southern Nevada all began to subside during Late Mississippian time, and were the dominant areas of negative mobility throughout Pennsylvanian and much of Permian times. Oquirrh Basin was the major downwarp in Utah, as evidenced by some 16,000 ft of interbedded orthoquartzites, a vast spectrum of carbonate types, and minor volumes of shale. Figure 4-6 is an isopach map of total Pennsylvanian sedimentary rocks for essentially the same area of the Cordilleran miogeosyncline as shown in Fig.4-5. Comparison of Fig.4-6 with Fig.4-5 indicates that the north—south elongated sedimentary basin in Utah west of Provo and Salt Lake City received about 7,000 ft of sediments during the Mississippian. This area developed into the Oquirrh Basin during Late Mississippian time and then became the northwesterly-aligned negative belt throughout the Pennsylvanian. During this period some 16,000 ft of shallow-water (dominantly marine) sediments were deposited in this basin; sedimentation kept a fairly even pace with the rate of subsidence, although at times and places substantial bathymetric depths may have developed. Synorogenic sedimentation typified some of the western depocenters of the Cordilleran miogeosyncline, because clastic sediments are interleaved with carbonates and shales of Lower Pennsylvanian age (see Fig.4-6).

An additional 10,000 ft of orthoquartzites, carbonates, and sandstones formed in the Oquirrh basin during the Wolfcampian time (Permian). Thinner accumulations formed in the Arcturus and other basins (Bissell, 1970). Continued negative mobility and great basement subsidence characterized the eastern half of the Cordilleran geosyncline during Leonardian and Guadalupian times (Permian). As a result, a total thickness of from 10,000 to 20,000 ft of sedimentary rocks accumulated in the Bird Spring, Oquirrh, and other basins in Utah and Nevada. The Sublett Basin of south-central Idaho was one of the major downwarps of the miogeosyncline at this time, because approximately 24,000 ft of orthoquartzites and carbonates, with minor volumes of siltstones and shales, formed during the Permian there (Cramer, 1971). Slightly more than 22,000 ft of this sedimentary prism is of Wolfcampian and Leonardian ages. This is thicker than the total Permian section exposed farther south in the Hogup-Terrace Mountains area of northwestern Utah, where the aggregate thickness of sediments is slightly in excess of 17,000 ft (Bissell, 1970, fig.2, p.290). Figure 4-7 shows the total thickness of Permian rocks in the same area as shown in Fig.4-5 and 4-6; the Sublett Basin of south-central Idaho between Burley and Malad experienced hypersubsidence. The northern part of a similar downwarp in northwestern Utah similarly underwent hypersubsidence. The data provided on

Fig.4-6. Isopach map of total Pennsylvanian sedimentary rocks in western Utah, eastern Nevada, and part of southern Idaho.

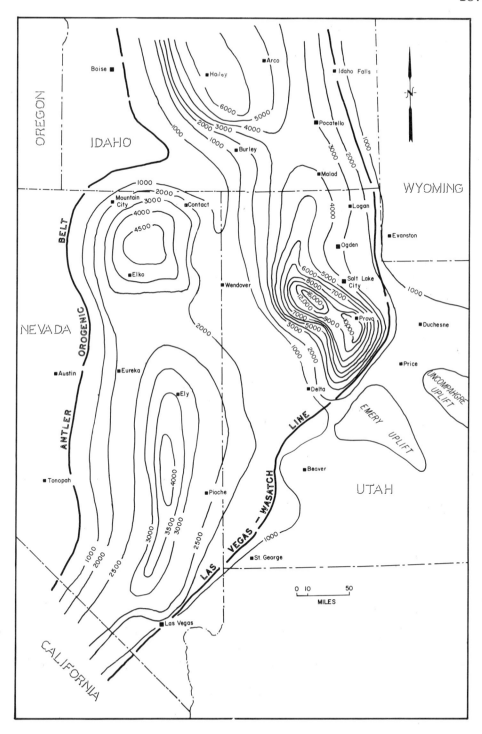

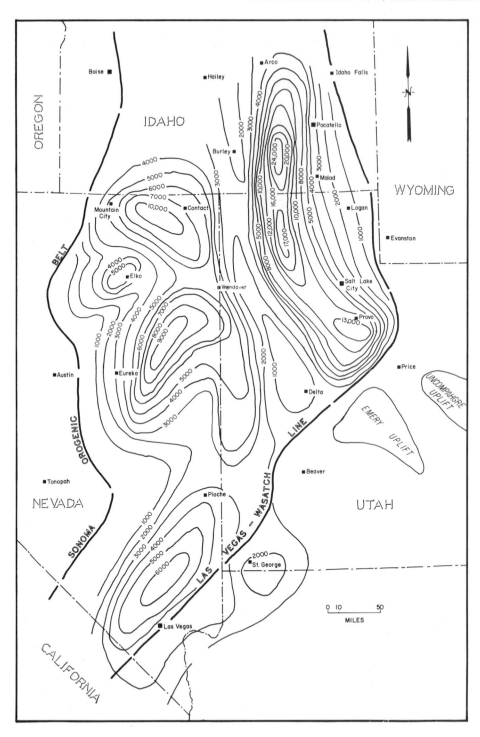

Fig.4-5—4-7 indicate excessive subsidence (hypersubsidence) in the Cordilleran miogeosyncline in northwestern Utah. This was one of the classic and major sedimentary basins for which fairly complete records are available for such a relatively short span of geologic time. Up to 36,000 ft of Mississippian, Pennsylvanian and Permian sedimentary rocks accumulated in this major downwarp. This exceeds the thickness of Permian down to Ordovician sedimentary rocks (30,050 ft) penetrated in the Anadarko Basin in western Oklahoma and of total Mississippian, Pennsylvanian, and Permian sedimentary rocks (approximately 30,000—32,000 ft), which accumulated in the Wood River—Sublett Basin of southern and south-central Idaho. Quite a contrast in subsidence of Anadarko Basin and Cordilleran miogeosyncline in northwestern Utah is in evidence, because in the Anadarko Basin the time interim was Ordovician, Silurian, Devonian, Mississippian, Pennsylvanian and Permian time, i.e., most of the Paleozoic Era. Both are examples of hypersubsidence. Depocenters in the Cordilleran miogeosyncline had that "let-down feeling" for a shorter period, but the total subsidence was more rapid.

The Cordilleran miogeocline "closed its doors" (so to speak) and ceased to perform as a negative mobile belt (a miogeosyncline composed of various sedimentary basins) by about mid-Triassic time. This event did not occur until an additional 1,000—4,000 ft of dominantly marine shales, carbonates, and some sandstones (Thaynes Formation) had accumulated. From about medial Triassic time on it ceased to subside and, in fact, at times was epeirogenically uplifted. Previously-accumulated sediments were then subjected to erosion and much material was removed, being transported eastward to aid in filling the Mesozoic Rocky Mountain geosyncline.

A quick glance at the figures of total sediment accumulation, that have been presented in previous pages, indicates that from 10,000 to 25,000 ft of sediments (now largely metasedimentary rocks) accumulated in the Beltian geosyncline, precursor to the Cordilleran geosyncline. This pile of sediments rests upon a crystalline basement of older Precambrian rocks. Above the Late Precambrian metasediments occur from 30,000 to as much as 70,000 ft of Paleozoic sedimentary rocks that represent all the Systems of this Erathem. Before the Cordilleran miogeosyncline was terminated by epeirogeny and thus changed from negative to positive mobility, about 1,000 to 4,000 ft of Early Triassic sediments accumulated in it. Had it been possible, in about medial Triassic time, to drill a borehole penetrating the aggregate thickness of Lower Triassic, all the Paleozoic, and all of the Late Precambrian sediments, that had accumulated up to that time in the miogeosyncline, this "hole" would have been from 12 to 15 miles deep. Accumulated thicknesses amount to as much as 100,000 ft, but it is rather doubtful that such a tremendous prism of sediment accumulated in one particular area of the ancient geosyncline. The sediments of Early Paleozoic age are thickest in

Fig.4-7. Isopach map of total Permian sedimentary rocks in western Utah, eastern Nevada, and part of southern Idaho. (Modified from Bissell, 1970, fig.2, p.290.)

western Utah and eastern Nevada, whereas those of Late Paleozoic age are thickest in the Oquirrh and Sublett basins of northwestern Utah—southwestern Idaho. The ancient geosyncline seemingly consisted of various sedimentary basins, dynamic basins, and other depocenters. At times and places, some were interconnected, but at other times some behaved independently. All in all, this represents an overthickened prism of sedimentary materials that documents the fact that tremendous basement subsidence occurred. When viewed from an analytical Dallmus' model of earth's curvature, one must conclude that the Cordilleran geosyncline was one of the major downwarps of the earth's crust; for some hundreds of millions of years the basement subsided to accommodate the great pile of sedimentary rocks. Truly, this miogeosyncline provides a classic example of geologic subsidence and hypersubsidence of the earth's crust. An isostatic or other adjustment, beginning in about Triassic time, reversed this process and the Cordilleran area was uplifted, whereas the region east of the Las Vegas— Wasatch Hinge Line subsided to accomodate the Rocky Mountain geosyncline.

The probability is very strong that during Late Precambrian, the entire Paleozoic, and into Early Mesozoic times the Cordilleran geosyncline lay along the western part of the North American plate. So interpreted, when this geosyncline was differentiated into its couplet of the miogeocline (discussed above) and a eugeocline, the latter characterized deeper waters west of the miogeocline in which shallow waters (at times "knee-deep") were typical. Accordingly, sediments were washed, dumped, turbidited, or otherwise transported to deeper waters where they accumulated with cherts, igneous materials, and deep-water accumulations in the eugeocline. Throughout Late Precambrian, Paleozoic, and Triassic times this depocenter differed from that of the miogeocline in that it was not a typical sedimentary basin. Contrarily, its accumulated rocks were not thickest in a *central* part of a basin. Rather, they thickened into deeper waters of the depocenter. In addition, since the accumulation of a thick prism of Precambrian sediments, metamorphism has affected the earlier-formed sediments and volcanics, thereby imparting various shades of post-depositional change to the materials.

Pre-Mesozoic age rocks in that part of the eugeocline that extends along western North America from Alaska southward across western Canada, Washington, Oregon, and California are from 15,000 to 31,500 ft in thickness (Eardley, 1951, pp.43—68). In addition to this Precambrian and Paleozoic prism of sediments and volcanics, one must add an additional 10,000 to 25,000 ft of Triassic sediments and volcanics. In recent years, personnel of the U.S. Geological Survey, and Professor N. J. Silberling and his associates and students at Stanford University have added a wealth of factual data concerning Permian and Triassic rocks in western Nevada, and have advanced some thought-provoking interpretations. It has been known for a number of years that the Sonoma orogeny occurred in Late Permian time and profoundly affected an area that extended from southern Nevada, across western and central Nevada, into western Idaho, eastern Oregon, and, probably, farther north into Canada (possibly, also to Alaska). According to Silberling and his co-workers, the Koipato Group, for years regarded as of probable Late Permian age, may be entirely of Early Triassic age, because ammonites of probable Spathian age have been collected from the Koipato, high upsection, in the Humboldt Range of western Nevada. This group consists predominantly of volcanic and volcanic-derived rocks. A great thrust — the Golconda thrust — carried what may be the sea-floor volcanic (mafic) and sedimentary rocks (bedded cherts, fine-grained siliceous and calcareous clastic rocks) eastward over partly correlative shallow-water marine carbonate and clastic rocks of the continental shelf. According to Silberling (1971, p.355): "Both before the Sonoma orogeny, during the Late Paleozoic, and after it, during the Early Mesozoic, continental-shelf carbonate and clastic deposition took place in west-central Nevada without orogenic interruption. Events of the Sonoma orogeny near the beginning of Triassic time invite speculative explanation invoking plate tectonic theory. The abrupt thrusting of ocean-floor rocks onto the continent, followed by the arc-like volcanism of the Koipato Group, might best be attributed to collision of an isolated, east-facing, oceanic arc-trench system against the west side of the continent situated on the North American plate. Such an event could reflect some more general change in the interaction of the world's lithosphere plates at about the same time and thus be of more than only local significance." Later he (1973) rejected this speculation.

Perhaps it is not so speculative after all; the writers believe that Silberling's hypothesis is well taken and that his interpretation is in close accord with the data that have been recently acquired by scientists working on the plate-tectonic hypothesis. If correct, therefore, this interpretation may explain better than theories that have been advanced to date, the causes for the excessive subsidence of the Cordilleran geosyncline and accumulation of an over-thickened prism of sediments and volcanics, and, most importantly, why this geosyncline reverted from negative to positive mobility before the Triassic Period ended. Additional statements of possible relationships of subsidence, hypersubsidence, and excessive sediment accumulation are presented later in this chapter. Before terminating the discussion on subsidence and hypersubsidence of the Beltian and Cordilleran geosynclines, the reader is invited to again examine Fig.4-3. The writers interpret the diastrophic setting for this part of North America as follows: a western continental crust formed much of the basement of the miogeocline, whereas oceanic crust was beneath the eugeocline. The sedimentary basins of the Beltian and

Cordilleran geosynclines were downwarped segments of the earth's crust, and were aligned generally north—south, along the western margin of the North American continental plate. A Late Precambrian separation is in evidence.

Appalachian geosyncline

It was pointed out previously in this chapter that the concept of a geosyncline is not that of this generation of geologists; rather, the definition is one century old (Dana, 1873). The part of the definition that is germane to this part of the discussion (Dana, 1873, p.717) is as follows: some mountains "... were the result of a slowly progressing geosynclinal, and consequently a very thick accumulation in the *trough* (italics added by the writers) of sedimentary beds, ending in an epoch of displacements and solidification, and often of metamorphism of the sedimentary beds, as in the case of the Alleghenies and other *synclinoria*". It should be remembered that Dana, Barrell, Gilbert, Schuchert, and many other geologists of past centuries did not have access to facts that corroborate Hess's (1972) theory of seafloor spreading, nor to the great body of information that now relates to the theory of lithosphere plate tectonics. Thus, as attention now focuses on the Appalachian geosyncline and the fact that subsidence and hypersubsidence characterize that great prism of sediments, and some igneous rocks, that accumulated in a trough-like depocenter, an attempt can be made to relate such geologic subsidence to the new concepts of geology.

Before taking a closer look at the Appalachian geosyncline, the term *orogen* is first examined because recent publications which relate to this ancient geosyncline discuss the evolution of the Appalachian orogen (see Bird and Dewey, 1970; and others). It was Gilbert (1890) who coined the term orogenesis for the episodic processes that operate along mobile belts, such as those that have produced great linear crustal disturbances of mountain chain proportions. Thus, if a single movement or a series of related movements occur within or adjacent to a linear depocenter, one is dealing with an orogeny. In an orogeny, the strain result of the stress is expressed in considerable tangential component of movement, and this movement is manifested adjacent to the long geosynclial troughs of the crusts. Accordingly, the orogeny results in folding and thrusting of the crust, accompanied by crustal shortening. Stated otherwise, the orogen identifies the most mobile and linear belts of the earth. Also, it is to be remembered that it was Gilbert who coined the term epeirogenesis for the more general vertical movements, or changes of level, over broad areas up to continental proportions; such a movement is termed an *epeirogeny*.

Because Kamen-Kaye stated the crux of all this so succinctly (1967, pp.1833—1834), he is quoted here: "The end of the first phase of the basin tectonic cycle may be conceived alternatively as the end of the phase of epeirogeny or as the beginning of the phase of orogeny. The change of phases presumably occurs when the basin is no longer able to protect itself from the onset of overbalancing or even catastrophic instability. Especially in the asymmetric basin the second phase may occur when *mature subsidence* (italics added by the writers) develops into the compressive downbuckle of tectogenic theory. If the tangential forces developed in the downbuckle are sufficiently great, their presumed effect is to squeeze the contained sediments into a core and to force this core upward so that it emerges in the catastrophic form of the orogen, commonly with injection and metamorphism, and in every case with structural failure by thrusting and imbrication."

Bird and Dewey (1970, p.1032) stated: "Although the Appalachian orogen lies along almost the entire eastern continental margin of North America, it clearly formed part of a continuous Appalachian/Caledonian orogen... extending in Late Paleozoic times from at least Florida to Spitzbergen on the Bullard, Everett and Smith (1965) reconstruction of the North American continents." They then proceeded to outline and discuss the various zones of crustal mobility of the Appalachian orogen. In addition, Bird and Dewey (p.1036) pointed out that it was Hall (1859) who was first to advance the concept that sediments which make up the Appalachian region accumulated in a furrow, giving credit to Dana for coining the name "geosynclinal" for the Allegheny region in 1873. In reviewing the history of investigation of the Appalachian geosyncline by various geologists, Bird and Dewey (p.1037) emphasized the importance of Kay's (1944) classification of its western, nonvolcanic belt (his miogeosyncline), and the eastern, volcanic and clastic belt (the eugeosyncline). According to Kay (1944), the two geosynclines were separated by a volcanic uplift along what is now the Green Mountains axis. Thus, Kay was first to envision that a sequence of facies belts related in some manner to a continental margin-island arc complex that bordered an ocean. The next step was taken by Drake et al. (1959) when they called attention to the close analogy between Kay's geosynclinal couple (that is, inner miogeosyncline and an outer eugeosyncline) and the paired continental rise geosynclines of the present eastern margin of the North American continent. Still later, Dietz and Holden (1966) argued that the term miogeosyncline should be replaced by "miogeocline" and that eugeosyncline should now be termed "eugeocline". Since then, various scientists have argued the case, some favoring an Early Mesozoic opening of the Atlantic Ocean, others favoring an opening-and-shutting followed by an opening of continental plates to account for the facies types and for Early Paleozoic orogenies (Taconian, Humberian, Acadian). For example, Dewey (1969) proposed a plate model for the evolution of the orogen based on the concept of an expanding and then contracting Proto-Atlantic Ocean. So interpreted, Acadian deformation of Devonian times occurred during continental collision.

Bird and Dewey (1970) examined the Appalachian orogen for the Late Precambrian to Middle Devonian interim, stating (p.1047): "In Late Precambrian times, a continuous North American/African continent began to distend along a narrow zone with the progressive rise of the base of the lithosphere associated with the early development of a tensional plate margin." According to their interpretation, distension resulted in development of a graben into which the early coarse clastics were dumped. Volcanicity accompanied the distension, followed by establishment of an accreting plate margin along an Appalachian Atlantic mid-oceanic ridge. Accordingly, new oceanic lithosphere was produced during an expanding—spreading phase that lasted until Ordovician times. They contended that in Early Ordovician time, when the Appalachian Atlantic had expanded to its maximum width, plate loss was initiated and a trench was formed. Later in the Ordovidian, an island arc developed on the oceanward side of the Piedmont. During progressive westward spreading of volcanic activity, the block-faulted miogeocline sagged and an exogeosyncline formed. For the purpose of stressing subsidence and hypersubsidence of a prism or prisms of sediments in the Appalachian geosyncline, the classical work of Bird and Dewey (1970, p.1031) is cited: "In terms of plate tectonic theory, and by analogy with modern continental margins, the Appalachian orogen evolved through a sequence of interrelated sedimentation—deformation—metamorphism patterns within a tectonic belt situated along the eastern margin of the North American continent. As exemplified by the northern part of the orogen, Appalachian stratigraphic—tectonic zones and deformation sequences are related to Late Precambrian to Ordovician expansion, followed by Ordovician through Devonian contraction, of a Proto-Atlantic ocean. This oceanic opening and closing was achieved by initial extensional necking of a single North American/African continent and by lithosphere plate accretion, followed by contractional plate loss along a trench, or complex of trenches, marginal to the drifted North American continent." It was the contention of these geologists that pre-orogenic Appalachian realms and patterns of sedimentation were much the same as those that exist along modern continental margins. This would be the continental shelf/slope/rise/abyss quadrad. Continental margin subsidence—sedimentation patterns for the Cenozoic and Holocene are discussed later in this chapter.

Continental margins

Atlantic Coastal Plain

The Cordilleran and Appalachian areas that were discussed in previous pages have been cited by geologists of the North American continent as constituting so-called "classic" examples of orogens. Numerous textbooks of geology allude to these areas as having been the sites of typical geosynclines from Late Precambrian through Paleozoic times. Furthermore, each depocenter consisted of a couplet of an outer volcanic—clastic eugeocline and an inner nonvolcanic, carbonate miogeocline. Numerous paleogeographic maps that were constructed by Willis, Schuchert, Eardley, and many more authors depicted both Americas more or less as a single entity, a block not necessarily juxtaposed against another landmass. The idea of continental drift is not new, but the application of mechanics of the plate tectonic theory to the evolution of geosynclines relates mostly to the past decade. Perhaps the hypothesis that ocean basins have been permanent since the formative stages of the earth is not tenable to most geologists. The writers shall not become completely entangled in this web of controversy, although they favor the theory of continental drift. This bias stems from the fact that plate tectonics is a reality today; ocean basins are not fixed in size or shape, but contrarily are either opening or closing. The Atlantic Ocean is opening and the Pacific Ocean is closing. The question arises, however, whether the continents always drifted or not? Dietz (1972) stated that: "The drifting of continents is another theme; every continent must have a leading edge and a trailing edge. For the past 200 million years the Pacific coast of North America has been the leading edge and the Atlantic coast the trailing edge. The trailing margin is tectonically stable, and since the continental divide is near the mountainous Pacific rim, most of the sediments are ultimately dumped into the Atlantic Ocean, including the Gulf of Mexico. Therefore, it is primarily along a trailing edge that the great geosynclinal prisms are deposited." So interpreted, the northern Gulf of Mexico throughout the Cenozoic and up to and including the present is a geosyncline — the Gulf Coast geosyncline. The eastern Atlantic coast and adjacent area is another modern example. Quoting again from Dietz (1972): "Now that the ocean floor is becoming better known, one need not look far to find an example of the geosynclinal couplet in the process of formation. A probable example of a 'living' eugeocline is the continental rise that lies seaward of the continental slope off the eastern United States of America. Landward of the rise and capping the continental shelf is a wedge of sediments that becomes progressively thicker as it extends toward the shelf edge. This wedge seems to be a living miogeocline."

How, the reader may justifiably inquire, does this relate to geologic subsidence? Dietz had an answer to this question (1972, pp.129—130): "An important aspect of geosynclines requiring explanation is that they are laid down on foundations that are *continuously subsiding* (italics added by the writers). This aspect is particularly evident in miogeoclines, which can attain a total thickness of their seaward edge of five kilometers, even though they are entirely composed of beds deposited in shallow water. This phenomenon is nicely accounted for by plate tectonics; the margins of rift oceans inherently have, as one geologist has expressed it, a 'certain sinking feeling'." Dietz then proceeded to cite the Atlantic Ocean between the U.S.A. and the bulge of Africa as an example of this phenomenon. This basin came into being about 180 million years ago through the mechanics of the insertion of a spreading rift that split North America away from Africa. Accompanying this rifting was the swelling of the mantle, which caused arching of the continents along a rift line to the tune of about 2 km. Attendant and subsequent erosion transported copious quantities of sediments as the uparched blocks were beveled, thus thinning the two continental plates. An abundant supply of sediments was assured and the top of the accumulating prism was maintained near sea level. Rate of sediment accumulation more or less kept pace with subsidence. Excess clastic sediment bypassed the shelf, was temporarily dumped on the continental shelf, and then cascaded onto the continental rise by turbidity currents. Thus, the continental shelf edge and the continental-rise prism make up the couplet of the geosyncline. Accelerated subsidence within either or both depocenters accompanied by excessive sediment supply from the continent could account for lithologic assemblages within the geosyncline. Jostling of various continental plates against each other would give rise to various changes of sediments within the prisms, which were caught in the jaws of the vise. Diastrophism in the form of epeirogenic subsidence could change to orogenic folding. Dietz (1972, p.132) stated it thusly: "The hypothesis that geosynclines are deposited along a continental margin and then crushed against the continent as a result of plate tectonics seems to explain satisfactorily how geosynclines are transformed into folded mountains. The close relation between the eugeoclines and foldbelts is not one of cause and effect but a simple consequence of location: geosynclines are laid down along continental margins and such margins are the locus of interaction between continents and subduction zones." The writers are of the opinion that because of a much improved understanding of the mechanics of plate tectonics, geologists are now in a position to understand, and perhaps explain, some of the possible reasons for geologic subsidence and hypersubsidence. Accordingly, a model has been constructed that is designed largely after a block-diagram of Dietz (1972), to show the geosynclinal couplet of the Atlantic coast of the U.S.A. (Fig.4-8). Again, the writers emphasize the importance and necessity of diastrophism to set the stage for subsidence, to be accompanied by and followed by sedimentation. Obviously, the weight of added sediment has considerable effect. Dietz (1972, p.130) explained it this way: "Additional subsidence, however, is caused by the steadily growing mass of the sedimentary apron,

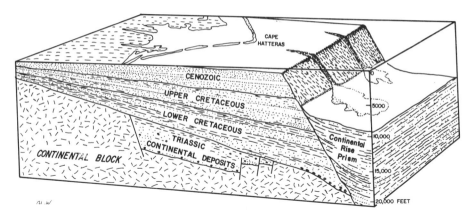

Fig.4-8. Geosynclinal couplet off the Atlantic Coast of the U.S.A. depicting the miogeocline with its shallow-water sediments, which accumulated on the shallow continental shelf, and the eugeocline consisting of sand and mud that were deposited by turbidity currents. Material in the Triassic basin represents continental deposits that accumulated before the foundering of the continental margin under tension 190 million years ago, before the opening of the Atlantic Ocean. (Modified from Dietz, 1972.)

which must be isostatically compensated because the earth's crust is not sufficiently strong to sustain the load. For every three meters of sediment deposited, the crust sinks about two meters. This crustal failure, however, is spread over a large geographic area, so that the growth and subsidence of a huge continental-rise prism causes a sympathetic downward flexing of the adjacent continental margin. As the continental shelf slowly tilts, wedges of shallow-water sediments are deposited." So interpreted, this provides a working hypothesis applicable to the Beltian, Cordilleran, Appalachian, and many other geosynclines and the multitude of sedimentary basins.

Gilluly (1964) constructed an isopach map of sediments of the Atlantic Coastal Plain, Continental Shelf, and Continental Slope, modifying it after fig.29 of Drake et al. (1959). He arrived at a figure of 190,000 sq miles for the mapped area of Triassic and younger sedimentary rocks for this area, and he computed the volume at approximately 280,000 cu miles from the 0-, 1,000-, 5,000-, 10,000-, and 15,000-ft isopachs. One is obviously dealing with hypersubsidence in this area, which Dietz (1972; also see previous discussion) termed a "living" geosynclinal couplet. Gilluly seemingly did not at that time enthusiastically embrace the postulate of westward drift of America from the Mid-Atlantic Ridge; in addition, he believed in the reality of a former "Appalachia". In his words (p.489): "The shelf, without doubt, was variably emergent during parts of Paleozoic time and a source of sediment for the Appalachian geosyncline." As far as geologic subsidence is concerned, Gilluly (1964, p.489—490) considered that available facts and attendant interpretations shed light on the origin of the Continental Shelf and stated: "Inasmuch as no sediment is likely to be more than two thirds as dense as the subcrustal material displaced, patently no amount of accumulating sediment can depress a surface of active sedimentation below its original level by purely isostatic forces. The old surface of subaerial erosion of Appalachia was thus not changed into a submarine surface of deposition merely by being loaded with sediment. Some subcrustal process must have brought about the submergence; isostatic adjustment to a growing sedimentary load would of course operate to continue the sinking." Herein lies the crux of the problem that relates to subsidence and hypersubsidence in continental-margin geosynclines: weight of wet, low-density (ca. 2.3) sediments can hardly be considered the principal operative force to *initiate* subsidence. It is obvious, however, that once subsidence has begun, taking the weight (sediment) from one "pan" of the balance and transferring it to the already sinking pan on the other arm will enhance subsidence. Gilluly (p.490) stressed this, as he added: "The Continental Shelf, therefore, is not due primarily to subsidence of an offshore area under a load of fluvial sediments discharged at the shore, as conceived by Kuenen (1950). The subsidence preceded much of the sedimentary loading. The loading must indeed have induced further subsidence, but it cannot account for the 5,000 ft of differential sinking of the landward troughs with respect to the median rise, on the seaward side of which even deeper basins occur. Five thousand feet of sediment cannot account for the difference in basement elevation between basin and rise when both columns are overlain by the same depths of water. The still greater subsidence of the troughs of the median rise is even harder to attribute to isostasy, for here the upper surface of the sediment in the troughs is in deep water, well down the Continental Slope." Gilluly believed that in order that former Appalachia could subside, the sialic crust beneath it must have been thinned; the thinning, he argued, must have been subcrustal. It was his contention that thinning was due to subcrustal processes that involved mass transfer, not just phase transformations. He argued for subcrustal flowage and erosional thinning of the sial by movements at its base. According to his interpretation (1964, p.490): "... the Continental Shelf is due to a complex of factors — sedimentation, isostatic response to loading, and subcrustal flow. I suggest that the thinning of the crust toward the foot of the Continental Slope is a result of subcrustal erosion by the mantle current, localized by the sedimentary load." Gilluly's thesis rests upon his interpretation of depth of the Moho discontinuity; he said (p.490): "The continuity of the upper surface of the basement and the general similarity of the seismic properties of the crust from the Piedmont to the Continental Slope suggest that the whole crust is sial. The M discontinuity here separating this sial from the mantle is almost surely not a phase transition, but a compositional boundary."

Determination of sedimentary volumes is of great significance in subsidence and hypersubsidence studies, particularly for the Atlantic Coastal Plain and the geosyncline adjacent to it (see also Fig.4-8). Gilluly et al. (1970, p.353) stressed the importance of studying volumes of sediments and stated: "Sedimentary volumes are of prime interest in many fields of geology: as measures of erosional rates, of geochemical balance, and recently, with the virtual demonstration of continental drift, as measures of movement of the continental and oceanic plates." Their studies led them to the conclusion that the volume of sedimentary rocks and unconsolidated sediments offshore of the Atlantic Coast of the United States, that can reasonably be attributed to erosion of the continent, amounts to 10×10^6 km^3. They contended that the sediment of the Appalachian geosyncline ". . . was supplied largely from the now drowned former land of Appalachia — perhaps from Africa. . . .". They believed that sediments of the Atlantic have probably been accumulating since the early Jurassic time, whereas those in the Pacific have been forming for a shorter period of time, perhaps since the middle Tertiary. They believed that North America has been over-riding the Pacific floor on a Benioff fault zone, as it drifted westward from a widening Atlantic. Their final statement is worthy of repeating (p.368): "The tremendous accumulation of sediment in the Gulf of Mexico need imply nothing as to continental drift; it is sufficiently explained by the contribution of the huge Mississippi drainage system." This geosyncline is discussed later in this chapter.

Gulf Coast

Central Gulf Coastal Plain area. The Gulf Coastal Plain is coextensive with the Florida Platform and the Atlantic Coastal Plain; a distance of 3,000 miles is involved from Cape Cod on the east to Tampico, Mexico. The three principal divisions of the Gulf Coastal Plain are: Tampico—Rio Grande region of east Mexico, the Mississippi embayment delta, and the Florida Platform. The writers discuss the central area that includes the Mississippi embayment delta region first, inasmuch as this is a major part of the Gulf Coast geosyncline and evidently has experienced hypersubsidence. Geosynclines have received the great bulk of sediments during the past and behave in a similar manner today. One can safely conclude, therefore, that these sedimentary basins experience hypersubsidence and, accordingly, accumulate the tremendous volumes of sediments. The central area of the Gulf Coast, which has been such a downwarped part of the earth's crust, still behaves as a negative mobile belt. The average increment of sediment that is transported to the Gulf by the Mississippi and other rivers amounts to about 2 million tons per 24-hour period. With this in mind, the writers briefly examine some of the sedimentary volumes, followed by a perusal of sediment thickness and significance of geologic subsidence.

So important is this area to geologists that one issue of the Bulletin of the Geological Society of America (63: 12-II, 1952) was devoted wholly to papers dealing with sedimentary volumes in the Gulf Coastal Plain. Objectives were clearly spelled-out: to study the distribution, thickness, volume, and general character of Mesozoic and Cenozoic sedimentary rocks of the Gulf Coastal Plain province rimming the western and northern shores of the Gulf of Mexico. Subsequently, the magnificent tome by Murray (1961), entitled "*Geology of the Atlantic and Gulf Coastal Province of North America*", has appeared. The interested reader should consult the latter for details and bibliography; one should also make reference to the "*Tectonic Map of the United States*" for additional aid.

In the 1952 paper mentioned above, Murray (pp.1177—1192) defined the Central Gulf Coastal Plain as the area between the Texas—Louisiana and Alabama—Georgia State lines south of the approximate latitude of Memphis, Tennessee, and Alabama between 85° and 94°W and 29° and 35°N. The Sabine River on the west and the Chattahootchee—Apalachicola River system on the east are natural boundaries. The subaerial part of the area constitutes about 145,000 sq miles, whereas the submerged portion to a depth of 10,000 ft includes about 140,000 sq miles. The gulfward-dipping post-Paleozoic sedimentary rocks underlie the entire area to an average thickness of about 2 miles. In some depocenters subjected to hypersubsidence, Mesozoic and Cenozoic deposits are from 7 to 8 miles in thickness. Murray (p.1178) said: "The rate of sedimentation and the rate of subsidence maintained approximate equality for extended periods of time as evidenced by the shallow depositional environment of the sediments." His interpolations, based on regional studies, indicated that the total volume of Mesozoic and Cenozoic deposits in the Central Gulf Coastal Plain exceeds 300,000 cu miles. By interpolation and extrapolation, utilizing well and geophysical data, Murray (p.1188) estimated that at least 200,000 cu miles of Mesozoic and Cenozoic deposits are present offshore. According to him, if the basement rocks slope up evenly from 7 miles down at the modern shore to a junction with the continental slope at 10,000 ft, a safe estimate can be made that about 500,000 cu miles of sediments are present beneath the submerged plain.

One of Murray's summary statements is noteworthy (1952, p.1188): "Regional studies reveal that Mesozoic and Cenozoic geosynclinal deposits accumulated in arcuate belts of varying thicknesses rimming the northern and western shores of the Gulf of Mexico. These sedimentary belts have a general shape of lenticularly flattened link sausages. The thicker lenses are areas or loci of maximum deposition (depocenters)." He interpreted the arcuate depositional belts from a structural standpoint as zones of regional subsidence and accumulation. The thicker lenses, he argued, appear to be areas of greater than normal subsidence, resulting from concentrated deposition at or near the mouths of major drainages. His studies indicated that strata have accumulated repeatedly in the depocenters to thicknesses of more-or-less 3,000 ft, before shifting of depositional locale. That is, depocenters are not known to occur vertically above immediately older ones. Murray (1952, p.1188) stated: "Such repeated shifts in sedimentary loci, after the accumulation of approximately 3,000 ft of strata, suggest this thickness of deposits is related to the maximum load and, accordingly, to the maximum depression that can be accommodated by an elastic crust in a specific area and time." He noted the economic significance of thick sedimentary sections: up to the time of his studies (more than a decade ago), the Central Gulf Plain had produced 12,500 barrels of oil and 36,000,000 cu ft of gas per cu mile, with proven reserves (then) slightly exceeding 13,000 barrels of oil and 111,900,000 cu ft of gas per cu mile.

Western Gulf Coastal Plain area. In the 1952 symposium alluded to above — *G.S.A. Bull.*, 1952, 63(II) — Colle et al. (pp.1193—1200) discussed the volume of Mesozoic and Cenozoic sediments for the Western Gulf Coastal Plain of the United States, whereas Guzmán (1952) similarly investigated sediments of the same age for the Mexican Gulf Coastal Plain. Considered together, this would involve an area rimming the west and northwest portions of the Gulf of Mexico from just north of Campeche at the south to the Texas—Louisiana State line on the east. Colle et al. (1952, p.1193) calculated a volume of 453,000 cu miles under a land area of 90,000 sq miles and 271,000 cu miles under the continental shelf area of 29,000 sq miles for the Western Gulf area. Thus, the total volume of Mesozoic and Cenozoic sediments is equal to 724,000 cu miles, representing a mean thickness of about 6 miles. Their two restored dip sections, one from Houston to the Gulf and the other from the Little Colorado district to the Gulf (their plates 2 and 3, respectively), show excessive downdip thickening southward into the depoaxis of the Gulf Coast geosyncline. Their Wilcox, Cockfield, Frio and other prisms seemingly aggregate from 25,000 to 30,000 ft of sediments. This represents a good example of hypersubsidence within various sedimentary basins. The reader is also referred to the work of Wilhelm and Ewing (1972).

In his paper relating to the volumes of Mesozoic and Cenozoic sediments, which accumulated in the Mexican Gulf Coastal Plain, Guzmán (1952) gave an estimate of 303,900 cu miles, excluding the Yucatàn

Peninsula. It was his interpretation that during the Mesozoic Era most of the Mexican Gulf Coastal Plain was a submerged shelf or foreland area along the eastern border of the great Rocky Mountain geosyncline, which extended from Central America through Mexico, the United States, and Canada. Guzmán noted that it was during the Lower and Middle Cretaceous and, later, in the Eocene time that the most significant volumes of sediments were dumped into the sedimentary basins. For example, some 65,800 cu miles of thick reefs, arkoses, and other clastics formed during the Cretaceous, whereas 92,400 cu miles of shales, sands and conglomerates accumulated during Eocene time. Guzmán stated (p.1201): "During Eocene time, several basins or foredeeps along the western margin of the present Coastal Plain received thick deposits of geosynclinal sediments (flysch and molasse), derived by erosion of the uplifted Mesozoic geosynclinal block." The isopach maps assembled by Guzmán for the Cretaceous sedimentary rocks clearly prove the prior existence of strong subsidence, whereas those for the Eocene indicate hypersubsidence (his fig.5, 7, 9). As shown in his fig.11 and 13, strong subsidence (just short of hypersubsidence) prevailed during the Oligocene and post-Oligocene.

It is not an easy matter to assess what effects subsidence and hypersubsidence have upon sediments that are at the bottom of a 10,000—30,000 ft prism. Some geologists do not consider that the weight of overlying sediments by itself can cause significant incipient metamorphism (i.e., "load" metamorphism). Inasmuch as the entire Gulf Coast province that is a part of the Gulf Coast geosyncline was a complex of sedimentary basins during Mesozoic and Cenozoic times (and still is) and in which oil and gas were generated and accumulated, the importance of compaction, cementation, and induration during subsidence cannot be overemphasized. In this connection, Jam et al. (1969) investigated subsurface temperatures in wells in south Louisiana. Temperature—depth curves were obtained for 123 fields. These authors reported that temperature gradients range from 18 to 36° C/km of depth, but most were between 22 and 24° C/km. They pointed out (p.2141) that: "At a depth of 3,048 m (10,000 ft) there is a belt of high temperature near the present coastline. This hot belt is located approximately where the sedimentary strata are believed to be of maximum thickness, which is estimated to be about 15,000 m. It is suggested that metamorphism and recrystallization already have begun in the lower part of the sedimentary section, and the consequent increased thermal conductivity may account for the high temperatures along the belt of maximum thickness." They were puzzled about the location of the prominent hot belt, for they made this observation (p.2149): "Most authorities (Meyerhoff, 1968) agree that the maximum thickness of Tertiary rocks and the greatest depth to basement should be near the present coastline. The thick sedimentary section should have lower thermal conductivity than the basement, and should have a blanketing effect on heat flow. The zone near the present shoreline thus should be cool, rather than hot." Distribution of temperatures at a depth of 10,000 ft in south Louisiana is shown in Fig.4-9. There is a rapid downdip thickening of Cenozoic sediments towards the south in the northern part of the Gulf Coast geosyncline. Apparently, oil and gas are generated below depths of about 7,500 ft in this depocenter. There is an abruptness in the prism thickening here. The tectonic hinge which separated the craton on the north from the geosyncline south of it was the belt of high temperature gradients as reported by Jam et al. (1969).

Bahama—Florida Platform

The vast Bahama Platform and much of peninsular Florida are discussed separately from either the Atlantic Coastal Plain or the Gulf Coastal Plain, because this vast area was not part of either geosyncline during Mesozoic and Cenozoic times, nor was its mode of subsidence and hypersubsidence comparable to the continental-edge depocenters. In addition, the Bahama—Florida Platform was almost wholly a carbonate depocenter (or series of related depocenters) that received most of its sediments *from the sea* and not from an adjacent continent or continents. The Bahama—Florida Platform is unique in its tectonic, sedimentotectonic, and subsidence patterns that differ markedly from the two great geosynclinal areas discussed in previous pages. Thus, it is treated as an important, discrete geologic entity in time and space.

Newell (1955) pointed out that the Bahamas include several more-or-less detached banks or platforms, each bounded by a steep submarine escarpment; he referred to them as the "Bahamian Platforms". He said (p.303): "Since the Early Cretaceous pure carbonate deposits have been accumulating in the area to a depth of more than 14,500 ft. It may be inferred from this that the region has been isolated from sources of terrigenous sediments. There is no compelling evidence of folding or faulting in later geologic times, and there is little reason to think that interruptions in the general subsidence have been frequent during the past 130 million years or so. Probably this downwarping has resulted from isostatic adjustment to a steady accumulation of carbonates." Newell concluded that the submarine profile in the Bahamas is a Late Tertiary coral-reef profile and that the banks probably originated during Cretaceous times as oceanic coral atolls, which were gradually incorporated into the North American continent by the spread and coalescence of calcareous deposits. He noted that (p.304): "Although separated from Florida by the moderately deep (300—500 fathoms) Florida Straits, the Bahamian Platforms are an extension of the Floridian Plateau."

Rocks of the Coastal Plain floor in the north peninsular Florida and the Florida panhandle are Precambrian and Early Paleozoic crystallines; they

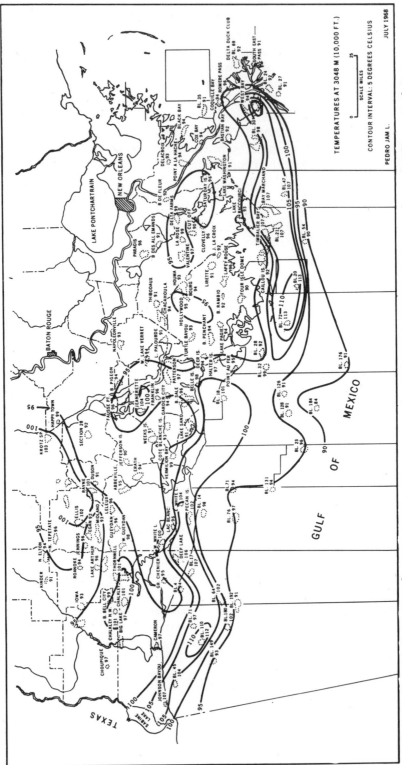

Fig.4-9. Temperature distribution in South Louisiana at a depth of 10,000 ft (3,048 m). Contour Interval (CI) equals 5° C. Hot belt with temperatures between 100 and 113°C is near the present shoreline. (From Jam et al., 1969. Courtesy of the American Association of Petroleum Geologists.)

have been encountered in holes drilled for oil there at depths around 2,600 ft. Oil well tests in south peninsular Florida, the Keys, and in the Bahamas have not been drilled sufficiently deep to penetrate pre-Coastal Plain rocks. In South Florida, the Fort Pierce Formation of Late Jurassic (?) or earliest Cretaceous (?) age is the oldest of the Coastal Plain rocks (Applin and Applin, 1965, p.1). In a test by Amerada Petroleum Corporation near Fort Pierce northeast of Lake Okeechobee, this formation was topped at a depth of 10,460 ft; the hole bottomed at 12,748 ft in igneous rock. Base of the Fort Pierce Formation there is at a depth of 12,680 ft; the formation consists of oolitic and other limestones, dolostones, and anhydrite, with red beds in the lower 250 ft. Five other tests by other companies encountered the Fort Pierce Formation progressively south, along the peninsula near Lake Okeechobee, West Palm Beach, Key Largo, and Big Pine Key. The test well drilled by The Gulf Oil Corporation at Big Pine Key (the western part of the Florida Keys) penetrated 15,455 ft of sediments, essentially below sea level. About 1,115 ft of carbonates and evaporites of the Fort Pierce Formation comprise the basal units drilled in this test. Applin and Applin (1965) provided an excellent discussion, maps, sections, and other data relating to the Comanche Series and associated rocks tested to 1965 in the subsurface in central and south Florida. Earlier, Pressler (1947) discussed occurrence of oil in Florida and presented a north—south restored section of Cretaceous and Cenozoic sedimentary rocks in peninsular Florida. He regarded south Florida and adjacent areas as an embayment from the Gulf of Mexico sedimentary basin. Accordingly, he proposed the name South Florida Embayment, considering it to have accumulated from 15,000 to 20,000 ft of sedimentary fill. Spencer (1967) published certain details of a deep test that was made in the northern part of Andros Island by The Bahamas Oil Company Ltd.; the Andros No.1 was drilled in 1946—1947 to a total depth of 14,585 ft, before being abandoned. Spencer indicated that Pleistocene, Pliocene, Miocene and Oligocene limestones and dolostones were penetrated in the upper 2,200 ft, Eocene in the next 2,440 ft, and Paleocene in the next 4,120 ft. Part of the latter, however, was questioned as being of Paleocene age. Upper Cretaceous rocks, which were penetrated from a depth of 8,760—10,660 ft, are mostly dolostones with minor limestones. The bottom 3,925 ft of rocks penetrated during the test are mostly dolostone, which was referred to the Lower Cretaceous. Spencer (1967, pp.266—267) and Kraft et al. (1971, fig.A-1, p.134) showed correlation between the rocks penetrated during the Bahamas deep test and the Cretaceous and Cenozoic rocks in South Florida. Winston (1971) stressed the petroleum possibilities of this huge carbonate platform, whereas Applin and Applin (1965) provided a wealth of data on subsurface carbonates and evaporites for south peninsular Florida. Thus, Pressler's (1947) South Florida Embayment was probably an area of hypersubsidence for accumulation of mostly nonclastics. Figures 4-10 and 4-11 represent some of the data accumulated by Applin and Applin (1965) relative to what is termed the Florida—Bahama Platform, a distinctive carbonate depocenter. Viewing this huge platform of dominantly marine carbonates from the standpoint of petroleum potential should provide food for thought among personnel of oil companies. It has been pointed out (Chilingar et al., 1972) that as of early January of 1968, the estimated proved oil reserves for the world were 399.5 billion barrels. Figures are not available indicating precisely what percentage of those reserves lie pooled in carbonate rocks; however, the authors estimate that somewhere between 50 and 60% of the world petroleum reserves are contained in carbonate rocks. The writers believe that large amounts of oil will be found in carbonate rocks in the future. The Felda Field, Sunniland Field, and Forty Mile Beds Field of South Florida (all south of Lake Okeechobee) produce oil from carbonates of Cretaceous age (Winston, 1971).

When viewed in the light of geologic subsidence and hypersubsidence, the Bahama—Florida Platform ranks with many geosynclines of the earth as a region of great negative mobility. It was a subsiding platform upon which shallow-water carbonates accumulated to a thickness in excess of 19,000 ft; this alone sets it apart from some other depocenters as a unique example of subsidence. Various geologists have investigated this area in an attempt to arrive at a reasonable explanation for its sediment types and reason(s) for subsidence. Dietz et al. (1970) concluded that the Bahama platform and, probably, the adjacent Blake Plateau and southern tip of Florida were laid down on an oceanic crust. The excessive section of more than 5 km of flat-lying Cretaceous carbonates implies an unusual geotectonic history during which great subsidence was characteristic. Dietz et al. (1970, p.1915) stated: "We suggest that this platform is underlain neither by sial nor by a volcanic foundation creating a 'mega-atoll'. Instead we propose a basement of oceanic crust about 11 km down which has undergone slow subsidence. Triassic rifting in the Atlantic probably initially created a closed small ocean basin, or mediterranean, in the Bahama region. This was accomplished by a rotational movement of the North American plate away from North Africa accompanied by shearing across the top of the South American plate which remained stationary and attached to Africa. A wedge-shaped sphenochasm resulted which became a sediment trap within Pangaea and quickly filled to sea level with turbidites. With renewed continental drift, the Bahama platform became a subsiding marginal plateau attached to the North American craton upon which algal-coral growth explosively flourished under holo-oceanic conditions, providing sufficient upbuilding to offset subsidence and maintain a sea level freeboard." These geologists pointed out that, based on some Cuban—Soviet geophysical data, the inference can be made that there is a thick sedimentary section beneath the Bahamian carbonates. In their words (p.1926): "We infer the presence of a hidden or sequestered Lower Mesozoic clastic section for which we suggest the name Bahama Cryptobasin. If its existence can be verified, this basin would seem to offer bright prospects for oil and gas with deep drilling. We infer that the clastics would be largely Triassic. The ±35,000-ft-thick sequence of Jurassic—Triassic red beds of the Cayetano Formation of northwestern Cuba may represent an upthrust and westward-transported (with Cuba) outcrop of the inferred Bahama Cryptobasin."

Uchupi et al. (1971) investigated the structural framework of the southeastern Bahamas through seismic profiler, magnetic and gravity data. In their scheme of interpreting the structure and origin of the southeastern Bahamas, the 20-km crust beneath this area would consist of a lower section more than 10 km thick made of oceanic crust (including oceanic basement and seamounts). They stated (p.702): "Above this basement is a carbonate section 5—10 km thick. If the continents began to separate in the Middle Mesozoic, no sedimentary rocks older than Early Jurassic, or possible very Late Triassic, should be present in the southeastern Bahamas. In the northwestern Bahamas the crustal section would consist of an oceanic crust about 10 km thick capped by a 20-km terrigenous—carbonate section. If the trough on which the northwestern Bahamas lie existed before the continents spread apart, then part of the terrigenous sediment fill is pre-Mesozoic." Obviously, the oceanic model presented by Uchupi et al. (1971) differs from that proposed by Dietz et al. (1970) as presented above. The model of Dietz et al. calls for the entire Bahamas to be formed in an intercratonic basin. Under the scheme of Uchupi et al. (1971) only the northwestern Bahamas were formed in this fashion, and the southeastern Bahamas were formed along a fracture zone generated during rifting.

To date, no well data are available to test the validity of the two postulated origins for the Bahama—Florida Platform presented above. Both hypotheses, however, suggest that the Bahama Escarpment, like the Blake, Campeche, and Florida Escarpments, was formed through carbonate accretion by reefs and other processes that prevailed during the Mesozoic and Cenozoic Eras. If the carbonate apron in the northwestern Bahamas is as much as 10 km thick (a reasonable postulate) and the crust beneath this apron is only 10 km thick, then hypersubsidence occurred throughout Mesozoic and Cenozoic times. Some geologists believe that the crust here is thicker, possibly at least 20 km. The schools of thought championed by Dietz et al. (1970) and by Uchupi et al. (1971) are only two among those prepared recently by several groups of geologists who postulate that formation of this unusual and unique depocenter, with an excessive carbonate pile, was related in some manner with mechanics of plate tectonics.

Krivoy and Pyle (1972) discussed the anomalous crust beneath the West Florida shelf with the assistance of a new Bouguer gravity anomaly map. They pointed out that 20,000 sq km of the shelf area is characterized by anomalies greater than +30 mgal. This, they thought, indicated that this part of the shelf is underlain by a crust having a thickness intermediate between that of continents and that of oceans. They stated (p.107): "A transition from oceanic toward continental crust in this area may have been accomplished by reef progradation across an ancient oceanic embayment. Alternately, a transition from continental toward oceanic crust may have been produced by rotation of Florida and consequent rifting. The reef-progradation hypothesis is most consistent with what is known of the deep structure and tectonic setting of the Florida Platform." These scientists did not believe that the area of high positive Bouguer gravity anomaly west of Florida has subsided any more than the surrounding parts of the Florida Platform, nor did they subscribe to "oceanization" of the area. They did, however, point out that possibly the great relative subsidence has not yet occurred, and that the area of high positive Bouguer anomalies suggests a region of denser basement and mantle rocks that eventually will undergo significant relative subsidence. The anomaly, they noted, has been associated with evidence of arching of Lower Cretaceous strata (Winston, 1971), which, in turn, suggests additional oil and gas possibilities in the deeper buried Cretaceous carbonates. Further geophysical investigations, followed by offshore drilling, should throw light on the origin of this anomalous crust beneath the west Florida shelf. It is significant to remember that the area beneath the shelf west of the State of Florida is much greater than the emerged peninsula. Thus, this huge submerged block and its flanks may provide a setting for economic oil and gas accumulation, possibly of greater dimensions than the few producing fields in South Florida.

Ouachita geosyncline

Geologists have differed in their interpretation and discussion of the geology of the Ouachita Mountains in relation to geosynclinal and/or other settings. For example, Eardley (1951, pp.206—218) discussed the Ouachita, Marathon, and Coahuila Systems in one chapter. On his coloured plate 5 (tectonic map of the Mississippian), Eardley depicted a basin for Mississippian and earliest Pennsylvanian times for eastern Texas, western Louisiana, and southern Arkansas, in which at least 15,000 ft of "Ouachita facies" accumulated. So interpreted, this would have been a dynamic basin that experienced hypersubsidence. By Pennsylvanian time (Eardley, 1951, plate 6) the Arkansas Basin occupied northern Mississippi and Arkansas, and extended westerly into eastern Oklahoma. At least 10,000 ft of sediments accumulated in this basin.

Cline (1970) examined the sedimentary features in the Oklahoma portion of the Ouachita Mountains, focusing particular attention on the Late Paleozoic rocks. He indicated that from the Late Cambrian to Middle Mississippian time the Ouachita geosyncline was a starved basin, accumulating only 5,000 ft of graptolitic shale, chert and novaculite, at the same time that thicker carbonates were accumulating in the Arbuckle area to the west. Cline pointed out that during Osage Mississippian time rapid subsidence in the Ouachita geosyncline, coincident with active tectonism in Llanoria to the south, resulted in the rapid deposition of perhaps as much as 25,000 ft of alternating sandstones and shales. As a result of his studies of numerous sedimentary features preserved within this thick prism of Meramecian and Chesterian age clastics, Cline concluded that these are turbidite deposits. Sandstones were derived from marginal sources, but once in the geosyncline they were transported axially. Cline (1970, pp.86—87) stated that: "Subsidence in the Ouachita geosyncline, accompanied by flysch sedimentation, continued until at least Mid-Pennsylvanian time. Sometime later, possibly at the end of the Missourian Epoch and/or perhaps in the Permian, the sediments of the geosyncline were deformed and moved northward in a series of imbricating thrust sheets ... This is a pattern repeated in many places in the world where geosynclinal flysch sediments are thrust against the adjacent craton ...". In reconstructing the depositional environment, Cline (p.100) concluded that: "The Ouachita environment is, therefore, interpreted as having been one of sedimentation in a deep trough in which the deposition of dark gray muds was frequently interrupted by density currents which debouched sands from lateral sources to the south and from a shelf area to the north. There is evidence that the lateral margins of the trough were steep and that subaqueous gliding and slumping brought in exotic limestone masses to be deposited along with the Johns Valley turbidites."

One of Cline's graduate students, Chamberlain (1971), also studied the bathymetry and paleoecology of the Ouachita geosyncline in southeastern Oklahoma and used trace fossils as well as other data in his interpretations. He summarized (p.34) the findings as follows: "The persistence of widespread, distinctive trace-fossil assemblages in the central Ouachitas, mainly feeding trails and burrows, through approximately 25,000 ft of Mississippian—Pennsylvanian rocks, attests to the continuous hospitality of the flysch environment in a basin that remained deep despite isostatic and eustatic fluctuations." Cline, Chamberlain, and others who have studied the Ouachita geosyncline have not explained the mechanics of subsidence of this depocenter. They believed in the formation of a deep (a foredeep north of Llanoria, perhaps?), which was a starved basin from Cambrian into Mississippian times and then was progressively filled (in part at least) with flysch facies that were rapidly dumped into the trough. Does the Ouachita geosyncline fit the model of Dietz (1972) discussed earlier in this chapter? The problem, of course, is that there is a lack of objective information concerning the time and mechanics of opening of the Gulf of Mexico and just how Llanoria fits into the picture. If interpreted in the light of mechanics of plate-tectonic theory, Llanoria (akin to Appalachia) was part of the Africa—South America plate that was juxtaposed against the North America plate, and was a source for the flysch sediments of the Ouachita geosyncline.

The Ouachita geosyncline was not the only depocenter of hypersubsidence in Oklahoma. The Anadarko foredeep of western Oklahoma was recently tested when the deepest well in the world (to date) was drilled, with the total depth being slightly more than 30,000 ft (T. D. = 30,050 ft). The hole was spudded in Permian sedimentary rocks and bottomed in the Viola (Ordovician) carbonates. In this part of the Anadarko Basin, which evidently is a dynamic basin, it would appear that one is dealing with a great prism of shallow-water sediments. The basin subsided during sediment accumulation. Bathymetry was not that of the area of study by Cline, Chamberlain, and others. Contrarily, the sediments penetrated in this deep test formed in shallow, mostly marine waters of the Anadarko basin. Kamen-Kaye (1972, p.13) was impressed by this world's deepest test and stated: "... I submit that it was ... a dramatic demonstration of one of the simplest geological phenomena we know — subsidence." Kamen-Kaye continued by saying (p.13): "... what we have in the Anadarko sedimentary foredeep of western Oklahoma is an absolute subsidence which could be almost as great as 30,000 ft. The subsidence would be less if a few degrees of dip should be involved at the wellsite, but the amount would almost certainly exceed 25,000 ft. Also, as the definition of a sedimentary foredeep requires, there is an asymmetric front. This brings the Viola up in the Wichitas, no more than 10 miles to the south. Reversing the direction of view we could say that subsidence on the asymmetric side of the Anadarko sedimentary foredeep takes place at the rate of almost 3,000 ft per mile. The crust of part of the Anadarko area has gone down abruptly, as though a relatively small cork had been pulled at some semi-fluid level of the crust of upper mantle."

Kamen-Kaye (1972) then hypothesized concerning possibility of subsidence up to 30,000 ft for the shallow-water sediments of the northern Gulf of Mexico, calling attention to the fact that some 19,000 ft of carbonates accumulated on the subsiding "platform" south of Florida. He added (p.13): "The super-deep well in western Oklahoma remains for the present the only actual proof we have of tremendous subsidence, and at best can apply only to a relatively small epeirogenic element such as the Anadarko downwarp. Deep wells in the Valverde downwarp of West Texas show that similar conditions may exist in another relatively small epeirogenic element. There appear to be others like these two within the continental United States. For great subsidence over great area we must wait for the results of widespread deep drilling completely across the Gulf of Mexico. When these results are in, one could expect to find that subsidence varying from considerable to tremendous had taken over an area as large as 600,000 square miles. Within such an area subsidence probably would not be less than 15,000 ft, or not less than almost 3 miles." In looking for a cause to such excessive subsidence, Kamen-Kaye asked some thought-provoking questions concerning the oceanic crust, and in conclusion he stated (p.14): "As I ponder the meaning of the super-deep well drilled in western Oklahoma I wonder where there is room for reconstruction of the mobility of oceanic crust in the vertical direction." Kamen-Kaye's comments may not be received with favor among some geologists and geophysicists, but then during the past decade many scientists have found it expedient to abandon, or at least modify, some of their theories.

Paleozoic sedimentary rocks of Texas

Most of the Paleozoic Erathem is documented in prisms of sediments which seemingly accumulated in shallow-water marine depocenters over portions of the earth's crust that experienced excessive subsidence. A few examples are cited and brief mention is made of the sedimentary record in this section.

First, attention is focused on the Texas Panhandle and the so-called Amarillo Arch. During most of the Paleozoic Era, this area was a dynamic basin, possibly an epeirogen. It was the depocenter of a prism of shallow-water sediments in excess of 20,000 ft in thickness. Carbonates, shales, and sands accumulated in this subsiding basin from Cambrian into Wolfcampian Permian time. During the remainder of the Permian, salt, gypsum, anhydrite, and some redbeds also formed. Some geologists interpret this region to have been the southern, and more mobile, side of the Ardmore—Anadarko geosyncline. Furthermore, it may have been connected in an east—southeasterly direction with the Ouachita geosyncline. Thus, what some geologists term the "Amarillo Arch" is part of the Wichita trend system. Accordingly, from the southeast to the northwest the latter would include the Criner Hills, Wichita Mountains, Amarillo Arch of the Texas Panhandle, and then would extend westerly to the Wet Mountains and Uncompahgre area of Colorado. It is not entirely unlikely that the "Amarillo Arch" area began to subside as a graben or a half graben, with the master fault on the south. Seemingly, subsidence of the trough was accompanied by upthrusting of the southern block. This would hardly be classed as a zeugogeosyncline, or "yoked-basin" in the concept of some geologists. Yet, it is akin to a yoke of oxen pulling a load, i.e., as the head of one ox goes down, that of the other goes up. Some 20,000—22,000 ft of Cambrian through Permian sediments accumulated in this sedimentary basin. Whether this was (1) a graben or half graben throughout essentially most of the Paleozoic, (2) part of a geosyncline (call it zeugogeosyncline, taphrogeosyncline, or whatever), (3) an asymmetric basin, or (4) some sort of combination of belts of vertical mobility, the fact remains that this is another excellent example of geologic subsidence on an excessive scale, i.e., hypersubsidence.

Any discussion of geologic subsidence and hypersubsidence would be incomplete were the Midland and Delaware Basins of West Texas to be omitted. Accordingly, Fig.4-12 has been constructed as a restored cross section in a general east-to-west direction across part of West Texas. Numerous geologists refer to this area as the "Permian Basin of Texas and New Mexico", which is a justifiable name because it identifies basins in which petroleum has been generated and then accumulated. Galley (1958) discussed the evolution through time of this area of basins, noting that approximately 16,000 ft of Upper Cambrian through Upper Permian sedimentary rocks comprise the prism of this downwarp. He observed that Pre-Mississippian strata consist chiefly of carbonate deposits that formed in shallow, clear marine waters. Near the end of Mississippian time the tectonic environment changed from one consisting of broad structures having gentle relief on the craton to an almost closed basin surrounded by mountainous areas, the largest of which was the Ouachita—Marathon complex. For the remainder of the Paleozoic, the area assumed the tectonism of the Marathon geosyncline; shales and sandstones accumulated in this subsiding depocenter. Permian times were characterized by deep (at times and places) and areally restricted marine basins of clastic deposition. They were stagnant at depth and surrounded by shoal platforms on which thick masses of carbonates accumulated; shallow lagoons extended to the shorelines. As shown in Fig.4-12, there was a broad "platform ridge" over which Permian carbonate reefs grew and flourished. These reefs store most of the oil found to date in the Permian basin of West Texas and southeastern New Mexico. When it is realized that the petroleum reservoirs of the Golden Lane of Mexico are found in similar reefal deposits (in Lower Cretaceous carbonates), and that similar reef development took place (and still is going on) in the Bahamas, Florida Keys, Yucatan and other places, the economic importance becomes clear. Furthermore, the extensive Florida Platform, that extends westerly into the Gulf of Mexico from emerged peninsular Florida, may also contain just such reef trends in the Jurassic and Cretaceous carbonates of the continental shelf there.

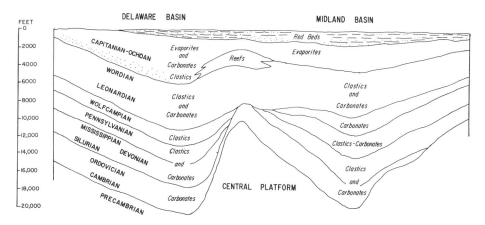

Fig. 4-12. Restored cross section, east—west, in part of West Texas, showing magnitude of subsidence in the Midland and Delaware Basins. (From various sources.)

In discussing the geologic subsidence in Texas, it is apropos to mention the downdip sediments of the Houston area. Sands and other clastics of the Wilcox, Cockfield, Frio (all Eocene), and basal Miocene units demonstrate that productivity trends occur in a pattern. That is, each productive zone occurs successively southward of the preceding one, as the axis of each successive basin and related hinge line shifted southward. There are numerous other examples of this progressive shift of depocenters with time in the Gulf Coast geosyncline. A substantial body of objective data, which relates to the Cordilleran miogeosyncline, was presented previously in this chapter, showing the shift in basins of maximum subsidence and sedimentation throughout Mississippian, Pennsylvanian, and Permian times (refer to Fig. 4-5—4-7).

The relationship of Texas clastic and carbonate sediments in subsurface is apparent: with subsidence and concomitant draping of reefs (particularly the pinnacle reefs), petroleum was generated and forced to migrate and accumulate in reservoirs. Chilingar et al. (1972, p.19—20) pointed this out for Texas and other areas of carbonate—clastic couplets.

Subsidence within the various cratonic basins, foredeeps, geosynclines, and other depocenters in the western half of Texas and eastern New Mexico was quite varied. Hypersubsidence occurred in some basins. Bathymetry evidently ranged from "knee-deep" to bathyal marine depths. As Galley (1958, p.395) pointed out: "The principal sites of oil generation probably were in the basins; reservoirs include basinal sandstone and platform carbonate and sandstone rocks." He noted that the Guadalupian-age rocks produced more than half of all the oil that had been found up to about 15 years ago in the Permian basin. Petroleum companies are continuing to probe deeper stratal units in basins of hypersubsidence in Texas and other areas in search for oil and gas. Kamen-Kaye (1972, p.13) stated: "Depending upon the interplay of events and circumstances, we may see in the relatively near future wells drilled as deep as 30,000 ft in the northern waters of the Gulf of Mexico, that is, in the tract of terrigenous lithofacies of the Gulf Coast 'geosyncline'." In view of the above discussion, the writers coined the following credo: *Better things through geologic subsidence!*

Four Corners area

Paradox Basin

The Four Corners area of Utah—Colorado—Arizona—New Mexico was part of the craton throughout the Paleozoic—Early Mesozoic times. Some areas were depressed into intracratonic basins, whereas others were uplifted forming source areas. The Paradox Basin (one such depressed area) was a southeast—northwest trending half-graben (taphrogenic basin), situated mostly in southwestern Colorado and southeastern Utah (the interested reader is referred to the papers of Wengerd, 1962; Baars, 1966; Peterson, 1966; Peterson and Ohlen, 1963; and others, for a wealth of details relating to maps, thicknesses of sedimentary rocks, sediment types, oil accumulation, etc.). During Late Precambrian times, certain fault lineaments were operative, thereby setting the stage for intracratonic basin subsidence and concomitant sedimentation in the Paradox Basin, as well as in the Black Mesa Basin south of it in northeastern Arizona and the San Juan Basin of northwestern New Mexico. The Uncompahgre uplift was an upfaulted block east of the Paradox basin, whereas the Nacimiento, Defiance—Zuni, and Kaibab uplifts farther south were positive areas from which clastic sediment was washed into the depocenters. Paleozoic carbonates, arkoses and other clastics, black shales, reef rocks, and evaporites total more than 10,000 ft in this thick sedimentary prism. This basin of subsidence was fault-controlled in large measure. The major subsidence of the Paradox Basin occurred during Pennsylvanian time concomitantly with upfaulting of the Uncompahgre block.

The sedimentary basin for Pennsylvanian and Permian times in southwestern Colorado and northern New Mexico, in a northeasterly direction across the Uncompahgre Highland, was a *zeugogeosyncline* (Brill, 1952). It was a unique trough among similar basins in western continental United States in that it was a rapidly subsiding basin with upfaulted blocks (highlands) on both sides. It was truly a yoked basin; that is, as the southeast—northwest aligned basin subsided, the adjacent blocks were uplifted to form highlands. Brill (1952) measured a total thickness of 13,170 ft of Late Paleozoic marine and nonmarine sedimentary rocks in this trough; they are Morrowan through Virgilian (Pennsylvanian) and Wolfcampian through

Guadalupian (Permian) ages. When a postulated thickness of some thousands of Pre-Pennsylvanian sedimentary rocks is added to the 13,170 ft of Permo-Pennsylvanian sediments, it becomes readily apparent that this taphrogenic basin, or zeugogeosyncline of some authors, experienced rapid subsidence, eventually amounting to hypersubsidence of yet another type of depocenter.

San Juan Basin

Wengerd (1958) discussed the San Juan Basin as "... an ovate, frontal, intermontane, structural basin slightly less than 20,000 sq miles in area, encompassing almost 36,000 cu miles of sedimentary rocks within the Dakota outcrop and above the Precambrian basement. The basin occupies most of northwestern New Mexico and a part of southwestern Colorado. The area of the basin was a southeastern shelf and positive area genetically related to the Early Paleozoic Cordilleran geosyncline... In Late Cretaceous time, the area subsided gently as a highly oscillatory sedimentational shelf related to the Rocky Mountain geosyncline, with sediments moving into the shelf environment from westerly directions." Something of the order of approximately 15,000 ft of Paleozoic, Mesozoic, and Early Tertiary sediments accumulated in the San Juan Basin, thereby qualifying it for a basin of hypersubsidence. The San Juan Basin seemingly fits the description of a dynamic basin as given by Dallmus (1958, p.884—898).

Rocky Mountain geosyncline

Reference to one of the current textbooks of Historical Geology will indicate that the Rocky Mountain seaway, or geosyncline, had its inception subsequent to the termination of the Cordilleran geosyncline. Its western shoreline, however, was approximately in the same geographic position as the eastern shoreline of the Cordilleran miogeosyncline. Stated otherwise, the tectonic hinge line (Las Vegas—Wasatch line), which separated the Cordilleran geosyncline and the shelf and craton east of it, now during the Mesozoic and into Early Cenozoic time was the dividing line between the Rocky Mountain geosyncline and the so-called Mesocordilleran geanticline (and Sevier orogenic belt) that existed west of it. By Mid-Cretaceous time, the Rocky Mountain geosyncline divided the North American continent into two portions. Earlier, in Early Cretaceous time, some 15,000 ft of sediments accumulated in the Mexican geosyncline that formed south of the Rocky Mountain seaway (see Eardley, 1968, plates 11 and 12).

Chapter 1, "The Mobile Belts", in Bucher's "*Deformation of the Earth's Crust*" (1933, pp.8—10, fig.1) dramatically emphasizes facts of subsidence and hypersubsidence of the earth's crust beneath a trough-like depocenter of the Rocky Mountain geosyncline. Almost 30 years ago,

Reeside (1944) presented some additional factual data proving that excessive geologic subsidence characterized this Mesozoic geosyncline, far in excess of the Mexican geosyncline farther south. Reeside compiled 10 maps; although geologists have revised some of his isopach maps, the changes are largely due to differences in interpretation and do not markedly alter the fact that hypersubsidence took place. Significant facts that these and other isopach maps reveal are that some subsidence occurred in foredeep-like troughs or other depressions directly east of the Wasatch Hinge Line during Paleozoic time and that tremendous quantities of coarse- and medium-textured clastics were derived from a westerly source. Considerable quantities of sediments were derived from the Sevier orogenic belt (see Armstrong, 1968, 1972). Isopach maps reveal that from 10,000 to 25,000 ft of conglomerates, sandstones, shales, coal beds, and minor carbonates accumulated in various rapidly-subsiding depocenters within the Rocky Mountain geosyncline.

Numerous geologists since the days of Bucher (1933) have wondered about this "teeter-totter" phenomenon (up-and-down movement) of the western continental United States. In examining some of the facts relating to vertical movements in the western Wyoming overthrust belt, and pondering over possible isostatic effects, Crosby (1968) made this observation (p.2000): "The miogeosyncline in southeastern Idaho subsided slowly and filled with about 50,000 ft (15,210 m) of Paleozoic and Mesozoic sediments, whereas the Wyoming shelf remained relatively positive. During destruction of the miogeosyncline by mountain building, more than 20,000 ft (6,100 m) of Cretaceous sediment, derived from some of the oldest formations in the rising highlands, accumulated in a foredeep in western Wyoming." Crosby noted that during destruction of the miogeosyncline, a linear or semilenticular basin characteristically develops at the edge of the craton and is progressively filled with debris shed from rising tectonic lands. He commented that this subsiding surface has been termed a "marginal basin" by European geologists, and is the "exogeosyncline" of Kay (1951, p.17). Eardley (1967, p.39) termed such a feature a "foredeep basin". As Crosby pointed out (p.2000—2001), the detrital sediments include much coarse syntectonic material, and aggregate thickness of sedimentary rocks ranges from a few thousand feet to more than 5 miles (8 km). Crosby (p.2001) stated: "Subsiding foredeep basins are complementary to the tectonic lands rising from the miogeosyncline; thus, adjacent areas are moving vertically, but in opposite sense. Net horizontal movement during overthrusting and folding is from lands rising to areas subsiding, and occurs during the time of most pronounced vertical movement. Much of the foredeep sedimentary rocks eventually may be overlapped and included in the youngest compressional structures." Such phenomena were stressed by Spieker (1946) in his discussion of Late Mesozoic and Early Cenozoic history of central Utah, and also by Armstrong (1968; 1972) for synorogenic events along and east of the Nevada—Utah—Idaho Sevier orogenic belt. Both of these geologists investigated syntectonic events along the western margin of the Rocky Mountain geosyncline.

Crosby (1968, p.2007) believed that the Cordilleran miogeosyncline began to break up in Late Jurassic time, and by the end of the Mesozoic Era the area occupied by the basin in southeastern Idaho had attained considerable height. He stated (p.2007): "The Pre-Paleozoic surface that had subsided nearly 50,000 ft (15,220 m) rebounded beyond its former position to become a marked positive feature. Coincident with uplift in southeast Idaho, a marginal trough in western Wyoming subsided rapidly and filled with more than 20,000 ft (6,100 m) of sediment. This is a relative vertical uplift of more than 70,000 ft (21,360 m) between the two adjacent areas." Isostatic considerations, based on the analysis of 166 gravity data from both sides, led Crosby to the interpretation that the latest phase of opposed vertical movements is still active. He did not theorize, however, at great length concerning causal processes, but he did suggest more than one possible processes, for he added (p.2013): "Whatever the process that produced the vertical movements of the surface in this region, whether it be phase changes at the crust—mantle boundary . . ., horizontal transfer of deep crustal material . . ., migration of partial melts . . ., subsidence due to ascent of magmas to the surface . . ., mantle diapirs . . ., or some unknown process, it must be slow and progressive through time." He made no mention linking this area to mechanics of plate tectonics. Eardley (1968) described the major structures of Colorado and Utah and presented a theory of origin of vertical uplifts. He stated (p.79): "It proposes that the ancestral Rockies and the more modern ones of Cretaceous and Early Tertiary age of both the shelf of Colorado and eastern Utah and the miogeosyncline of western Utah are the result of vertical uplifts of the silicic crust. The uplifts are caused by the rise, from the upper mantle, of basalt in scattered places to the base of the silicic crust. This rise domed the silicic crust and the overlying sedimentary veneer." He also noted (pp.84—85): "The evidence seems convincing that the uplifts of the shelf province are caused by vertical pressures and this, together with the general broad oval shape of the uplifts, leads directly to the postulate of a large intrusion beneath each in the silicic crust."

Literature pertaining to various basins in Wyoming, Montana, and the Dakotas is so voluminous that any attempt to summarize in this chapter the timing, extent, and amount of subsidence and hypersubsidence would fall short of adequate discussion. Some excellent papers for this area appear in the symposium edited by Weeks (1958), two particular Bulletins (Vol. 50(10), 1966; Vol. 51(10), 1967) of the American Association of Petroleum Geologists, numerous guidebooks, and other publications. Noteworthy are the three papers by the following authors: Armstrong and Oriel (1965), Furer (1970), and Wilson (1970). The first-named geologists suggested that the Gannett Group (Lower Cretaceous) is a synorogenic deposit in the Idaho—Wyoming belt. This may indicate that a "foredeep basin" (see Crosby, 1968, pp.2000—2001) was forming along the western margin of the Rocky Mountain geosyncline at least by Early Cretaceous time. The writers believe that it experienced subsidence in Jurassic time, which continued into Early Cretaceous time to accumulate the thick prism of conglomerates and other clastics of the Gannett Group. Furer (1970) studied nonmarine Upper Jurassic and Lower Cretaceous rocks of western Wyoming and southeastern Idaho. He provided much additional information relating to the petrology and some on stratigraphy of the Morrison (Jurassic) Formation and Gannett Group (Early Cretaceous); however, the writers do not agree with some of his interpretations of the paleotectonics. Seemingly, he envisioned that the Cordilleran miogeosyncline was perpetuated into Cretaceous time, because he stated (p.2283): "During the Paleozoic Era western Wyoming was near the 'hinge line' between the Cordilleran miogeosyncline on the west and the stable craton on the east. Generally, sediment was shed westward from the craton into the subsiding trough during numerous transgressions and regressions of the sea. During the Jurassic the sea invaded the area from the north and west. . . . After deposition of the Twin Creek Limestone, in Middle Jurassic time, the subsidence of the trough almost ceased. The overlying Preuss—Stump—Sundance sequence was deposited in a shallow sea that subsided only slightly. The basin remained stable until the deposition of the Gannett Group during Early Cretaceous time." Apparently Furer considered that the Cordilleran miogeosyncline continued into the Cretaceous by merely expanding as a trough eastward, because he stated (p.2298): "By the Early Cretaceous, either the pattern of winds changed or else new volcanic centers developed as volcanism in the Cordilleran geosyncline became more widespread in the north."

Wilson (1970) made an in-depth study of the Upper Cretaceous—Paleocene synorogenic conglomerates of southwestern Montana, which are slightly more than 10,000 ft in thickness. He concluded that the major source area was an elongate chain of uplifts in southwesternmost Montana and easternmost Idaho. His map (fig.1, p.1844) depicts a chain of comparable-age Upper Cretaceous and Paleocene synorogenic conglomeratic units from near Drummond, Montana, on the north, to central Utah on the south. The latter area is the one which was studied by Spieker (1946, 1949). Subsequently it has received the attention of numerous geologists and was recently studied by Armstrong (1968, 1972). Figure 4-13 consists of four block diagrams showing four stages in the evolution of a segment of the

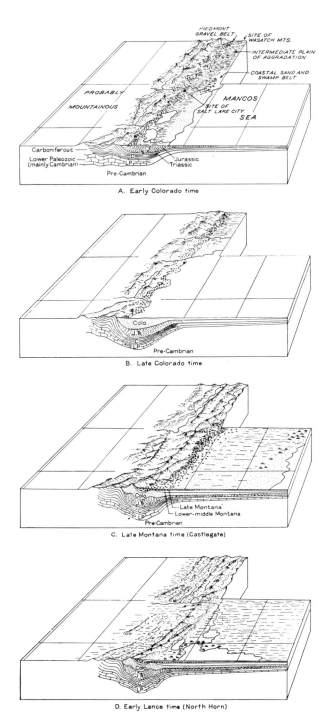

Fig.4-13. Block diagrams showing evolution of a segment of the western part of the Rocky Mountain geosyncline (Utah, north of latitude 39°30') during Late Cretaceous time. This figure demonstrates that geologic subsidence to hypersubsidence occurred along an area that during Paleozoic time was a craton. (After Spieker, 1946, fig.21, p.157.)

western part of Rocky Mountain geosyncline in central Utah, whereas Fig.4-14 is a diagrammatic section, west—east, across this geosyncline also in central Utah. Spieker (1949, pp.13—14) tabulated some thicknesses for Jurassic, Cretaceous, and Paleocene marine and nonmarine sediments that accumulated in this "foredeep basin". The minimum and maximum total sediment thicknesses are 15,850 ft and 42,700 ft, respectively. Of the latter total thickness, 13,000 ft belong to the Jurassic, 25,200 ft are assigned to the Cretaceous, and the remaining 4,500 ft of sediments are of Paleocene age. In this depocenter, which experienced hypersubsidence during the Jurassic, Cretaceous, and Paleocene times, the prism of sedimentary rocks is dominantly of shallow-water marine origin, with substantial thicknesses of fluvial deposits, mudflows, and other nonmarine materials. Uppermost 200—1,500 ft consist of the Flagstaff Formation (Late Paleocene) and a succession of lacustrine limestones, shale, sandstone, and minor volumes of volcanic ash. These deposits may be termed "frosting on the cake" (so to speak) and, according to some geologists, are not part of the Rocky Mountain geosyncline. It is also important to point out that the maximum thickness of 42,700 ft tabulated by Spieker does not occur in one single locality, but represents an aggregate of various stratigraphic sections that are exposed in the Spanish Fork Canyon, Wasatch Mountains, and then southerly in the Wasatch Plateau and adjacent areas on the west in San Pete and Sevier Counties, Utah. The significant conclusion that can be drawn from the wealth of data that this prism of sediments provides is that in an area of former hinge-line stability (near the Wasatch Line), which was part of the craton during Paleozoic into mid-Triassic time, a complete reversal, in the vertical sense, took place. At least by Jurassic time what had been the craton started to experience the early stages of negative mobility. This area continued as a foredeep basin on into the early Paleocene. One should not overlook the possibility that it persisted locally for a longer period of time, even into Late Paleocene time when the Flagstaff Formation accumulated. For example, Armstrong (1968, pp.448—449) questioned the Cretaceous age of clastics that were termed the Indianola (?) Formation by Christiansen (1952) for the Canyon Range in central Utah, some tens of miles southwest of Spanish Fork Canyon. Christiansen also considered variegated siltstones, conglomerates, sandstones and limestones that overlie the alleged Indianola (?) to represent the North Horn (?) Formation. Armstrong (1968, pp.448—449) termed this succession of 10,000-plus ft of conglomerates and other clastics and carbonates the Canyon Range fanglomerate, stating (p.338): "The writer suggests that the Canyon Range fanglomerate may be a lateral equivalent of the Paleocene Flagstaff Limestone of the Utah Plateaus. This correlation can be supported by evidence as strong as that supporting an Indianola (?) age, although either correlation may be proven to be correct

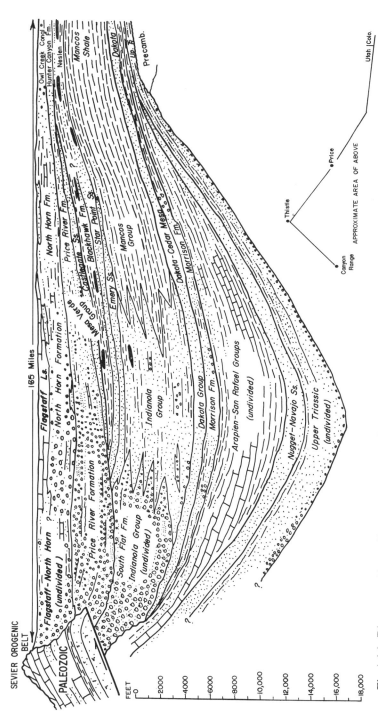

Fig.4-14. Diagrammatic section across a part of the Rocky Mountain geosyncline in central Utah. (Principal sources: Spieker, 1946; Armstrong, 1968.)

when fossil evidence is found. The important point is that the age of the fanglomerate is not well known." The writers of this chapter have visited the area mentioned by Armstrong (east of the Canyon Range) and concur with his suggestions.

The record is quite clear that an unusual thickness of mostly clastic sediments accumulated in some sort of north—south elongated depocenter east of what is now the Great Basin. Whether this Cretaceous depocenter is termed sedimentary basin(s), dynamic basin, foredeep basin, exogeosyncline, marginal basin, or any of a number of other names, a detailed sedimentary record indicates advances and retreats of the Cretaceous seas in response to eustatic changes, deformation, and numerous floods of clastic sediments from a westward source. The latter was the great Sevier orogenic belt (Armstrong, 1968; 1972), that was first regarded as the "Sevier Arch" by Harris (1959) who so named it. As shown by Armstrong (1968, fig.2, p.436), this orogenic belt extended from southern Nevada, west of Las Vegas, in a northeasterly direction into southwestern Utah and then through central Utah, leaving this state west of the Utah—Idaho—Wyoming corner. Armstrong suggested presence of a Hinterland to the west of the belt and a Foreland east of it. It was his interpretation that the Cordilleran and Rocky Mountain geosynclines were separate entities for he said (p.429): "In Nevada and Utah, sedimentation in the Cordilleran miogeosyncline began before the appearance of Cambrian fossils, and continued without orogenic interruption through the Triassic. During the Jurassic, deformation and regional metamorphism occurred in the western part of the miogeosyncline, and the source of sediment accumulation shifted onto the Colorado Plateau. A major source of clastic material appeared along the eastern margin of the Cordilleran miogeosyncline in Early Cretaceous time; this source supplied the sediments that filled the Cretaceous to Paleocene Rocky Mountain geosyncline.... This source area was the Sevier orogenic belt, which had a history of deformation through most of the Cretaceous (Sevier orogeny)." It is easy to believe that isostatic adjustments occurred as a result of the shift of excessive quantities of clastic materials from the westerly source to an easterly, nearby depocenter, but it is hard to consider that the wet unconsolidated sediment was the original cause of geologic subsidence. The Rocky Mountain geosyncline, like the Cordilleran geosyncline which was geographically to the west, is a classic example of geologic hypersubsidence. Both were *miogeosynclines.* Space in this chapter does not permit an in-depth study of the possible relationships of timing of sedimentation and deformation (Antler, Sonoma, and Sevier orogenies) in this large area of the western United States to the plate tectonics theory, although it does indeed invite speculative interpretation.

When traced in a southwesterly direction into southern Nevada, effects of the Sevier orogenic belt—Rocky Mountain geosyncline couplet are seen in various mountain ranges and in the Colorado Plateau. The Las Vegas—Wasatch hinge line here, as well as farther north, separated the easterly craton from the geosyncline which was situated west of it. A complete reversal took place, however, as the region west of the tectonic hinge experienced negative mobility (even to the extent of hypersubsidence), whereas the area to the east was a mildly positive shelf and craton from Late Precambrian into Mid-Triassic time. The Las Vegas—Wasatch "fulcrum" remained in about the same geographic position, but "what had gone down now came up" west of the hinge line. The Rocky Mountain geosyncline now experienced pronounced negative mobility. Hypersubsidence typified its western area near the Sevier orogenic belt as the foredeep basin came into being. Moore (1966) and Seager (1966) studied some of these features in the Virgin and Beaverdam Mountains of northwestern Arizona and adjacent southeastern Nevada. Still farther south in southern Nevada the record of synorogenic and other syntectonic accumulations during Cretaceous into Early Tertiary times has been assembled largely by Longwell (1949, 1963) and Longwell et al. (1965). There, the Cretaceous Willow Tank, Baseline, and Thumb Formations (thousands of feet thick) record a history of syntectonic sedimentation of conglomerates, sandstones and shales. In addition, the Cretaceous (?) or Tertiary (?) Overton Fanglomerate and Gale Hills Formation consist of coarse- to medium-textured clastic sediments, whereas the Tertiary Horse Spring Formation consists of from 4,200 ft to 6,200 ft of conglomerates and interbedded tuff breccia and tuffaceous sediments. In conclusion, one can quote Armstrong (1968, p.451): "A proper summary then is that the central Cordillera of the western United States was affected during the Mesozoic by at least the Nevada, Sevier, and Laramide orogenies. These orogenies, along with the Paleozoic Antler and Sonoma orogenies, are the principal Cordilleran orogenies which have been significant in the development of the central Cordillera of the western United States. Even this statement is an embarrassing oversimplification."

Pacific Coastal Plain of North America

Numerous Mesozoic and Cenozoic depocenters along the western part of North America, particularly those of continental United States, contain a detailed record of subsidence, hypersubsidence, and accompanying extensive sediment accumulation. Several of these depocenters are discussed here.

Northwestern Washington. Since the pioneer days of Weaver (1937), a substantial amount of field work has been directed towards mapping, measuring, and studying the Cenozoic rocks of Pacific Northwest in hopes of discovering economic hydrocarbon accumulations. For years, less attention was directed to the older rocks. Much information, however, has been gained relating to eugeosynclinal Permian rocks and also to some thick sections of Mesozoic sedimentary and associated igneous rocks. Brown and Hanna (1971) discussed the geologic structure of the northern Olympic Peninsula and Strait of Juan de Fuca in Washington, particularly the section exposed on the south limb of the Clallam syncline. They (p.1941) indicated that the marine Eocene and Oligocene sedimentary rocks are more than 20,000 ft in thickness. A scaled section on their fig.2 (p.1942) shows a thickness of 6 miles, i.e., in excess of 30,000 ft. This signifies hypersubsidence during only part of the Tertiary.

Sacramento Valley, California. Geologists in California have long pondered, and argued, about the age, terminology, and modes of accumulation of some of the thick piles of Jurassic and Cretaceous rocks in that state. Ojakangas (1968) investigated the thick Cretaceous section along the west side of the Sacramento Valley of California in an attempt to ascertain the framework in which the sediments were deposited. Upper Jurassic and Cretaceous sedimentary rocks total more than 35,000 ft in thickness in that area. Seemingly, the ancestral Klamath Mountains and Sierra Nevada were the sources of the detritus. The entire Upper Jurassic and Cretaceous sequence thickens westward from the Sierra Nevada. The Lower Cretaceous Series generally thickens to the north as well, whereas the Upper Cretaceous Series generally thickens to the south. In his concluding statements Ojakangas (p.1004) stated: "Stratigraphic and paleogeographic evidence indicate that much of the studied sequence was deposited a few tens of miles offshore in at least moderately deep water. The great thickness of the sequence, and other features, indicate a geosynclinal environment, and the general lack of volcanic rocks suggests a miogeosyncline rather than a eugeosyncline. Bedding characteristics, especially grading, are indicative of deposition below wave base by turbidity currents." Dietz (1963) suggested that the prisms of turbidity-current deposits at the base of the continental slopes may be modern eugeosynclinal accumulations, with the material at the edge of the shelves (continental terraces) representing the miogeosynclinal deposits. Ojakangas also made this observation (pp.1004—1005): "Dietz and Holden (1966) further suggested that turbidites, such as comprise the continental-rise prism at the bases of slopes, are confined to the eugeosynclinal facies. The evidence from this investigation suggests that miogeosynclinal deposits, which do not contain the volcanics and radiolarian cherts typical of the eugeosynclines, can also be composed of 'hemipelagic turbidite' beds."

Central California. A thick marine sequence of Upper Cretaceous and Early Tertiary clastic sedimentary rocks is exposed in the La Panza, southern Santa

Lucia, and Sierra Madre Ranges of central California. Chipping (1972) reported that the sedimentary rocks, which are Upper Cretaceous, Paleocene, and Eocene in age are more than 30,000 ft thick here. Chipping (1972, p.492) named this sequence, which consists of conglomerate, sandstone and shale, the Sierra Madre sequence. He indicated that it is composed of turbidites and fluxoturbidites derived from the north and east. Chipping termed this thick sedimentary section *flysch* and interpreted it to have accumulated continuously in a steep-sided, troughlike basin southwest of the present San Andreas fault zone. This taphrogenic basin experienced rapid subsidence with concomitant rapid filling of clastic sediments derived from uplifted nearby blocks. This is another excellent example of hypersubsidence. A modern analogue for such sediment accumulation possibly is to be found in the San Pedro and Santa Monica Basins of southern California. Gorsline and Emery (1959) pointed out that the floors of these two basins lie at a depth of nearly 1,000 m, yet they contain layers of clean sand interbedded with normal deep-water green muds. They noted that the sea floor off southern California is a block-faulted complex of basins separated by islands and banks. They believed that these basins are being supplied sediments via turbidity currents. Nearshore basins fill first because of turbidity currents, resulting in eventual spilling of sediments from basin to basin.

San Joaquin Valley, California. Rocks of Miocene or younger age produce most of California's oil and gas. Slightly less than half of the total oil and gas production in the state comes from sedimentary rocks of this age in the San Joaquin Valley. Structurally, this valley is a synclinorium that is about 250 miles long and 50—60 miles wide, lying between the Sierra Nevada Mountains to the east and the Coast Ranges to the west. Depth to the basement seemingly is in excess of 30,000 ft, which indicates that this dynamic basin experienced hypersubsidence. Simonson (1958) indicated that the San Joaquin Valley is one of the most petroliferous basins in California. Bandy and Arnal (1969) discussed the Middle Tertiary development of the San Joaquin Basin that lies in the southern half of the San Joaquin Valley. This basin, they pointed out, was already well-defined by the beginning of Oligocene (Zemorrian Stage). They also indicated (p.816) that of the more than 5,000 cu miles of rocks representing the marine sediments that were deposited during the Oligocene and Miocene times, 4,000 cu miles were deposited in water depths greater than 300 ft, and most of the oil fields of the basin produce from these strata. Oil production, they noted, is from strata that were laid down on or near the steeper slopes of the basin where there are rather rapid changes in biofacies. Some of these deposits appear to be subsea fan deposits related to the subsea valleys through which coarse sediment was transported. It was also noted that during the Oligocene and Lower Miocene times, maximum water depths of the marine basin were abyssal (in excess of 6,000 ft). Through time, the basin assumed progressively shallower depths, so that by latest Miocene and Pliocene times the basin had shoaled to neritic depths. Thus, the analysis of isopach data alone may give an incomplete concept of the magnitude of subsidence in a basin.

Los Angeles Basin, California. For its size, the Los Angeles Basin has proved to be one of the richest oil-producing areas in the world; average recovery is in excess of 100,000 barrels per acre. The Long-Beach field exhibited prolific oil recovery, amounting to about 480,000 barrels per acre. Barbat (1958) indicated that pre-basinal sediments in the Los Angeles Basin range from Upper Cretaceous through Lower Miocene; these sediments were deformed and beveled in mid-Miocene time. Following this event, this area was depressed relative to sea level and received sediments that Barbat (p.66) termed the basinal sediments. His isopach map (fig.3, p.66) indicates an aggregate thickness of approximately 18,000 ft of Upper Miocene, Pliocene, and Pleistocene sediments. He argued (p.72) in favor of water depths of 4,000—6,000 ft during the mid-cycle stages of basin development, with normal shale deposition being interrupted by turbidity current deposition of coarser clastics. He noted also that at times the basin filled more rapidly than it sank to accommodate the sediment. This dynamic basin presents another example of variable basin subsidence and hypersubsidence, with folding and faulting control.

The literature is full of other classic examples of geologic subsidence and hypersubsidence and excessive sediment accumulation. For example, the great Persian Gulf synclinorium experienced hypersubsidence (30,000—40,000 ft) through Phanerozoic time (Kamen-Kaye, 1970); ultimate producible reserves of crude oil in this area are believed to exceed 250 billion bbl. Kamen-Kaye (p.2371) suggested that as much as 500 billion bbl of in-place oil accumulated in major structures there. Another area which deserves mentioning is the Western Canada Sedimentary Basin; the reader is referred to the classic publication for that area, which was edited by Clark (1954). Another publication to which attention is directed is "*The Geology of Egypt*" by Rushdi Said (1962).

Isostatic subsidence

Bloom (1967) discussed the Pleistocene shorelines as a means of testing the theory of isostasy, pointing out that (p.1478): "Postglacial isostatic recovery of formerly glaciated regions has provided one of the best available tests for the theory of isostasy." The writers of this chapter direct attention to two examples in the western conterminus United States, which prove isostatic subsidence and subsequent rebound: (1) the Lake Mead area of Arizona and Nevada, and (2) the Pleistocene Lake Bonneville of Utah. Precise and repeated surveys were made prior to and after the filling of the Lake Mead reservoir behind Hoover Dam (Longwell, 1960). Between 1935 and 1950, a measurable depression developed, having a closure of 70 mm and a diameter of about 60 km; it centered over the area of maximum loading. The water load exerted a pressure of over 140 pounds per sq inch (psi) which is equal to nearly 10 bars (Longwell, 1960, p.35). In commenting on this area, Bloom (p.1482) stated that ". . . in spite of the complexities, it seems clear that the Lake Mead area was deformed by a water load of 10 bars in the short time of only 15 years. Such rapid deformation must surely be elastic, although Longwell (1960, p.37) did not rule out the possibility of eventual plastic deformation at great depth."

Lake Bonneville

During the Wisconsinian Stage of the Pleistocene Epoch, pluvial Lake Bonneville came into existence in western Utah and small parts of adjacent eastern Nevada and southern Idaho. In Monograph I of the U.S. Geological Survey, Gilbert (1890) presented a tremendous amount of information, and a map, relating to this ancient Ice Age lake. Workers since Gilbert's day have added some details, but have not materially changed his interpretations. This lake was the largest of the Pleistocene lakes of the Great Basin, being some 325 miles long and 125 miles wide, with a surface area of about 20,000 sq miles. Its maximum depth was slightly more than 1,100 ft in the main northern body near the west edge of the present Great Salt Lake. In his discussion of deformation of the Bonneville shore line (chapter 7), and on maps depicting magnitude of this deformation, Gilbert did not use the term "isostasy". The word, however, appears in the index of Monograph I. Actually, the term "isostasy" had been defined only 10 years before Gilbert completed his studies on Lake Bonneville and was not widely accepted. Crittenden (1963a, 1963b) presented new data on the isostatic deformation of Lake Bonneville, stating: "Domical upwarping of the area formerly occupied by Pleistocene Lake Bonneville is verified by 75 new measurements of elevation on the Bonneville shore line. Gilbert's conclusion that this uplift was an isostatic response to the removal of load is confirmed by maps which show that the deformation is closely correlated with the former distribution and average depth of water." Crittenden noted that the calculated viscosity of the subcrust in this area is 10^{21} poises, compared with 10^{22} poises in Scandinavia. Figure 4-15 consists of two maps showing deformation of the Bonneville shore line; one is after Gilbert (1890) and the other is that of Crittenden (1963b).

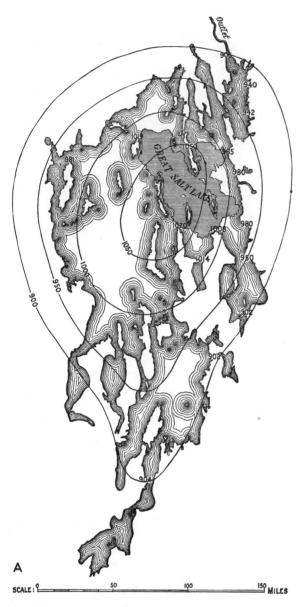

Fig.4-15. Maps showing deformation of Lake Bonneville shore line due to isostatic rebound during Holocene time. (A, after Gilbert, 1890, plate L; B, after Crittenden, 1963a, fig.3.)

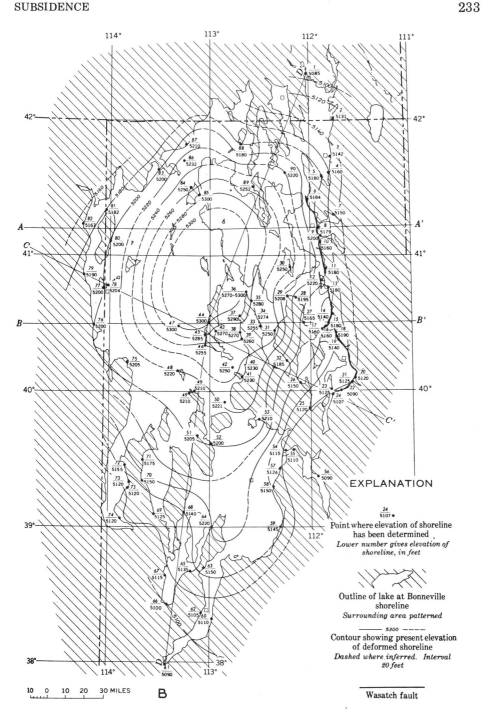

Brief theoretical considerations of geologic subsidence causes

It is tempting to speculate about causes of geologic subsidence for geosynclines and other depocenters. The writers indulged in the luxury of some speculations in previous pages. Wang (1972), for example, stated: "The concept of the geosyncline has been a controversial subject ever since it was proposed. It appears, however, that in the 'new global tectonics', geosynclinal theory could be rationalized into a sound unified concept of geosynclinal genesis and evolution." Page (1970) discussed continental margin tectonics of the Sur-Nacimiento Fault zone of California, and speculated on ocean-floor spreading as a causal mechanism. Papers by Larson et al. (1968), Chase et al. (1970), Coney (1970), Pitman III and Talwani (1972), and the Bulletin of the American Association of Petroleum Geologists (Vol. 56(2), February 1972) will provide the reader food for thought on the pros and cons of this exciting new concept, i.e., *plate tectonics*.

In all previous discussions in this chapter, the thicknesses of consolidated rocks were reported. One should not lose sight of the fact that thicknesses of the corresponding unconsolidated sediments were much greater. For example, Fig.4-16 shows the different theoretical stages of compaction of argillaceous sediment, whereas Fig.4-17 depicts compaction stages of a sand, as the interstitial fluids move out. In these figures, the final thickness of consolidated sediments (rocks) is 62.5% of the original thickness of unconsolidated sediment. The effects of cementation and other diagenetic changes were not considered in preparing Fig.4-16 and 4-17. In addition, the effect of crushing of sand grains under high overburden stresses is also not included. Figure 4-18 is a schematic diagram showing the difference between a sediment consisting of flat particles (e.g., clay particles) and having a porosity of 33.3% (Fig.4-18A) and a similar sediment with a porosity of 20% (Fig.4-18 B). In Fig.4-19, a sand having a porosity of 33.3% (Fig.4-19A) is compared with a better consolidated, similar sand having a porosity of 20% (Fig.4-19B).

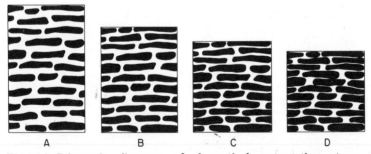

Fig.4-16. Schematic diagrams of theoretical compaction stages of an argillaceous sediment, having initial and final porosity of 50% (A) and 20% (D), respectively. Intermediate porosities are 40% (B) and 30% (C). All porosities were measured exactly.

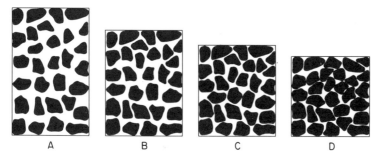

Fig.4-17. Schematic diagrams of theoretical compaction stages of a sand, having initial and final porosities of 50% (A) and 20% (D), respectively. Intermediate porosities are 40% (B) and 30% (C). All porosities were measured exactly in preparing these diagrams.

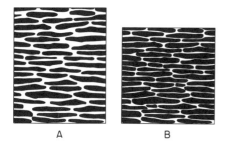

Fig.4-18. Comparison between clayey sediment having a porosity of 33.3% (A) and a similar sediment with a porosity of 20% (B). Porosities were measured exactly in preparing these theoretical diagrams.

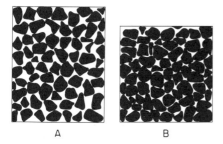

Fig.4-19. Comparison between degree of compaction of a sand having a porosity of 33.3% (A) and that of a similar sand with 20% porosity (B). Porosities were measured exactly in preparing these theoretical diagrams.

Thus, in the opinion of the writers the gravitational compaction of sediments can give rise to such great changes in the initial thicknesses and volumes of the sediments of the earth's crust, that quantitative studies of oscillatory movements, which are based on measurements of the thicknesses and volumes of sediments, are not very accurate if compaction is not taken into consideration. This opinion, however, is not shared by Ronov (1948).

PRESENT-DAY SUBSIDENCE

Subsidence due to fluid withdrawal

Tapping the earth for ground water and petroleum resulted in sinking of the land, i.e., subsidence, in many areas. Withdrawal of steam for geothermal power and of brines also have caused subsidence. Fluid pressures in petroleum reservoirs have been reduced by as much as 2,000—4,000 psi from the initial hydrostatic pressure, whereas those in ground-water reservoirs have been reduced by as much as 200—600 psi ($\approx$ 460—600 ft, head of water). Withdrawal of the fluids, and consequent removal of hydrostatic support, can cause sinking of the land surface.

The most dramatic and damaging effects of subsidence occurred in Long Beach, California, in the 1940's and 1950's, above the Wilmington Oil Field. Sinking over an area of 22 sq mi gave rise to a bowl up to 27 ft deep (see Fig.4-20), causing damages in excess of $150 millions. The vertical subsidence has been accompanied by a horizontal movement of as much as 9 ft, directed inward toward the center of subsidence. This area is underlain by 6,000 ft of sediments of Recent to Miocene ages that unconformably overlie a basement schist of Pre-Tertiary age. Seven oil-producing zones extend from a depth of about 2,500 ft to 6,000 ft. The average porosity of productive zones, containing 23—70% of sand, ranges from 24 to 34%.

Other areas of land subsidence related to exploitation of oil and gas fields include: (1) Goose Creek, Texas; (2) Lake Maracaibo, Venezuela; (3) Niigata, Japan; and (4) Po Delta, Italy.

The land subsidence in San Joaquin Valley in California, due to intensive pumping of underground water for irrigation, covers three different areas of 1,400, 800, and 300 sq miles. The rate of sinking of the land was up to a foot per year. In some places, the land subsided to more than 20 ft below its former level. Other areas of major subsidence related to ground-water removal include: (1) Denver, Colorado; (2) Eloy—Picacho area, Arizona; (3) Houston—Galveston area, Texas; (4) Las Vegas, Nevada; (5) London, England; (6) Mexico City, Mexico; (7) Savanna, Georgia; (8) Tokyo and Osaka, Japan.

According to Poland and Davis (1969, p.197), the amount of compaction that a confined aquifer system will experience depends on the compressibility of the sediments within the range of change in effective pressure

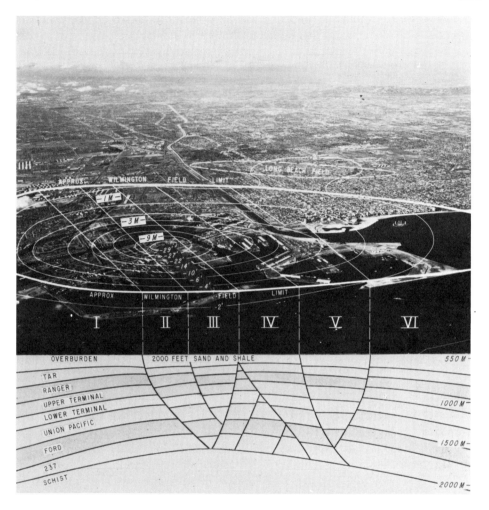

Fig.4-20. Subsidence of Long Beach area (1928—1962) and geologic section at Wilmington oil field, California. The subsidence in upper right is due to fluid withdrawal from Long Beach (Signal Hill) oil field, whereas that in foreground is owing to withdrawal from the Wilmington oil field. Lines of equal subsidence are in feet (—2', —4', —6', —10', —14', —22', —27', and —29'); only three contours are shown in meters (—1, —3, and —9 m). Approximate vertical scale in meters is added at the geologic section. (Courtesy of City of Long Beach Department of Oil Properties, Calif.)

(artesian-head decline) and the magnitude of that change. As shown in Chapter 2, the compressibility decreases with decreasing void ratio. The amount of subsidence also depends on the thicknesses and vertical permeabilities of the clayey interbeds and the time duration of pressure decline owing to production of fluids. The reader is referred to the classic paper on subsidence by Poland and Davis (1969) for further details.

Miscellaneous causes of subsidence

Various other causes of land subsidence include: (1) solution; (2) oxidation; (3) compaction of sediment under surface loading, vibration, or wetting; and (4) tectonic movements. Loose, dry, low-density deposits, that compact when they are wetted, cover extensive areas in Asia, Europe, and North America. Lofgren (1969) referred to this process as *hydrocompaction*, which produces widespread subsidence of the land surface.

In certain highly porous soils, which are commonly derived from mudflows, the individual grains are held together with a dry clay film. These films appear to prevent consolidation of such soils (high dry strength). As these soils become saturated with water, however, the clay loses its strength and the soil structure collapses. As much as 15—20 ft of subsidence has been recorded where more than 100—150 ft of highly compactible alluvium has been wetted. Heavy rains or water leaking into the soil from canals and pipelines brings about this dramatic collapse of the land. Bara (1960) reported that if the ratio of porosity at liquid limit to porosity at natural density is greater than 1, then the samples immersed in water compact from 10 to 25% under an applied load of 100 psi (7.03 kg/cm^2). The intergranular strength of some soil deposits is due to lime cementation, which also breaks down on wetting.

The major types of deposits susceptible to hydrocompaction are: (1) loose, moisture-deficient alluvial deposits, and (2) moisture-deficient loess and related eolian deposits (Lofgren, 1969).

Earthquakes also can cause subsidence. The possibility of earthquakes affecting the benchmark elevations in California was investigated by Green (1972). The two principal types of earthquakes are those causing (1) surface faulting and (2) shaking, with the former being uncommon at the present time because the majority of earthquakes are relatively deep-seated. Earthquakes are normally reported in terms of *magnitude* (M), which is a measure of the energy released at the source (1 to 10 on the Richter scale), and *epicenter*, i.e., location on the surface directly above the quake. *Intensity*, which is a measure of the severity of ground motion at a particular point on the earth's crust (I to XII on the modified Mercelli scale), is of prime importance in studying the effect of earthquakes on magnitude of subsidence.

According to Green (1972), the largest earthquake during the past 20 years in California (Kern County, 1952, M = 7.7) resulted in a maximum vertical displacement of 4 ft and a maximum horizontal displacement of

2—3 ft. The 1971 earthquake in San Fernando Valley in California (M = 6.6) resulted in maximum vertical and horizontal movements of 3 and 5 ft, respectively. Green (1972) showed that there is a direct relationship between the anomalous settling of benchmarks and increased shaking. The amount of energy released at a particular source does not appear to be as important as the secondary effect of ground shaking.

Techniques of measuring subsidence

The simplest method of detecting subsidence at the surface involves the use of firmly anchored posts (bench marks), which are spaced in a grid pattern over a wide area. Their relative positions and elevations are examined periodically with surveying instruments.

Marsden and Davis (1967, p.95) and Poland and Davis (1969) described several techniques of measuring the compaction of underground strata:

(1) Bullets of radioactive material are shot into the rock at measured intervals in the wellbore from a firing mechanism (gun) lowered down the hole. A gamma ray detector is then periodically lowered into the well on a cable to determine the changes (lessening) in distances between the radioactive markers.

(2) The vertical compression of casing (steel pipe) that lines the well is measured. Inasmuch as casing is usually cemented to the rock surrounding it, the vertical compression of the rock will shorten the pipe length. Casing comprises 30- or 40-ft pipe joints. Magnetic flux at the collar joining two pipe lengths is different from the flux along the length of the pipe. Thus, a magnetometer can be used to locate the successive collars and measure the shortening of individual pipe lengths. Shortening as low as ½ inch in a 40-ft joint can be detected. The shortening also can be measured by marking the pipe with radioactive pellets at definite intervals.

(3) Compaction recorders are used, which consist of a heavy weight emplaced in the formation below the bottom of a well casing. Cable, which is attached to the anchor weight (200—300 lb), extends to the land surface and is counter-weighted (in 50-lb increments) to maintain constant tension. The cable is free to move at its upper end. The amount of cable rise above the land surface is continuously measured by a recorder, as subsidence occurs. When amplified, the record can reveal subsidence as small as 1/10 mm.

(4) Floats, quartz rods and transducers are employed, which enable measurement of vertical and horizontal strains and tilts of the land. Ground movements of less than $0.1\,\mu$ can be measured with strain recorders and tiltmeters (Marsden and Davis, 1967, p.99).

In conclusion, one can state that probably a great deal can be learned concerning the geologic subsidence and compaction from the studies of the present-day subsidence, due to various causes.

REFERENCES

Applin, P. and Applin, E. R., 1965. Comanche Series and associated rocks in subsurface in central and south Florida. *U.S. Geol. Surv., Prof. Pap.*, 447: 84 pp.
Armstrong, F. C. and Oriel, S. S., 1965. Tectonic development of Idaho—Wyoming thrust belt. *Bull. Am. Assoc. Pet. Geol.*, 49: 1847—1866.
Armstrong, R. L., 1968. Sevier orogenic belt in Nevada and Utah. *Bull. Geol. Soc. Am.*, 79: 429—458.
Armstrong, R. L., 1972. Low-angle (denudation) faults, hinterland of the Sevier orogenic belt, eastern Nevada and western Utah. *Bull. Geol. Soc. Am.*, 83: 1729—1754.
Aslanyan, A. T., 1960. Dynamic problem of geotectonics. *Int. Geol. Congr., Rep. Sov. Geol., 21st Sess.*, pp. 5—16,
Baars, D. L., 1966. Pre-Pennsylvanian paleotectonics — key to basin evolution and petroleum occurrences in Paradox basin, Utah and Colorado. *Bull. Am. Assoc. Pet. Geol.*, 50: 2082—2111.
Babaev, A. G., 1956. Concerning one method of solution of problem of compensation of depressions with sedimentation. *Izv. Akad. Nauk Arm. S.S.R.*, 9(8): 31—35.
Bandy, O. L. and Arnal, R. E., 1969. Middle Tertiary basin development, San Joaquin Valley, California. *Bull. Geol. Soc. Am.*, 80: 783—820.
Bara, J. P., 1960. Laboratory studies in loessial foundations and embankment samples — Sherman Dam — Farewell Unit, Nebraska. *U.S. Bur. Reclam., Earth Lab., Denver, Colo., Rep.*, EM-572: 17 pp.
Barbat, W. F., 1958. The Los Angeles Basin area, California. In: L. G. Weeks (Editor), *Habitat of Oil*. Am. Assoc. Pet. Geol., Tulsa, Okla, pp. 62—77.
Beal, L. H., 1965. Geology and mineral deposits of the Bunkerville mining district, Clark County, Nevada. *Nev. Bur. Mines Bull.*, 63: 96 pp.
Bird, J. M. and Dewey, J. F., 1970. Lithosphere plate-continental margin tectonics and the evolution of the Appalachian orogen. *Bull. Geol. Soc. Am.*, 81: 1031—1060.
Bissell, H. J., 1962a. Pennsylvanian and Permian rocks of Cordilleran area. In: *Pennsylvanian System in the United States — a Symposium*. Am. Assoc. Pet. Geol., Tulsa, Okla., pp. 188—262.
Bissell, H. J., 1962b. Permian rocks of parts of Nevada, Utah, and Idaho. *Bull. Geol. Soc. Am.*, 73: 1083—1100.
Bissell, H. J., 1962c. Pennsylvanian—Permian Oquirrh basin of Utah. In: *Geology of the Southern Wasatch Mountains and Vicinity, Utah. Brigham Young Univ. Geol. Stud.*, 9(1): 26—49.
Bissell, H. J., 1970. Realms of Permian tectonism and sedimentation in western Utah and eastern Nevada. *Bull Am. Assoc. Pet. Geol.*, 54: 285—312.
Bloom, A. L., 1967. Pleistocene shorelines: a new test of isostasy. *Bull. Geol. Soc. Am.*, 78: 1477—1494.
Brill Jr., K. G., 1952. Stratigraphy of the Permo-Pennsylvanian zeugogeosyncline of Colorado and northern New Mexico. *Bull. Geol. Soc. Am.*, 63: 809—880.
Brown Jr., R. D. and Hanna, W. F., 1971. Aeromagnetic evidence and geologic structure, Northern Olympic Peninsula and Strait of Juan de Fuca, Washington. *Bull. Am. Assoc. Pet. Geol.*, 55: 1939—1953.

Bucher, W. H., 1933. *The Deformation of the Earth's Crust.* Princeton University Press, Princeton, N.J., 518 pp.

Bullard, E., Everett, J. E. and Smith, A. G., 1965. The fit of the continents around the Atlantic. *R. Soc. Lond. Philos. Trans.,* Ser. A, 258: 41—51.

Chamberlain, C. K., 1971. Bathymetry and paleoecology of Ouachita geosyncline of southeastern Oklahoma as determined from trace fossils. *Bull. Am. Assoc. Pet. Geol.,* 55: 34—50.

Chase, C. G. and others, 1970. History of sea-floor spreading west of Baja California. *Bull. Geol. Soc. Am.,* 81: 491—498.

Chilingar, G. V., Mannon, R. W. and Rieke III, H. H., 1972. *Oil and Gas Production from Carbonate Rocks.* Am. Elsevier, New York, N.Y., 408 pp.

Chipping, D. H., 1972. Early Tertiary paleogeography of central California. *Bull. Am. Assoc. Pet. Geol.,* 56: 480—493.

Christiansen, F. W., 1952. Structure and stratigraphy of the Canyon Range, central Utah. *Bull. Geol. Soc. Am.,* 63: 717—740.

Clark, L. M. (Editor), 1954. *Western Canada Sedimentary Basin.* Am. Assoc. Pet. Geol., Tulsa, Okla., 521 pp.

Cline, L. M., 1970. Sedimentary features of Late Paleozoic flysch, Ouachita Mountains, Oklahoma. *Geol. Assoc. Can., Spec. Pap.,* 7: 85—101.

Colle, J., Cooke Jr., W. F., Denham, R. L., Ferguson, H. C., McGuirt, J. H., Reedy Jr., F. and Weaver, P., 1952. Volume of Mesozoic and Cenozoic sediments in Western Gulf Coastal Plain of United States. *Bull. Geol. Soc. Am.,* 63: 1193—1200.

Coney, P. J., 1970. The geotectonic cycle and the new global tectonics. *Bull. Geol. Soc. Am.,* 81: 739—748.

Cramer, H. R., 1971. Permian rocks from Sublett Range, southern Idaho. *Bull. Am. Assoc. Pet. Geol.,* 55: 1787—1801.

Crittenden Jr., M. D., 1963a. New data on the isostatic deformation of Lake Bonneville. *U.S. Geol. Surv., Prof. Pap.,* 454-E: 31 pp.

Crittenden Jr., M. D., 1963b. Effective viscosity of the Earth derived from isostatic loading of Pleistocene Lake Bonneville. *J. Geophys. Res.,* 68: 5517—5530.

Crittenden Jr., M. D., Schaeffer, F. E., Trimble, D. E. and Woodward, L. A., 1971. Nomenclature and correlation of some Upper Precambrian and Basal Cambrian sequences in western Utah and southeastern Idaho. *Bull. Geol. Soc. Am.,* 82: 581—602.

Crosby, G. W., 1968. Vertical movements and isostasy in western Wyoming overthrust belt. *Bull. Am. Assoc. Pet. Geol.,* 52: 2000—2015.

Dallmus, K. F., 1958. Mechanics of basin evolution and its relation to the habitat of oil in the basin. In: L. G. Weeks (Editor), *Habitat of Oil.* Am. Assoc. Pet. Geol., Tulsa, Okla., pp.883—931.

Dana, J. D., 1873. On some results of the Earth's contraction from cooling. *Am. J. Sci.,* 3rd Ser., 5: 430; 6: 717.

Dewey, J. F., 1969. Evolution of the Appalachian—Caledonian orogen. *Nature,* 22: 124—129.

Dietz, R. S., 1972. Geosynclines, mountains, and continent-building. In: J. T. Wilson (Editor), *Continents Adrift. Sci. Am.,* pp.124—132.

Dietz, R. S. and Holden, J. C., 1966. Miogeoclines in space and time. *J. Geol.,* 74: 566—583.

Dietz, R. S., Holden, J. C. and Sproll, W. P., 1970. Geotectonic evolution and subsidence of Bahama Platform. *Bull. Geol. Soc. Am.,* 81: 1915—1928.

Drake, C. L., Ewing, M. and Sutton, J., 1959. Continental margins and geosynclines: The east coast of North America north of Cape Hatteras. In: *Physics and Chemistry of the Earth,* 5. Pergamon, New York, N.Y., pp.110—198.

Eardley, A. J., 1951. *Structural Geology of North America.* Harper, New York, N.Y., 2nd ed. 1962, 624 pp.
Eardley, A. J., 1967. Idaho and Wyoming thrust belt: its divisions and an analysis of its origin. In: L. A. Hale (Editor), *Anatomy of the Western Phosphate Field. Guideb., 15th Annu. Field Conf., Int. Assoc. Pet. Geol.,* pp.35—44.
Eardley, A. J., 1968. Major structures of the Rocky Mountains of Colorado and Utah. *Univ. Mo. J.,* 1: 79—99.
Furer, L. C., 1970. Petrology and stratigraphy of nonmarine Upper Jurassic—Lower Cretaceous rocks of western Wyoming and southeastern Idaho. *Bull. Am. Assoc. Pet. Geol.,* 54: 2282—2302.
Gabrielse, H., Roddick, J. A. and Blusson, S. L., 1965. Flat River, Glacier Lake and Wrigley Lake, district of Mackenzie and Yukon Territory. *Can. Geol. Surv. Pap.,* 64(52): 30 pp.
Galley, J. E., 1958. Oil and geology in the Permian Basin of Texas and New Mexico. In: L. G. Weeks (Editor), *Habitat of Oil.* Am. Assoc. Pet. Geol., Tulsa, Okla., pp.395—446.
Gilbert, G. K., 1890. Lake Bonneville. *U.S. Geol. Surv., Monogr.,* 1: 438 pp.
Gilbert, G. K., 1928. Studies of basin range structure. *U.S. Geol. Surv., Prof. Pap.,* 153: 92 pp.
Gilluly, J., 1963. The tectonic evolution of the western United States. *Q.J. Geol. Soc. Lond.,* 119: 133—174.
Gilluly, J., 1964. Atlantic sediments, erosion rates, and the evolution of the Continental Shelf: some speculations. *Bull. Geol. Soc. Am.,* 75: 483—492.
Gilluly, J., Reed Jr., J. C. and Cady, W. M., 1970. Sedimentary volumes and their significance. *Bull. Geol. Soc. Am.,* 81: 353—376.
Gorsline, D. S. and Emery, K. O., 1959. Turbidity-current deposits in San Pedro and Santa Monica Basins off Southern California. *Bull. Geol. Soc. Am.,* 70: 279—290.
Green, J. P., 1972. An approach to analyzing multiple causes of subsidence. *Annu. Fall Meet., Soc. Pet Eng. AIME, 47th, San Antonio, Texas, Oct. 8—11, Pres. Pap.,* SPE 4079: 12 pp.
Guzmán, E. J., 1952. Volumes of Mesozoic and Cenozoic sediments in Mexican Gulf Coastal Plain. *Bull. Geol. Soc. Am.,* 63: 1201—1220.
Hall, J., 1859. Description and figures of the organic remains of the Lower Helderberg group and the Oriskany sandstone. *N.Y. Geol. Surv., Paleontol.,* 3: 532 pp.
Harris, H. D., 1959. A late Mesozoic positive area in western Utah. *Bull. Am. Assoc. Pet. Geol.,* 43: 2636—2652.
Harrison, J. E., 1972. Precambrian Belt Basin of Northwestern United States: Its geometry, sedimentation, and copper occurrences. *Bull. Geol. Soc. Am.,* 83: 1215—1240.
Hess, H. H., 1962. History of the oceans. In: A. E. J. Engel et al. *Petrological Studies* (A Volume in honour of A. F. Buddington). Geol. Soc. Am., Boulder, Colo., pp.599—620.
Jam, L., Dickey, P. A. and Tryggvason, E., 1969. Subsurface temperature in South Louisiana. *Bull. Am. Assoc. Pet. Geol.,* 53: 2141—2149.
Johnson, R. H. and Huntley, L. G., 1916. *Principles of Oil and Gas Production.* Wiley, New York, N.Y., 355 pp.
Kamen-Kaye, M., 1967. Basin subsidence and hypersubsidence. *Bull. Am. Assoc. Pet. Geol.,* 51: 1833—1842.
Kamen-Kaye, M., 1970. Geology and productivity of Persian Gulf Synclinorium. *Bull. Am. Assoc. Pet. Geol.,* 54: 2371—2394.

Kamen-Kaye, M., 1972. Viola---voila! *Geotimes*, 17(7): 13—14.
Kay, M., 1944. Geosynclines in continental development. *Science*, 99: 461—462.
Kraft, J. C., Sheridan, R. E. and Maisano, M., 1971. Time-stratigraphic units and petroleum entrapment in Baltimore Canyon Basin of Atlantic Continental Margin geosyncline. *Bull. Am. Assoc. Pet. Geol.*, 55: 658—679.
Krivoy, H. L. and Pyle, T. E. 1972. Anomalous crust beneath West Florida Shelf. *Bull. Am. Assoc. Pet. Geol.*, 56: 107—113.
Kuenen, Ph. H., 1950. *Marine Geology*. Wiley, New York, N.Y., 551 pp.
Landes, K. K., 1951. *Petroleum Geology*. Wiley, New York, N.Y., 660 pp.
Larson, R. L., Menard, H. W. and Smith, S. M., 1968. Gulf of California: A result of ocean floor spreading and transform faulting. *Science*, 161: 781—784.
Lofgren, B. E., 1969. Land subsidence due to the application of water. In: *Reviews in Engineering Geology II*. Geol. Soc. Am., Boulder, Colo., pp. 271—303.
Longwell, C. R., 1949. Structure of the northern Muddy Mountain area, Nevada. *Bull. Geol. Soc. Am.*, 60: 923—968.
Longwell, C. R., 1960. Interpretation of the levelling data. In: Comprehensive survey of sedimentation in Lake Mead, 1948—49. *U.S. Geol. Surv., Prof. Pap.*, 295: 33—38.
Longwell, C. R., 1963. Reconnaissance geology between Lake Mead and Davis Dam, Arizona—Nevada. *U.S. Geol. Surv., Prof. Pap.*, 374-E: 51 pp.
Longwell, C. R., 1965. Geology and mineral deposits of Clark County, Nevada. *Nev. Bur. Mines, Bull.*, 62: 218 pp.
Longwell, C. R., Pampeyan, E. H., Bowyer, B. and Roberts, R. J., 1965. Geology and mineral deposits of Clark County, Nevada. *Nev. Bur. Mines Bull.*, 218 pp.
Marsden Jr., S. S. and Davis, S. N., 1967. Geological subsidence. *Sci. Am.*, 216(6): 93—100.
McGill, G. E. and Sommers, D. A., 1967. Stratigraphy and correlation of the Precambrian Belt Supergroup of the Southern Lewis and Clark Range, Montana. *Bull. Geol. Soc. Am.*, 78: 343—352.
Meyerhoff, A. A. (Editor), 1968. Geology of natural gas in South Louisiana. In: *Natural Gases of North America. Am. Assoc. Pet. Geol., Mem.*, 9(1): 376—581.
Milton, C. and Grasty, R., 1969. "Basement" rocks of Florida and Georgia. *Bull. Am. Assoc. Pet. Geol.*, 53: 2483—2493.
Misch, P. and Hazzard, J. C., 1962. Stratigraphy and metamorphism of Late Precambrian rocks in central northeastern Nevada and adjacent Utah. *Bull. Am. Assoc. Pet. Geol.*, 46: 289—343.
Moore, R. T., 1966. *A Structural Study of the Virgin and Beaverdam Mountains, Arizona*. Thesis, Stanford Univ., Calif., 99 pp.
Murray, G. E., 1952. Volume of Mesozoic and Cenozoic sediments in central Gulf Coastal Plain of United States. *Bull. Geol. Soc. Am.*, 63: 1177—1192.
Murray, G. E., 1961. *Geology of the Atlantic and Gulf Coastal Province of North America*. Harper, New York, N.Y., 692 pp.
Newell, N. D., 1955. Bahamian platforms. In: *The Crust of the Earth — A Symposium*. Geol. Soc. Am., Spec. Pap., 62: 303—315.
Newell, N. D. and Rigby, J. K., 1957. Geological studies on the Great Bahama Bank. In: *Regional Aspects of Carbonate Deposition. Soc. Econ. Paleontol. Mineral., Spec. Publ.*, 5: 15—79.
Ojakangas, R. W., 1968. Cretaceous sedimentation, Sacramento Valley, California. *Bull. Geol. Soc. Am.*, 79: 973—1008.
Page, B. M., 1970. Sur-Nacimiento fault zone of California: Continental margin tectonics. *Bull. Geol. Soc. Am.*, 81: 667—690.

Park Jr., C. F. and Cannon Jr., R. S., 1943. Geology and ore deposits of the Metaline quadrangle, Washington. *U.S. Geol. Surv., Prof. Pap.*, 202: 81 pp.

Peterson, J. A., 1966. Stratigraphic vs. structural controls on carbonate-mound hydrocarbon accumulation, Aneth area, Paradox Basin. *Bull. Am. Assoc. Pet. Geol.*, 50: 2068—2081.

Peterson, J. A. and Ohlen, H. E., 1963. Pennsylvanian shelf carbonates, Paradox Basin. In: *Shelf Carbonates of the Paradox Basin. Four Corners Geol. Soc. Symposium*, pp. 65—79.

Peterson, J. A. and White, R. J., 1969. Pennsylvanian evaporite—carbonate cycles and their relation to petroleum occurrence, southern Rocky Mountains. *Bull. Am. Assoc. Pet. Geol.*, 53: 884—908.

Pitman III, W. C. and Talwani, M., 1972. Sea-floor spreading in the North Atlantic. *Bull. Geol. Soc. Am.*, 83: 619—646.

Poland, J. F. and Davis, G. H., 1969. Land subsidence due to withdrawal of fluids. In: D. J. Varnes and G. Kiersch (Editors), *Reviews in Engineering Geology II*. Geol. Soc. Am., Boulder, Colo., pp.187—269.

Pressler, E. D., 1947. Geology and occurrence of oil in Florida. *Bull. Am. Assoc. Pet. Geol.*, 31: 1851—1862.

Reeside Jr., J. B., 1944. Maps showing thickness and general character of the Cretaceous deposits in the Western Interior of the United States. *U.S. Geol. Surv., Oil Gas Invest., Prelim.* map 10.

Roberts, R. J., 1968. Tectonic framework of the Great Basin. *Univ. Mo. J.*, 1: 101—119.

Roberts, R. J., 1972. Evolution of the Cordilleran fold belt. *Bull. Geol. Soc. Am.*, 83: 1989—2004.

Ronov, A. V., 1948. Consolidation of sediments and accuracy of methods used in studying history of oscillatory movements of the earth's crust. *Dokl. Akad. Nauk S.S.S.R.*, 62(5): 673—676.

Said, R., 1962. *The Geology of Egypt*. Elsevier, Amsterdam, 377 pp.

Scheidegger, A. E. and O'Keefe, J. A., 1967. On the possibility of the origination of geosynclines by deposition. *J. Geophys. Res.*, 72: 6275—6278.

Schuchert, C., 1923. Sites and nature of the North American geosynclines. *Bull. Geol. Soc. Am.*, 34: 151—230.

Seager, W. R., 1966. *Geology of the Bunkerville Section of the Virgin Mountains, Nevada and Arizona*. Thesis, Univ. Ariz., 124 pp.

Seager, W. R., 1970. Low-angle gravity glide structures in the Northern Virgin Mountains, Nevada and Arizona. *Bull. Geol. Soc. Am.*, 81: 1517—1538.

Simonson, R. R., 1958. Oil in the San Joaquin Valley, California. In: L. G. Weeks (Editor), *Habitat of Oil*. Am. Assoc. Pet. Geol., Tulsa, Okla., pp.99—112.

Silberling, N. J., 1971. Geological events during Permian—Triassic time along the Pacific margin of the United States. In: *International Permian—Triassic Conference, Calgary, Alta, 1971*. Program with abstracts, p.355 (abstr.).

Silberling, N. J., 1973. Geologic events during Permian—Triassic time along the Pacific margin of the United States. In: A. Logan and L. V. Hills (Editors), *The Permian and Triassic Systems and Their Mutual Boundary*, pp.345—362.

Spencer, M., 1967. Bahamas deep test. *Bull. Am. Assoc. Pet. Geol.*, 51: 263—268.

Spieker, E. M., 1946. Late Mesozoic and Early Cenozoic history of central Utah. *U.S. Geol. Surv., Prof. Pap.*, 205-D: 117—161.

Spieker, E. M., 1949. The transition between the Colorado Plateau and the Great Basin in central Utah. *Utah Geol. Soc. Guideb.*, 4: 106 pp.

Stewart, J. H., 1972. Initial deposits in the Cordilleran Geosyncline: Evidence of a Late Precambrian (< 850 m.y.) continental separation. *Bull. Geol. Soc. Am.*, 83: 1345–1360.

Uchupi, E., Milliman, J. D., Luyendyk, B. P., Bowin, C. O. and Emery, K. O., 1971. Structure and origin of southeastern Bahamas. *Bull. Am. Assoc. Pet. Geol.*, 55: 687–704.

Wang, C. S., 1972. Geosynclines in the new global tectonics. *Bull. Geol. Soc. Am.*, 83: 2105–2110.

Weeks, L. G., 1958. Habitat of oil and some factors that control it. In: L. G. Weeks (Editor), *Habitat of Oil*. Am. Assoc. Pet. Geol., Tulsa, Okla., pp.1–61.

Wengerd, S. A., 1958. Origin and habitat of oil in the San Juan Basin of New Mexico and Colorado. In: L. G. Weeks (Editor), *Habitat of Oil*. Am. Assoc. Pet. Geol., Tulsa, Okla., pp.366–394.

Wengerd, S. A., 1962. Pennsylvanian sedimentation in Paradox Basin, Four Corners region. In: *Pennsylvanian System in the United States*. Am. Assoc. Pet. Geol., Tulsa, Okla., pp.264–330.

Wilhelm, O. and Ewing, M., 1972. Geology and history of the Gulf of Mexico. *Bull. Geol. Soc. Am.*, 83: 575–600.

Wilson, M. D., 1970. Upper Cretaceous–Paleocene synorogenic conglomerates of southwestern Montana. *Bull. Am. Assoc. Pet. Geol.*, 54: 1843–1867.

Winston, G. O., 1971. The Dollar Bay Formation of Lower Cretaceous (Fredericksburg) age in South Florida: Its stratigraphy and petroleum possibilities. *Fla. Bur. Geol., Spec. Publ.*, 15: 99pp.

Young, J. C., 1960. Structure and stratigraphy in north-central Schell Creek Range: In: *Guidebook to the Geology of East-central Nevada. Int. Assoc. Pet. Geol. East. Nev. Geol. Soc.*, pp.158–172.

Ziegler, P. A., 1959. *Guidebook for Canadian Cordillera Field Trip, Internat. Symposium on the Devonian System*. Alberta Soc. Petroleum Geologists, 1055 pp.

Chapter 5

ROLE OF SEDIMENT COMPACTION IN DETERMINING GEOMETRY AND DISTRIBUTION OF FLUVIAL AND DELTAIC SANDSTONES

(CASE STUDY: PENNSYLVANIAN AND PERMIAN ROCKS OF NORTH-CENTRAL TEXAS)

L. F. BROWN Jr.

INTRODUCTION

The role of sediment compaction in determining the geometry and distribution of fluvial and deltaic sandstones was studied in the Upper Pennsylvanian and Lower Permian rocks of the Virgil and Wolfcamp Series on the Eastern Shelf in North-central Texas, which are fluvial, deltaic, interdeltaic, and open shelf facies (Fig.5-1). Ten to fifteen repetitive sequences contain limestones, coals, clays, shales, and sheet and elongate sandstones (Fig.5-2). Elongate sandstone bodies occur at more than 30 stratigraphic levels within the 1,200-ft section of predominantly nearshore facies. Sandstone bodies provide information concerning depositional environments

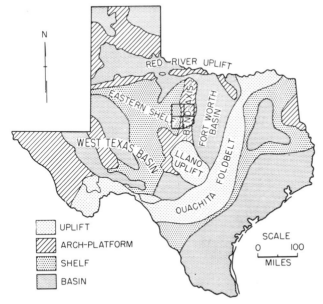

Fig.5-1. Index map and major structural features of Texas. (After Dallas Geological Society, Texas Highway Map.)

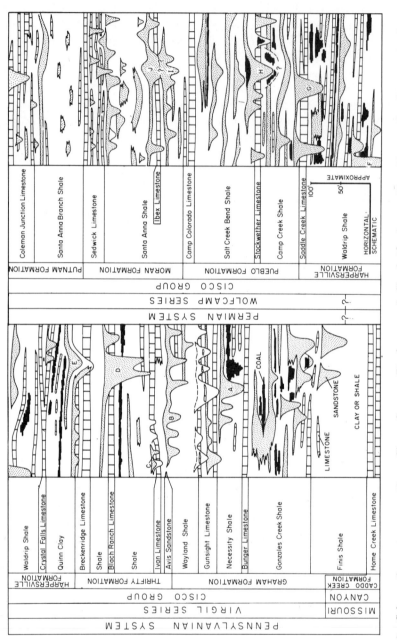

Fig.5-2. Schematic outcrop section, Cisco Group, Stephens and Shackelford counties, Texas. Sandstones A–J are principal sandstone systems. (After McGowen, 1964; Waller, 1966; and Ray, 1968.)

and paleoslope; more specifically, the spatial distribution and external geometry of these sandstone units provide a unique opportunity to analyze the effect of differential sand—mud compaction and tectonics on the deposition of fluvial and deltaic facies.

A depositional model for these rocks in North-central Texas has been proposed (L. F. Brown, 1969; Galloway and Brown, 1972). In this chapter an attempt is made to understand better the respective roles played by compaction and tectonics in controlling fluvial and deltaic depositional sites. The spatial distribution of elongate sandstones and the structural framework of the region provide principal data for interpretation.

In addition to standard stratigraphic and structural methods, decompaction of mudstones provides a tool to estimate the relative importance of differential compaction and differential shelf subsidence. Conclusions based on decompaction are necessarily speculative because of the status of research on compaction and the nature of assumptions which must be made (Chapman, 1972).

The external geometry of elongate sandstones and the factors which controlled their distribution in upslope areas on the Eastern Shelf were studied. More specifically, the investigation involved (1) construction of a stratigraphic framework which included the spatial distribution of principal sandstone units; (2) interpretation of dominant source direction, paleoslope, and depositional model from stratigraphic and sedimentary information; (3) development of a structural framework using both conventional and residual mapping techniques; (4) comparison of stratigraphic and structural data in order to evaluate possible structural control of elongate sandstone patterns; and (5) general approximation of paleosurface and subsidence trends using decompaction techniques.

Considerable attention has been given to the depositional nature and areal distribution of sedimentary facies, especially sandstones, in order to insure a valid paleogeographic and paleoenvironmental framework within which compaction and tectonic effects can be properly evaluated. The sequential history of deposition in the area, as interpreted from spatially arranged fluvial and deltaic facies, provides a means for testing the relative significance of factors that guided the shifting fluvial and deltaic environments.

The area investigated comprises approximately 2,000 sq miles in Stephens, Shackelford, Callahan, and Eastland counties, Texas. It is sufficiently large to contain significant parts of elongate sandstone systems, but the size minimizes regional facies and tectonic variations.

Earlier workers in this area include Cummins (1891), Drake (1893), and Plummer and Moore (1922). Lee (1938) first clearly recognized the complexity of the stratigraphy and contributed significant ideas on environmental conditions. Cheney (1929) and Cheney and Goss (1952) examined regional structural history. L. F. Brown (1960, 1962), Eargle (1960), Shankle (1960), Stafford (1960a, b), Terriere (1960), Rothrock (1961a, b), and Myers (1965) described Pennsylvanian and Permian elongate sandstones in the region. Adams et al. (1951), Rall and Rall (1958), Van Siclen (1958), Adams and Rhodes (1960), and Jackson (1964) considered deposition on the Eastern Shelf. L. F. Brown (1969) and Galloway and Brown (1972) proposed a general depositional model for Virgil and Wolfcamp rocks in the central part of the Eastern Shelf. Regional Pennsylvanian facies on the Eastern Shelf were discussed by Wermund and Jenkins (1968).

Elongate sandstones occur in rocks of many ages and areas. Investigations involving Paleozoic sandstones similar to Virgil and Wolfcamp facies on the Eastern Shelf were tabulated by Friedman (1960), Pettijohn (1962), and Potter (1963, 1967). Studies have been concentrated in the Appalachian Basin, Eastern Interior Basin, and Midcontinent region; except for the data presented by Galloway and Brown (1972), there is very little published information on Paleozoic sandstones in Texas.

A number of depositional models have been proposed for rocks similar to those in North-central Texas: for example, Pepper et al. (1954), P. Allen (1959), Feofilova (1959), Moore (1959), Pryor (1961), Wanless et al. (1963), Beerbower (1964), Swann (1964), Williams et al. (1964), Duff (1967), Wright (1967), and others. Several studies of Recent sediments, which bear on the depositional interpretation of these Upper Paleozoic rocks, include Kruit (1955), Treadwell (1955), Fisk et al. (1954), Welder (1959), Coleman and Gagliano (1964), J. R. L. Allen (1965), Kolb and Van Lopik (1966), Bernard and LeBlanc (1965), and Frazier (1967).

Stratigraphic and structural control was based on surface and subsurface data. Thin, relatively persistent limestone beds were key stratigraphic units. Elongate sandstones outline the skeletal framework of delta lobes and fluvial channels. Surface control was based on 800 described localities, 300 measured sections, limestone facies maps, and reconstructed elongate sandstone patterns tied by maps of all members and key beds at 1:20,000 scale. Subsurface control included 250 wells correlated with 12 dip and strike sections tied to the outcrop section.

VERTICAL SEQUENCES: STRATIGRAPHIC UNITS

Repetitive sequences

Virgil and Wolfcamp rocks in upslope areas of the Eastern Shelf are composed of thin, persistent limestones that are interstratified with thicker mudstone or shale units containing sheet sandstones, elongate sandstones, and less common coal or bituminous shales (Fig.5-2). Ten to fifteen mudstone and sandstone sequences (or "cycles") are separated by regionally persistent limestone beds within the 1,200-ft section.

Sequences normally display an orderly vertical arrangement of facies (Fig.5-2). A generalized sequence of facies (upward) includes: (1) thin, persistent limestone beds; (2) extensive clay—shale facies, which contain marine fossils near the base but become unfossiliferous and silty near the top; (3) local elongate sandstone bodies, which are commonly oriented east—west and laterally are equivalent to clay—shale facies; in places clay—shale contain limestone lenses, coals, bituminous shales, and lenticular sandstones; and (4) clay—shale facies overlying elongate sandstones, which contain some coal and bituminous shale, sheet and bar sandstones. Although ten to fifteen such repetitive sequences occur, many local variations exist in the vertical succession of facies. Aside from minor variations, the most significant regional variation is the presence or absence of dip-fed elongate sandstone facies.

Formal stratigraphic classification developed by early workers was unsatisfactory (Plummer and Moore, 1922). Formation names rarely coincide with significant stratigraphic units. Names applied to individual limestone and sandstone beds were utilized where applicable. Most major sandstone systems are unnamed and are denoted by letters $A-H$ (Fig.5-2).

Formats

Superposed sequences of limestone-bounded, dominantly terrigenous clastic rocks are extensively distributed throughout approximately 25 counties on the Eastern Shelf (Fig.5-3). Bounding limestone units are not necessarily time-stratigraphic but represent the best time-markers in the section (L. F. Brown, 1969). Limestone-bounded stratigraphic units or sequences are persistent and mappable rock units at the outcrop and in the subsurface, and have genetic significance important in understanding the origin of the repetitive shelf facies (Fig.5-2 and 5-3). In this chapter, these subdivisions are called *formats* (Forgotson, 1957). These marker-defined, operational units are informal stratigraphic subdivisions designated by the names of bounding limestones (e.g., Home Creek—Bunger format). Virgil and Wolfcamp formats are commonly 100 ft thick in upslope areas, but several superposed sequences may be combined into thicker formats useful in solving specific problems of stratigraphic analysis. Format boundaries used herein are regionally persistent limestones — Home Creek, Bunger, Gunsight, Blach Ranch, Breckenridge, Crystal Falls, Saddle Creek, Stockwether, and Camp Colorado (Fig.5-3).

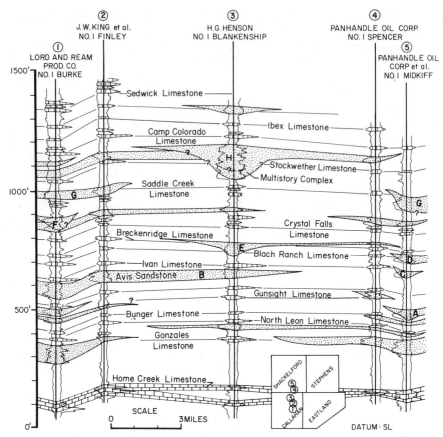

Fig. 5-3. Subsurface reference section, Cisco Group, Callahan and Shackelford counties, Texas. Formats are arbitrarily-defined units bounded by limestones. (After Seals, 1965.)

The term Cisco in this chapter refers to nearshore, primarily terrigenous rocks of the Virgil and Wolfcamp Series that occur between the underlying open marine Canyon facies and the overlying Wichita—Albany facies (Fig. 5-2). Cisco rocks crop out in the Brazos and Colorado Valleys; Cisco Group is not necessarily synonymous with Cisco Series (L. F. Brown, 1959). Dominantly nonmarine rocks, which are approximate upslope equivalents of the Cisco Group, crop out in the Trinity Valley. These fluvial and associated facies are informally designated "Bowie rocks".

SANDSTONE FACIES: GEOMETRY AND ORIGIN

Two general classes of sandstones occur within Virgil and Wolfcamp

Series in North-central Texas: elongate and sheet sandstones. More common elongate sandstones can be mapped from outcrop westward into the subsurface of the Eastern Shelf; sheet sandstone facies are thin and difficult to map. Interpretations of origin of individual sandstone bodies are based on sedimentary structures, internal geometry and stratigraphic relationships, and more importantly on external geometry, sand-body distribution, and facies relationships.

Elongate sandstones

Most elongate sandstones within the area are oriented northeast–southwest (Fig.5-4A). Some sandstones are of channel origin, displaying prominent erosional contacts with subjacent rocks. Other sandstones display gradational contacts; locally gradational units may contain small channel bodies or extensive channels may be superimposed on the nonchannel facies. Sedimentary and stratigraphic evidence indicates that most elongate sandstones in the area are segments of dip-fed fluvial and/or deltaic facies deposited on a surface sloping southwestward at less than 5 ft/mile. Few significant strike-fed, elongate bar sandstones have been observed.

Elongate sandstones at the outcrop are parts of major fluvial and delta complexes which normally extend into the subsurface for tens of miles. Individual sandstone bodies commonly include more than one facies. For example, a sandstone may be composed of superposed delta front, channel-mouth bars, distributary channels, and superimposed fluvial sandstones, as well as destructional sheet and bar sandstone bodies deposited along the periphery of the deltas, all of which might be mapped in outcrop or subsurface as a single sand unit. Progradation of fluvial and delta facies resulted in sandstone geometry ranging from relatively simple distributary bodies to complex belt sandstones. Although many sandstones may locally display erosional bases, they are principally gradational with underlying strata.

Fluvial sandstone facies

Channel deposits include sandstones, conglomerates, and rarely mudstones. Sandstone channels range from bodies deposited contemporaneously with associated rocks to those which cut tens of feet into subjacent strata (Fig.5-4B). Few levee and overbank deposits have been recognized. Channel sandstones and conglomerates at the outcrop contain sedimentary structures which confirm westward to southwestward paleoslope.

Abrupt contacts of channel deposits with regionally persistent limestone marker beds clearly outline channel boundaries. In the subsurface the

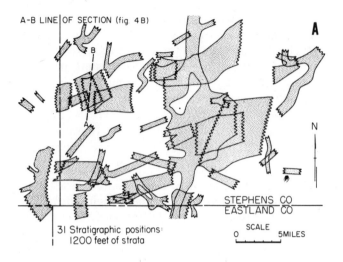

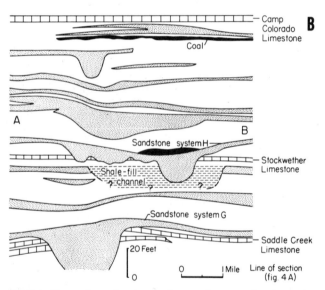

Fig. 5-4. Geometry of outcropping sandstones, Stephens County, Texas. A. Restored areal distribution of elongate sandstones. B. Cross section of elongate sandstones. (After McGowen, 1964; Waller, 1966; and Ray, 1968.)

local replacement of a limestone bed with a sandstone along a linear belt indicates channel erosion.

Fluvial sandstones are medium- to fine-grained. Lenses of chert conglomerate are common, especially near the base of channels. Locally derived limestone conglomerates occur near the base and along the flanks of some channel bodies. Simple fluvial channels display asymmetric cross sections with maximum erosion near one bank of the channel. Most fluvial sandstones are composed of numerous superimposed channels resulting in complex internal structure.

Plant fragments and clay clasts are common constituents. Marine fossils occur in the upper part of some channel sequences. Upward fining of sediment characterizes many fluvial channel units. Festoon crossbedding is more common near the base of channels, and ripple cross stratification typifies uppermost sandstones. Some channels, especially those containing extensive chert gravel, show horizontal, foreset and trough crossbedding typical of braided stream deposits; others exhibit an upward decrease in grain size, as well as an upward decrease in the scale of sedimentary structures within the point bar sequences.

Fluvial sandstones grade upward into overbank mudstones, marine sandstones, or marine limestones (Fig.5-5). Bituminous shales or impure coal beds occur within some channels, indicating cut-off and filling with suspended sediment (Fig.5-4B). Fluvial channels commonly cut into or through underlying deltaic facies; larger channels commonly cut deeply into marine facies of the previous depositional sequence. At the outcrop, lower Cisco fluvial channels are normally incised into thick subjacent delta facies. Upper Cisco channels cut into thin deltaic facies or rest directly on marine facies, suggesting erosion of deltaic facies in upslope areas or, more probably, rapid progradation across nearshore marine facies.

The lack of extensive subaerial erosion indicates that most fluvial sandstones were deposited within meander belts on a low-lying plain near regional base level, probably the upper part of a delta plain. Fluvial channels in the area normally do not fit a braided piedmont or upper coastal plain fluvial model, although such models explain Bowie fluvial rocks exposed upslope in the Trinity Valley.

Deltaic sandstones

Sandstones of deltaic origin exhibit vertical sequences composed of constructional delta front sheet sands, distributary mouth bars, and distributary channel facies (Fig.5-5). In addition, some local sandstone bars of destructional origin occur along the fringes of the delta. Fluvial channel deposits, which prograded over the upper delta plain and occupied many previous distributary channels, are difficult to differentiate from distributary channel deposits, especially in the subsurface.

Delta front facies are normally well-bedded and well-sorted sandstones which contain parallel laminae, some ripple cross laminae, and symmetrical

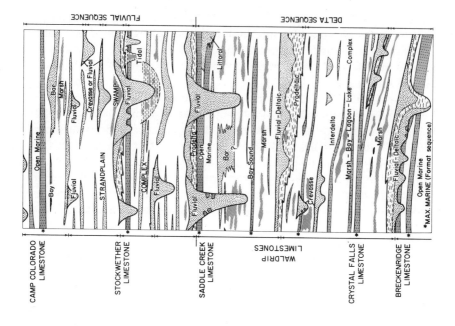

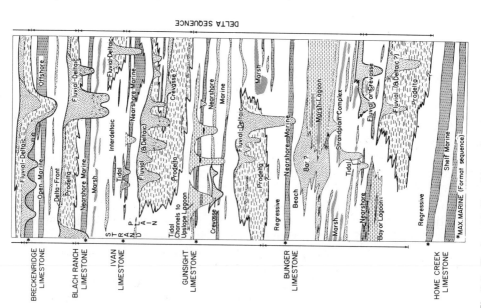

Fig.5-5. Facies relationships of outcropping Cisco and Bowie rocks, Stephens County, Texas. Compare sequences between bases of fluvial or deltaic facies and sequences bounded by transgressive limestones (format sequences).

ripple bedforms on upper surfaces. The base of delta front sandstones commonly exhibits highly-contorted bedding. Delta front sandstones are gradational below with laminated, plant-rich prodelta clays and siltstones. Distributary mouth bars are composed of well-sorted, highly-contorted sand. Relict parallel laminae and some trough crossbedding may be preserved. Injection of sand from below, along with rolled and squeezed structures, indicates extreme compactional distortion of the facies resulting from progradation of deltaic sands over water-saturated prodelta muds. Shallow, symmetrical distributary channel sandstone facies occur near the top of the bars. These channels may be cut into the underlying bar, or they may occur within overlying delta plain mudstones. Erosion within the delta sequence is restricted to the base of channels. Distributary channels are commonly narrow; smaller channels display crevasse splay characteristics, including climbing ripples and subaqueous levees. Later fluvial channels may cut into or through the delta facies, leaving a strong imprint of superimposed fluvial character on the sandstone facies complex.

Sheet sandstones

Sheet sandstones compose a small part of the total volume of sandstone within the Cisco Group. Sheet and thin bar sandstones primarily occur as fringing deltaic facies and within interdeltaic areas.

Delta front sands were reworked and redistributed along strike during deltaic deposition, and this redistribution of sand continued after abandonment and during destruction of various lobes. At the outcrop, these extensive, thin (2—8 ft thick), well-bedded sandstones normally occur within interdeltaic areas at the approximate stratigraphic position of fluvial and deltaic sandstones (Fig.5-5). Sheet sandstones contain burrowing pelecypods and wave ripples on upper surfaces. At some localities, distributary channels cut the sandstones, indicating delta front origin. It is probable, however, that sheet sandstones flanking elongate deltaic sandstones represent intergradational delta front and marine destructional deposition.

Along the flank of deltas, thin bar sandstones and sandy limestones (1—4 ft thick) were deposited during marine reworking of deltaic sands. These bars are very local, rippled, commonly repetitive and highly burrowed. Bars may grade upward into clastic limestones.

Within interdeltaic areas, sheet and bar sandstones (1—20 ft thick), interbedded with relatively unfossiliferous mudstones, compose strike-fed strandplain or chenier facies (Fig.5-5). These sandstones, which commonly display beach characteristics, were derived from abandoned deltas during marine destruction. They were transported along strike into delta flank basins and other interdeltaic embayment areas. Sheet sandstones are genetically important facies within the Cisco, but in the subsurface these units are normally too thin to map.

Sandstone distribution

Depositional patterns

In plan view, dip-fed sandstones display three general areal patterns: distributary, belt, and composite. These patterns are similar to Eastern Interior Pennsylvanian sandstones (Friedman, 1960; Potter, 1963).

Distributary sandstone patterns bifurcate down paleoslope from single trunk streams (Fig.5-6A). Similar patterns in Illinois (Swann, 1964), in Oklahoma (Busch, 1953, 1959), and in Kansas (S. L. Brown, 1967) were interpreted to be of deltaic origin. These bifurcating sandstone bodies are principally barfinger sandstones (Fisk, 1961). Distributary sandstones commonly terminate downslope within about 20 miles of the principal trunk stream. They represent relatively small deltas restricted to upslope areas. Inasmuch as they failed to prograde far downslope, deeply-eroding fluvial channels did not significantly modify the deltaic facies.

Belt sandstone patterns characterize more extensive elongate sandstone systems which extend far down paleoslope (Fig.5-6B). At the outcrop, these sandstones display deltaic sequences, but prominent fluvial channels commonly cut the deltaic sandstones, modifying and eroding these upslope deltaic facies as the delta—fluvial complex shifted basinward. Deposition was maintained by relatively steady sediment supply. Individual sandstone facies within a belt complex are not necessarily contemporaneous.

Within the belt sandstones there are many small coalescing lobes composed of delta front sheet sands and distributary mouth bars, distributary channels, and destructional bars and sheet sands. Superimposed fluvial channels at the outcrop are commonly conglomeratic and deeply erosional.

Composite sandstone patterns are intermediate between distributary and belt geometry (Fig.5-6C, D). Within some formats, relict distributary sandstones in upslope areas were bypassed by belt sandstones which extend farther down paleoslope. Composite patterns reflect a period of deltation in upslope areas during which distributaries switched from site to site from a single trunk fluvial stream, followed by extensive westward progradation of a major distributary of the system.

Paleoslope

Most Cisco sandstone systems display dominant northeast—southwest orientation. Detailed mapping and reconstruction of sandstone geometry at

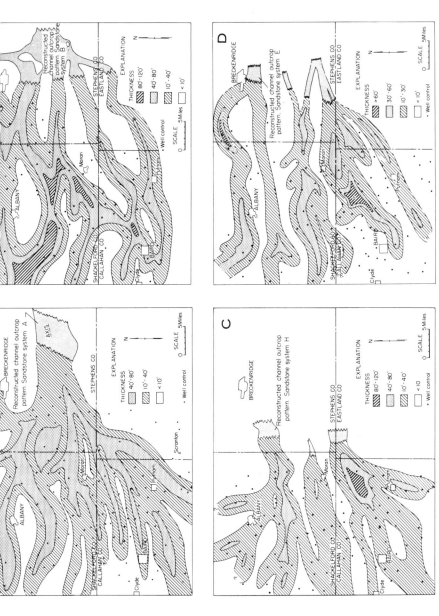

Fig.5-6. Cisco and Bowie fluvial–deltaic sandstones, North-central Texas. A. Sandstone system A. B. Sandstone system B. C. Sandstone system H. D. Sandstone system E. Refer to Fig.5-2 and 5-3 for stratigraphic position. Pattern represents general trends of delta sand complex; interdistributary mudstone facies and local subdelta sandstone variations may occur within pattern area at this scale and well control. (After Seals, 1965.)

31 stratigraphic positions illustrate west—southwest paleotransport routes (Fig.5-4A). Sedimentary structures within these linear sandstone facies confirm this transport direction. Principal sediment source was to the northeast, probably in the Ouachita Mountains and associated piedmont.

The persistence of depositional sites of sandstone facies is a distinctive feature of Cisco sedimentation (Fig.5-7). Concentration of sandstone bodies within 10 principal systems points to an average paleotransport direction of S65°W. Inferred paleoslope was in the westward direction in the northwestern part of the map area. A discontinuity separating southwestward trends from westward trends coincides with inferred structural axes discussed below.

Gradients during deposition of dip-fed sandstones were significantly less than 5 ft/mile. Regionally parallel sandstones, limestones, and coal beds indicate little difference in gradient during deltaic progradation and limestone deposition. Prodelta mud slopes probably locally increased the gradient at which distributary mouth bars and channels were deposited.

Present structural configuration at the base of elongate sandstones is evidence of erosional and/or compactional nature of basal channel surfaces (Fig.5-8). Structural contours at the base of sandstones define V-patterns with axes plunging westward to southwestward at about 50 ft/mile. Asymmetry of structural contours at the base of sandstones, especially those with northeast—southwest orientation, confirm post-depositional tilting to the northwest during the Permian (Fig.5-8B). When regional structure (50 ft/mile northwest) is subtracted from structure at the base of a sandstone system, resulting structural residual values outline the sandstone trend

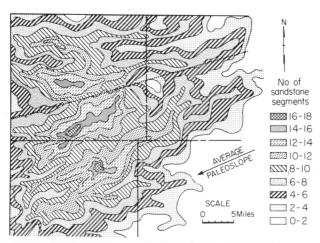

Fig.5-7. Concentration of Cisco and Bowie elongate sandstones, North-central Texas. Based on number of sandstone segments per 4 sq mi within ten major sandstone systems (A—J, Fig.5-2). Dotted line lies along discontinuity in sandstone concentration.

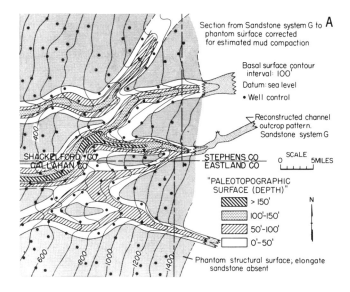

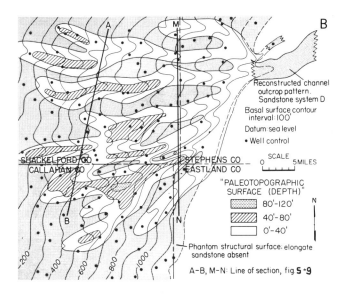

Fig.5-8. Structure and inferred paleotopography at base of elongate sandstones. A. Sandstone system *G*. B. Sandstone system *D*. Paleosurface configuration is isopach map from base of sandstone to overlying phantom structural surface near top of sandstone system. (After Seals, 1965.)

and indicate that regional gradients were consistently less than 5 ft/mile. Sandstone was, therefore, deposited upon surfaces with very low gradients. Variations probably reflect local scouring or subsidence of sandstones into water-saturated prodelta muds.

Isopach maps were prepared of intervals between the base of sandstones and overlying or underlying stratigraphic datum surfaces in order to estimate the configuration of basal sandstone surfaces at or soon after deposition (Siever, 1951; Andreson, 1961, 1962). These methods, commonly referred to as "paleotopographic mapping", eliminate some distortion resulting from post-depositional mud compaction. Maps of distributary sandstone patterns display closed contour patterns in distributary branches (Fig. 5-8B). Because unusually deep channel erosion in distributary lobes is unlikely, closed patterns probably reflect (1) distal bar-finger sandstones that are laterally gradational with prodelta mudstones and (2) differential compaction of bar-finger sands and interdistributary muds resulting in closed thick areas. Decompaction of mudstones within the isopach interval (see discussion of decompaction, p.283) did not significantly change or open the isopach patterns in a downslope direction, confirming that distributary patterns most likely represent gradational bar-finger sandstones rather than deep-cutting, valley-fill sandstones.

Belt sandstones display fluvial patterns on similarly constructed isopach maps. Closed thick areas occur where the sandstone is thickest, suggesting some compactional distortion. When the mudstones within the interval are decompacted, however, the isopach patterns open down paleoslope (Fig. 5-8A). Such patterns are typical of uppermost Cisco sandstones which occur within deeper erosional channels at the outcrop. Far downslope these sandstones exhibit less basal channel erosion where they grade into dominantly deltaic facies.

The nature of sandstone facies along dispersal routes within a single sandstone system varies markedly from upslope deltaic areas to distal depositional sites. In upslope areas, fluvial channels commonly cut subjacent shelf limestones (Fig. 5-9, section $M-N$). Downslope in the same sandstone system bar-like sandstones rarely cut subjacent limestone beds and are generally gradational with adjacent strata (Fig. 5-9, section $A-B$). Greater compactional distortion also occurs in downslope bar-finger bodies where thicker, water-saturated prodelta muds surrounded deltaic sands.

Use of structural and stratigraphic methods to estimate paleogradients in distal parts of distributary sandstone systems is impractical because of the gradational nature of the basal contact and the effects of differential compaction of bar-finger sands and prodelta muds (Fig. 5-8B). Paleogradients of upslope fluvial channels, however, can be approximated using structural residual or paleotopography (isopach) mapping techniques. By subtracting

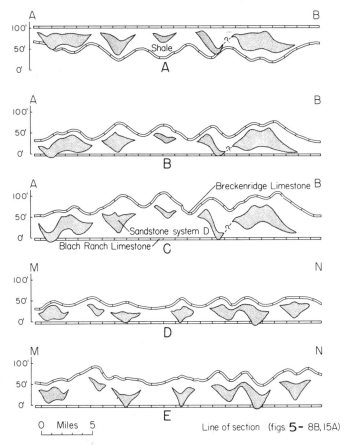

Fig. 5-9. Cross sections, Blach Ranch—Breckenridge format. A. Breckenridge Limestone datum (downdip). B. Blach Ranch Limestone datum (downdip). C. Decompacted downdip section. D. Blach Ranch Limestone datum (updip). E. Decompacted updip section.

regional structure from structural configuration at the base of erosional sandstones, gradients at the base of Cisco fluvial sandstones can be inferred to have been less than 1—2 ft/mile; this supports other evidence of delta plain fluvial origin rather than piedmont or upper coastal plain fluvial deposition (Fig. 5-8A).

Spatial distribution of sandstones: vertical patterns

Spatial distribution of principal sandstone bodies is a key to factors controlling depositional sites. It was noted above that Cisco sandstones were deposited within distributary, belt, and composite systems (Fig.5-6). The vertical relationship between these superposed sandstone systems provides evidence of the respective roles played by compaction and structure in determining the distribution of sand depositional sites within each repetitive sequence. Most Cisco sandstones display vertically offsetting relationships with overlying and underlying sandstone systems (Fig.5-10A). Along several narrow east—west belts, however, superposed sandstones are commonly stacked in vertical multistory arrangement (Fig.5-10B). Offset and multistory vertical patterns persist throughout 1,200 ft of fluvial and deltaic facies.

Coastal plain fluvial systems, which occurred updip from the present outcrop, must have maintained relatively permanent routes. The distribution of ten Cisco sandstone systems indicates that most deltaic systems originated at common point sources, indicating little shifting of river channels upslope from their junction with widely shifting delta distributaries.

Offsetting sandstones

The common occurrence of vertically-offset sandstones is compatible with deposition on a slowly subsiding, slightly tilting shelf where dip-fed deltaic sands were deposited on water-saturated prodelta muds. Exceedingly slight topographic depressions or troughs that developed above highly-compacted interdistributary and interdeltaic muds provided an efficient path for subsequent delta progradation. Differential compaction of relatively non-compactible deltaic sands and highly compactible muds should, therefore, result in vertically offset deltaic depositional routes.

Following delta deposition in the area, delta destructional and marine transgressive facies were deposited over the abandoned and subsiding delta platform. The effect of differential compaction of delta facies was transmitted to these subsequent facies, for delta sands of the next overlying sequence commonly offset underlying sandstones. Offset sandstone deposition is further discussed in the section entitled "Sandstones and paleosurface trends" (p.285). Throughout most of the area, superposed Cisco sandstones are arranged in offset patterns (Fig.5-10A). This common spatial distribution is compatible with deltaic deposition of sand and water-saturated muds on a slowly subsiding shelf. Not all sandstones in the section, however, display offset patterns, e.g., multistory patterns.

Multistory sandstones

Vertically stacked or multistory sandstones occur in narrow belts oriented along paleoslope. These belts, illustrated by charting areas where maximum thickness axes of superposed sandstones overlap (Fig.5-10B),

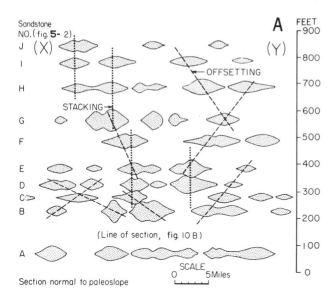

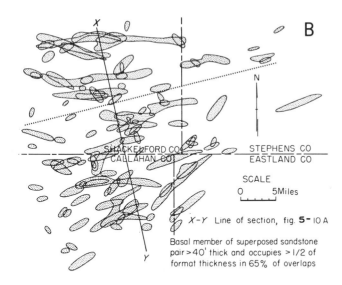

Fig.5-10. Vertical arrangement of Cisco—Bowie sandstones. A. Offset and multistory patterns. Sandstone geometry is schematic. B. Areal distribution of multistory sandstones. Each stippled area outlines one superposed pair of multistory sandstones. Dotted line lies along discontinuity in multistory orientation. (After Seals, 1965.)

correspond closely with belts of maximum sandstone concentration (Fig.5-7).

In 65—70% of multistory sandstone occurrences, the underlying sandstone displays greater than average thickness. Subsidence of thick bar-finger sands into subjacent muds provided subtle overlying paleotopographic depressions which guided progradation of later, superposed delta systems. It is difficult to explain the persistence of multistory belts throughout 1,200 ft of Cisco section by differential compaction. Subsidence of unusually thick linear sand bodies into underlying muds might locally have produced overlying paleotopographic depressions responsible for diverting a subsequent delta system. Compactional subsidence sufficient to stack multiple superposed sandstones does not appear likely within thin shelf sequences. Slight structural weakness or instability, coupled with loading of these less stable belts, is more likely a dual mechanism for multistory deposition in North-central Texas. The relationship of unstable structural belts and multistory sand deposition is discussed in the section entitled "Sandstones and subsidence trends" (see p.281).

Spatial distribution of Cisco fluvial and deltaic systems points to interplay between differential compaction of deltaic sand and mud facies, as well as slight contemporaneous subsidence related to structurally unstable belts. Differential compaction exerted greater control, except in local belts where overriding tectonics produced multistory relationships.

DEPOSITIONAL MODEL

The Eastern Shelf is a relatively stable tectonic area developed on the Early Pennsylvanian Concho Platform (Cheney and Goss, 1952), which was obscured by the Late Pennsylvanian and Early Permian westward tilting. The north—south axis of westward tilting that approximately coincides with the western Flank of the Fort Worth Basin has been designated the Bend Axis (idem). Cisco strata display little evidence of the axis, but slight structural adjustment along the feature may have been responsible for localized structural activity on the shelf. East of the shelf lay the Ouachita Mountains and the exposed Fort Worth Basin rocks termed piedmont. North of the shelf was the Wichita structural system in southern Oklahoma, whereas southward the shelf apparently deflected around the Llano structural complex (Fig.5-11).

Southeastward Post-Triassic tilting was related to development of the Gulf of Mexico. Lower Cretaceous strata in the region dip southeastward at about 30 ft/mile over a dissected Sub-Cretaceous surface cut into Permian and Pennsylvanian rocks.

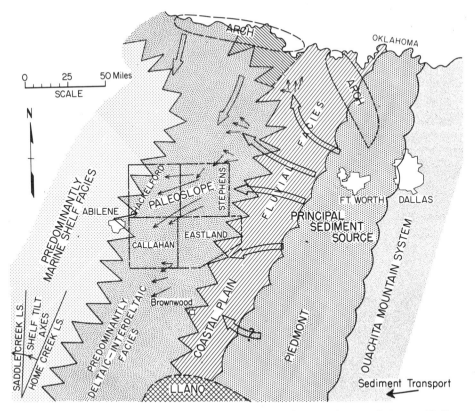

Fig.5-11. Paleogeographic map, Cisco and Bowie rocks, North-central Texas. (Sediment transport directions are after Lee, 1938; Terriere, 1960; Stafford, 1960; Rothrock, 1961; and Brown, 1960, 1962.)

Virgil and Wolfcamp rocks dip northwest at approximately 50 ft/mile. Strike migrates from N 25° E at the base of the Virgil Series to N 10° E at the base of Wolfcamp rocks. Wermund and Jenkins (1964) reported counterclockwise shift in strike from N 45° E in older Desmoines rocks to north—south strike in lower Wolfcamp rocks, indicating accelerated development of the West Texas Basin and gradual decline of the Fort Worth Basin subsidence.

Depositional systems

Virgil and Wolfcamp facies are composed of fluvial, deltaic, interdeltaic, and shelf depositional systems (Fig.5-12). Depositional systems are stratigraphic packages of genetically related facies comparable to modern facies

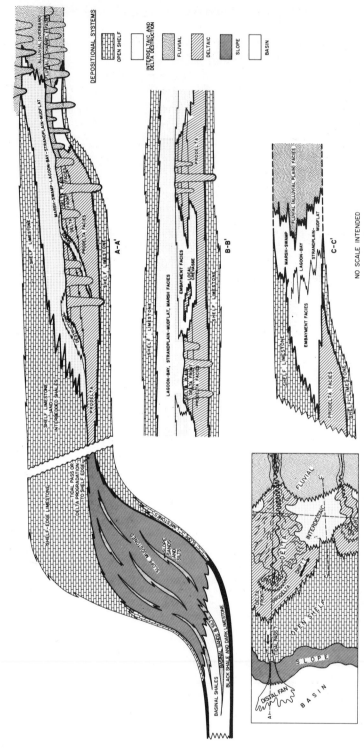

Fig.5-12. Distribution of component facies, Cisco and Bowie depositional systems, North-central Texas.

complexes readily apparent from physiographic characteristics, such as barrier, delta, and fluvial systems (Fisher and McGowen, 1967, p.106).

Fluvial systems contain channel and overbank facies primarily upslope and northeast of the area in the Jack and Montague counties. Coastal plain and piedmont fluvial facies upslope within the Bowie rocks (Fig.5-13) have rarely been recognized within the Brazos or Colorado River valley outcrop belt where fluvial channels were deposited on lower coastal plains and delta plains.

Prograding Cisco delta systems contain prodelta mudstones and siltstones, delta front sheet sandstones and distributary mouth bars, and delta plain facies (distributary channel sandstones, interdistributary coals, mudstones, and crevasse splays) deposited during delta construction (Fig.5-12). Many bar and sheet sandstones, mudstones, coals, and impure limestones were deposited during destruction of the delta. Bar-finger sandstones are composed of various sand facies deposited during delta construction. These

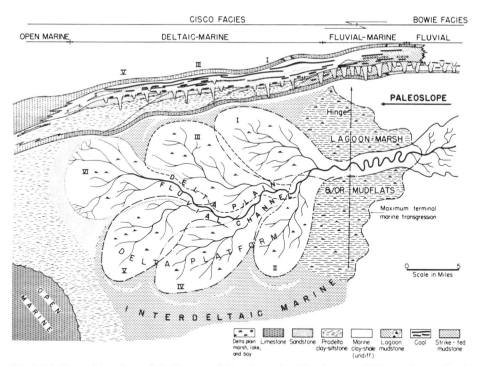

Fig.5-13. Depositional model, Cisco and Bowie rocks, North-central Texas. Many fluvial systems display braided stream geometry and sedimentary structures.

facies, along with superimposed fluvial sandstones, constitute the elongate sandstone facies of this chapter.

Most mudstone facies that are interstratified downslope with shelf limestones are probably of prodeltaic origin. Delta progradation is the only significant mechanism capable of transporting mud across the broad, shallow-water Cisco shelf. Prodelta mud and silt transported to the shelf edge by deltas constitute the major volume of slope sediment responsible for prograding the shelf (Galloway and Brown, 1972).

Cisco interdeltaic systems are principally upslope mudstone facies deposited within embayments flanking delta systems (Fig.5-12). Deltaic sediments redistributed by marine currents were deposited as sheet and bar sands and nearshore muds. Specific facies recognized within interdeltaic systems include lagoon, bay-sound, marsh, and strandplain and mudflat deposits. Interdeltaic sandstones are primarily thin, strike-fed facies.

Cisco shelf systems are composed principally of various limestone facies (Fig.5-12). Possibly some thin shelf mudstones occur in downslope areas, but most muds interstratified with shelf limestones are of prodeltaic origin. It is possible that with accumulation of additional data, it can be demonstrated that some downslope muds were derived by longshore drift from major deltas to the north. Shelf limestones in the lower Cisco are commonly thin, but some upper Cisco shelf-edge limestones are very thick and may occupy an entire format. Shelf limestones were deposited in shallow water in the absence of local terrigenous clastic supply. Limestones on the shelf edge thin upslope into transgressive tongues which separate delta sequences. Downslope, shelf limestones pinch out or thin into slope and basinal facies.

Depositional summary

Delta systems prograded rapidly across the slowly subsiding Eastern Shelf (Fig.5-13). Sediments from the east supplied crevassing delta lobes until avulsion of over-extended systems occurred. Delta construction, accompanied by deposition of complementary facies within nearby interdelta areas, restricted open-shelf limestone facies to downslope areas beyond the effects of local deltaic deposition. Upper delta plain facies were cut by meandering to slightly sinuous fluvial channels. Shelf edges migrated basinward primarily in response to offlapping mudstone deposits supplied by delta systems.

Marine processes slowly modified abandoned, compacting, and subsiding deltas. Winnowed sediments were swept onto interdeltaic mudflats and strandplains; complex mudstone facies occupied coastlines and foundering delta plains. Widespread open-shelf limestone environments, in the absence of local delta deposition, transgressed upslope and along the coast to coalesce over marsh-stabilized, subsiding deltas and interdelta areas.

Within the proposed shelf model, all environments occur simultaneously, shifting with distribution of delta sites to produce repetitive sequences composed of thin, superposed depositional systems, each displaying a more-or-less homotaxial sequence of component facies (Fig.5-13). At any point on the shelf, each limestone-bounded sequence or format contains some variation of the following vertical arrangement of facies: (1) open-shelf limestone system; (2) delta system (prodelta, delta front and distributary mouth bar, distributary channel, mudstone and coal, sheet and bar sandstone); (3) fluvial system (channel sandstone or overbank mudstone facies); and/or (4) interdelta system (marsh, lagoon—bay—sound, strandplain and mudflat); and again (1) open-shelf limestone system. These format sequences appear to be diachronous and aperiodic, but they represent the most synchronous stratigraphic package on the shelf. Constructional fluvial and delta facies occupy short discrete time intervals within the format, whereas most of the total deposition time was consumed by the deposition of destructional and transgressive facies. Cisco and Bowie depositional systems (and component facies) shifted southwestward through time as the eastern flank of the West Texas Basin was filled by the westward prograding shelf.

FACTORS CONTROLLING SANDSTONE DISTRIBUTION

The geometry and spatial distribution of Cisco sandstones indicate that the configuration of surfaces over which deltas prograded was controlled principally by differential sand—mud compaction of subjacent deltaic facies. Multistory sandstones point to subtle, local control of delta sites by differential rates of structural subsidence. Within such a model, each deltaic episode should inherit a paleosurface resulting from compactional and minor structural effects during deposition of the subjacent delta sequence. The relationship of each of these factors — tectonics, compaction, and inherited paleotopography — is considered independently.

Structure and sandstone trends

Cisco rocks on the Eastern Shelf dip northwest at about 50 ft/mile (Fig.5-14A). Northwest-trending minor structural axes commonly have local closure of less than 50 ft. Multistory sandstone belts and axes of maximum density of sandstone bodies locally coincide with regional synclinal trends (Fig.5-10B and 5-7). Elongate sandstone trends and inferred paleoslope deviate about 30°—35° from present structural axes.

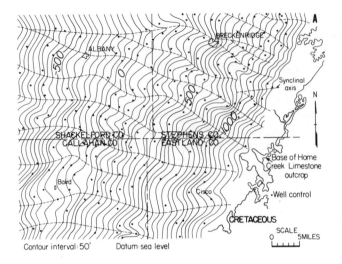

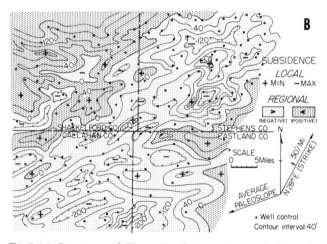

Fig.5-14. Structure of Home Creek Limestone, North-central Texas. A. Structure at base of limestone. (After Seals, 1965.) B. Residual structure at base of limestone. Structural residuals are derived by subtracting regional structure from Fig.5-14A. Note discontinuity of residual values along zero contour in northwest quadrant.

A meaningful display of structural data is obtained by mapping structural residual values. When regional structure is subtracted from structure at the base of key limestone beds, resulting positive and negative residual values define strong northeast—southwest trends. Approximately 70% of all Cisco elongate sandstones in the area, including multistory belts, occurs within a broad negative residual belt trending northeast—southwest

GEOMETRY/DISTRIBUTION OF SANDSTONES: ROLE OF COMPACTION 273

(Fig. 5-14B). Elongate sandstones within positive residual areas normally coincide with minimum positive values; Cisco sandstones are commonly absent in areas of maximum positive residual values. Coincidence of sandstone multistory trends and maximum sandstone density belts with negative structural residual trends suggests persistent but subtle structural control of some deltaic sandstones.

Residual structural patterns clearly delineate broad belts where shelf subsidence was considerably more or less than average for the region. These structural trends apparently affected the general location of the Cisco delta deposition and, therefore, other facies related to delta deposition. Boundaries separating major positive and negative residual belts probably represent subtle flexures that were contemporaneous with deposition. A prominent flexure (zero residual contour) trends northeast—southwest across the northwestern part of the area (Fig. 5-14B). This structural discontinuity coincides with discontinuities in multistory and sandstone density trends (Fig. 5-10B and 5-7). The flexure probably exerted some persistent control of delta distribution in this part of North-central Texas.

Importance of compaction

Because Cisco rocks in North-central Texas are primarily clays and shales, the effects of mud compaction must be considered when analyzing the configuration of erosional or depositional surfaces. The original geometry of fluvial and deltaic facies is distorted by differential compaction of sandstone facies and water-saturated mudstone facies. Insufficient data limit the validity of decompaction techniques to restore original geometry. The potential value of decompaction methods in combined stratigraphic—structural problems, however, warrants its cautious application, if for no other reason than to demonstrate the need for continued research on compaction phenomena.

Evidence of compaction

Isopach patterns of Cisco formats reflect differential sand—mud compaction. Thick isopach trends consistently coincide with the distribution of elongate fluvial and deltaic sandstone bodies within the format (Fig. 5-9 and 5-15A).

Regionally, Cisco shelf formats are tabular to slightly wedge-shaped stratigraphic units which display local thickness variations. Significant regional erosional unconformities have not been recognized. Local channels within fluvial and delta systems, which removed minor volumes of sediment, were immediately filled with sand or mud. Reefs or limestone bank deposits are absent in upslope Cisco areas where thin limestones and sheet sandstones, along with elongate fluvial—deltaic sandstones, compose the relatively noncompactible part of the section. Thickness variations within upslope areas resulted principally from differential compaction of fluvial and deltaic sediments.

Thickness of an interdistributary mudstone section is commonly as little as 50% of adjacent areas containing bar-finger sandstones (Fig.5-15A). Similar compaction estimates were obtained by Mueller and Wanless (1957) for Pennsylvanian rocks in Illinois. Such compactional distortion is extremely significant in producing local structural anomalies.

Fracture patterns based on aerial photograph lineations commonly coincide with fluvial and deltaic sandstone bodies. Fracture density patterns (Fig.5-15B) suggest intense differential compaction along multistory belts and maximum sandstone concentration belts (Fig.5-10B and 5-7). At the outcrop, many local structural anomalies result from drape of strata over relatively noncompactible* sandstones.

Cisco rocks have been significantly distorted by differential sand—mud compaction, as well as moderate contemporaneous subsidence. Elimination of compactional distortion should ideally provide avenues for restoring original geometry and for estimating differential structural subsidence.

Problems of decompaction

Early work by Athy (1930) and Hedberg (1936), among others, provided empirical data related to decreasing shale porosity with depth of burial. Weller (1959) summarized much of the Athy and Hedberg data and graphically illustrated the potential of decompaction in stratigraphic analysis.

Within the past fifteen years, a growing number of workers have been investigating fundamental aspects of fluids, pressures, and mineralogical changes with increasing depth (Dickinson, 1953; Powers, 1967; Rochon, 1967; Jones, 1968; Burst, 1969; and others). These studies have been primarily concentrated in Gulf Coast Tertiary rocks where porosity and depth data are available from thousands of logs and cores obtained from boreholes. Martin (1966) and Conybeare (1967) recently showed the importance of decompaction in stratigraphic restoration studies in Canada.

---

*Data on relative compactibility of sands and clays, which is presented in Chapter 2, does not support this statement. (Editorial comment.)

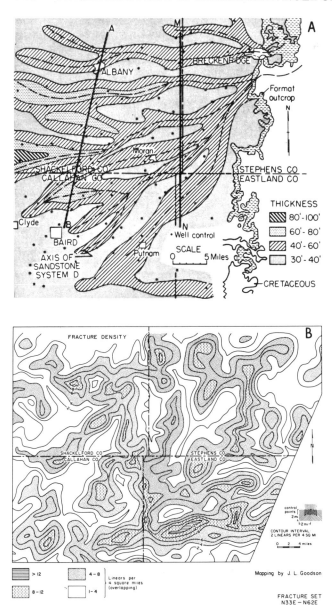

Fig.5-15. Evidence of differential sand—mud compaction, Cisco rocks, North-central Texas. A. Isopach map, highly compacted Blach Ranch—Breckenridge format. (After Seals, 1965.) Sections $A-B$ and $M-N$ on Fig.5-9. B. Inferred compactional fracture density patterns. Based on aerial photographic linears within 4 sq mi, centered on square-mile grid. Compare with trends on Fig.5-7, 5-10B, 5-14B, 5-17 and 5-19.

There is general agreement among workers (Burst, 1969, p.90) that dehydration and compaction of clays occur as a three-stage process. To depths of about 3,000 ft pore water and excess water layers (third layer and higher) are expelled. The second to last water monolayer is in part thermally removed at about 3,000—15,000 ft. During the final stage, below 15,000 ft the expulsion of the last water layer occurs. Dickinson (1953, fig.15), Weller (1959, fig.1), and Conybeare (1967, fig.1) presented porosity—compaction—depth curves which are generally similar to and, in part, based on those presented by Athy (1930) and Hedberg (1936). These graphs are relatively consistent, at least within a general order of magnitude. Application of these curves (or derived curves) to Cisco data resulted in a reasonably close spread of inferred compaction percentages (Fig.5-16). Use of such compaction data and derived curves is made, however, with full understanding of limitations inherent in the quality and quantity of data on compactional processes, as noted by Chapman (1972).

Regional geology indicates that Cisco rocks probably were never buried below 3,000 or 4,000 ft. Fluvial and deltaic sandstones were conduits through which fluids were expelled during compaction. It is assumed that deltaic Cisco rocks have reached compaction equilibrium. It is not known whether or not the time factor has resulted in significant volume reduction in these Paleozoic rocks beyond that of burial pressure and temperature. Possible volume increase by rebound during regional uplift and erosion may be insignificant, but this phenomenon has not been investigated.

In addition to basic problems of mud compaction phenomena and adequate observational and experimental data, Cisco strata present other unique problems: (1) effect of thin, tabular limestone and sandstone beds on fluid migration during compaction; (2) variable clay mineral facies (e.g., prodelta, delta plain, interdeltaic) possessing different compaction properties; (3) limitation of estimating composition and thickness by electric logs; and (4) the arbitrary decision to ignore sandstone and limestone compaction. In the opinion of the writer, the decompaction procedures at present must be strictly qualitative and relative. Qualitative estimates of mud compaction, however, require the use of numbers just like paleoecological or paleoenvironmental reconstruction. Decompaction is a second order interpretation or approximation; therefore, until adequate research on compaction has been accomplished, maps resulting from decompaction procedures should not be interpreted quantitatively. Numbers simply illustrate relative, approximate values. Thus, if they are related to a datum, the datum is also arbitrary and interpretative. Compaction data presented by Athy (1930), Hedberg (1936), Dickinson (1953), Weller (1959), and Conybeare (1967) provided the basis for calculating theoretical decompaction of Cisco rocks (Fig.5-16).

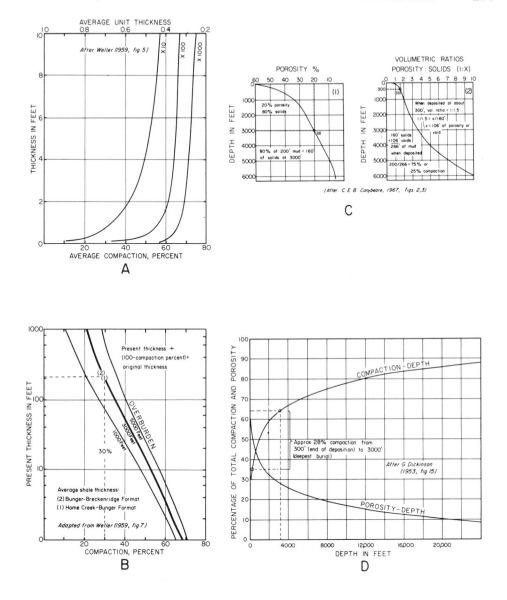

Fig.5-16. Mud compaction data. A. Average compaction of mud columns. B. Mud compaction from overburden. C. Compaction of mud with depth of burial. D. Relation of porosity and compaction of shales with depth of burial. (After Dickinson, 1953; Weller, 1959; and Conybeare, 1967. Published with permission from the *Bulletin American Association of Petroleum Geologists* (Fig.5-16A, B, C); and *Bulletin Canadian Petroleum Geology* (Fig.5-16D).)

Control of depositional surfaces

Inherited paleotopography

The orientation of Cisco fluvial and delta facies was apparently influenced by paleotopography inherited from subjacent limestone bank deposits of the Canyon Group. Initial Cisco deltas prograded over the Home Creek Limestone, uppermost of several bank-like shelf limestones of the Canyon Group (Fig.5-2, p.248). An isopach map of the Home Creek in the area exhibits distinct northeast—southwest thick and thin trends reflecting the irregular limestone bank (Fig.5-17A).

Regional structure was subtracted from the structural configuration at the top of the Home Creek Limestone. The residual surface was reoriented along paleoslope (based on sandstone orientation) with an arbitrary 5 ft/mile southwest dip and recontoured using a phantom datum. Strong northeast—southwest grain appeared that may reflect the paleotopography which controlled initial Cisco progradation routes (Fig.5-17B). The first fluvial and deltaic sequence deposited during Cisco regression (Home Creek to Bunger format) displays several maximum sandstone percentage axes, which coincide with inferred paleotopographic depressions or troughs on top of the underlying Home Creek Limestone. Prograding deltas apparently followed subtle interbank low areas on the upper Home Creek surface, where thicker sections of prodelta muds compacted to provide more efficient progradational routes.

West to southwest, paleosurface grain persisted throughout Cisco deposition in this area. Deltaic deposition within each successive sequence followed the most favorable paleosurface routes inherited from the succeeding sequence. The configuration of paleosurfaces at the end of each Cisco sequence (format) apparently resulted from two principal factors: variable rates of shelf subsidence and differential sand—mud compaction.

Differential subsidence

Isopach maps of conformity-bounded units are commonly used for estimating contemporaneous structural activity when there are no significant depositional breaks in the section. Compactional distortion of Cisco deposits precludes direct use of isopach maps for approximating relative rates of subsidence during deposition (Fig.5-18). A hypothetical noncompaction model of formats that theoretically reflects format geometry, assuming that no compaction occurred during deposition and burial, provides a method for estimating relative rates of shelf subsidence.

Estimation of subsidence. Noncompaction models were constructed by decompacting all mudstones within each format (Fig.5-18A) to hypothetical

GEOMETRY/DISTRIBUTION OF SANDSTONES: ROLE OF COMPACTION   279

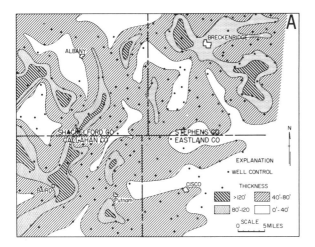

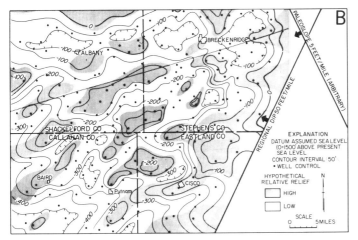

Fig.5-17. Geometry of Home Creek Limestone. A. Isopach map of Home Creek Limestone. (After Seals, 1965.) B. Inferred paleotopography, upper surface of Home Creek Limestone. Inferred paleosurface is structure map following arithmetic tilting to paleoslope. Dotted line lies along discontinuity in contour pattern.

precompaction (or initial depositional) thickness (Fig.5-18C). Limestones and sandstones were arbitrarily assumed to be noncompactible. For example, 235 ft of mudstones within a format, which has theoretically undergone 30% mud compaction during burial, was 335 ft thick at the end of format deposition (Fig.5-16B). Hypothetically, this 335 ft of mudstone represents about 930 ft of mud having 80% porosity and which underwent compaction during and immediately after deposition (Fig.5-16A). Empirically, a column of mud 150 ft thick will undergo about 60% average compaction, whereas

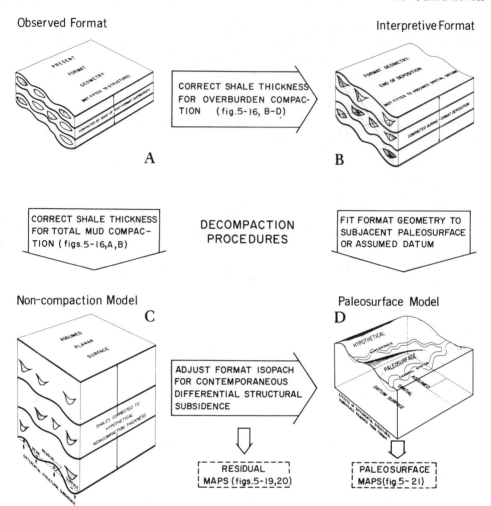

Fig.5-18. Decompaction procedures. A. Observed format. Isopach map based on well data. B. Interpretative format. Isopach map decompacted to pre-overburden thickness. C. Noncompaction model. Isopach map decompacted to hypothetical thickness assuming no compaction. D. Paleosurface model. Isopach map is decompacted, corrected for structural subsidence, and fitted to arbitrary subjacent surface; thickness reflects relative paleotopographic relief.

700 ft of mud will theoretically display 65% average compaction (Fig.5-16A). Formats containing more than 150 ft of mudstone were selected to minimize the percentage error.

If the geometry of shelf sand and mud formats resulted primarily from differential rates of structural subsidence and from differential sand—mud compaction, removal of compactional distortion should produce geometry reflecting structural effects. Residual thickness (isopach) values of non-compaction models (based on mean thickness), which theoretically have had the effects of compaction removed, should approximate relative (positive and negative) magnitude of shelf subsidence (Fig.5-18C). For convenience in applying residuals to paleosurface (isopach) restoration, the values were arithmetically returned to hypothetical thickness at the end of format deposition.

Residual map patterns of various formats based on this technique define northeast—southwest-trending belts of negative and positive values (Fig.5-19). If such residual patterns reflect relative rates of structural subsidence, a relationship should exist between sandstone distribution and subsidence patterns.

Sandstones and subsidence trends. Areas containing high concentrations of elongate sandstones and multistory belts commonly coincide with northeast—southwest-trending negative isopach residual belts (Fig.5-20). Non-compaction isopach residuals are interpreted to reflect relative rates of shelf subsidence. Residual patterns of successive formats display progressive changes which reflect slight structural evolution, but the fundamental orientation of negative and positive belts persists throughout deposition of the Cisco sequence (Fig.5-19, 5-20).

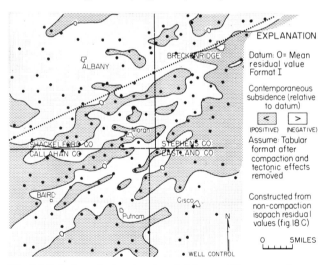

Fig.5-19. Inferred structural subsidence, Home creek—Bunger format. Isopach residual map using positive or negative values is based on mean thickness of noncompaction isopach map (Fig.5-18C). Contours are omitted. Dotted line lies along discontinuity in residual patterns.

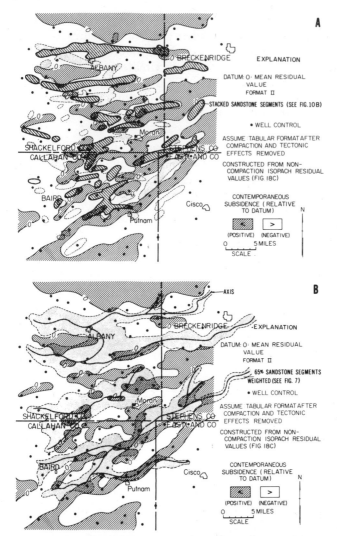

Fig.5-20. Relationship between inferred structural subsidence and distribution of elongate sandstones, Bunger—Breckenridge format. A. Structural subsidence and multistory sandstones. Heavy lines outline multistory sandstones within this format. B. Structural subsidence and concentration of elongate sandstones. Structural subsidence map is noncompaction isopach residual map (Fig.5-18C). Contours are omitted.

A northeast-trending discontinuity displayed by isopach residual patterns in the northwest quadrant of the area (Fig.5-19) coincides with similar features on stratigraphic maps and residual structural maps (Fig.5-7, 5-10B, 5-14B on p.260, 265, 272 resp.). This similarity strengthens arguments for some local structural control of fluvial and deltaic depositional sites. The area northwest of the discontinuity near Albany, for example, was predominantly negative during deposition of the Home Creek—Bunger format (Fig.5-19), but during deposition of the Bunger—Breckenridge format an east—west positive belt divided the earlier negative area (Fig.5-20). Each sandstone system deposited within the latter format reflects the local positive belt by splitting around the more structurally stable area (Fig.5-6A, B and 5-8B on p.259 and 261).

Areas where sandstones are absent commonly coincide with unusually positive isopach residual areas, whereas channel sandstone concentration and multistory belts generally coincide with extremely negative residual areas. Persistent nondepositional belts, such as the one trending northeast—southwest through the extreme southeastern corner of Shackelford County (Fig.5-7, p.260) coincide with strong positive residual trends (Fig.5-19) and areas of maximum paleosurface relief inferred from restored isopach maps (Fig.5-21).

In summary, fluvial and deltaic sandstone geometry and inferred unstable structural trends indicate that (1) unusually high or low rates of structural subsidence along narrow west to southwest-trending belts controlled location of highly concentrated and multistory sandstone belts (negative) or persistently nonsandstone belts (positive), respectively; and (2) dominant differential sand—mud compaction in larger areas with intermediate subsidence rates apparently resulted in common offsetting vertical patterns.

Differential compaction

Offset vertical relationships exhibited by sandstone facies point to control by differential sand—mud compaction. Although multistory sandstone belts coincide with inferred trends of maximum shelf subsidence, most delta deposition occurred within areas underlain by interdistributary and interdeltaic mud facies of the previous sequence.

Decompaction procedures. If mudstones within a format containing fluvial and deltaic facies can be decompacted theoretically to approximate thickness at the end of deposition (before burial under an overburden), geometry or thickness variations displayed by the format could logically be a guide to paleosurface configuration on top of the format (Fig.5-18B). For example, if thin areas or trends coincide with interdistributary facies within the format and if thick trends coincide with fluvial and deltaic sandstone facies, it is likely that these thickness trends were reflected by subtle differences in relative paleotopographic relief. This assumption is further

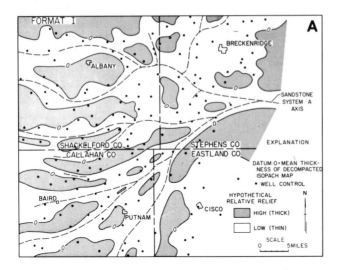

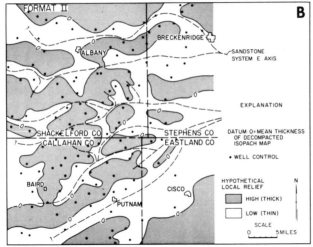

Fig. 5-21. Inferred paleosurface map. A. Decompacted isopach map, Home Creek–Bunger format (pre-sandstone system A surface). B. Decompacted isopach map, Bunger–Breckenridge format (pre-sandstone system E surface). Superjacent fluvial and deltaic sandstone trends indicated by dashed line. Refer to Fig. 5-18D. Contours are omitted.

confirmed when an overlying sandstone system coincides with thin areas or inferred paleosurface lows on the underlying format.

Compaction of mud begins with deposition, but the degree of compaction following deposition is critical when attempting to restore approximate geometry. It is assumed that the Cisco rocks were buried by at least 3,000 ft of strata, although precise thickness of overburden makes little significant difference in the average percent compaction (Fig.5-16B, C, D). Graphs (idem) illustrating percentage of mud compaction with increasing depth show that under approximately 3,000 ft of overburden, 250-ft mud columns have been compacted by approximately 25—30%. Cisco mudstones at 200 well locations were decompacted using a 30% compaction estimate.

Differential rates of shelf subsidence, especially within a terrigenous clastic province, would result in local thickness variations which would not necessarily affect the paleosurface. For this reason, isopach residual values were algebraically added to format thickness values for each control point (Fig.5-18D). Resulting format thickness trends theoretically represent only the effects of differential sand—mud compaction during deposition of the format.

When the decompacted isopach map is fitted to a horizontal, tilted, or assumed subformat surface, the isopach values theoretically reflect areas of relative paleotopographic relief at the end of deposition (Fig.5-21). Each format inherits some paleosurface relief from subjacent formats, so that a decompacted isopach map is not quantitative but supplies additional confirmation that interdistributary or interdeltaic mudstone areas were prime depositional sites for superjacent deltas. The best and final test of the validity of such inferred paleosurface (isopach) maps is comparison with the distribution of superjacent sandstone systems.

Sandstones and paleosurface trends. Fluvial and deltaic sandstones, which prograded over each Cisco format, generally followed paleotopographic depressions or troughs inferred from decompacted isopach maps. For example, axes of maximum thickness for sandstone A generally coincide with interdistributary depressions displayed on the subjacent Home Creek—Bunger format (Fig.5-21A). Distributary sandstone bodies clearly deflect around areas of greater than average thickness or inferred paleosurface relief. Compaction of the underlying format provided sufficient relative relief to control depositional sites of prograding distributaries that deposited deltaic sandstone system A. Similarly, the distribution of sandstone system E, which is composed dominantly of fluvial channel deposits, closely follows a series of closed depressions on the decompacted isopach map of the underlying format (Fig.5-21B).

The close fit between fluvial and deltaic sandstones and thin trends (or inferred paleosurface lows) displayed by subjacent, theoretically decompacted formats adds confirmation of the role played by sand—mud compaction in controlling progradational routes. Greater paleosurface relief over relatively noncompactible deltaic sandstones may have provided higher energy sites that affected the nature of limestone facies deposited during subsequent transgressions (Waller, 1969).

Use of decompaction techniques for restoring cross sections to approximate geometry at the end of deposition provides another means for studying sandstone geometry. Proper datum surfaces for restoring as much original sandstone geometry as possible have been explored by various workers (Andreson, 1962; Potter, 1963). Application of decompaction methods to mudstones within a section containing elongate sandstones, restores the sandstones to more probable cross-sectional shape, especially when the section is fitted to a subjacent limestone datum (Fig.5-9C, E, p.263). Such restorations assume that most distortion resulted from compactional drape over relatively noncompactible sands, but obviously some subsidence of sand bodies into underlying muds also occurred.* Sections restored in this manner (idem) display fluvial channel shape in upslope areas and delta barfinger geometry in downslope areas.

The study of terrigenous clastic deposition should benefit from conservative use of various decompaction procedures. With caution and continued research, decompaction may eventually provide a relatively reliable technique for investigating paleosurfaces and differential subsidence.

Evolution of paleosurfaces: summary

Cisco deltaic deposition developed initially over limestone bank deposits of the underlying Canyon Group, probably in response to an increasing sediment supply related to Ouachita uplift. Deltaic and associated facies built westward rapidly over subtle paleotopographic relief on the subjacent limestones. Compaction of prodelta muds over bank and interbank relief provided west-to-southwest paleosurface depressions along which delta lobes prograded. Following a decrease in sediment supply, increase in subsidence, or a possible eustatic rise in sea level, a long period of deltaic destruction, compaction, and marine transgression ended deposition of the fluvial–deltaic sequence.

Continued differential compaction of delta sand and interdistributary mud of the first Cisco deltaic sequence resulted in slight compactional depressions overlying former mud facies, while sites of earlier fluvial and deltaic sand deposition stood slightly higher (Fig.5-22). The heavy load of terrigenous clastic sediments apparently overloaded certain narrow, structurally weak belts, resulting in local but relatively persistent subsidence.

When the second period of delta deposition occurred in the area, caused by upstream avulsion, increased sediment supply, or filling of earlier drowned channels, deltas prograded principally across available compactional

---

*In the opinion of the editors, the compaction of sand itself should also be taken into consideration.

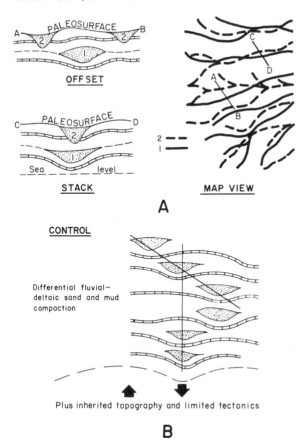

Fig.5-22. Schematic relationship of sandstone geometry and inferred controlling factors.

depressions. Compactional control was responsible for offset vertical sandstone patterns. Greater than average structural subsidence along certain belts locally countered the effect of compaction, resulting in some multistory sand deposition. During each subsequent Cisco fluvial and deltaic episode, the distribution of sand depositional sites was similarly controlled.

Paleosurface grain was maintained throughout the Cisco, but successive predelta surfaces displayed progressive changes in configuration that shifted subsequent delta depositional sites. The configuration of surfaces over which successive delta systems prograded evolved through the interplay of compaction and structural instability. The distribution of each sandstone system inherited its pattern from the previous system and, in turn, controlled the paleosurface on which the next deltaic sandstones were deposited.

SUMMARY AND CONCLUSIONS

Upper Pennsylvanian and Lower Permian rocks of the Eastern Shelf in North-central Texas are composed of 10—15 repetitive sequences including open shelf, deltaic, fluvial, and interdeltaic facies. Sediments derived from the Ouachita Mountains and associated piedmont were transported westward across a narrow coastal plain. Fluvial and deltaic sandstone facies define a southwest paleoslope less than 5 ft/mile. Position and orientation of individual elongate sandstone bodies were significantly dictated by the effects of differential mud—sand compaction during and immediately following deposition.

Sandstones displaying distributary patterns represent distal deposition in the upslope area. Belt sandstones, typified by uncommonly thick fluvial channel deposits, prograded far downslope. Composite patterns include distributary and belt sandstones representing complex progradational history. Rocks presently display one-half degree northwest regional dip; negative structure residuals outline a broad area within which 70% of the deltaic facies were deposited.

Elongate sandstone facies are generally arranged parallel to the paleoslope in vertically offset patterns controlled by differential compaction of fluvial and deltaic sands and interdistributary muds. Multistory sandstone bodies were deposited along narrow, structurally unstable belts that were periodically overloaded and later reoccupied by prograding deltas. Initial Cisco deltas followed a paleosurface grain controlled by underlying differential compaction of prodelta mud over paleotopography displayed by bank limestones; this fundamental orientation was maintained during deposition of 1,200 ft of Cisco strata. As a result of differential sand—mud compaction, each fluvial—deltaic system inherited its distribution pattern from previous systems and, in turn, provided control for the next deltaic depositional episode. Stratigraphic and structural mapping utilizing mud decompaction techniques, though qualitative and highly subjective, confirms the roles played by compaction and structure in controlling the geometry of sandstone bodies.

ACKNOWLEDGEMENTS

The writer is indebted to the following individuals for their help. J. H. McGowen, T. H. Waller, M. J. Seals, and J. R. Ray contributed much basic data. Bureau of Economic Geology staff members, W. L. Fisher, P. U. Rodda, and P. T. Flawn, critically read the manuscript and contributed ideas. J. L. Goodson computed much of the decompaction data and provided critical evaluation of the results. Miss Josephine Casey and Mrs. Elizabeth T. Moore processed the manuscript; drafting was done under the supervision of J. W. Macon. The West Texas Geological Society kindly permitted the use of several illustrations from "Cyclic Sedimentation in the Permian Basin" (L. F. Brown, 1969).

Publication of this chapter was authorized by the Director, Bureau of Economic Geology, The University of Texas at Austin. It was reprinted with changes from a report originally entitled "Geometry and Distribution of Fluvial and Deltaic Sandstones (Pennsylvanian and Permian), North-central Texas" in *Trans. Gulf Coast Assoc. Geol. Soc.*, 19(1960): 23—47. Permission was granted by the Association in January, 1973.

REFERENCES

Adams, J. E., Frenzel, H. N., Rhodes, M. L. and Johnson, D. P., 1951. Starved Pennsylvanian Midland basin. *Bull. Am. Assoc. Pet. Geol.*, 35: 2600—2607.

Adams, J. E. and Rhodes, M. L., 1960. Dolomitization by seepage refluxion. *Bull. Am. Assoc. Pet. Geol.*, 44: 1912—1920.

Allen, J. R. L., 1965. Late Quaternary Niger delta and adjacent areas: sedimentary environments and lithofacies. *Bull. Am. Assoc. Pet. Geol.*, 49: 547—600.

Allen, P., 1959. The Wealden environment: Anglo—Paris basin. *Philos. Trans. R. Soc. Lond.*, Ser. B, 242: 283—346.

Andreson, M. J., 1961. Geology and petrology of the Trivoli Sandstone in the Illinois basin. *Ill. State Geol. Surv., Circ.*, 316: 31 pp.

Andreson, M. J., 1962. Paleodrainage patterns: their mapping from subsurface data and their paleogeographic value. *Bull. Am. Assoc. Pet. Geol.*, 46: 398—406.

Athy, L. F., 1930. Density, porosity, and compaction of sedimentary rocks. *Bull. Am. Assoc. Pet. Geol.*, 19: 1—36.

Beerbower, J. R., 1964. Cyclothems and cyclic depositional mechanisms in alluvial plain sedimentation. In: *Symposium on Cyclic Sedimentation, I. Kans. Geol. Surv., Bull.*, 169: 31—42.

Bernard, H. A. and LeBlanc, R. J., 1965. Resumé of Quaternary geology of the northwestern Gulf of Mexico province. In: *The Quaternary of the United States* — Review volume for the 7th Congr. of the Int. Assoc. for Quaternary Res. Princeton University Press, Princeton, N.J., pp. 137—185.

Brown Jr., L. F., 1959. Problems of stratigraphic nomenclature and classification, Upper Pennsylvanian, north-central Texas. *Bull. Am. Assoc. Pet. Geol.*, 43: 2866—2871 (discussion).

Brown Jr., L. F., 1960. Stratigraphy of the Blach Ranch — Crystal Falls section (Upper Pennsylvanian), northern Stephens County, Texas. *Bur. Econ. Geol., Univ. Tex., Rep. Inv.*, 41: 45 pp.

Brown Jr., L. F., 1962. A stratigraphic datum, Cisco Group (Upper Pennsylvanian), Brazos and Trinity valleys, north-central Texas. *Bur. Econ. Geol., Univ. Tex., Rep. Inv.*, 46: 42 pp.

Brown Jr., L. F., 1969. Virgil—Lower Wolfcamp repetitive depositional environments in north-central Texas. In: *Cyclic Sedimentation in the Permian Basin.* West Tex. Geol. Soc., Midland, Texas, pp.115—134.

Brown, S. L., 1967. Stratigraphy and depositional environment of the Elgin Sandstone (Pennsylvanian) in south-central Kansas. *Kans. Geol. Surv., Bull.*, 187(3): 9 pp.

Burst, J. F., 1969. Diagenesis of Gulf Coast clayey sediments and its possible relation to petroleum migration. *Bull. Am. Assoc. Pet. Geol.*, 53: 73—93.

Busch, D. A., 1953. The significance of deltas in subsurface exploration. *Tulsa Geol. Soc., Dig.*, 21: 71—80.

Busch, D. A., 1959. Prospecting for stratigraphic traps. *Bull. Am. Assoc. Pet. Geol.*, 43: 2829—2843.

Chapman, R. E., 1972. Primary migration of petroleum from clay source rocks. *Bull. Am. Assoc. Pet. Geol.*, 56: 2185—2191.

Cheney, M. G., 1929. Stratigraphic and structural studies in north-central Texas. *Univ. Tex., Bull.*, 2913: 27 pp.

Cheney, M. G. and Goss, L. F., 1952. Tectonics of central Texas. *Bull. Am. Assoc. Pet. Geol.*, 36: 2237—2265.

Coleman, J. M. and Gagliano, S. M., 1964. Cyclic sedimentation in the Mississippi River deltaic plain. *Trans. Gulf Coast Assoc. Geol. Soc.*, XIV: 67—80.

Conybeare, C. E. B., 1967. Influence of compaction on stratigraphic analyses. *Can. Pet. Geol.*, 15: 331—345.

Cummins, W. F., 1891. Report on the geology of northwestern Texas. *Tex. Geol. Surv., 2nd Annu. Rep.*, 1890: pp.357—552.

Dickinson, G., 1953. Geological aspects of abnormal reservoir pressures in Gulf Coast Louisiana. *Bull. Am. Assoc. Pet. Geol.*, 37: 410—432.

Drake, N. F., 1893. Report on the Colorado coal field of Texas. *Tex. Geol. Surv., 4th Annu. Rep.*, 1892: pp.355—446.

Duff, P. McL. D., 1967. Cyclic sedimentation in the Permian coal measures of New South Wales. *J. Geol. Soc. Aust.*, 14(2): 293—307.

Eargle, D. H., 1960. Stratigraphy of Pennsylvanian and Lower Permian rocks in Brown and Coleman counties, Texas. *U.S. Geol. Surv., Prof. Pap.*, 315-D: 55—77.

Feofilova, A. P., 1959. Facies environment of Lower Carboniferous coal measures accumulation in the Donets basin. *Bull. U.S.S.R. Acad. Sci. (Geol. Ser.)*, 5. (English translation published by Am. Geol. Inst., pp.28—29.)

Fisher, W. L. and McGowen, J. H., 1967. Depositional systems in the Wilcox Group of Texas and their relationship to occurrence of oil and gas. *Trans. Gulf Coast Assoc. Geol. Soc.*, XVII: 105—125. (Reprinted as *Bur. Econ. Geol. Univ. Tex., Geol. Circ.*, GC 67-4.)

Fisk, H. N., 1961. Bar-finger sands of Mississippi delta. In: *Geometry of Sandstone Bodies*. Am. Assoc. Pet. Geol., Tulsa, Okla., pp.29—52.

Fisk, H. N., McFarlan Jr., E., Kolb, C. R. and Wolbert Jr., L. J., 1954. Sedimentary framework of the modern Mississippi delta. *J. Sediment. Petrol.*, 24: 76—99.

Forgotson Jr., J. M., 1957. Nature, usage and definition of marker-defined vertically segregated rock units. *Bull. Am. Assoc. Pet. Geol.*, 41: 2108—2113.

Frazier, D. E., 1967. Recent deltaic deposits of the Mississippi River — their development and chronology. *Trans. Gulf Coast Assoc. Geol. Soc.*, XVII: 287—311.

Friedman, S. A., 1960. Channel-fill sandstones in the Middle Pennsylvanian rocks of Indiana. *Indiana Geol. Surv., Rep. Prog.*, 23: 59 pp.

Galloway, W. E. and Brown Jr., L. F., 1972. Depositional systems and shelf-slope relationships in Upper Pennsylvanian rocks, north-central Texas. *Bur. Econ. Geol., Univ. Tex., Rep. Inv.*, 75: 62 pp.

Hedberg, H. D., 1936. Gravitational compaction of clays and shales. *Am. J. Sci., 5th Ser.*, 31: 241—287.

Jackson, W. E., 1964. Depositional topography and cyclic deposition in west-central Texas. *Bull. Am. Assoc. Pet. Geol.*, 48: 317—328.
Jones, P. H., 1968. Hydrodynamics of geopressure in the northern Gulf of Mexico basin. *Soc. Pet. Eng. AIME, Paper*, SPE 2207: 12 pp.
Kolb, C. R. and Van Lopik, J. R., 1966. Depositional environments of the Mississippi River deltaic plain — southeastern Louisiana. In: *Deltas in Their Geologic Framework*, Houston Geol. Soc., Houston, Texas, pp.17—61.
Kruit, C., 1955. Sediments of the Rhone delta, I. Grain size and microfauna. *K. Ned. Geol. Mijnbouw Gen. Verh.*, 15: 397—499.
Lee, W., 1938. Stratigraphy of the Cisco Group of the Brazos basin. In: *Stratigraphic and Paleontologic Studies of the Pennsylvanian and Permian Rocks in North-central Texas. Univ. Tex. Pub.*, 3801: 11—90.
Martin, R., 1966. Paleogeomorphology and its application to exploration for oil and gas (with examples from western Canada). *Bull. Am. Assoc. Pet. Geol.*, 50: 2277—2311.
McGowen, J. H., 1964. *The Stratigraphy of the Harpersville and Pueblo Formations, Southwestern Stephens County, Texas.* Thesis, Baylor Univ., Waco, Texas, 440 pp., unpublished.
Moore, D., 1959. Role of deltas in the formation of some British Lower Carboniferous cyclothems. *J. Geol.*, 67: 522—539.
Mueller, J. C. and Wanless, H. R., 1957. Differential compaction of Pennsylvanian sediments in relation to sand/shale ratios, Jefferson County, Illinois. *J. Sediment. Petrol.*, 27: 80—88.
Myers, D. A., 1965. Geology of the Wayland quadrangle, Stephens and Eastland counties, Texas. *U.S. Geol. Surv., Bull.*, 1201-C: 1—63.
Pepper, J. F., DeWitt Jr., W. and Demarest, D. F., 1954. Geology of the Bedford Shale and Berea Sandstone in the Appalachian basin. *U.S. Geol. Surv., Prof. Pap.*, 259: 111 pp.
Pettijohn, F. J., 1962. Paleocurrents and paleogeography. *Bull. Am. Assoc. Pet. Geol.*, 46: 1468—1493.
Plummer, F. B. and Moore, R. C., 1922. Stratigraphy of the Pennsylvanian formations of north-central Texas. *Univ. Tex., Bull.*, 2132: 237 pp.
Potter, P. E., 1963. Late Paleozoic sandstones of the Illinois basin. *Ill. Geol. Surv., Rep. Inv.*, 217: 92 pp.
Potter, P. E., 1967. Sand bodies and sedimentary environments: A review. *Bull. Am. Assoc. Pet. Geol.*, 51: 337—365.
Powers, M. C., 1967. Fluid-release mechanisms in compacting marine mudrocks and their importance in oil exploration. *Bull. Am. Assoc. Pet. Geol.*, 51: 1240—1254.
Pryor, W. A., 1961. Sand trends and paleoslope in Illinois basin and Mississippi embayment. In: *Geometry of Sandstone Bodies.* Am. Assoc. Pet. Geol., Tulsa, Okla., pp.119—133.
Rall, R. W. and Rall, E. P., 1958. Pennsylvanian subsurface geology of Sutton and Schleicher counties, Texas. *Bull. Am. Assoc. Pet. Geol.*, 42: 839—870.
Ray, J. R., 1968. *Stratigraphy of the Moran and Putnam formations, Lower Permian, Shackelford and Callahan Counties, Texas.* Thesis, Baylor Univ., Waco, Texas, 200 pp., unpublished.
Rochon, R. W., 1967. Relationship of mineral composition of shales to density. *Trans. Gulf Coast Assoc. Geol. Soc.*, XVII: 135—142.
Rothrock, H. E., 1961a. Origin of the Zweig Sandstone lens. In: *A Study of Pennsylvanian and Permian Sedimentation in the Colorado River Valley of West-central Texas.* Abilene Geol. Soc., Guideb., Abilene, Texas, pp.33—35.

Rothrock, H. E., 1961b. Deposition of the Coon Mountain Sandstone. In: *A study of Pennsylvanian and Permian Sedimentation in the Colorado River Valley of West-central Texas.* Abilene Geol. Soc., Guideb., Abilene, Texas, pp.36—38.

Seals, M. J., 1965. *Lithostratigraphic and Depositional Framework Near-surface Upper Pennsylvanian and Lower Permian Strata, Southern Brazos Valley, North-central Texas.* Thesis, Baylor Univ., Waco, Texas, 128 pp., unpublished.

Shankle, J. D., 1960. The "Flippen" sandstone of parts of Taylor and Callahan counties, Texas. *Abilene Geol. Soc., Geol. Contrib., Abilene, Texas,* pp.168—201.

Siever, R., 1951. The Mississippian—Pennsylvanian unconformity in southern Illinois. *Bull. Am. Assoc. Pet. Geol.,* 35: 542—581.

Stafford, P. T., 1960a. Geology of the Cross Plains quadrangle, Brown, Callahan, Coleman, and Eastland counties, Texas. *U.S. Geol. Surv., Bull.,* 1096-B: 39—72.

Stafford, P. T., 1960b. Stratigraphy of the Wichita Group in part of the Brazos River valley, North Texas. *U.S. Geol. Surv., Bull.,* 1081-G: 261—280.

Swann, D. H., 1964. Late Mississippian rhythmic sediments of the Mississippi valley. *Bull. Am. Assoc. Pet. Geol.,* 48: 637—658.

Terriere, R. T., 1960. Geology of the Grosvenor quadrangle, Brown and Coleman counties, Texas. *U.S. Geol. Surv., Bull.,* 1096-A: 1—35.

Treadwell, R. C., 1955. Sedimentology and ecology of southeast coastal Louisiana. *La. State Univ., Coastal Stud. Inst., Tech. Rep.,* 6: 78 pp.

Van Siclen, D. C., 1958. Depositional topography — examples and theory. *Bull. Am. Assoc. Pet. Geol.,* 42: 1897—1913.

Waller, T. H., 1966. *The Stratigraphy of the Graham and Thrifty Formations, Southeastern Stephens County, Texas.* Thesis, Baylor Univ., Waco, Texas, 370 pp., unpublished.

Waller, T. H., 1969. Lower Cisco carbonate deposition in north-central Texas. In: *Late Pennsylvanian Shelf Sediments, North-central Texas.* Dallas Geol. Soc., Guideb., Dallas, Texas, pp.34—39.

Wanless, H. R., Tubb Jr., J. B., Gednetz, D. E. and Weiner, J. L., 1963. Mapping sedimentary environments of Pennsylvanian cycles. *Bull. Geol. Soc. Am.,* 74: 437—486.

Welder, F. A., 1959. Processes of deltaic sedimentation in the lower Mississippi River. *La. State Univ., Coastal Stud. Inst., Tech. Rep.,* 12: 90 pp.

Weller, J. M., 1959. Compaction of sediments. *Bull. Am. Assoc. Pet. Geol.,* 43: 273—310.

Wermund, E. G. and Jenkins Jr., W. A., 1964. Late Missourian tilting of the Eastern Shelf of the West Texas basins. *Geol. Soc. Am., Spec. Pap.,* 82: 220—221.

Wermund, E. G. and Jenkins Jr., W. A., 1968. Late Pennsylvanian sand deposition in north-central Texas. *Geol. Soc. Am. Program, 1968 Anu. Meet. South-central Sect.,* p. 40.

Williams, E. G., Ferm, J. C., Guber, A. L. and Bergenback, R. E., 1964. Cyclic sedimentation in the Carboniferous of western Pennsylvania. In: *Guideb. 29th Field Conf. Pa. Geol.* Pa. State Univ., Dep. Geol., 35 pp.

Wright, M. D., 1967. Comparison of Namurian sediments of the central Pennines, England, and Recent deltaic deposits. *Sediment. Geol.,* 1: 83—115.

Chapter 6

THE EFFECT OF COMPACTION ON VARIOUS PROPERTIES OF COARSE-GRAINED SEDIMENTS

O. G. INGLES and K. GRANT

INTRODUCTION

The effect of compaction, either by natural or anthropomorphic processes, on the various properties of coarse-grained materials (>0.1 mm in diameter), that have not been subjected to normal rock-forming processes, is discussed in the present chapter. The major properties discussed include: (1) porosity, (2) strength, (3) permeability, (4) compressibility, (5) liquefaction, (6) wave transmission, (7) collapse, and (8) erosion susceptibility. In addition, modern engineering methods for compacting and densifying the sediments, or controlling their properties to meet specific requirements of usage, are briefly discussed.

Porosity and pore structure are the principal factors which control the strength, permeability, compressibility, and other properties of coarse-grained sediments whether in a loose or a compacted condition. The properties of a compacted material depend mainly upon the origin, composition, texture and fabric, and subsequent diagenetic history, which are discussed in detail in Chapter 1 of this book and Chapter 3 of Vol. II.

In any discussion of the properties of sediments, it is convenient to adopt, or to refer to, the extended Casagrande classification (Casagrande, 1948) for coarse-grained materials. An abridged version of his classification is presented in Table 6-I.

COMPACTION AND SHAPE FACTORS

The density achieved in, and consequently the porosity of, compacted coarse-grained materials depends upon a number of factors related to particle morphology and the fabric of the particles within the material mass.

Methods of measuring and describing quantitatively the fabric of sediments have been extensively developed in recent years, and are described by a number of authors (e.g., McMahon, 1972, for rocks; Lafeber, 1965, for soils) to whom reference should be made whenever detailed characterization is required.

In considering detailed morphology, all particles may be divided into shape classes in accordance with standard procedures given by either (1)

TABLE 6-I

Extended Casagrande (1948) classification for coarse-grained materials (abridged)

Sediment type	Grain diameter	Symbol	Unit dry weight (g/cm^3)	Engineering properties		Permeability (cm/sec)
				potential frost action	drainage properties	
Boulders and cobbles	>3 inch	—	2.0 —2.25	almost none	good	>1
Gravel and gravelly soils	7 mesh to 3"	GW*	2.0 —2.15	none to very slight	excellent	>10^{-2}
		GP**	1.85–2.00	very slight	excellent	>10^{-2}
Sands and sandy soils	200 mesh to 7 mesh	SW*	1.75–2.10	none to very slight	excellent	>10^{-3}
		SP**	1.60–1.95	very slight	excellent	>10^{-3}

*W signifies well-graded, few fines.
**P signifies poorly-graded, few fines.

(Note: the terminology "well-graded" in engineering practice corresponds to "poorly-sorted" in sedimentary usage; conversely, "poorly-graded" is equivalent to "well-sorted".)

American Society for Testing Materials (1966) — rounded, angular, sub-rounded, and sub-angular particles; or (2) British Standards Institution (1967) — rounded, irregular, angular, flaky, elongated and flaky, and elongated particles, as shown in Fig.6-1.

Within these classes, a further subdivision may be made on the basis of the dimensions, i.e., the length, width, and breadth of the particles. Moss (1962) discussed methods of measuring particle shapes in relation to stream saltation. Krumbein (1941) differentiated between shape and roundness and presented several methods for measuring these properties. Standard methods for measuring particle size and shape are given in British Standards Institution (1967) publication.

Imposed upon morphology is the surface texture of the particles. Within any subdivision of any class of particle morphology, the surface texture may be glassy, smooth, granular, rough crystalline, or honeycombed, as discussed in British Standards Institution (1967) publication. This specification also includes particle characteristics associated with each surface texture as shown in Table 6-II.

Additional information on the texture and fabric of coarse-grained sediments is presented in Chapter 1 of this book.

Regardless of other factors, wholly symmetrical structures can only be assumed for particles that are spherical or close to spherical. As the angularity or the ellipticity of particles increases, the degree of anisotropy introduced into the packing fabric also increases. The influence of tectonic

TABLE 6-II

Surface texture

Group*	Surface texture	Characteristics
1	glassy	conchoidal fracture
2	smooth	water-worn, or smooth due to fracture of laminated or fine-grained rock
3	granular	fracture showing more-or-less uniform rounded grains
4	rough	rough fracture of fine- or medium-grained rock containing no easily visible crystalline constituents
5	crystalline	containing easily visible crystalline constituents
6	honeycombed	with visible pores and cavities

*Surface texture has been described above under six headings. This grouping is broad, being based on the impression which would be gained by a visual examination of hand specimens. It does not purport to be a precise petrographical classification. Different specimens of the same rock type may not fall into the same group in Table 6-III, or subsequent tables.

Fig. 6-1. Typical particle shapes; the standard descriptions for granular materials. (After British Standards Institution, 1967, fig. 2, p. 19.)

forces may also introduce anisotropy into the fabric. Packing of irregular, angular particles may be random with no elements of symmetry being present in the structure. On the other hand, the packing of flat and highly elliptical particles may have one axis of n-fold symmetry, i.e., the packing is cross-anisotropic. Figures 6-2 and 6-3 illustrate the role and the representation of symmetry elements in a granular mass. In addition, Lafeber and Willoughby (1971) have determined that samples of soil and beach sand taken from two localities in Australia have fabrics which possess only one axis of symmetry, i.e., the fabric is monoclinic with the oblique axis controlled by consolidation in relation to the topographic slope at the site. They further showed that this anisotropic condition is reflected in the porosity of anisodimensional specimens for the soil (ϕ = 0.55, for specimens taken with their long axis vertical to the ground surface; ϕ = 0.565, for specimens with their long axis horizontal to the ground surface, oriented north—south; ϕ = 0.585, for specimens with their long axis horizontal to the ground surface and oriented east—west). Likewise strength properties for sand, such as the secant modulus ($5.54 \cdot 10^2$ kg/cm^2 vertical; $4.10 \cdot 10^2$ kg/cm^2 parallel to the coastline; $3.93 \cdot 10^2$ kg/cm^2 horizontal, 30° to the coastline; $3.84 \cdot 10^2$ kg/cm^2 horizontal, 60° to the coastline), show that the porosity and strength as measured by the secant modulus may differ along each of the three axes (cf. Fig.6-4).

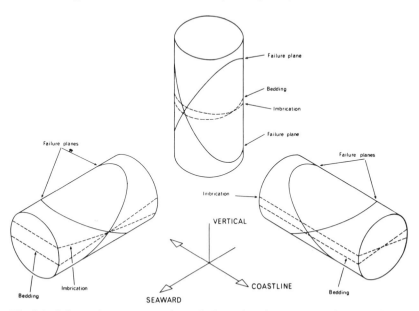

Fig.6-2. Schematic representation of the role of symmetry elements in determining the failure planes of a beach sand. (After Lafeber and Willoughby, 1970, fig.6, p.84.)

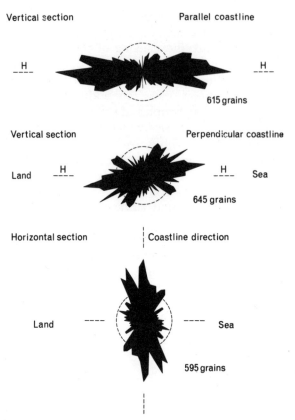

Fig.6-3. Representation of the symmetry elements of a granular mass. (After Lafeber and Willoughby, 1971, fig.4, p.167.)

Another good experimental investigation has recently been reported by Oda (1972), who showed that for several mixed mineral sands the *initial fabric* (as expressed by the spatial relationship of the constituent particles, for example, by the probability density of contact), *but not the particle orientation*, has considerable effect on the deformation behavior (the secant moduli varied as much as threefold). The shear strength at failure, however, is at least partly determined by the shape of the constituent grains; though for near-spherical particles, the mobilized shear strength at failure is almost independent of the initial fabric.

The intervention of factors of morphology and fabric effectively reduces the symmetry of a compacted material. The more particles deviate from spherical shape, the less it can be expected that the density of the close-packed (rhombohedral) structures (see p.301) will be the compaction limit. Planar particles pack more closely than the spherical particles in an

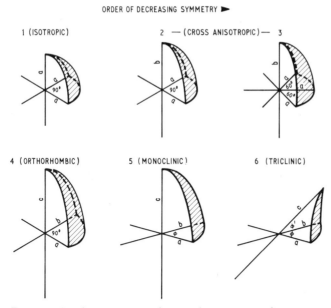

Fig. 6-4. The six symmetry classes; the measure of any property in a given axial direction being denoted by the length of that axis (a, b, or c of arbitrary magnitude).

ideal rhombohedral packing, whereas acicular particles pack less densely than the latter. The more the departure from a condition of isotropic packing, the more an orientation factor tends to modify such properties as porosity, permeability, and strength, as discussed in later sections.

The roughness of the particles, which does not significantly affect the final compacted density, can appreciably increase the compactive effort that is required to achieve the final density. This is, of course, due to the necessity for overcoming increased frictional forces between the particles. The degree of roughness may also affect the strength of the compacted material. Rough particles that are capable of interlocking, form stronger fabric structures than do smooth particles which do not interlock.

POROSITY, DENSITY AND PORE STRUCTURE

The voids of any compacted sediment (better referred to as pores, because they may or may not be liquid-filled) can vary considerably in volume, tortuosity, size distribution, shape, surface roughness and other factors of significance to the physical and mechanical behavior of the sediment as a whole. Their importance in affecting the bulk properties of a coarse sediment cannot be overemphasized.

For example, the magnitude of measured permeabilities is usually determined by the pore-size distribution, because the major portion of the flow occurs through the largest pores. In recent years it has also been recognized that the strength parameters of a porous mass are primarily dependent on the properties of the pores (Knudsen, 1959; Ryskewitch, 1953; Ingles and Frydman, 1963).

The accuracy of measurement of the true density of a porous solid depends both on the structure of the solid particles or particulate mass, i.e., volume of intercommunicating pores, and on the penetrative power of the immersion medium used for the determination. It is usual to consider the true density of a material as that measured on the finely powdered solid (to ensure destruction of any closed pores, as may occur in a scoria, tuff, or vesicular rock, for example) using helium gas as a displacement medium (to achieve maximum penetration). Less rigorous methods often give good approximations, however, depending on the material under examination. Test 14B of British Standards Institution (1967) defines one such method (sand replacement), which for coarse-grained soils has a permissible standard deviation of 0.16 g/cm^3.

If the bulk, or apparent, density is measured using the geometrical bounds of the porous mass, the porosity, ϕ, can be defined as:

$$\phi = 1 - (\rho_a/\rho_t) = V_p/(V_p + V_s) \tag{6-1}$$

where ρ_a and ρ_t denote the apparent and true densities, respectively; whereas V_p and V_s denote the volumes of pores and solids, respectively.

In engineering use, particularly the discussion of compression and settlement behavior, a parameter related to the porosity but based on the volume of pores relative to the volume of solids is frequently used. It is called the void ratio, e_v or e, and is related to the porosity as:

$$e_v = V_p/V_s = \phi/(1 - \phi), \text{ or:}$$

$$\phi = \frac{e_v}{1 + e_v} \tag{6-2}$$

Because of its imprecise nomenclature, use of the parameter void ratio is avoided as far as possible in the present treatment in favor of the parameter porosity.

For an ideal system of equal-sized spherical particles, a relationship between porosity and packing density (or mean coordination number) can be readily shown (Graton and Fraser, 1935). This relationship is supported by the experiments of Smith et al. (1929) and Kiselev (1954) and is presented in Fig.6-5. Precise values of porosity (as a fraction) for various regular packings are as follows:

rhombohedral	$\phi = 0.2595$
body-centred tetragonal	$\phi = 0.3018$
body-centred cubic	$\phi = 0.3198$
simple orthorhombic	$\phi = 0.3954$
simple cubic	$\phi = 0.4764$

Although many natural processes, such as dune formation by wind or water, can establish sediments of a fairly uniform grain size, even these seldom approximate the ideal systems. The apparent density of a coarse-grained sediment is greatly increased when it possesses a wide range of particle sizes, i.e., good grading. Indeed, artificial compaction of selectively graded granular material can yield very high bulk densities (and correspondingly low porosities), to as much as 95% of the true density ($\phi \leq 0.05$; McGeary, 1961). It is commonly found that for a well-graded, natural sand—gravel mixture, the porosity lies below 0.20, i.e. lower than that of the densest regular packing. It has been further found that certain optimum gradings in natural materials allow the greatest densities to be achieved in compaction. These optimum gradings are shown by what are known as Fuller curves (Fuller and Thompson, 1907), which are a series of grading plots for various size distributions having different maximum grain sizes. In construction practice, some latitude is allowed on these curves, as specified

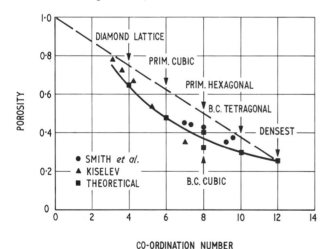

Fig.6-5. Variation of porosity with packing density (spheres). Co-ordination number is equal to the number of spheres touching any given sphere. (After Kruyer, 1958, fig.7, p.1766.)

in ASTM Standards D556-40T and D557-40 (1944). A typical series of curves is presented in Fig.6-6. A recent illustration of the interrelationship of grading, compaction, and strength for some natural surface sediments of low clay content has been given by Ingles (1971).

Apart from total porosity, the most important factors in whether a natural or artificial coarse-grained mass can be further compacted or not are (1) degree of saturation and (2) environmental and internal drainage conditions. This is considered in detail in the section on the strength of compacted sediments. The general principles involved in densification, however, can be illustrated by reference to Fig.6-7, in which a given compactive effort is being considered to be applied to a granular mass (sandy gravel) of varying moisture content. To the right of the optimum density point, densification is limited by the pore pressure of water opposing the compactive force. This restriction can only be overcome *if* free drainage can occur within and without the granular mass. To the left of the optimum density point, frictional resistance opposes interparticle movement and hence densification. At sufficiently high compactive forces, however, crushing of the primary particles will result in a further densification. If no crushing occurs, the porosity of the densest state obtainable by compaction of a coarse-grained sediment is normally below that of the loose uncompacted material by less

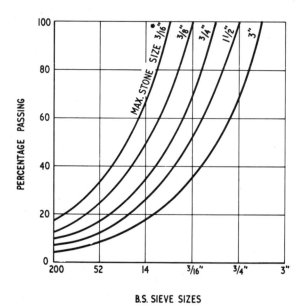

Fig.6-6. Grading curves for optimum density packing for various maximum stone sizes (sizes in inches as per the original specifications). (Based on the specifications of Fuller and Thompson, 1907.) *Curve designations indicate the permissible latitude.

than 0.15. For rounded gravels and sands in an intermediate density condition, the porosity can be estimated from the grading according to the formula proposed by Winterkorn (1970):

$$\phi = 0.385 - 0.08 \log_{10} D_{max}/d_{min} \qquad (6\text{-}3)$$

where d_{min} and D_{max} represent the smallest and largest particle sizes present, respectively. The actual porosity of most coarse-grained sediments is better measured in situ, however, using hydraulic methods, because the undisturbed sampling of unconsolidated coarse-grained sediment is indeed very difficult. Porosity limits for the loosest and densest conditions readily attainable in a laboratory can be conveniently determined by the method of Kolbuszewski (1965).

The crushing of porous primary particles such as those which originated from the non-plastic clay sands, volcanic scoria and tuffs, or materials with planes of weakness (e.g., micaceous rocks), will result in considerable reductions of porosity at higher stress levels due to the production of fines and the consequent increase of the term D_{max}/d_{min} in eq.6-3. Figure 6-8 shows also the increase in surface area per unit of pore volume with increasing size range. This is even further enhanced whenever crushing of primary particles occurs, because crushing increases the size range of the material and the surface roughness of particles.

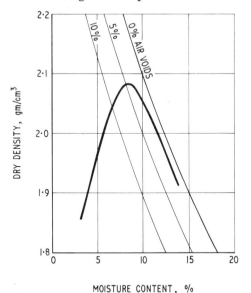

Fig.6-7. Typical relationship between dry density (g/cm³) and moisture content (%) for a sandy gravel. Curves showing the residual percentage of air voids are superimposed.

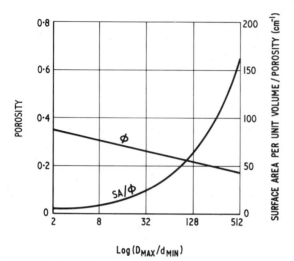

Fig. 6-8. Porosity and surface area (per unit of pore volume) of granular material as a function of the grain size range. (After Winterkorn, 1971, fig. 4.)

Maximum densities (minimum porosities) in a given cohesionless granular material are obtained by vibration under load, and in engineering practice the vibratory roller is a popular machine for the compaction of loose, dry sands. Shock waves have also been used for the compaction of sands and gravels at depth, the surcharge load being provided by the depth of overburden. In saturated materials, the effect of shock waves may be enhanced by liquefaction phenomena, which are discussed later.

The pore structure within any loose, granular mass consists essentially of a series of comparatively large cavities interconnected by narrow channels near the points of grain contact. In addition to this irregularity of pore shape, the usual anisotropy of grain shape also leads to an anisotropic pore structure during the formation of the sediment (Lafeber and Willoughby, 1971). This anisotropy of pore structure may be either diminished or enhanced during further compaction. In general, the application of an *isotropic* stress field will reduce the initial anisotropy, whereas an *anisotropic* stress field will modify the pore structure according to the relative anisotropies of the applied stress and of the porous sediment.

For practical calculations, especially of fluid flow, the porosity of coarse-grained sediments is frequently reported in terms of an effective porosity or a mean pore size (hydraulic radius). This conceals the physical reality of the wide range of pore sizes, pore angularity, anisotropy, surface roughness, etc., which are characteristic of most natural materials. It should be recognized, however, that the use of such quantities leads only to approximate or averaged descriptions of the properties of the sediment mass.

TABLE 6-III

Porosity, density, and surface area for various coarse-grained sediments[*1]

Type of material	Class[*2]	Surface area (m^2/g)	Porosity		Density[*3] (g/cm^3)	
			loose	compacted	loose	compacted
Poorly-graded gravel	GP	<0.1	0.25–0.40	0.15–0.30	1.60–2.00	1.90–2.25
Well-graded gravel; gravel—sand	GW	<0.1	0.20–0.35	0.10–0.25	1.70–2.10	1.95–2.30
Poorly-graded sand	SP	0.1–1	0.35–0.50	0.25–0.40	1.35–1.70	1.60–2.00
Well-graded sand; sand—silt	SW	0.3–3	0.30–0.45	0.20–0.35	1.50–1.90	1.80–2.15
Lime sands	SP/SC	—	—	0.25–0.35	—	1.70–2.00
Collapsing sands (granite)	SC	—	0.40–0.55	0.25–0.35	1.25–1.55	1.70–2.10
Glacial till	GW	<0.1	0.15–0.20	0.10–0.15	2.10–2.25	2.20–2.35
Scoria and tuffs	GM/SM	1–10	0.60–0.75	0.40–0.70	0.70–1.10	0.80–1.60[*4]
Sub-plastic clay sands	SP/SC	1–10	0.40–0.50	0.20–0.40	1.30–1.60	1.65–2.15
Cemented sands; oil sands	—	—	0.05–0.35	0.05–0.35	1.75–2.50	1.75–2.50

[*1] The range of values shown is the common range, but occasionally extreme values will be found. See, for example, Ueshita and Nonogaki (1971).
[*2] In these usual class symbols of engineering practice, G denotes gravel, S = sand, C = clay, W = well-graded, P = poorly-graded, M = medium-graded (see Glossary).
[*3] The tabulated values are for apparent density; the true density of all these materials varies from that of normal silicates (about 2.65 g/cm^3), principally according to the iron content. For example, iron enrichment raises the true density of scorias to about 2.75 g/cm^3, and of lateritic gravels and sands to as high as 3.0 g/cm^3.
[*4] The wide range of compacted densities for volcanic ejecta reflects the varying degrees of hardness (hence crushing resistance) of these porous rocks.

Cementation processes in coarse-grained sediments, especially where the cementation arises from evaporative processes associated with the rise and fall of a fluctuating water table, have a pronounced tendency to cement the mass around the points of particle contacts. Thus, there is only a relatively small effect on the total porosity or the volume of large pores, but a substantial effect on the fine pore volume, with ultimate blocking of the natural constrictions in the flow path and great reduction in permeability.

Some typical values of porosity, density, and surface area of natural coarse-grained sediments are presented in Table 6-III, together with the approximate apparent density to which they can be artificially compacted by modern heavy mechanical equipment or vibrational shock.

By pressure or vibration alone, even tectonic forces do not normally achieve greater densities than those shown in Table 6-III, and only the processes of diagenesis (as described in detail in Larsen and Chilingar, 1967) can further reduce the porosity or increase the density of a buried sediment.

COMPRESSIBILITY

The compressibility of a sand or sediment has been found to be directly related to the porosity (or void ratio). In engineering practice, it is common to describe the compressibility of a material under static load by means of a plot of porosity versus effective pressure (i.e., overburden pressure minus pore pressure), the slope of this line being termed the coefficient of volume compressibility, m.

Nevertheless, the relationship is not a simple linear one and, if plotted with a logarithmic pressure scale, may show several regions of different behavior (Fig.6-9): A — an elastic behavior region dominated by the cohesion and friction of the grain skeleton; B — a region of inelastic permanent compaction due to internal shear and consolidation of the mass; and C a new elastic region of dense structure dominated by the very low compressibility of the solid grains themselves. If the solid grains are themselves porous (e.g., scoria, tuffs, sub-plastic clayey sands) or brittle (lime sands), a further compression due to crushing of the grains may occur at the pressure level appropriate to the porous or brittle grain strength.

It is not unusual for coarse-grained sediments to show a well-defined region A as a result of earlier deep burial and subsequent removal of the overburden, as for instance by melting of an icecap. Such behavior is known as preconsolidation, and would be represented on Fig.6-9 by a compressibility curve such as that labelled E.

The above remarks, and the following discussion, apply to the material behavior in response to short-time loadings. Creep and creep compliance associated with loads of long duration (in the sense of geological time) have not been measured for these materials. Most elastic deformations are in fact only pseudo-elastic, with long-term creep components. If a load is maintained for a considerable period of time, then paths such as that shown by D on Fig.6-9 will be followed.

Because of the conditions of deposition, weathering, or folding (for deeply-buried sedimentary strata), the porosity of a sediment is often anisotropic, as already discussed. The compressibility thus can also be anisotropic. In particular, a sand mass is usually stiffer vertically than horizontally, because the long axes of particles of irregular shape tend to align in the horizontal direction when deposited, resulting in a higher apparent density in the vertical direction than in the horizontal direction (Kallstenius and Bergau, 1961).

The higher lateral compressibility of sand is taken practical advantage of in the process of lateral consolidation (Itoh, 1969) developed in Taiwan and Japan. As the ambient stress level rises, however, initially anisotropic coarse-grained material tends to become increasingly isotropic (Gerrard, 1967). On the other hand, if a deviator stress is applied, the reverse behavior applies, i.e., sediment becomes more anisotropic (Onas, 1970). Onas' curves showing relationship between stiffness and relative density of sand and rock aggregates are presented in Fig.6-10. Relative density is defined as:

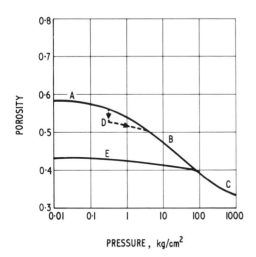

Fig.6-9. Compaction curve for a sand as a function of applied pressure. Solid line = continuous loading path; dashed line = effect of time delay at a fixed load. A = elastic behavior region; B = inelastic permanent compaction region; C = new elastic region of dense structure; D = effect of creep; E = effect of precompression, or unloading from a compressed state.

$$\frac{(\phi_{max} - \phi_0)(1 - \phi_{min})}{(\phi_{max} - \phi_{min})(1 - \phi_0)}$$

where ϕ_{max}, ϕ_{min}, and ϕ_0 denote maximum, minimum, and initial porosities, respectively. It thus describes the density relative to somewhat arbitrary conditions, namely, those used to determine the maximum and minimum porosities; however, it covers the range encountered in normal engineering practice and conveys some idea of how much further a given material can be compacted by normal means.

The compressibility of a sand, gravel, or other coarse-grained sediment depends also on other factors besides porosity; however, porosity usually has the most important influence. Such other factors include degree of cementation, water content, surface roughness and grain shape.

Following the ordinary laws of soil mechanics, a saturated coarse-grained sediment cannot be compressed at all unless drainage or vibration occur, because the pore-water pressure opposes the compaction forces. On the other hand, the compaction of a sediment under either static or dynamic forces may be greatly assisted by the presence of small amounts of water, which lubricates the motion of the grains over one another, i.e., reduces the intergranular friction.

Friction, arising from surface roughness, etc., and cohesion arising from cementation, etc., are both factors with considerable influence on the compressibility of a coarse-grained sediment. Thus, compressibility values

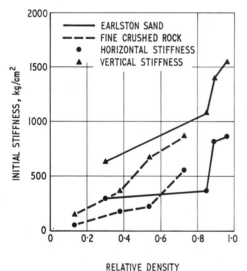

Fig.6-10. Stiffness as a function of relative density, showing the influence of anisotropy. (After Onas, 1970, fig.5.1, p.116.)

can be small or great and, due to preconsolidation effects, may differ substantially even at the same porosity (compare the slopes of curves E and B at porosity 0.42 on Fig.6-9, for example).

The measurement of compressibility of many coarse-grained sediments requires field experiments, because the problems of sampling loose strata without disturbance are complex and, therefore, laboratory measurements commonly provide only a rough guide. The compressibility is usually measured in the field with a cone penetrometer (Terzaghi and Peck, 1948), utilizing a shaft which is driven into the material by a static or dynamic effort. Some limitations of the dynamic cone method in sands in which liquefaction can occur have been discussed by Yamanouchi (1970). Inasmuch as sand compressibility is strongly dependent on both density and overburden stress (cf. Fig.6-9), it is generally closely correlated with the standard penetration test. Curves showing this relationship have been given by D'Appolonia and D'Appolonia (1970) and are reproduced in Fig.6-11. For coarser sediments (gravels), field settlement observations of large loaded areas are the only means by which compressibility can be determined. For these sediments, the effect of overburden stress has been studied by Pellegrino (1965), whose compressibility measurements are presented in Fig.6-12.

It has been shown by Schultze (1970), however, that the modulus of compressibility of non-cohesive sands (in region B of Fig.6-9) can be estimated without previous compression tests simply from the porosity parameters, provided the natural (in situ) porosity and those of the loosest

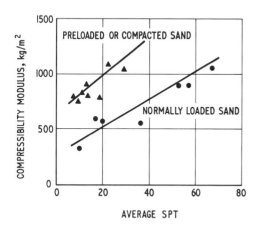

Fig.6-11. Relationship of sand compressibility modulus (M) to the standard penetration test value (*SPT*). $M = E/(1-\nu^2)$, where E is Young's modulus and ν is Poisson's ratio. (After d'Appolonia and d'Appolonia, 1970, fig.6, p.21.)

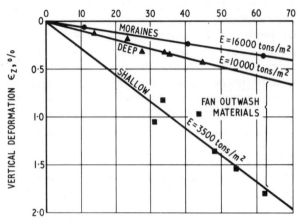

Fig.6-12. Compressibility of sediments of glacial origin. (After Pellegrino, 1965, fig.6, p.90.)

and densest states are all known. In such circumstances, the modulus of compressibility (K) is given by:

$$K = V(\sigma'_e)^w \tag{6-4}$$

where σ'_e is the effective compressive stress per unit stress ($=\dfrac{\sigma'}{\sigma_e}$, where σ' is the effective compressive stress and σ_e is the unit stress) and V (kg/cm²) and w are parameters of the material which are defined by the following two equations:

$$V = \frac{2.3 \, e_{max}^m \, \sigma_e}{c \, e_0^4 \, \ln(b e_{max}/e_0)} \tag{6-5}$$

$$w = 1 - \frac{1}{2.3} \ln(b e_{max}^n/e_0) \tag{6-6}$$

TABLE 6-IV

Values of coefficients in eq. 6-5 and 6-6

$(e_{max} - e_{min})$	b	c	m	n
0.262~0.366	2.49	0.73	6.46	1.40
0.366~0.420	2.43	1.29	6.27	1.22
0.420~0.476	2.00	2.34	8.93	2.29

where $e_0 = \phi_0/(1 - \phi_0)$, $e_{max} = \phi_{max}/(1 - \phi_{max})$, and the subsidiary coefficients b, c, m and n can be found in Table 6-IV.

The grain size distribution effect is accounted for in the above ranges of $(e_{max} - e_{min})$. Values of w lie between 0.3 and 0.8, and those of V between 80 and 1,300 kg/cm². A pressure-settlement line (overburden pressure versus surface settlement) can thus be readily constructed.

For *elastic* settlements arising from the plate loading of coarse sediments (e.g., regions A or C of Fig.6-9), the formula $S = PBI/M$ can be applied, in which S is the settlement, P is the bearing stress (compressive stress), B is the loaded area, I is an influence factor dependent on the geometry of the loaded area, given by D'Appolonia and D'Appolonia (1970), $M = E/(1 - \nu^2)$, E is Young's modulus, and ν is Poisson's ratio. Although for an ideal material (non-dilatant, frictionless, etc.) the theoretical value for Poisson's ratio is 0.5, for natural, non-cohesive sands it may be taken as 0.2—0.3. It should be noted that M is a modulus *defined by a plate-bearing test* and is not identical with the true bulk modulus K, to which it is related by the equation $M = K^3(1 - 2\nu)/(1 - \nu^2)$.

Once the inelastic structural compressibility of region B (Fig.6-9) has been taken up, the very much smaller elastic compressibility of the hard-grain skeleton region C (Fig.6-9) is followed. Nevertheless, for some materials such as lime sand, scoria, etc., the low intrinsic crushing strength leads to a gradual change from inelastic to elastic behavior, and the latter is approximated only at the highest stress levels. The density level at which structural compressibility ceases to occur for some natural, coarse-grained materials is indicated in Table 6-III by the compacted density value. At greater densities than these, the compressibility is essentially elastic except when particle crushing or long-term creep occur.

Although deeply-buried, coarse-grained sediments are generally in the elastic and dense condition prior to being subjected to tectonic forces, the application of such forces often results in anisotropy of fabric with consequent anisotropy of such properties as compressibility. The greater the tectonic forces, the greater will be these effects. Strata folded as a result of the application of tectonic forces are generally partly in a state of compression and partly in a state of tension. In such strata, material adjacent to the curve of longer radius is in a state of tension and hence has higher compressibility, whereas material adjacent to the curve of shorter radius is in a state of compression with the resultant lower compressibility (Fig.6-13). Measured values for the compressibility of coarse-grained natural sediment, are incorporated in Table 6-V together with other related strength and elastic properties. A further general account of the compressibility of sediments, with special reference both to the geological factors and engineering properties associated with them, has been given by Terzaghi (1955).

TABLE 6-V

Moduli, elastic constants, and shear parameters for various coarse materials*

Type of material	Young's modulus E (kg/cm^2)	Modulus of compressibility K (kg/cm^2)	Modulus of rigidity G (kg/cm^2)	Poisson's ratio ν	Cohesion c (kg/cm^2)	Angle of internal friction** θ (degrees)	Unconfined compressive strength (kg/cm^2)
Poorly-graded gravel (loose)	200–800	—	—	—	0–2.0	35–40	15–30
Well-graded gravel; gravel–sand (dense)	500–3,000	100–1,000	1,000–2,000	—	0–4.0	40–45	25–65
Poorly-graded sand (loose)	300–2,000	200–800	100–800	0.05–0.15	0	25–35	≈15
Well-graded sand; sand–silt (dense)	1,000–10,000	500–1,000	300–3,000	0.15–0.30	0–2.0	30–40	≈30
Lime sands	≈4,000	—	≈1,600	≈0.30	—	—	5–30
Glacial tills	500–2,000	300–1,000	—	—	≈3	35–45	≈27
Volcanic scoria, tuffs, etc.	100–800	—	100–300	0.05–0.15	5–25	—	≈5

* Unless otherwise specified, the values of the parameters tabulated are secant values for the first loading cycle. Resilient (i.e., repeated load) secant values are normally slightly greater than those quoted.
** These values should be reduced up to 50% for saturated conditions. Considerable controversy on the magnitude of the friction angle has arisen from the fact that its observed value is highly dependent on the method adopted for test.

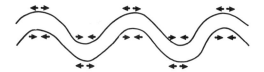

Fig.6-13. Schematic diagram showing tension and compression zones resulting from folding.

PERMEABILITY

Even in coarse-grained sediments, permeability is largely controlled by flow in the largest channels. If the sediment is poorly graded, tending to be monosized, permeabilities are higher than those for a well-graded material of similar maximum grain size, but of which the grain interstices are substantially filled with smaller grains.

Nevertheless, all coarse-grained sediments possess a higher permeability compared to fine-grained sediments. Some typical permeability values for natural sediments are presented in Table 6-VI.

On the basis of measurements on many loose sands, Schultze (1970) found a very high correlation coefficient (0.999) for the equation:

$$k = \frac{\mu_{20}}{\mu_t} (1.767 \, d_w - 0.334) \left(\frac{1}{1-\phi}\right)^3 \qquad (6\text{-}7)$$

where k is the permeability coefficient, μ_t is the viscosity of the fluid at temperature t, ϕ is the porosity, and d_w is the effective grain diameter in mm (as per Kozeny, 1927). The lower limit of grain size for eq.6-7 is 0.2 mm. Inasmuch as d_w is found to be almost the same as d_{50}, the mean grain size, the Schultze equation relates permeability to grain size. The possibility of an even more fundamental and general relationship of permeability to certain petrologic and geometric parameters deduced from thin sections

TABLE 6-VI

Some typical permeability values for natural sediments

Sediment type	Normal permeability range (cm/sec)
Heavy clays	$10^{-8} - 10^{-12}$
Loam soils	$10^{-5} - 10^{-8}$
Silts, tills, loess	$10^{-3} - 10^{-5}$
Sands, eskers	$1 \ \ -10^{-3}$
Gravels	$10^2 \ - 1$

(Teodorovich method) has been indicated by the work of Aschenbrenner and Chilingar (1960).

As a coarse-grained sediment is densified, its permeability falls, often to a marked degree, because densification is accompanied primarily by removal of the *largest* voids which results in a considerable restriction of flow. This situation can be illustrated by reference to known compaction—permeability relationships in soils (Ingles, 1968), where quite small increases in density can lead to extremely large decreases of permeability. For example, a 10% variation in density can result in a ten thousand-fold change in permeability for some soils (cf. Fig.6-14). Where the sediment is saturated, however, and particularly in coarse-grained sands, such effects are very much less pronounced, because the particles can move easily over one another to assume configurations giving approximately uniform pore size. Thus, pores or channels of large dimension become less common. Even so, compaction of these sediments also results in permeability reduction.

The high permeability of coarse-grained sediments can be a property of considerable practical advantage. For example, it is very desirable in drainage

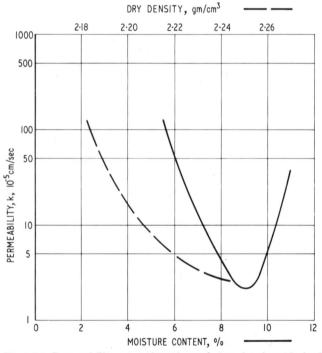

Fig.6-14. Permeability as a function of dry density (dashed line) and moisture content (solid line) for compacted crushed basalt. (After York, 1970, fig.5, 6, p.52.)

ways, to ensure the stability of natural slopes and earthworks, etc., and is important within natural underground reservoirs of oil, gas, and artesian or other waters. In the case of liquid or gas withdrawal from such reservoirs, substantial settlements can ensue due to reduction of the pore pressure in the sediment. An extensive discussion of this behavior is contained in the Proceedings of the International Conference on Land Subsidence (1969).

In saturated sediments, the flowing fluid usually contains entrained or suspended solids which may be filtered out by the coarse-grained sediment. Thus, the permeability will probably change over a period of time whenever bulk flow occurs in a sediment. The magnitude of the permeability reduction due to filtering action depends on the total impressed or natural flow volume. The effectiveness of the filter action is determined not only by the mean grain size of the sediment, but also has an inverse proportionality to the flow velocity because the latter determines the size of particle which can be deposited in the grain interstices. Under prolonged flow conditions a coarse-grained sediment may become clogged with fines, and this becomes equivalent in its effect to a densification by grading.

It is difficult to reverse this clogging process by simple means, because high hydraulic heads are needed to scour an already impeded drainage path, and reversal of the flow direction is seldom practical, especially for underground water reservoirs. One practical method developed by Chilingar and co-workers (Chilingar et al., 1970; Anbah et al., 1965) is based on the flocculation by electrochemical means of the clay infilling, which in its non-flocculated (deflocculated) condition constitutes a serious flow-restriction. These researchers showed that application of direct electric current may also result in destruction of highly-swelling clays, such as montmorillonite. The possibility of this occurring in nature during diagenesis is not excluded because natural electric currents do flow through sediments (Serruya et al., 1967).

Some diagenetic and authigenic processes, e.g., silicification, calcitization and sulphatization, are very effective in reducing the permeability of coarse-grained sediments (see Larsen and Chilingar, 1967, for details). Other diagenetic and epigenetic processes such as dolomitization and solution may increase porosity and permeability of carbonate rocks. Diagenetic processes can be reversible, however, e.g., dedolomitization, which can reduce porosity and permeability of dolomitized rocks, and decementation which may increase porosity and permeability.

In contrast to the hydraulic permeability of saturated sediments, permeability of unsaturated coarse-grained sediments is very low and, in certain instances, is even lower than that of dense clays. The reason for this apparent paradox lies in the fact that in an unsaturated, coarse-grained sediment the water films are discontinuous and no capillary transfer can

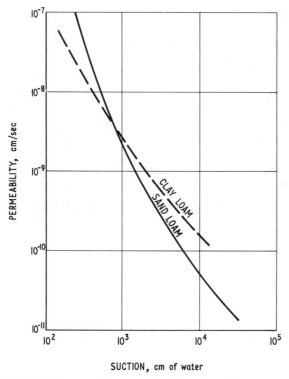

Fig.6-15. Permeability as a function of suction.

occur. Water movement is then possible only by surface ion migration or by vapor-phase transport, and for low humidities (i.e., high suctions) these transport modes are extremely slow. Hydraulic permeabilities are, therefore, very low. Figure 6-15 illustrates this situation and shows why it is possible to use a coarse-grained material (sand, gravel, etc.) as a *moisture barrier*. Provided no significant amount of water (bulk water) is allowed to enter, no significant water transfer will occur across the sand barrier.

In considering the permeability of coarse-grained sediments, it must be remembered that the coarser the sediment, the more likely is the occurrence of non-Darcy and even turbulent flow conditions. Turbulent flow is most likely to occur in gravels, where the inertial forces become large compared to the viscous forces.

Numerous and extensive studies have been published on the flow of fluids in porous media. For a modern treatment in depth, the reader is referred to the work by Lee et al. (1968), which includes the mathematical analysis of aquifers.

The basic equation for streamline (Darcy) flow is:

$$v = k \nabla h \tag{6-8}$$

where v is the discharge velocity, ∇ is the gradient symbol, h is the piezometric head, and k is the proportionality constant known as the coefficient of permeability. The latter (k) has the dimensions of velocity and is a function of the material properties of both the porous solid and the fluid. A specific permeability independent of the fluid properties can be defined, if the specific weight and viscosity of the fluid are incorporated into the equation. Where only water flow at normal temperatures is involved, however, this is unnecessary and thus not commonly used.

By combining the Darcy equation with the continuity equation (mass inflow = mass outflow + change of mass storage in time), it can be shown that for reasonably flat flows:

$$\nabla^2 h = \frac{S_s}{k} \frac{\delta h}{\delta t} \tag{6-9}$$

where S_s is the specific storage [$S_s = \rho g(\alpha + \phi\beta)$, where ρ is the fluid density, g is the gravitational acceleration, ϕ is the porosity, α is the vertical compressibility of the solid skeleton (i.e., reciprocal of the bulk modulus), and β is the fluid compressibility]. For confined aquifers of thickness b, eq.6-9 is better written as:

$$\nabla^2 h = \frac{S}{T} \frac{\delta h}{\delta t} \tag{6-10}$$

where S ($= S_s b$) and T ($= kb$) are the storage and transmissivity coefficients, respectively.

For turbulent flows, the earliest expressions were represented by the equation:

$$\nabla h = a v^n \tag{6-11}$$

where a and n are constants. This equation does not take into sufficient account such material properties as grain angularity, grading, degree of cementation, etc., which affect the physical arrangement of the medium. Thus, it does not allow an extrapolation of laboratory tests to variable conditions in the field with the accuracy currently desired. Other expressions have, therefore, been advanced (e.g., Cohen de Lara, 1956; Wilkins, 1956). In general, however, their advantages may not be sufficient to justify departure from the simplicity of eq.6-11, provided the utmost care is taken to simulate field conditions in testing.

For all flow and permeability problems in coarse-grained porous media, it is probably desirable to establish a friction factor versus the Reynolds number correlation diagram for the material, under conditions as close as possible to field conditions. Reynolds number is a dimensionless number which indicates whether the flow is laminar or turbulent. It is equal to $\frac{vd\rho}{\mu}$, where v is velocity of the fluid, d is diameter of the pore channel, ρ is the density of fluid flowing, and μ is the viscosity of fluid. The head loss due to friction, h_f, is equal to $f\frac{l}{d}\frac{v^2}{2g}$, where f is the friction factor, l is the length of flow channel, d is the diameter of pore channel, v is the velocity, and g is the gravitational acceleration.

York (1970), using compacted granular basalt of various gradings found that the flow regime, and hence the permeability coefficient, changed markedly at an hydraulic gradient of 1. Above this value, turbulent flow resulted in sensibly lower (and constant) values of permeability than in the laminar flow regions below 1. York also observed that there is a variation in the value of k depending on the conditions of compaction in the manner shown in Fig.6-14.

Some reference should be made to the measurement and the control of permeability of coarse-grained sediments, in view of the importance of permeability control in solving many practical engineering problems; for example, those associated with rockfill dams, sand blankets and filter zones, foundation settlement, mine safety (including water entry into shafts, tunnels, etc.), and drainage and stability of slopes.

Laboratory tests are not recommended whenever an in situ field measurement can be made. Such field measurements normally consist of measuring the volume rate of inflow to a well-point in the sediment. Some care must be taken in collecting these data. For example, a drawdown recovery test or an outflow test is much to be preferred over an infiltration test or injection test under pressure head, because (a) it ensures that surface blockage of the hole is less likely and (b) the water withdrawn is in full thermodynamic equilibrium with the strata, the permeability of which is being measured. Also, the test must be continued for a sufficiently long period of time for a quasi-steady state to be established. Some good field methods are described in the "Earth Manual" of the U.S. Bureau of Reclamation (1963).

Reference has already been made to methods of *increasing* the permeability of a sediment (Chilingar et al., 1970), but the most common practical method of *decreasing* the permeability of a coarse-grained sediment in the field is grouting. The cheapest available material suitable to the mean pore size of the sediment should be used. For example, cement can be used for sands, whereas a cement—bentonite mixture can be used for silty sands. Mine wastes, ash, sand, etc., mixed with a small amount of cement are used for grouting gravels. For grouting work, it is usually considered desirable to reduce the permeability of the strata to about 10^{-5} or 10^{-6} cm/sec. A good summary of current grouting methods is given by Fujii (1968).

STRENGTH

As in the case of permeability, the strength of coarse-grained sediments is also a function of their density and pore size distribution. The large pores and channels, which determine the overall permeability of the sediment, also constitute the major weaknesses which effectively control its mechanical properties. This mechanical control exercised by major pores and discontinuities in sediments was recognized by Ingles (1961) and has been subsequently studied by Lafeber and Willoughby (1970). Their studies on beach sands show also the role of pore anisotropy in determining the bulk strength of coarse-grained sediments.

The effects of fabric on deformation and strength of sands and the important contributions of Oda (1972) on the subject have been discussed previously. Porosity remains nevertheless the principal determinant of strength and deformation. It is well to emphasize here that what is usually measured and referred to by the term *strength*, is more truly a measure of *weakness*, due to the great difference between the measured and theoretical material strengths.

Figures 6-16 and 6-17 show typical interrelationships among moisture content, density, and strength for two coarse-grained materials artificially compacted. The difference between the two sets of curves in these two figures illustrates the difference between the properties of a rounded grain aggregate (Fig.6-16) and of an angular grain aggregate (Fig.6-17).

In theory, uncemented, coarse aggregates should be cohesionless, but in practice this is not the case owing to mechanical interlocking of the irregular particle shapes and surfaces. Even perfectly spherical, smooth-surfaced particles are not wholly cohesionless, because the condensation of small amounts of water at the points of grain contact lend a certain capillary cohesion, albeit small (Ingles, 1968). For materials of the smallest grain sizes, residual electrostatic charges can also lead to some particle agglomeration and cohesion.

The mechanical effects of grain size are also substantial. On theoretical grounds, strength should be proportional to the inverse square root of grain diameter, but in practice somewhat smaller exponents (0.25—0.52) are often found. As a result, at low porosities material of smaller grain size shows the higher strength, whereas at high porosities material of large grain size is stronger. The porosity at which strength is independent of grain size for clean,

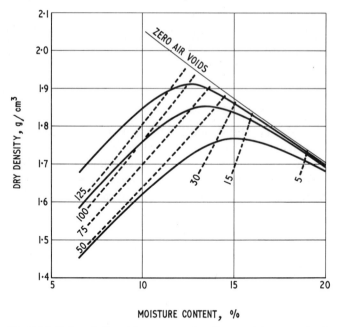

Fig.6-16. Interrelationship among density, moisture content, and strength of a rounded coarse sand (strength isobars expressed in terms of the penetration resistance). (After Tanimoto, 1957.)

coarse-grained sediments is not known, but probably is around 0.35—0.4.

The basic empirical equation governing the shear strength of granular materials and which assumes that failure occurs on the plane in which sliding first becomes possible (i.e., shear failure is assumed), can be presented as follows:

$$\tau_f = c' + (\sigma - p_{pw} - p_{pa}) \tan \theta' \qquad (6\text{-}12)$$

where τ_f is the shear stress on the plane of sliding; σ is the normal stress on the plane; p_{pw} and p_{pa} are the pore water and pore air pressures, respectively; c' is the apparent cohesion; and θ' is the apparent angle of internal friction of the sediment.

The strains which develop during the application of stress may be either elastic or inelastic. In any elastic or pseudo-elastic region the relationship between stress and strain can be expressed by the various sediment moduli. Young's modulus, E, is the ratio of axial stress to axial strain; the bulk modulus, K, is the ratio of hydrostatic pressure to the dilation it produces (this modulus was also discussed earlier as a modulus of compressibility); and the shear modulus, G also known as the modulus of rigidity, is the ratio of

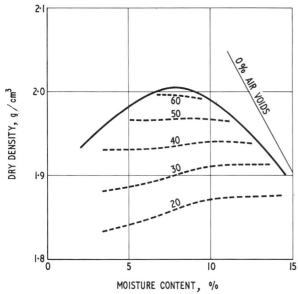

Fig.6-17. Interrelationship among density, moisture content, and strength for an angular gravelly sand (strength isobars expressed in terms of the California bearing ratio). (After Tanimoto, 1967, fig.28, p.24.)

shear stress to shear strain in simple shear. Some values of these three moduli for coarse-grained sediments are given in Table 6-V, p.312. Whenever the material behavior is not strictly linear elastic, it is necessary to specify whether a tangent or a secant value of the modulus is being measured, and whether a loading, unloading, or resilient (repeated loading) value of the modulus is referred to (see Fig.6-18).

The values for Poisson's ratio, ν, which is the ratio of longitudinal contraction to lateral extension are also presented in Table 6-V. As in the case of moduli, it is necessary to specify the method of measurement (secant, tangent, resilient, etc.), although this has seldom been done hitherto in the literature. Because granular materials are seldom truly elastic, the Poisson's ratio increases with increasing stress level. The quoted values in Table 6-V are applicable from zero to about 60% of peak stress, after which the value rises rapidly to 0.5 and often even higher (values up to 1.6 have been recorded at very high strains). Inasmuch as most design calculations are based on the application of stresses well below failure stresses, the quoted figures are those of practical value. Resilient values are higher by about 0.1 than those shown in Table 6-V. In saturated conditions the value approaches 0.5 (Ishihara, 1971). Sandstones show a linear increase in Poisson's ratio with increase of stress levels.

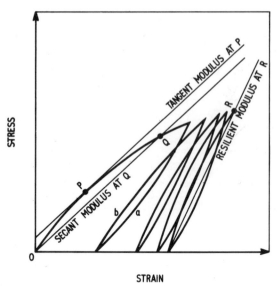

Fig.6-18. Various moduli approximately describing the real material behavior O—P—Q—R; a = an unloading curve, b = a loading curve.

As mentioned before, increased confining pressure tends to reduce the initial anisotropy of a sediment, which greatly affects the strength properties. Inasmuch as the properties of a sediment change with the application of stress, measurement of strength properties of anisotropic materials is difficult and generally unsuccessful (Onas, 1970). Gerrard (1967, 1969) has discussed the relationships of stress and strain in a transversely isotropic material (also known as cross-anisotropic material), particularly for sands. Gerrard has shown that for stress paths in which the octahedral shear stress increases monotonically, i.e., undergoes a steady progression, either increase or decrease, without reversal, then the lowest value of $\int_{\text{initial}}^{\text{final}} (\sigma_{\text{oct}} - \sigma_c) d(\tau_{\text{oct}})$ corresponds to the greatest vertical strain and most dilatant behavior. The application of any deviator stress is known to increase the anisotropy of all coarse-grained materials, except when applied in the plane of an existing anisotropy; and stress-induced anisotropy is an important factor in determining the strength of these materials (Ingles and Lafeber, 1967; Onas, 1970).

Chaplin (1965) showed that for cohesionless sands the strain should increase linearly with the square root of any of the stresses (σ_1, σ_2, σ_3, σ_{oct}, etc.). In 1962, Rowe published a complete theory on the behavior of rigid rotund particles in a stress field, giving a stress-dilatancy solution for granular materials as follows:

$$\frac{\sigma'_1}{\sigma'_3} = \left(1 - \frac{\delta V}{V} \frac{1}{\delta \epsilon_1}\right) \tan^2\left(\frac{\pi}{2} + \frac{\theta_u}{2}\right) \qquad (6\text{-}13)$$

where $\frac{\delta V}{V}$ = volumetric strain; θ_u = true friction angle between the particles; σ'_1, σ'_3 = major and minor principal effective stresses; and ϵ_1 = the axial strain in the σ'_1 direction. Hence a plot of σ'_1/σ'_3 versus $\left(1 - \frac{\delta V}{V.\delta \epsilon_1}\right)$ should be a straight line through the origin with a slope of $\tan^2\left(\frac{\pi}{2} + \frac{\theta_u}{2}\right)$. Although this has been confirmed in a large number of cases, it was subsequently shown by El Sohby (1964) that with constant stress ratios in sand, there is a significant component of elastic deformation. Consequently, in the case of constant stress ratio, Rowe's theory, which is based on non-elastic sliding deformation, applies best at high stress ratios ($\sigma_1 \gg \sigma_3$), because these favor non-elastic sliding.

The observed strength of sand and other granular materials largely depends on the test method adopted, one of the most important variables being the time taken to reach peak load. The latter is defined as failure, though a coarse granular mass will often carry substantial loads in the post-peak region. Lee et al. (1973) descibed this property for a sandstone. Decreasing the time to failure increases the effective strength of a sand, i.e., a sand is stronger to shock (impact) loading than to slow loading. This is particularly true of saturated sands, because the pore pressure ($p_{pw} + p_{pa}$ in eq.6-12) is unable to dissipate. For a completely dry sand, however, the effect becomes insignificant. It is for the above reasons that dynamic moduli differ, sometimes very appreciably, from static moduli, as will be seen by reference to the values of sonic modulus presented in the following section.

Referring back to eq.6-12, it can be seen that where any cementation of the grains occurs, the strength of a granular mass is enhanced (c' increases). When the component grains are rough, as a result of interlocking, etc., the strength is also enhanced ($\tan \theta'$ increases). The "cohesion" (c') and "angle of internal friction" (θ') are frequently used as parameters to quantify the strength of coarse granular materials, and some of their recorded values are shown in Table 6-V, p.312. It is important to remember, however, that these values are obtained by using a mathematical equation based on idealized material properties and, therefore, should not be confused with any real physical property such as cohesion or friction in the true sense. Indeed, there has been considerable dispute on the correct method for determining θ', and there is a wide range in values reported in the literature. Many attempts have been made to correlate friction with real physical parameters; Orchard et al. (1966) have described one such method for correlating friction with surface roughness. Winterkorn (1971) has claimed that the angle of friction as measured in triaxial cells is related to the porosity by the following equation:

$$\tan \theta = \frac{A(1-\phi)}{\phi - B(1-\phi)} \qquad (6\text{-}14)$$

where A and B are constants. A somewhat simpler relation has been established by Schultze (1970) for sands and gravels:

$$\tan \theta = \frac{0.46(1-\phi_i)}{\phi_i} \qquad (6\text{-}15)$$

where ϕ_i is the initial porosity.

The response of sands to repeated cycles of loading is densification to a stable condition, as has been described by Morgan (1966).

In engineering practice, a number of aspects of the strength of coarse granular materials have great practical importance. These include the coefficients of active and passive earth pressure (which determine the security of retaining walls, etc.), the calculation of safe slopes (to determine angles of

TABLE 6-VII

Design specifications for safe slopes and bearing pressures

(a) Safe slopes (U.S. Atomic Energy Commission, Division of Materials Licensing requirements, except as indicated otherwise)

Type of material	Horizontal to vertical ratio
Sand—gravel mixture, dry	2:1
Sand—gravel mixture, wet	3:1
Coarse-grained sand, dry	2½:1
Coarse-grained sand, wet	3:1
Fine-grained sand, dry	2½:1 (British specification)
Coarse tailings, dry	2:1
Coarse tailings, wet	2½:1

(b) Bearing pressures (Road Research Laboratory, United Kingdom specifications)

Type of material	Bearing pressure (tons/ft²)
Well-graded gravel, sand—gravel mix	10
Loose gravel, loose sand—gravel mix	3
Well-graded sand, coarse-grained	4
Loose sand, coarse-grained	2
Well-graded sand, fine-grained	3
Loose sand, fine-grained	1

cut and repose), and allowable bearing pressures (to avoid undue settlements, etc.). In many cases, their importance is so great that codes of practice have arisen to enable safe designs to be calculated and specified. These codes are, therefore, quoted here (Tables 6-VII and 6-VIII) as essential guides to any estimated behavior of coarse granular masses under load. They are necessarily very conservative and, hence, may indicate lower strengths than some of those previously quoted in Table 6-V from actual field or laboratory experiments.

Inasmuch as cohesion is relatively low in coarse-grained sediments, with the exceptions noted for cemented and angular materials, the internal stress redistributions occasioned by application of any external force system are largely controlled by interparticle movements, i.e., shear. A most important property of coarse-grained sediments, undergoing shear deformation, is the *critical void ratio*. This is the void ratio at which the bulk volume of the sediment remains unchanged on application of a shearing force. All coarse-

TABLE 6-VIII

Coefficients of earth pressure (Civil Engineering Code of Practice, 1951)

(a) Values of active pressure coefficients, vertical wall, horizontal ground surface, cohesionless soil

δ*	θ				
	25°	30°	35°	40°	45°
0°	0.41	0.33	0.27	0.22	0.17
10°	0.37	0.31	0.25	0.20	0.16
20°	0.34	0.28	0.23	0.19	0.15
30°		0.26	0.21	0.17	0.14

(b) Values of passive pressure coefficients, vertical wall, horizontal ground surface, cohesionless soil

δ*		Code		Revised**	
	Triaxial θ	33°	40°	33°	40°
	Plain strain θ			34°	42°
0°		3.4	4.6	2.5	4.6
10°		4.5	6.5	3.0	5.8
20°		5.6	8.8	3.6	7.2
30°		6.7	11.4	4.3	8.8

* Wall friction angle.
** Rowe and Peaker (1965) have demonstrated that the original code figures may be on the unsafe side, and proposed revisions as shown.

grained sediments have their own characteristic critical voids ratio, such that if the sediment has an initial void ratio higher than this value (less dense), the application of shear will result in densification of the mass until a void ratio equal to the critical one is reached, at which time a steady state will ensue. Conversely, if the mass initially has a void ratio lower than the critical void ratio, application of shear will lead to a dilation of volume (loosening of the mass) until the critical void ratio is attained (Fig.6-19). It has been shown by Morgenstern (1963) that the critical void ratio state is the state of maximum entropy.

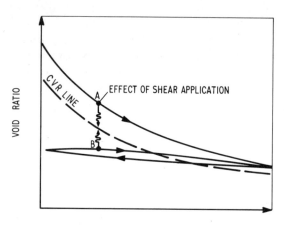

Fig.6-19. Relationships between void ratio and pressure, showing the critical void ratio line and the effect of shear (wavy arrows) applied from points A and B.

LIQUEFACTION AND COLLAPSE IN SANDS

Liquefaction

Under certain conditions, cohesionless granular materials and particularly sands when subjected to an applied load can behave as a liquid for the duration of the load. The relevant conditions are chiefly (1) saturation and (2) impeded drainage.

As described in the previous section, if cohesion and pore-air pressure are neglected, the shear strength of a saturated cohesionless granular material is expressed by the equation:

$$\tau_f = (\sigma - p_{pw}) \tan \theta \qquad (6\text{-}16)$$

where the pore-water pressure, p_{pw}, includes both static and dynamic components. If the pore-water pressure cannot readily dissipate, then as the applied load causes a rearrangement of the grains and densifies the mass, the pore-water pressure may rise sufficiently to equal the total normal stress, σ, in the sediment mass. According to the eq.6-16, the shear strength then becomes zero and the system behaves as a viscous liquid. In fact, the total load on the system is then borne by the pore water, which has become the continuous phase. The solid particles behave as though they were suspended in this medium, giving rise to its viscous nature. This condition prevails until the pore-water pressure is dissipated by water movement out of the loaded area, or by removal of the excess load.

Liquefaction of cohesionless granular materials has caused surface structures, such as embankments, buildings, etc., to sink and underground structures, such as tanks, pipes, piles, etc., to rise until they are in hydrostatic equilibrium with their surrounding environment (Kawasumi, 1968). Also, natural sand or silt beds, or structures such as embankments or fills formed from sand or silt, may flow when the angle of internal friction is exceeded. Bjerrum et al. (1961) cited examples of flow on fine loose sand in the beds of Norwegian fiords. They showed that the friction angle decreases very rapidly when the porosity exceeds a certain value and may fall as low as 11°.

Nonuniform loading on structures may cause rotational movements resulting in tilting, etc. A particularly common and serious problem is flow movement on steep slopes in the form of landslides, mud and gravel flows, etc., where infiltration of the slope by water at a higher level causes excessive pore pressure at lower levels, unless good slope drainage is available.

Liquefaction may be caused either by static or dynamic loading. The liquefaction arising from static loading is most frequently caused by raising the water content of a poorly-drained material, increasing the surcharge load on it, or both. This case is well illustrated by the Aberfan (Wales) disaster of 1968 (H.M.S.O. Welsh Office, 1969). In general, however, it is usually possible to apply loads sufficiently slowly to allow pore-water pressures to dissipate or, else, to provide effective avenues for drainage.

This is not the case with dynamic loading, such as that applied by the seismic forces of an earthquake or resulting from underground explosions. As a result of rapid accretion of load, even in otherwise well-drained situations the pore-water pressure cannot be dissipated at a sufficiently fast rate, and liquefaction occurs. During repeated loadings by an earthquake, pore water may be expelled into surrounding strata or liberated at the surface in the form of "sand-boils", producing the condition popularly known as quicksand. The concomitant effects of superincumbent structures, the general environment, etc., are well illustrated in the General Report on the Niigata Earthquake of 1964 and the Alaska Earthquake of the same year

(U.S. Department of Interior, 1970).

Liquefaction is dependent on the ability of the pore-water pressure to exceed the normal stress. Several factors within the material, however, may critically affect the normal stress and, hence, the susceptibility to liquefaction. One such factor is the degree of mechanical interlock between the solid particles as determined by grain size, shape, and roughness, which give an effective cohesion to the otherwise cohesionless grains. It is also obvious that the higher the relative density of the mass (lower porosity), the less susceptible it will be to liquefaction, because there will be less possibility of further densification or water infiltration necessary to initiate the pore-pressure rise. Prakash and Gupta (1970) stated that for relative density greater than 52% in coarse-grained sands and 62% in fine-grained sands, no significant pore pressure can be produced. They have concluded from this that for depths of burial greater than about 4 m, liquefaction probably cannot occur.

Yamanouchi (1970) has shown that materials susceptible to liquefaction include those in which the mechanical interlocking of the skeletal grains is weak and those that already possess appreciable porosity. Figure 6-20 illustrates that materials which (a) are composed of very large particles, (b) are well graded (i.e., have relatively high bulk density), and (c) have good mechanical interlock, will liquefy less readily than those that do not exhibit these characteristics. Hence good compaction or the addition of

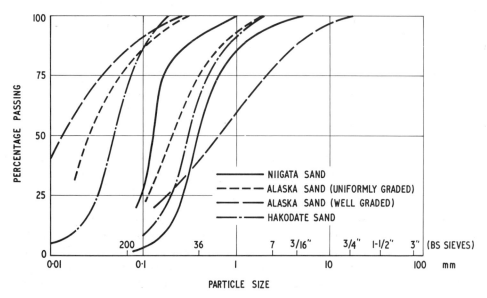

Fig. 6-20. Grading limits for sands which liquefied during major earthquakes. (After Yamanouchi, 1970, fig. 7, p. 7.)

material to densify the mass or produce a better-graded material alleviates the tendency to liquefaction in potentially dangerous materials. In a liquefied material, as the pore-water pressure dissipates, particles settle into a new position of equilibrium generally at a greater density than that of the original material, resulting in compaction. D'Appolonia (1953) has described a method known as vibro-flotation for compacting loose sands by artificial liquefaction. An increase of up to 45% in the density of a sand mass was reported using this method.

Florin and Ivanov (1961) discussed in detail the mechanics of the liquefaction process. They showed that coarse-grained materials are susceptible to liquefaction, but because the duration of liquefaction is so much less than for fine-grained material the effects are not as noticeable, although mud flows emanating from talus slopes can be cited as examples of the results of liquefaction. Such flows may carry rock fragments up to many cubic yards in size. They tend to occur when natural drainage is insufficient to carry away all water entering the material. This often occurs during periods of heavy rainfall or the spring thaw of snow. Florin and Ivanov discussed the influence of entrapped gas that decreases the tendency for coarse-grained materials to liquefy, because the presence of gas tends to decrease the velocity of shock waves. Kurzeme (1970) has also reviewed some recent knowledge of liquefaction.

Collapse

Some sands, on wetting under load, will decrease markedly in volume, i.e., will *collapse*. This phenomenon is due to the fact that in their natural condition the granular skeleton of these materials is very porous, with the points of contact between the grains being in a metastable equilibrium (Ingles and Aitchison, 1969). Water entry can disturb this equilibrium in a number of ways, and acts as a lubricant to assist the sliding of the grains over each other until dynamic equilibrium is achieved at a higher bulk density. Indeed, most differential settlements on sand arise from density differences.

Typical coarse-grained sediments in which collapse has been observed include (1) sands formed in situ in tropical areas as a result of the leaching of constituents other than quartz from granites or felspathic sandstone masses and (2) sand deposits formed by aeolian action in arid areas (Brink and Kantey, 1961). In the latter case, the clearing of vegetation from the surface can seriously reduce evapo-transpiration, increase the depth of subsequent water penetration, and cause collapse beneath the surface structures.

Collapse phenomena can be observed in North Queensland (Australia) and Rhodesia (Africa), on roads built on granite subgrades that have been insufficiently compacted (Holden, 1957). Under traffic, these roads lose their grade line and assume a wavy appearance. Collapse under foundations has been reported by Jennings and Knight (1957) and, in the vicinity of irrigation canals, by Prokopovich (1969).

Sands subject to collapse are those that are poorly graded, have low or very low bulk densities (< 1.60 g/cm^3), or have cohesive bonds of a water-soluble nature (e.g., slowly soluble minerals, or dispersive clays). Holden (1957) cited examples of Rhodesian granite sands with a natural density of 1.20 g/cm^3 capable of compaction to a density of 1.55 g/cm^3. Collapsing sands are, of course, quite susceptible to liquefaction and can be compacted by this technique.

The correct procedure to avoid collapse is to preconsolidate these materials. In some cases this can be done simply by flooding, whereas in others this can be achieved by preloading, shock treatment, vibration, etc.

CEMENTATION AND ITS EFFECT ON PROPERTIES

Natural materials may experience cementation, i.e., undergo diagenesis when subjected to the action of groundwater containing certain dissolved salts. The effect of such cementation is to produce cohesion between the discrete grains. Ultimately, this may yield a cementing matrix so strong that the whole mass becomes rock-like. This matrix may consist of silica, iron oxides, calcium carbonate, or gypsum (all hydrated to some extent), and more rarely aluminum, manganese, or nickel oxides. Materials cemented by silica are known as silcretes, by iron oxides as ferricretes, by calcium carbonate as calcretes, and by gypsum as gypcretes. The terms silcrete, ferricrete and calcrete have been proposed by Lamplugh (1902).

Each type of cementation is the result of different physical and chemical processes. In the acidity range of natural groundwaters, silica possesses a low but finite solubility which is not greatly affected by changes in pH (Fig.6-21). It can be deposited from solution only as a result of evaporation of the solvent (water), i.e., by exceeding the solubility concentration. Similarly, gypsum ($CaSO_4.2H_2O$) has a significant solubility ($\approx 0.02\%$) and can be deposited by evaporation of the solvent water. Iron, on the other hand, although quite soluble in the ferrous state below pH 8, is insoluble in the ferric state above pH 3. Hence, when an aqueous solution of ferrous salts encounters conditions sufficiently oxidizing for the ferrous—ferric ion transition to occur, iron oxide will be precipitated. Oxygen from the air is sufficient to promote this reaction, which is probably responsible for the iron-staining rings in some sandstones. Natural limes (mostly calcium carbonate) are often quite soluble in the presence of acid waters containing appreciable amounts of dissolved carbon dioxide from the air, forming a

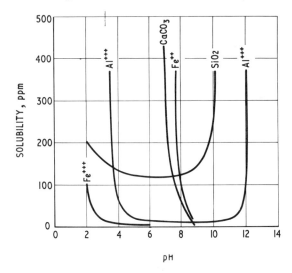

Fig.6-21. Variation with pH of the solubility of Fe^{2+} and Fe^{3+} (after Mattson and Gustafsson, 1923), Al^{3+} (after Raupach, 1963), and amorphous SiO_2 (after Morey et al., 1964.)

soluble calcium bicarbonate. When the dissolved carbon dioxide content is reduced by solubility effects resulting from a change of solution temperature, or the solubility product of the bicarbonate is exceeded as a result of the evaporation of water, calcium carbonate will again precipitate, especially at the grain-to-grain contact points. Evaporation which results in deposition of cementing materials, especially silica, leads to stronger bonding in a granular mass.

These cementing agents tend to deposit wherever water comes into contact with air, or where a seasonally fluctuating water table occurs. Thus, they are often deposited close to the surface, and their deposition is favored by high ambient temperatures which remove water by evaporation. This occurs in tropical climatic conditions (summer rainfall, winter drought), where ferricretes tend to be produced; in Mediterranean climatic conditions (winter rainfall, summer drought), where calcretes tend to be produced; and in arid conditions (irregular rainfall, high temperatures), where silcretes and gypcretes tend to be produced. In the latter case, temporary cementation with salt (NaCl, chiefly) may also be produced at or near the surface. These cemented sediments may be later buried under younger deposits. The cementation process is not restricted to any geological period of time and is active today in areas with the suitable environment (Grant and Aitchison, 1970).

The cemented aggregate is either hard or soft, depending on the degree of cementation and the type of cementing agent, and the range of strengths in such sediments is very great. For example, the recorded sandstone strengths range from 300 to 3,000 kg/cm^2. Silica is initially deposited from solution as a soft hydrated gel, which on dehydration first becomes opal and then crystalline silica (agate or quartz), producing successively harder materials. Similarly, the initial deposit of iron oxide may be quite a soft hydrated gel, whereas after oxidation on exposure to air the mass hardens to a brick-like consistency. Buchanan (1807), who originally noted this phenomenon in India, coined the term "laterite" for ferricrete materials.

Another effect of cementation is the bonding of fine clay particles into stable aggregates, which form a pseudo-coarse-grained material. The aggregate produced may be either sand- or silt-sized. Ackroyd (1967) has discussed this phenomenon and its significance for road building in Nigeria. Norrish (1964), who noted its occurrence in Queensland, Australia, has ascribed it to cementation with aluminum compounds. These "sub-plastic clays" appear to have no tendency to cementation into a single mass, however, and remain as more or less cohesionless sands and silts.

In their soft condition, cemented granular sediments may be compacted by normal means and their breakdown under rolling, for instance, often leads to the desirable size distributions for very dense grading and good compaction. Consequently, they became popular for low-cost road construction in semi-arid areas, provided certain elementary engineering precautions were observed (Andrews and Vlasic, 1968; Pryor, 1966). An added advantage is that material compacted in the soft condition may harden after exposure to atmospheric conditions, improving its performance in load-bearing earthen structures. Netterberg (1967) has discussed the hardening after repeated stress applications to calcrete, whereas Grant and Aitchison (1970) discussed the similar effect for silcrete and ferricrete. With the more strongly cemented materials, however, compaction is impossible without crushing.

Another form of cementation is lithification due to diagenetic processes on deep burial or under the influence of tectonic pressures. Elevated temperatures and pressures can have significant effects on the solubility products of some minerals. As a result of chemical reactions within the interstitial solutions, new minerals are formed, cementing the grains of the originally incoherent material into a coherent rock mass.

The main effect of cementation on coarse-grained materials is the great increase in the bulk strength of the material and the resistance against water-softening, which causes slope stability problems, liquefaction, collapse, etc. Even weakly-cemented materials are far more resistant to weathering and erosion than uncemented sediments. This is well illustrated by the mesa formations in arid zones (Fig.6-22) and the stream erosion in contiguous

Fig. 6-22. Mesa, showing cemented silcrete caprock and typical concave talus slope, near Winton, Queensland, Australia.

Fig. 6-23. Erosion of a deep coarse granular sediment with lightly-cemented interlayers, Fujiyama, Japan.

cemented and uncemented sediments (Fig.6-23). Not to be ignored, however, is the infilling of pores, which effectively reduces the porosity of the mass, with substantial effects on permeability, compressibility, and the angle of shearing resistance (internal friction). Cementation also effectively reduces the plasticity of any fine-grained material incorporated within the coarse-grained sediment. By reducing the number of discontinuities in the granular framework, cementation substantially alters the acoustic (and other wave propagation) properties of the sediment.

SUSCEPTIBILITY OF COMPACTED COARSE-GRAINED SEDIMENT TO EROSION

Compacted coarse-grained sediments are susceptible to erosion in the same ways as are other natural materials such as soil. The subject of erosion has been discussed at length by Termier and Termier (1963).

Erosion is dependent upon three factors. Firstly, the nature of the eroding medium must be considered; this medium may be ice, water, or air. Secondly, the energy of the medium in any given situation is critical; the amount and nature of the erosion increases with increasing energy of the eroding medium. For instance, the water of a quietly flowing stream may have essentially no sediment load, whereas a raging torrent may carry blocks of rock literally as large as a house. Thirdly, the ability of the material to resist disintegration is very significant, i.e., the harder the material, the less is the erosion that occurs in a given situation. Thus, cemented materials erode less easily than do those that are not cemented.

Coarse-grained materials, unless they are cemented, are "soft" materials and hence are prone to erosion. They have little resistance to ice and are readily removed. In ice, the sediment particles are transported to be emplaced eventually in the terminal moraine of the glacier, or deposited in a sea or lake if the glacier terminates there. Alternatively, the particles may be ground against rocks or debris either in or under the ice to produce rock flour that is usually carried away by streams of melt water issuing from the glacier. In water, the energy of the eroding medium plays the dominant role. In high-energy environments, the pressure of the water between particles may rise sufficiently to cause liquefaction and the particles literally float away. This type of action is particularly applicable to wave and high-velocity stream erosion. In a very high-energy environment, particles transported by this means may be very large. Although liquefaction may not occur, in low-energy environments there may be sufficient energy available to transport particles by a rolling action. In general, the smaller and more rounded the particle, the easier it is to transport it by this means. In the case of erosion by air, both the energy available in the air and the size of the coarse-grained particles are important. Air rarely has sufficient energy to move particles larger than 2 mm in diameter. Hence, it is sand- and silt-sized particles that are prone to aeolian erosion. Chepil (1958) noted that the aeolian erosion of 60-mesh sand required a wind velocity of 21 km/h, 0.3 m above the surface. Aeolian erosion tends to transfer unprotected compacted sands to sand dunes.

Sand dunes are usually in a state of dynamic equilibrium with their surroundings. Interference with natural dunes may have a profound effect on the state of dynamic stability. For instance, a coastal beach with its backing dune may be in a state of dynamic stability. The dune acts as a sand reservoir supplying sand to the beach during periods of sand depletion and accepting sand from the beach during periods of sand accretion. Interference with the dune by levelling, covering, building, etc., upsets the system of dynamic equilibrium and can cause partial or, in the extreme case, total loss of the beach.

The weathering of a porous mass is greatly accelerated when air and water action are combined and, especially, if assisted by thermal extremes. Thus, sands formed in situ as a result of decomposition of granites tend to show density gradients, with the lowest densities being present near the upper surface. These deposits are particularly prone to repeated shallow landslips, the "Masa soils" of Japan being a good example.

The erosion of cemented coarse-grained materials is dependent upon the degree of cementation. Weakly-cemented materials tend to erode by processes similar to those which cause erosion of uncemented materials, although they are more resistant to erosion than are the latter. Additional energy is, of course, required to break the weak cementing bonds. Materials cemented by lime or gypsum may be eroded by water, which is capable of removing the cementing material by dissolution. Strongly-cemented materials, on the other hand, are essentially rock-like and tend to erode in a manner similar to rocks. In elevated locations, strongly-cemented materials may be eroded by the preferential removal of soft material below the cemented horizon. Such a process undercuts the harder cemented material, which falls away forming a cliff. Silcretes, ferricretes and gypcretes are susceptible to this type of erosion. The typical escarpments with concave talus slopes, capped by vertical rock of mesas and breakaways in arid and semi-arid regions, are produced by this process (Fig.6-22).

The efficiency of methods designed to protect coarse-grained sediments from erosion depends upon their ability to lower the energy content of the eroding medium. Dykes, groynes, breakwaters, etc. are used extensively for this purpose. Similarly rip-rap or gabions, placed on the sloping surface of dam and dyke walls, lower the energy of either water or air impinging on the particles and thus tend to prevent erosion. In the simplest case, a mat of grass and other vegetation can be quite effective in stabilizing sands and gravels. Compaction is *not* an effective technique of combating the erosion of coarse-grained sediments, because of their high porosity.

EFFECT OF ARTIFICIAL COMPACTION ON PROPERTIES OF SEDIMENTS

All loose sediments may be compacted by external forces; and dense and cemented coarse-grained sediments can be densified or cemented still further by various artificial as well as natural means. Again, this deliberate modification of the properties of coarse-grained sediments greatly depends upon the initial porosity of the mass for its effectiveness.

Where it is desired to reduce water flow in a coarse-grained sediment (as in excavation or mining activities), grouting techniques can be adopted, whereby various quantities of cement, cement—bentonite, or other grouts are injected to fill the voids. The selection of the most economical shut-off grout largely depends on the grain size and permeability of the sediment. Discussion of the criteria and methods available for the selection of grouting materials is given by Ingles and Metcalf (1972). Generally, bulked cements (i.e., cements with any suitably-sized, cheap, inert filler) are considered best for gravels; whereas for sands, either cements, cement—bentonite mixes, or chemicals (including the new water-setting types) are most suited.

Many engineering methods are available for increasing the strength of a surface gravel or sand. These include improvement of the grading by mixing and recompacting with material of a suitable size to ensure a closer approach to the grading curves shown in Fig.6-6, p.302. If the gravel or sand is close to being monosized, the added material can be simply spread on the surface and vibrated in by a vibratory roller. Another method is to improve the cohesion of the sediment by mixing with either cement or bitumen and then recompacting. Whenever compaction or recompaction of a sand or gravel is desired, vibratory compaction (using rollers) is always the most efficient technique. When compaction of the sediment must be extended to considerable depths, however, then rolling is not applicable, because its depth of worthwhile effect is less than 25 cm owing to the difficulty in transmitting stress through a deep granular sediment. Consequently, other methods must be adopted which consist of some kind of impact rammer, as in the Vibroflotation process (Greenwood, 1965) or the new Direct Power Compaction process (Morimoto et al., 1969). Another method which has found application in USSR and Japan is the electric discharge compaction method (Lomise et al., 1963), whereby a shock wave is used to achieve compaction at depth. Other techniques using shock waves have been proposed and occasionally used, employing conventional explosive charges properly placed. All these methods for compaction of deep, coarse-grained sediments are most applicable to sands. Additional strength is seldom required in a gravelly deposit and is not readily achieved.

If the artificial compaction is intense, grain crushing can occur, especially in the case of porous materials such as the volcanic ejecta and brittle rocks such as granites. In these circumstances, compaction leads to the production of fines within the mass, which not only serve to densify the mass rapidly, but also may add to its cohesion. In engineering practice, this effect is usually measured by examining the plasticity index of the fines fraction after compaction. A small amount of plastic fines will greatly increase the strength of the mass by increasing its cohesion. An optimum amount of fines exists, however, above which the plasticity of these fines can lubricate the movement of the large grains and reduce the shear strength. Hence, in any compacted sand, gravel, or sand—gravel mix, which is used for construction purposes (e.g., roads, airfields), it is normal practice to specify an upper limit of the plasticity index (P.I.). Usually, a P.I. of around 6 is adopted; however, sometimes for particular materials, especially those cemented materials that harden on exposure, this requirement can be relaxed appreciably. Where the sand or gravel is to be used in dam construction, etc. as a filter blanket (i.e., where permeability and bulk density rather than strength properties are critical), then plasticity should be very low because it reduces the effectiveness of the filter. Materials like pumices, granites, etc., which can produce fines during the compaction process, are thus generally unsuitable and to be avoided for filter construction.

Compaction of sands liable to collapse, or subject to large settlements because of very high initial porosity, has conventionally been carried out by surcharging or heavy rolling. From the previous discussion of the properties and the fact that friction angles of dry sands are higher compared to wet sands, it is obvious that settlements in these materials will be best accelerated if the compactive effort is accompanied or preceded by some irrigation of the site. This should not be so great, however, as to saturate a poorly-draining sediment and thereby create positive pore pressures. Only sufficient amount of water is used to lubricate the grains.

ACOUSTIC, THERMAL AND OTHER MISCELLANEOUS PROPERTIES

Wave transmission

The velocity of sound in coarse-grained materials is low. The sonic velocity values for various types of deposits are presented in Table 6-IX.

TABLE 6-IX

Sonic velocity values for various types of deposits

Type of deposit	Sonic velocity (km/sec)	
Dry gravel	0.6 —0.8	
Wet gravel	1.2	
Sand	0.6 —1.85	lower values are for the
Sand and gravel	0.38—1.67	non-saturated condition;
Tuff	2.16	higher values are for the
Silt	0.43—1.73	saturated condition
Sandstone	1.4 —4.3 (soft) (hard)	

These figures can be compared with values in excess of 3.5 km/sec for most of the harder rock types.

Sonic velocities are higher in cemented materials and increase with increasing degree of cementation. Velocity in strongly-cemented materials, such as silcrete may approach that in hard rock.

Hardin and Richart (1963) showed that wave velocities in sands vary with the approximate inverse fourth power of the confining pressure and linearly with porosity, independently of grain size, gradation, and relative density. They deduced that $v_s = k_1 \sigma_0^{k_2}$ where v_s is the velocity of propagation of the travelling wave, σ_0 is the confining pressure, and k_1 and k_2 are constants with k_2 approximately equal to 0.25. Relationship between porosity and shear wave velocity is illustrated in Fig.6-24.

Another variable with a noticeable effect on the velocity of propagation of acoustic waves is the water content of the sediment. Whitman (1970) discussed this relationship with reference to sonic waves from nuclear explosions. Theoretically, in unsaturated conditions, the mineral grain framework is the continuous medium and is responsible for the conduction of acoustic waves. The mass of the initially dry material slowly increases with increasing moisture content, causing the velocity of wave propagation gradually to decrease. In the saturated condition, water is the continuous phase and effectively transmits the waves at a velocity of about 2.0 km/sec. Figure 6-25 shows theoretical relationship between dilation velocity and saturation. In practice, the change from mineral framework to water transmission has been found to be more gradual than Fig.6-25 indicates. There is still much uncertainty about the actual mechanism involved.

The effect of compaction on a coarse-grained sediment is essentially to decrease the void ratio and thus to increase the sonic velocity. If compaction of wet sediments results in an increase in their saturation, the velocity of

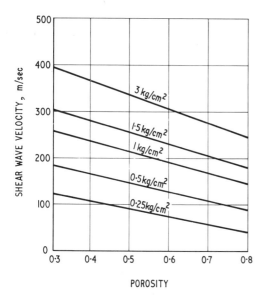

Fig.6-24. Shear wave velocity in sands as a function of porosity for various confining pressures. (After Hardin and Richart, 1963, fig.15, p.59.)

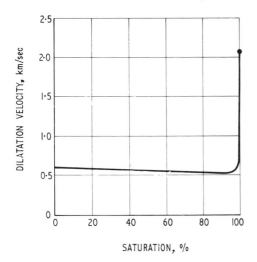

Fig.6-25. Relationship between dilation velocity in sand and water content. (After Whitman, 1970, fig.5.27, p.185.)

wave propagation should gradually increase and reach that in water. Ishihara (1971), however, has shown that the longitudinal wave velocity in saturated sands and gravels is about 10—20% greater than in pure water.

As with shock loadings, so with higher sonic frequency the apparent modulus, E, of the sediment rises substantially. Young's modulus for Ottawa sand determined from static loading tests lies between 2,000 and 10,000 kg/cm^2, whereas the modulus deduced from ultrasonic measurements on the same sand lies between 4,000 and 11,000 kg/cm^2.

Thermal properties

The thermal expansion of a poorly-graded sand has been quoted at 10^{-5} cm/cm °C (Ingles and Metcalf, 1972).

The thermal conductivity of coarse-grained sand and of pumice has been reported as $0.9-2.7 \cdot 10^{-3}$ and $0.6-1.2 \cdot 10^{-3}$ cal/cm sec °C, respectively (Clark, 1966). Other reported values of thermal conductivity for gravels and for sands are $3.4-6.9 \cdot 10^{-3}$ and $1.7-6.9 \cdot 10^{-3}$ cal/cm sec °C, respectively (Kersten, 1949); however, few measurements of coarse granular sediments have been reported in the literature.

Contact angle

Measurement of the contact angle of water in sands is useful in the computation of capillary rise, entry pressures, etc. (see Langnes et al., 1972, for detailed discussion). Emerson and Bond (1963) have found that some natural sand sediments have contact angles in excess of 90°, i.e., sands are not water-wet except under pressure. Although this is a somewhat unusual extreme (attributed in that case to grain coatings), the contact angles in natural sands should not be assumed to be zero (as for clear silica surfaces). In water-wet sands contact angles are less than 90°.

Resistivity

Measurements of ground resistivity may be and have been used to locate beds of sand and gravel at moderate depths below the surface (Bartlett and Foote, 1966). This method permits estimation both of the depth of the sand—gravel layer and the depth of water table (if any). It is based primarily on the large differences in resistivity between dry granular strata and wet granular strata, because saturated gravels or sands have low resistivities (often less than 3,000 ohm cm), whereas clean (low clay content), dry gravels and sands possess quite high resistivities (above 30,000 ohm cm). The presence of clay leads to intermediate values. The recorded values of resistivity for glacial tills range from 18,000 to 50,000 ohm cm and for granular tuffs from 1,300 to 5,900 ohm cm (Clark, 1966).

Evaluation of other aggregate properties

Crushing, impact, and abrasion tests for coarse aggregates have all been specified into standard procedures. For the two former tests the procedures of the Road Research Laboratory, U.K. (1952) are recommended; whereas for the latter test, the procedure of A.S.T.M. C131—69 of the American Society for Testing Materials is considered the best.

GLOSSARY

Anisotropy: exhibiting properties with different values when measured in different directions.
Anisodimensional: inequality in the dimensions of length, breadth, and depth of a material.
Brittle: possessing the property of sudden rupture under stress with little or no prior deformation.
Bulk modulus: see Modulus, bulk.
Creep: a time-dependent plastic deformation under constant load. (Usually considered to be slow, but no time scale is defined.)
Creep compliance: the ratio of the strain to the applied stress, where the ratio is a function of time.
Cross-anisotropy: anisotropy (see above) in which the particular measured property is the same for all directions in a given plane, but different when measured orthogonal to that plane.
Darcy flow: fluid flow of a laminar or smooth streamline type, for which the velocity of the fluid is given by the product of the coefficient of permeability and the hydraulic gradient ($v = ki$), the former being a constant characteristic of the porous medium in which flow takes place.
Deviator stress: the difference between the major and minor principal stresses in a triaxial test.
Dilatancy: the property of granular masses of expanding in bulk with change of shape. (Caused by changes in the relative orientation of particles.)
Dynamic modulus: see Modulus, dynamic.
Effective grain diameter: that grain diameter which, for an assemblage of smooth uniform spheres, would be hydraulically equivalent to the actual assemblage of non-uniform, irregular grains. (Note that in engineering usage a different definition, namely, the diameter corresponding to 10% finer on the grain size curve, is commonly used. These two definitions should not be confused.)
Effective stress: the average normal force per unit area transmitted from grain to grain in a soil mass.

Elastic: capable of sustaining stress without permanent deformation. In engineering usage it means that the strain is directly proportional to the stress (see also Modulus, elastic).

Eskers: long ridges of gravel deposited by streams flowing in channels beneath or in the lower part of a glacier. They form mounds at right angles to the ice front.

Grading: the degree of mixing of size classes in sedimentary or granular material. This term is a standard engineering term with approximately the same meaning as "sorting" in sedimentology. *Note*, however, that the common descriptive adjectives in engineering and sedimentology usage have almost precisely opposite sense, i.e.,

well-sorted ≈ poorly-graded and poorly-sorted ≈ well-graded. The engineering term "gap-graded" is used to describe materials in which particles of certain intermediate size range are substantially or wholly absent.

Gravel: small stones and pebbles from 2 mm to 60 mm in effective diameter.

Heavy clays: clays and clay soils of very dense texture, such that the permeability of the mass (uncracked) is less than 10^{-8} cm/sec.

Loam: a mixture of sand, silt, clay or a combination of any of these, with minor amounts of organic matter.

Loess: a uniform aeolian deposit of silty material having an open structure and relatively high cohesion due to cementation by clay or calcareous material at the grain contacts.

Modulus:
  bulk: the measure of a material's resistance to elastic changes in volume (for example, the force required to cause a specific change in volume). In engineering practice, this can be expressed in various ways according to the means adopted for measurement.
  dynamic: the elastic or Young's modulus as determined by a rapid (dynamic) loading test (for example, a sonic or wave transmission method).
  elastic: see Young's.
  resilient: the elastic or Young's modulus measured after many previous cycles of loading and unloading.
  shear: the ratio of the shearing stress to the shear strain.
  static: the elastic or Young's modulus as determined by a static (slow) loading test (for example, the progressive weighting of a specimen).
  Young's: the ratio of the stress (force per unit area) to the strain (deformation per unit length). Inasmuch as the strain is dimensionless, the dimensions are those of stress only. If the stress-strain proportionality is non-linear, it is usual to take a tangent or secant value from the curve.

Monoclinic: a system in which there is one twofold axis, one symmetry plane, or both. Monoclinic systems have three axes of unequal length, two of which are at right angles and one oblique.

Monosized: Of a single or uniform grain size.

Monotonic: A progression (or regression) without reversal of direction. The slope of a monotonic function does not change sign within the designated range.

Non-dilatant: see Dilatancy.

Pseudo-elastic: having a stress- strain proportionality over only a limited range of stress or strain.

Rotund: rounded, approximately spherical.

Sand: Granular material varying between 0.06 mm and 2 mm in effective diameter.

Shear modulus: see Modulus, shear.

Silt: Granular material varying between 0.002 mm and 0.06 mm in effective diameter.

Static modulus: see Modulus, static.

Sub-plastic: soil materials which do not exhibit the plasticity that would normally be expected from their known clay content, but which can be induced to do so by artificial means (for example, chemical treatment).

Till: non-sorted, non-stratified earth and rock materials carried or deposited by glaciers.

Turbulent flow: fluid motion such that various parts of the fluid are moving in random directions other than in a straight line parallel to the axis of bulk flow.

Young's modulus: see Modulus, Young's.

REFERENCES

Ackroyd, L. W. 1969. Formation and properties of concretionary and non-concretionary soils in Western Nigeria. *Proc. Reg. Conf. Afr. Soils. Mech. Found. Eng., 4th, Capetown,* 1967, 1: 47–52.

American Society for Testing Materials, 1944. Tentative specifications for materials for stabilized base and surface courses. In: *Book of ASTM Standards Including Tentative Standards, II.* ASTM, Philadelphia, Pa., D556-40T: 1357–1358, and D557-40T: 1359–1360.

American Society for Testing Materials, 1967. Description of soils (visual-manual procedure). In: *Book of ASTM Standards with Related Material, II.* ASTM, Philadelphia, Pa., D2488-66T: 772–780.

American Society for Testing Materials, 1969. *Standard Method of Test for Abrasion of Coarse Aggregates by Use of the Los Angeles Machine.* ASTM, Philadelphia, Pa., C131-69: 87–90.

Anbah, S. A., Chilingar, G. V. and Beeson, C. M., 1965. Application of electrical current for increasing the flow rate of oil and water in a porous medium. *J. Can. Pet. Tech.*, 4: 81–88.

Andrews, J. H. and Vlasic, Z. E., 1968. Decomposed lithic sandstone as a feasible pavement material. *Proc. Conf. Aust. Road Res. Board, 4th, Melbourne, 1968*, 4: 1083—1100.

Aschenbrenner, B. C. and Chilingar, G. V., 1960. Teodorovich's method for determining permeability from pore-space characteristics of carbonate rocks. *Bull. Am. Assoc. Pet. Geol.*, 44: 1421—1424.

Bartlett, A. H. and Foote, J. N., 1966. Resistivity method for location of sand and gravel. *Proc. Conf. Aust. Road Res. Board, 3rd, Sydney, 1968*, 3: 1196—1203.

Bjerrum, L., Kringstad, S. and Kummeneje, O., 1961. The shear strength of a fine sand. *Proc. Int. Conf. Soil Mech. Found. Eng., 5th, Paris, 1961*, 1: 29—37.

Brink, A. B. A. and Kantey, B., 1961. Collapsible grain structure in residual granite soils in southern Africa. *Proc. Int. Conf. Soil Mech. Found. Eng., 5th, Paris, 1961*, 1: 611—614.

British Standards Institution, 1967a. *Methods of Testing Soils for Civil Engineering Purposes*. B.S.I., London, 1377: 234 pp.

British Standards Institution, 1967b. *Methods for Sampling and Testing of Mineral Aggregates, Sands and Fillers*. B.S.I., London, 812: 104 pp.

Buchanan, F., 1807. *A Journey from Madras Through the Countries of Mysore, Canava, and Malabar, etc.* East India Company, London, 3 Vols.: 424, 556, 543 pp.

Casagrande, A., 1948. Classification and identification of soils. *Trans. Am. Soc. Civil Eng.*, 113: 901—992.

Chao, K.-H. and Chiu, K.-Y., 1962. The study of improving the bearing capacity of Taipei silt by using quicklime piles, 1. *Bull. Coll. Eng., Nat. Taiwan Univ.*, VII: 53—66 (in Chinese).

Chaplin, T. K., 1965. A fundamental stress-strain pattern in granular materials sheared with small or no volume change. *Proc. Int. Conf. Soil Mech. Found. Eng., 6th, Montreal, 1965*, 1: 193—197.

Chepil, W. A., 1958. Soil conditions that influence wind erosion. *U.S. Dep. Agric. Tech. Bull.*, 1185: 40 pp.

Chilingar, G. V., El-Nassir, A. and Stevens, R. G., 1970. Effect of direct electrical current on permeability of sandstone cores. *J. Pet. Tech.*, 22(7): 830—836.

Civil Engineering Code of Practice, 1951. *No. 2, Earth Retaining Structures*. Institute of Structural Engineers, London, 224 pp.

Clark, S. P., 1966. *Handbook of Physical Constants*. Geol. Soc. Am., Mem., 97: 587 pp.

Cohen de Lara, G., 1956. Etude d'infiltration dans les barrages en enrochements, applications au cas des batardaux de coupure. *Hydraul. Conf., 4th, Soc. Hydrotech. France, Paris*, 4: 7.

Correns, C. W., 1949. *Einfuhrung in die Mineralogie*. Springer, Berlin, 414 pp.

D'Appolonia, D. J. and D'Appolonia, E., 1970. Use of the SPT to estimate settlement of footings on sand. In: *Symposium on Foundations on Interbedded Sands*. CSIRO, Melbourne, pp.16—22.

D'Appolonia, E., 1953. Loose sands — their compaction by vibroflotation. *ASTM Symp. Dyn. Testing Soils, Spec. Tech. Publ.*, 156.

Department of Scientific and Industrial Research, Road Research Laboratory, 1952. *Soil Mechanics for Road Engineers*. H.M.S.O., London, 147 pp.

El-Sohby, M. A., 1964. *The Behaviour of Particulate Materials under Stress*. Thesis, Manchester Univ., Manchester.

Emerson, W. W. and Bond, R. D., 1963. The rate of water entry into dry sand and calculation of the advancing contact angle. *Aust. J. Soil Res.*, 1: 9—16.

Florin, W. A. and Ivanov, P. L., 1961. Liquefication of saturated sandy soils. *Proc. Int. Conf. Soil Mech. Found. Eng., 5th, Paris, 1961*, 1: 107—111.

Fujii, T., 1968. *An Introduction to Rational Grouting.* Tokyo, 27 pp.
Fuller, W. B. and Thompson, S. E., 1907. The laws of proportioning concrete. *Trans. Am. Soc. Civil Eng.*, 59: 67—143. Discussion, 59: 144—172.
Gerrard, C. M., 1967. Some aspects of the stress-strain behaviour of a sand. *Aust. Road Res.*, 3: 67—90.
Gerrard, C. M., 1969. *Theoretical and Experimental Investigations of Model Pavement Structures.* Thesis, Univ. of Melbourne, Melbourne, Vic., 528 pp.
Grant, K. and Aitchison, G. D., 1970. The engineering significance of silcretes and ferricretes in Australia. *Eng. Geol.*, 4: 93—120.
Graton, L. C. and Fraser, H. J., 1935. Systematic packing of spheres, with particular relations to porosity and permeability, I—II. *J. Geol.*, 43: 785—909.
Greenwood, D. A., 1965. Strengthening sand. *Consulting Eng.*, Oct., 28: 39—42.
Hardin, B. O. and Richart, F. E., 1963. Elastic waves in granular soils. *Proc. Am. Soc. Civil Eng., J. Soil Mech. Found. Eng.*, 89 SMI: 33—65.
H.M.S.O. Welsh Office, 1969. *A Selection of Technical Reports Submitted to the Aberfan Tribunal.* H.M.S.O., London, 218 pp.
Holden, A., 1957. Granitic Sandveld, Part II. Its stability, cementing and drainage characteristics. *Rhod. Eng.*, 2: 33—36.
Ingles, O. G., 1961. Microstructure in binderless briquetting. In: *Int. Symp. Agglom., 1st, Philadelphia.* Interscience, New York, N.Y., pp. 29—53.
Ingles, O. G., 1968. Brittle failure of unsaturated soils, stabilized soils and rocks. In: I. K. Lee (Editor), *Soil Mechanics, Selected Topics.* Butterworths, London, pp. 259—260, 267—268.
Ingles, O. G., 1971. Statistical control in pavement design. In: *Int. Conf. Applications of Statistics and Probability in Civil and Structural Eng., 1st, Hong Kong*, pp. 268—278.
Ingles, O. G. and Aitchison, G. D., 1969. Soil-water disequilibrium as a cause of subsidence in natural soils and earth embankments. In: *Land Subsidence, IASH, Publ.*, 89(2): 342—353.
Ingles, O. G. and Frydman, S., 1963. An examination of some methods for strength measurement in soils. *Proc. Aust.—N.Z. Conf. Soil Mech. Found. Eng., 4th, Adelaide, 1963*, pp. 213—219.
Ingles, O. G. and Lafeber, D., 1967. The initiation and development of crack and joint systems in granular masses. In: *Proc. Symp. Stress and Failure around Underground Openings. Univ. of Sydney, Pap.*, 7: 22 pp.
Ingles, O. G. and Metcalf, J. B., 1972. *Soil Stabilization, Theory and Practice.* Butterworths, Sydney, N.S.W., 380 pp.
International Conference on Land Subsidence, 1969. *Proc. Int. Conf. Land Subs. IASH*, Brussels, 1: 324 pp.; 2: 337 pp.
Ishihara, K., 1971. On the longitudinal wave velocity and Poisson's ratio in saturated soils. *Asian Reg. Conf. Soil Mech. Found. Eng., 4th, Bangkok, 1971*, 1: 197—201.
Itoh, N., 1969. *Foundation Improvement by Quicklime.* Nikan Kogyo Shimbunsha, Tokyo, 150 pp.
Jennings, J. E. B. and Knight, K., 1957. The additional settlement of foundations due to collapse of structure of sandy sub-soils on wetting. *Proc. Int. Conf. Soil Mech. Found. Eng., 4th, London, 1957*, 2: 151—153.
Kallstenius, T. and Bergau, W., 1961. Research on the texture of granular masses. *Proc. Int. Conf. Soil Mech. Found. Eng., 5th, Paris, 1961*, 1: 165—170.
Kawasumi, H., 1968. *General Report on the Niigata Earthquake of 1964.* Tokyo Electrical Engineering College Press, Tokyo, 693 pp.
Kersten, M. S., 1949. Thermal properties of soils. *Univ. Minn., Inst. Tech., Bull.*, 52(28): 226 pp.

Kiselev, A. V., 1954. Structure of certain xerogels, pores and particles. *Dokl. Akad. Nauk S.S.S.R.*, 98: 431—434.
Knudsen, F. P., 1959. Dependence of mechanical strength of brittle polycrystalline specimens on porosity and grain size. *J. Am. Ceramic Soc.*, 42: 376—387.
Kolbuszewski, J., 1948. An experimental study of the maximum and minimum porosities of sand. *Proc. Int. Conf. Soil Mech. Found. Eng.*, 2nd, Rotterdam, 1948, 1: 158—165.
Kolbuszewski, J., 1965. Sand particles and their density. Lecture to Materials Science Club. *Symp. on Densification of Particulate Materials*, London.
Kozeny, J., 1927. Über kapillare Leitung des Wassers im Boden (Aufstieg, Versickerung und Anwendung auf die Bemässerung). *Sitzungsber. Akad. Wiss. Wien., Math. — Naturwiss. Kl.*, 136(IIa): 271—306.
Krumbein, W. C., 1941. Measurement and geological significance of shape and roundness of sedimentary particles. *J. Sediment. Petrol.*, 11: 64—72.
Kruyer, S., 1958. The penetration of mercury and capillary condensation in packed spheres. *Trans. Faraday Soc.*, 54: 1758—1767.
Kurzeme, M., 1970. Foundation response to seismic loading. In: *Symp. on Foundations on Interbedded Sands*. CSIRO, Melbourne, Vic., pp.52—60.
Lafeber, D., 1965. The graphical representation of planar pore patterns in soils. *Aust. J. Soil Res.*, 3: 143—164.
Lafeber, D. and Willoughby, D. R., 1970. Morphological and mechanical anisotropy of a recent beach sand. *Proc. Symp. on Foundations on Interbedded Sands*, Perth, 1: 80—86.
Lafeber, D. and Willoughby, D. R., 1971. Fabric symmetry and mechanical anisotropy in natural soils. *Proc. Aust.—N.Z. Conf. Geomech.*, 1st, Melbourne, 1971, pp.165—174.
Lamplugh, G. W., 1902. Calcrete. *Geol. Mag.*, 9: 575.
Langnes, G. L., Robertson Jr., J. O. and Chilingar, G. V., 1972. *Secondary Recovery and Carbonate Reservoirs*. Am. Elsevier, New York, N.Y., 304 pp.
Larsen, G. and Chilingar, G. V., 1967. *Diagenesis in Sediments*. Elsevier, Amsterdam, 551 pp.
Lee, I. K., Lawson, J. D. and Donald, I. B., 1968. Flow of water in saturated soil and rockfill. In: I. K. Lee (Editor), *Soil Mechanics, Selected Topics*. Butterworths, London, pp.82—194.
Lee, I. K., Ingles, O. G. and Neil, R. C., 1973. Controlled deformation of a cemented soil and sand. *Proc. Int. Conf. Soil Mech. Found. Eng.*, 8th, Moscow, 1973, Sess.1, in press.
Lomise, G. M., Moshcheryakov, A. N., Gilman, Y. D. and Fedorov, B. S., 1963. Compaction of sandy soils by electric discharger. *Gidrotekh. Stroitel.*, 7: 9—13 (in Russian).
Mattson, S. and Gustafsson, Y., 1937. The laws of soil colloidal behaviour (XIX). *Soil Sci.*, 43: 543—473.
McGeary, R. K., 1961. Mechanical packing of spherical particles. *J. Am. Ceramic Soc.*, 44: 512—522.
McMahon, B., 1972. Estimation of risk in rock engineering by analysis of rock defects. In: *A Short Course in Rock Mechanics*. Univ. of New South Wales, Sydney, 2: 1—29.
Morey, G. W., Fournier, R. O. and Rowe, J. J., 1964. The solubility of amorphous silica at $25°C$. *J. Geophys. Res.*, 69: 1995—2002.
Morgan, J. R., 1966. The response of granular materials to repeated loading. *Proc. Conf. Aust. Road Res. Board*, 3rd, Sydney, 1966, 3: 1178—1191.

Morgenstern, M., 1963. Maximum entropy of granular materials. *Nature*, 200: 559—560.
Morimoto, T., Fukuzumi, R. and Itoh, T., 1970. The direct power compaction method and its application. *South-East Asian Regional Conf. Soil Eng.*, 2nd, Singapore, 1970, pp.189—197.
Moss, A. J., 1962. The physical nature of common sandy and pebbly deposits, I. *Am. J. Sci.*, 260: 337—373.
Netterberg, F., 1967. Some roadmaking properties of South African calcretes. *Proc. Reg. Conf. Afr., Soil Mech. Found. Eng.*, 4th, Capetown, 1967, 1: 77—81.
Norrish, K. and Tiller, K. G., 1964. Sub-plastic clays. *Proc. Aust. Clay Miner. Conf.*, 2nd, Adelaide, p.66 (suppl. p. 72).
Oda, M., 1972. Initial fabrics and their relations to mechanical properties of granular material. *Soils Found. (Japan)*, 12: 17—36.
Onas, J., 1970. *Anisotropy and the Stress-Strain Behaviour of Soil.* Thesis, Civil Eng. Sch., Melbourne Univ., Vic., 147 pp.
Orchard, D. F., Lye, B. R. X. and Yandell, W. O., 1970. A quick method of measuring the surface texture of aggregate. *Proc. Conf. Aust. Road Res. Board*, 5th, Canberra, 1970, 5(5): 325—340.
Pellegrino, A., 1965. Geotechnical properties of coarse-grained soils. *Proc. Int. Conf. Soil Mech. Found. Eng.*, 6th, Montreal, 1965, 1: 87—91.
Prakash, S. and Gupta, M. K., 1970. Liquefaction and settlement characteristics of loose sands. In: *Proc. Conf. on Dynamic Waves in Civil Engineering.* Wiley Interscience, pp.229—246.
Prokopovich, N., 1969. Prediction of future subsidence along Delta-Mendota and San Luis Canals, Western San Joaquin Valley, California. In: *Land Subsidence. IASH, Publ.*, 89(2): 600—610.
Pryor, A. J., 1966. The use of sandstone as pavement material in the Wimmera. *Proc. Conf. Aust. Road Res. Board*, 3rd, Sydney, 1966, 3: 900—930.
Raupach, M., 1963. Solubility of simple aluminium compounds expected in soils. *Aust. J. Soil Res.*, 1: 28—62.
Road Research Laboratory, D.S.I.R., 1952. *Soil Mechanics for Road Engineers.* H.M.S.O., London, 541 pp.
Rowe, P. W., 1962. The stress dilatancy relation for static equilibrium of an assembly of particles in contact. *Proc. R. Soc. London*, A269: 500—527.
Rowe, P. W. and Peaker, K., 1965. Passive earth pressure measurements. *Geotechnique*, 15: 57—78.
Ryskewitch, E., 1953. Compression strength of porous sintered alumina and zirconia. *J. Am. Ceramic Soc.*, 36: 65—68.
Schultze, E., 1970. *Soil Mechanics Problems in Sand.* Lectures to Central Building Research Institute, Roorkee, India; also *Bodenmechanische Problems bei Sand.* Institut für Verkehrswasserbau, Grundbau und Bodenmechanik, Aachen, VGB50: 1—124.
Serruya, C., Picard, L. and Chilingarian, G. V., 1967. Possible role of electrical currents and potentials during diagenesis ("electrodiagenesis"). *J. Sediment. Petrol.*, 37: 695—698.
Smith, W. O., Foote, P. D. and Busang, P. F., 1929. Packing of homogeneous spheres. *Phys. Rev.*, 34: 1271—1274.
Sundborg, A., 1956. *The River Klarälven; A Study of Fluvial Processes.* (Esselte Aktiebolag, Stockholm, 1: 165—219). *Upps. Univ. Geogr. Ann.*, 2/3(1956): 127—516.
Tanimoto, K., 1957. Doro Kensetsu, 116: 1—6 (in Japanese).
Tanimoto, K., 1967. *Research and Investigation of Subgrade Characteristics of the Kobe*

Highways. Special report of the Construction Engineering Research Foundation, Kobe, pp.1—26 (in Japanese).

Termier, H. and Termier, G., 1963. *Erosion and Sedimentation.* Van Nostrand, London, 433 pp.

Terzaghi, K., 1955. Influence of geological factors on the engineering properties of sediments. *Econ. Geol., Anniv. Vol.,* 50: 557—618.

Terzaghi, K. and Peck, R. B., 1948. *Soil Mechanics in Engineering Practice.* Wiley, New York, N.Y., 304 pp.

Ueshita, K. and Nonogaki, K., 1971. Classification of coarse soils based on engineering properties. *Soils Found. (Japan),* II: 91—111.

United States Atomic Energy Commission, undated. *Information and Criteria Pertinent to Evaluation of Embankment Retention Systems.* U.S. At. Energy Comm., Div. Mater. Licensing, Washington, D.C.

United States Department of the Interior, 1963. *Earth Manual.* U.S. Dep. Inter., Bur. Reclam., Denver, Colo., 783 pp.

United States Department of the Interior, 1970. *The Alaskan Earthquake, 27-3-1964.* U.S. Geol. Surv., Washington, D.C.

Whitman, R. W., 1970. The Response of Soils to Dynamic Loadings. *U.S. Army Eng. Waterways Exp. Stn. Vicksburg, Miss., Contract Rep.,* 26: 236 pp.

Wilkins, J. K., 1956. Flow of water through rockfill and its application to the design of dams. *Proc. Aust.—N.Z. Conf. Soil Mech. Found. Eng.,* 2nd, Christchurch, 1956, pp.141—149.

Winterkorn, H. F., 1971. Analogies between macromeritic and molecular systems and the mechanical properties of sand and gravel assemblies. In: *Chemical Dynamics (Papers in Honor of H. Eyring).* Wiley Interscience, New York, N.Y.

Yamanouchi, T., 1970. *Shear Characteristics of Sands under Static and Repeated Loads.* Guest Lecture, Monash Univ. Sch. of Civil Eng., Melbourne, Vic., 8 pp.

York, K. I., 1970. Permeability of granular materials. *Aust. Road Res.,* 4: 45—54.

Chapter 7

IDENTIFICATION OF SEDIMENTS — THEIR DEPOSITIONAL
ENVIRONMENT AND DEGREE OF COMPACTION — FROM WELL LOGS

D. R. ALLEN

INTRODUCTION

Great progress has been made in the fields of stratigraphy and sedimentation within the last decade as a result of the application of geophysical well logs to studies of depositional environments and compaction history. The application of well logs in these areas broadens the scope of data available to the investigator, because most rocks are found only in the subsurface and are not exposed on the surface as outcrops. Differential compaction over sand bodies and structures is often related to sand body geometry (see Chapter 5). Determination of degree of compaction and sand body geometry are interrelated, and both are dependent upon the environmental history of the sediment.

Geophysical well logging utilizes many different tools, each of which measures some particular petrophysical characteristic of the rocks penetrated by a borehole. Data from well logs have been utilized for many years to provide information on structure, lithology, and interstitial fluid content. The term "subsurface geologist" refers to a person skilled in the interpretation of well logs to determine structure, lithology and perhaps stratigraphy. Many geologists working with logs have noted that log data indicate various rock fabric changes that can be used for correlation purposes. Some also were aware that certain subtle logging patterns were characteristic of a particular sediment type and depositional environment. As these uses for well log data became known, various oil companies prepared manuals on the subject; however, much of these data were not published until the past few years.

This chapter emphasizes the data that may be obtained from well logs concerning lithology, depositional environment, and degree of compaction, particularly in the case of coarse-grained sediments. The skills involved in geologic log interpretation are obtained by experience, which has tended to be limited to the subsurface geologist. It is hoped that recognition by the sedimentologist and stratigrapher of the data available to them from these sources will encourage a more widespread use of well logs.

Recognizing that many logging terms and measurements may not be familiar to the reader, an elementary discussion of the principles of the major logging tools is given herein for convenience. This discussion is intended only to familiarize a person with the various logging techniques that enable the determination of various petrophysical properties. Quantitative log analysis is a specialty in itself and further study of logging principles is imperative for anyone who wishes to achieve an expertise in lithology identification and the determination of depositional environments from well logs.

DETERMINATION OF LITHOLOGY FROM WELL LOGS

Development of logging techniques*

The determination of lithology from well logs is not new and may have been the first qualitative use of borehole electrical measurements. The first measurements of the variations in natural (self potential or SP) and induced electrical response (resistivity to induced current) by different rock types were made by C. and M. Schlumberger about 1928. These early investigators referred to the records obtained as "electrical coring" and suggested that this type of measurement could be substituted for conventional coring and sampling procedures. These electric well logs were quickly adopted by the drilling industry and soon proved to be a tool which virtually eliminated coring and sampling for purposes of formation correlation and, in some cases, stratigraphic control. This was particularly true in areas where formations or marker beds of a visibly distinct character were not present. Coring and sample logging are still widely used for paleontological investigations and for determination of reservoir rock type and interstitial fluid content.

Variations in natural radioactivity (gamma radiation) of different rock types were recorded in boreholes about 1938. This logging technique (the gamma log) was offered as a commercial service in 1940. The neutron log, which involves formation bombardment by high-energy neutrons and recording the consequent excitation, was developed about the same time as the gamma log and was soon combined with it as a companion tool. These logging methods offered distinct advantages over the electric log in some rock types. SP—resistivity measurements often are not definitive when logging carbonate sections, whereas radioactivity logs are very responsive to carbonate lithology and fluid content variations. In addition, radioactivity logs can be run in cased boreholes, whereas electric logs can be run only in boreholes without casing. It is a common practice today to run both log

---

*Several logging service companies offer similar logging tools today, but often with different trade names. In this chapter, an attempt is made to use either generic or widely used terms, rather than those used exclusively by one company.

types whenever possible because each supplies a unique petrophysical measurement.

The two basic log types, i.e., the SP—resistivity and the gamma ray— neutron, are the logs commonly used for lithology determinations. They are also the logs normally available from older boreholes. In addition to the basic logging systems, various special-information logs are now routinely run, the type varying from one geographic area to another. They are usually used as a supplement to one or more of the basic log types. Many of the specialty logs record rock properties or responses from which porosity and fluid content data can be inferred. Among these are the scattered gamma ray or gamma-gamma density log, various neutron porosity logs, acoustic wave propagation logs, and focused current and multiple-depth investigation electric logs. Various tool configurations and methods of portrayal are used, dependent upon the particular use for which the log data is intended. There are several other logging tools that are available, such as the nuclear magnetic log, the pulsed neutron log, and various neutron activation logs. These logs are not discussed further because, as yet, they are of limited application as far as the identification of clastics and degree of compaction are concerned. For further details, the reader is referred to an excellent book by Pirson (1963).

The first step in any lithologic determination from well logs should be the comparison of core and/or other sample descriptions with available logs and the establishment of a reference base. Most log responses vary from one area or geologic province to another, and there is no substitute for experience. Even the most experienced observer needs to familiarize himself with the logging problems and responses, that might be limited to a certain locality, when working in unfamiliar territory.

Logging principles

Two basic electric log techniques have been used in the past: (1) conventional resistivity and (2) induction resistivity. Induction resistivity logging, since its inception about 1950, has virtually supplanted the conventional resistivity log in usage. Both resistivity and induction logs are normally combined with an SP curve when borehole conditions permit. A gamma ray (GR) curve is usually either substituted for the SP curve or run in addition to it when drilling fluids having either very high or very low resistivities are used (i.e., when oil or salt water constitute the major portion of the drilling fluid). The gamma ray log also may be used in combination with the resistivity curves in areas where dense rocks such as carbonates or volcanics are present. Because of the wide use of electric and radioactivity logs for formation correlation, which requires a readily recognizable format, both log types were standardized in the manner shown on Fig.7-1. This portrayal was called the "hour glass" log when it was first used, as opposed to recordings where the SP—resistivity and the GR—neutron deflections were all in the same direction. In the format used today, shales are seen as excursion towards the depth tract on both log types, whereas sands are seen as excursions away from the depth tract.

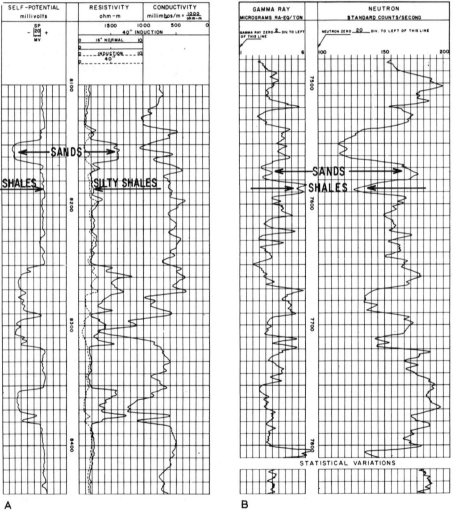

Fig.7-1. Induction—electrical log (A) and gamma ray—neutron log (B) in sand—shale sequences. (Courtesy of Schlumberger Well Services, Houston, Texas.)

IDENTIFICATION OF SEDIMENTS FROM WELL LOGS 353

Self potential (SP) log

The self or spontaneous potential curve is a recording of the natural electrical potentials that exist in a borehole owing to electrochemical, electrokinetic, and junction potential forces. These borehole potentials are complex in nature, involving battery-like cells of unequal electrical potential that exist between adjacent rock formations and between the rocks and the borehole fluid. These potentials are small and are easily influenced by fluid character, both in the borehole and in the formations. They also are influenced non-uniformly by formation permeability. Figure 7-2 illustrates the theoretical flow of the SP current in a sand—shale sequence. As shown in Fig.7-2, a negative potential is developed opposite a sand when there is a fresh-water drilling fluid in the borehole standing opposite a formation containing saline fluid. The following formula is often used to calculate the theoretical SP magnitude:

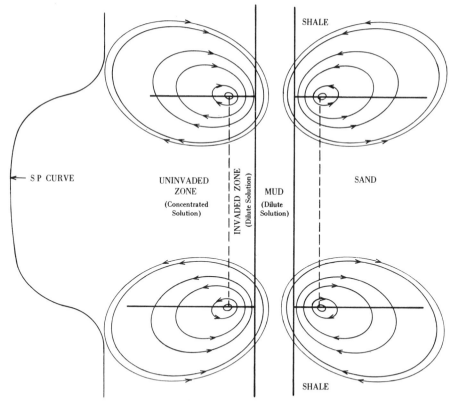

Fig.7-2. Hypothetical self potential current flow in a sand—shale sequence. (Courtesy of Dresser Atlas, Houston, Texas.)

$$E_c = -K \log(R_{mf}/R_w) \tag{7-1}$$

where E_c = the electrochemical component of the SP, K = a temperature-dependent constant, R_{mf} = resistivity of the drilling fluid filtrate, and R_w = resistivity of the formation water.

The constant K is determined from charts supplied by the logging service companies, R_{mf} is measured at the well, and R_w is estimated from other log data or obtained directly by analyzing formation water samples. This formula (7-1) is valid when the salts in both formation and drilling fluids are essentially of the same type, e.g., sodium chloride. The SP values may be positive or negative depending upon the R_{mf}/R_w ratio. The SP is measured from the shale base line and not from the center tract. A reduction is seen in the negative SP deflection opposite a sand that contains shale or clay either in a dispersed or particulate form. This reduction is approximately proportional to the shale or clay content and is an indication of sand purity; thus, it can be used as a very rough measure of degree of sorting or winnowing. It may also show the presence of graded bedding. Inasmuch as the borehole potentials form a curving magnetic field at each bed junction, a sharp boundary is not recorded on the log. In a sand—shale sequence, the sand boundary is considered to be at the inflection point where the negative SP excursion becomes convex. No SP current is developed opposite impermeable shales because there is only negligible diffusion between the borehole and formation fluids. In dense carbonates and similar rocks, the SP curve may have very little distinguishing character. Consequently, the SP log is not a good lithology tool in these rocks. The self potential current flow between dense formations is sharply restricted, forcing the currents deep into the formations which gives rise to a spurious reading (Fig.7-3).

Conventional resistivity curves

In all early electric log tools, some combination of electrode spacing and arrangement was used to produce what are termed the "normal curves". These resistivity measurements are made by passing current through the formations penetrated by the borehole and measuring the voltages (potentials) that develop at some particular point. A theoretical electrode configuration is shown in Fig.7-4A. Current is passed from electrode A to electrode B and the potential between electrodes M and N is measured. The reference point is the midpoint O, which is located between electrodes A and M. The electrode spacing, which determines the depth of investigation into the formation, is the distance between electrodes A and M. Deeper investigation devices (Fig.7-4B), which produce the lateral curves, are usually run in combination with the tools that produce the normal curves as an aid in

IDENTIFICATION OF SEDIMENTS FROM WELL LOGS

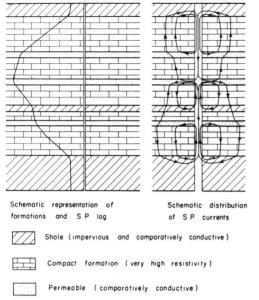

Fig.7-3. Hypothetical current flow and self potential log in dense formations. (Courtesy of Schlumberger Well Services, Houston, Texas.)

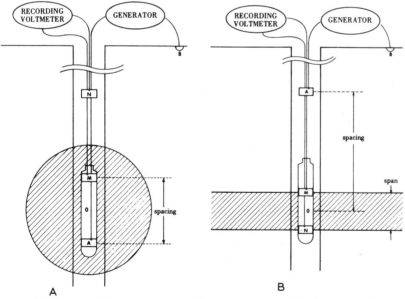

Fig.7-4. Theoretical electrode schematic diagram of logging tools for measuring formation resistivity. A. Normal curves. B. Lateral curves. (Courtesy of Dresser Atlas, Houston, Texas.)

determining the amount of drilling fluid invasion and the fluid content of the formations. The short normal curves have an electrode spacing of 16 or 18 inches. They primarily measure the resistivity of the mud cake and the shallow invaded zone around the borehole. Because of the short spacing and shallow depth of investigation, these curves usually will record bed boundaries with good definition and will show some indication of the presence of permeable beds owing to filtrate invasion. This is in contrast to the deeper investigation curves (laterals) which will not clearly define thin beds and lack bed definition in salt-water bearing, permeable zones (Fig.7-5A).

The configuration (envelope shape) of the SP and short normal curves are most often used for investigating lithology and depositional environment, whereas the long normal or lateral curves are used for fluid content determinations. Owing to the long spacing between the electrodes of the deep investigation logging tools, bed thickness has a large effect on the appearance of the lateral curves.

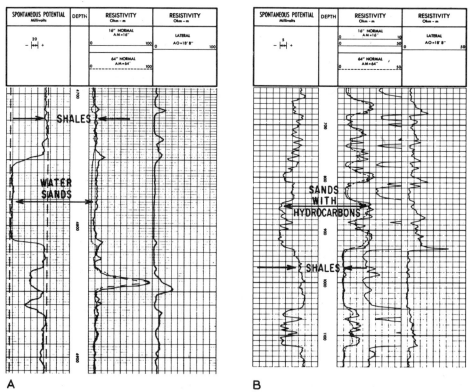

Fig.7-5. SP and resistivity logs showing both normal and lateral curves. (Courtesy of Dresser Atlas, Houston, Texas.)

Induction resistivity curves

Induction resistivity logging has almost replaced the conventional deep investigation devices for most purposes. The log format and appearance are essentially the same as the conventional resistivity curves (Fig.7-1 and 7-5). In the standard induction log recording, the lateral curve is replaced by a recording of formation conductivity. This curve is useful in determining lithology in very low-resistivity (high conductivity) intervals. Induction logs can be run in boreholes containing non-conductive fluids (i.e., oil or gas), where conventional resistivity curves cannot be obtained. The short normal curve is usually recorded along with the induction log when borehole conditions permit. A schematic diagram of an induction log tool configuration is shown in Fig.7-6. A high-frequency, alternating current is transmitted into the formations, where secondary currents are induced by the alternating magnetic field. These secondary currents in turn form magnetic fields which induce currents in a receiver coil. The currents that flow in the receiver coil are essentially proportional to formation conductivity (the reciprocal of resistivity). Standard spacing in most areas is 40 inches. Multiple-spaced induction and lateral curves also can be recorded on a logarithmic scale. When this format is used, a conventional presentation is usually made also. Bed thickness also affects the induction curves and corrections must be made to obtain true formation resistivity when logging beds that are between 4½ and 10 ft thick. True formation resistivity cannot be obtained for thin beds using conventional logging equipment.

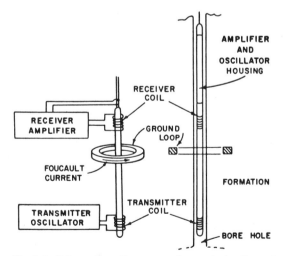

Fig.7-6. Schematic diagram of an induction log recording system. (Courtesy of Schlumberger Well Services, Houston, Texas.)

Resistivities measured by both the normal and induction log tools are recorded in ohm-meters (or ohm-meters^2/meter). Resistivity values are highest in dense (well cemented, non-porous) formations and are very sensitive to pore fluid content and salinity. Formation resistivities also are sensitive to gross changes in particle size, i.e., a change from sand to gravel or cobbles. Fine-grained, conductive material such as clay or shale tends to lower the resistivity of a formation, whereas coarse-grained conglomeratic material increases resistivity.

Gamma ray—neutron log

Although the gamma ray and neutron logs are separate and distinct logging methods, they often are run together. The resultant log resembles the SP—electric log in a shale—sand sequence (Fig.7-7 and 7-8). Both the gamma ray and neutron logs record some form of radiation by the use of sensing and counting devices such as the Geiger counter and scintillometer. Several different scales for recording radiation intensity have been used by the various logging service companies in the past. Thus, great care must be exercised when comparing radiation logs because of the possibility that different scales were used. Current logs, however, all use a standard scale based on deflections recorded in a standard test pit that contains various rocks. This scale is divided into arbitrary divisions designated as API units.

The gamma log

The gamma ray logging tool measures the natural radioactivity of the formations traversed. In sedimentary formations, the natural gamma ray intensity usually reflects the shale or clay content. Shales and clays have a relatively high content of radioactive material compared to most other rock types. The gamma ray log can be recorded in fluids in which SP measurements cannot be made. Natural gamma rays are bursts of high-energy electromagnetic radiation, which are continually emitted as a result of the decay of natural radioactive materials. These rays are gradually absorbed as they traverse the formations. Although their intensity can be recorded by several types of counters, the scintillation counter is the type most often used today. Inasmuch as the energy emitted is sporadic in nature, statistical averaging circuits (called time constants) are used to convert the random signals into a usable form. Due to this circuitry, both logging speed and the time constant can alter the appearance of a log. Slow logging speeds and increased time constants give a more accurate measurement of the average natural formation radioactivity. Because dense formations absorb the gamma rays more rapidly than do less dense strata, a lower counting rate is observed at the detector, assuming equal amounts of radioactive emissions from both.

The neutron log

The neutron log is often called the "hydrogen" log because the counting rate is dependent upon the amount of hydrogen atoms in the formations traversed. Inasmuch as the pore space in most rocks is filled with water or oil, both of which are rich in hydrogen, the neutron log can be used as a porosity indicator. In neutron logging, the formation is bombarded with high-energy neutrons from a radioactive source contained in the logging sonde. The neutrons are slowed down in the formation as a result of collisions with hydrogen atoms. Secondary gamma rays of capture are emitted by the capturing nucleus. The log recorded can be either a measurement of the capture gamma ray intensity or of the neutron population itself, both of which are indicative of the same formation characteristic, i.e., the hydrogen content. Neutron log response is very sensitive to the mechanical configuration of the tools, type and strength of the neutron source, spacing between the source and the counter, and borehole effects. Each individual logging tool is calibrated under standard conditions; yet, it still may respond differently from other tools when borehole and formation conditions are different. Recent improvements have been made in neutron porosity logging by the use of dual detectors and a sonde that presses against the borehole wall. This type of tool eliminates a large part of the problems previously noted.

A gamma ray—neutron log is compared with an induction log in a sand—shale sequence in Fig.7-7. The "hour glass" shape formed by the sand—shale—sand intervals is pronounced. The gamma log has a high counting rate in shales due to the relatively high content of radioactive, carbonaceous material usually contained therein. In a shaly or silty sand, the gamma count is higher than in a "clean" sand. This effect can be used to estimate the shale content of a sand interval. In shales, the neutron log, responding to the hydrogen of both porewater and adsorbed water, has a low counting rate owing to a rapid absorption of the bombarding neutrons. In a porous section containing natural gas, a very high counting rate is observed on the neutron log because gas contains smaller amounts of hydrogen than do either water or oil. In gas-bearing zones, therefore, the apparent porosity as calculated from the neutron log is less than the true porosity.

Both the gamma ray and neutron logs are widely used in carbonate sections for lithology determinations. The neutron log is a good porosity indicator in carbonates due to their normally low content of shale and other materials that contain hydrogen. The neutron log porosity calculation assumes that all hydrogen response is related to the volume of the pores containing fluid. Hypothetical electric and radioactivity log response to various lithologies is shown in Fig.7-8. These stylized logs were drawn to

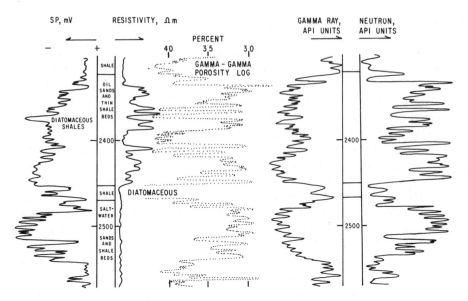

Fig.7-7. Induction—electric log, gamma ray—neutron log, and gamma—gamma porosity log response to a shale—sand sequence containing both oil and water-bearing sands.

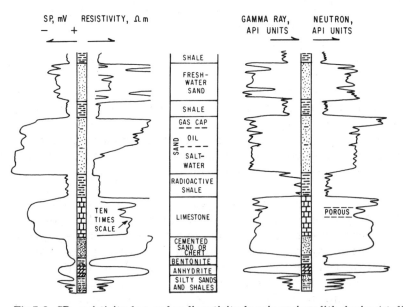

Fig.7-8. SP—resistivity log and radioactivity logs in various lithologies (stylized response).

show typical response patterns, and the relative log measurements do not represent quantitative comparisons.

Determination of lithology from porosity logs

There are three widely-used logging systems that are usually referred to as porosity logs. These devices do not actually measure porosity directly; instead, each one measures a petrophysical rock property from which porosity can be derived. The three systems are: (1) the neutron log, (2) the gamma—gamma density log, and (3) the sound propagation log. Inasmuch as each tool measures a different rock property, variations in the amount and type of the matrix, the degree of cementation, and the type of pore fluids may cause each log to record a response indicating different porosities. By comparing these different responses, lithology and matrix content often can be determined with fair accuracy.

The degree of compaction of a sediment is often related to its porosity; thus, by determining porosity, the degree of compaction can be estimated. Shale porosities are more likely to be representative of the degree of compaction of an interval than are sand porosities, because of the more pronounced diagenetic and epigenetic effects in sands, e.g., cementation, mineral neoformation, and solution. Overpressured zones (those with pore fluid pressures greater than hydrostatic) that were caused by a rapid deposition and subsidence are usually undercompacted and can be detected by shale bulk specific gravity measurements and by various porosity logs (see Chapter 2).

Neutron porosity log

The neutron log can be used for porosity determinations. The measurements made are of either secondary gamma rays of capture or of the neutron population at some fixed distance from the neutron source. Thus, the strength of the signals received is related to the amount of hydrogen present within the sphere of influence of the emitter—detector system. Because the hydrogen in porous rocks is contained almost entirely within the pore space (in water or hydrocarbons), the signal strength at the detector can be related to porosity. All neutron tools are calibrated against materials of known porosities. These data are used to prepare charts or determine equations that allow porosity to be approximated from the log measurements. Porosities calculated from neutron log data are usually in units calibrated as though the formation material were limestone. These data have to be converted, either by prepared charts or by instrumentation, when the matrix rock is not limestone. Siltstones, shales and clays are seen as porous rocks because of the water contained within their pore spaces. They may have a very low permeability and effective porosity, but the logging tool does not measure these properties. Gas in the pore space is indicated by an abnormally high counting rate. Erroneously low porosities may be calculated if corrections are not made. Neutron porosity measurements give best results in carbonate sections that are relatively shale free and have low to moderate porosities. The newer logging systems mentioned earlier have greatly enhanced the value of the neutron porosity log in sand—shale sequences.

Gamma—gamma porosity log

The gamma—gamma logging tool currently used by most companies contains a radioactive source that bombards the formations with gamma rays. The tool body is mounted on a skid pad and is pressed against the borehole wall. Two detectors are used, which are mounted at different distances from the source (about 1—2 ft). The bombarding gamma rays interact with the electrons of the materials surrounding the borehole and are eventually absorbed. Some of the gamma rays are back-scattered rather than being immediately absorbed and their intensity is measured at the two receivers. The signal strength at the detectors is related to the electron density of the surrounding material in an exponential manner. By using two detectors and a hole caliper recording simultaneously, the instruments can be calibrated to compensate for borehole and mud cake problems. The log shows the corrected bulk density* of the formations traversed. Older logging equipment did not automatically compensate for the problems of borehole size and rugosity and these corrections had to be made by using charts. If the grain and fluid densities are approximately known or can be estimated, a calculated porosity can be recorded directly at the logging truck.

The following equation is used for calculating porosity from the density log:

$$\rho_b = \phi \rho_f + (1 - \phi) \rho_{ma} \tag{7-2}$$

or rearranging:

$$\phi = \frac{\rho_{ma} - \rho_b}{\rho_{ma} - \rho_f} \tag{7-3}$$

where ρ_b = bulk density (log measurement), ρ_{ma} = density of the matrix, ρ_f = density of the pore fluid, and ϕ = porosity fraction.

The gamma—gamma (density) log usually is a reliable source of information as to the degree of sediment compaction and/or cementation.

---

*In gamma—gamma logging, the terms density and bulk density are universally used rather than specific gravity.

Acoustic porosity log

The acoustic logging tool measures the interval transit time (Δt) of compressional sound waves expressed in microseconds per foot ($\mu\text{sec}/\text{ft}$) of formation traversed. It is the reciprocal of the speed of sound through a formation usually reported in ft/sec. The interval transit time is a function of matrix composition, porosity, type and amounts of pore fluids, and the degree of compaction and cementation. The acoustic travel time log is a useful porosity tool in the case of consolidated formations and for lithology identification when used in combination with other porosity devices. Current instrumentation commonly utilizes two alternately pulsing sound transmitters, which are spaced a set distance apart, with either two or four receivers located in between. The signal recorded is compensated automatically for errors generated by sonde tilt, borehole rugosity and mud cake. Wyllie et al. (in: Guyod and Shane, 1969) concluded that a linear relationship exists between Δt and porosity in clean, consolidated formations. The empirical formula which relates Δt to porosity, ϕ, can be expressed as:

$$\Delta t_{\log} = \phi \Delta t_f + (1-\phi) \Delta t_m \qquad (7\text{-}4)$$

or rearranging:

$$\phi = \frac{\Delta t_{\log} - \Delta t_m}{\Delta t_f - \Delta t_m} \qquad (7\text{-}5)$$

where $\Delta t_{\log}$ = log reading in $\mu\text{sec}/\text{ft}$, Δt_m = matrix travel time in $\mu\text{sec}/\text{ft}$, Δt_f = fluid travel time in $\mu\text{sec}/\text{ft}$, and ϕ = fractional porosity.

Equation 7-5 is often referred to as the Wyllie time-average formula.

Transit times (reciprocals of acoustic velocities) for various rocks and interstitial fluids are given in Table 7-I.

In unconsolidated, undercompacted sands and in overpressured reservoirs, Δt does not accurately reflect porosity. In the cases of unconsoli-

TABLE 7-I

Transit times for various media

Medium	Transit time ($\mu sec/ft$)
Sandstone	55.5—57
Limestone	47.5
Dolomite	43.5
Anhydrite	50.0
Salt	67.0
Fluids	189—192

dated and undercompacted sands, the values of calculated porosity will be higher than the actual values. A rule-of-thumb method for correcting for this error is to multiply the right side of eq.7-5 by a correction factor whenever shale travel time exceeds 100 μsec/ft. This empirical correction is best obtained by comparison with the results obtained from other porosity logging tools or with core porosities.

Lithology determinations from electric logs

The electric log is an excellent lithology identification tool in shale and sandstone formations. These two rock types usually give rise to a distinctive hour-glass log pattern, i.e., the shales show low resistivity and no SP deflection, whereas sands develop a high SP deflection and may or may not show a high resistivity dependent upon fluid content (Fig.7-7 and 7-8). The shallow-investigation resistivity curve usually exhibits at least a small deflection in an opposite direction to the SP curve in sands, regardless of fluid content. In some cases, where the formations are not invaded by the drilling fluid filtrate, the short normal curve will give the same reading as the curves of the deeper-investigation devices. The sand—shale contrast on the electric log is important, both in lithology determination and in pattern recognition of depositional environments. The magnitude of the SP deflection, in a salt-water or an oil-bearing sand, is a qualitative measurement of the clay or silt content of the sand, i.e., the greater the shale percentage, the smaller the SP deflection. When the pore spaces are at least partially saturated with hydrocarbons, the measured resistivity decreases with increasing clay—silt (conductive solids) content similar to the SP reduction. This resistivity effect is not readily observed in the case of water-bearing sands.

When the pore fluid consists of either fresh or brackish water, great care must be exercised in attempting a lithologic determination because both the SP and resistivity measurements are sensitive to fluid resistivity. Figure 7-9 illustrates the relationship between the water salinity and log response in a fresh-water to salt-water sequence. The log response of clay or silt saturated with fresh water can be exceptionally misleading because of development of apparent SP and resistivity values that normally are associated with a permeable sand. As shown in Fig.7-9, the SP curve is neutral or reversed opposite the fresh-water intervals. The self potential gradually increases in a negative direction as the pore water becomes more saline.

The self-potential (SP) development opposite a relatively nonpermeable section also can be very misleading as to the lithology of the formations logged. Figure 7-10 is a log sample of a clean siltstone that could easily be mistaken for a permeable sand. It has all of the characteristics of a thick sand section in that sufficient porosity and permeability are present to

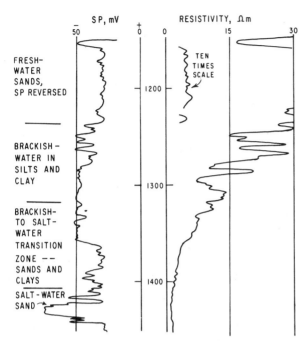

Fig. 7-9. Fresh to salt water transition in a clay—sand sequence as shown by SP and resistivity logs.

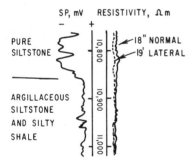

Fig. 7-10. SP and resistivity log response to siltstones. (Courtesy of Standard Oil Co. of California.)

develop a good SP curve. The separation between the short- and long-spaced resistivity curves is similar to that seen in the case of permeable sands. The slightly higher resistivity and the lack of curve separation in the siltstone—silty shale section indicates a lesser depth of invasion and perhaps a greater induration. Due to the somewhat lower water salinities usually found in shales, as compared to the nearby sands, this is a normal pattern for a water-bearing sand—shale sequence. Obviously, data from drill cuttings or cores may be necessary to distinguish between clean siltstones and sands, although the log response of the argillaceous siltstone of the lower part of Fig.7-10 is typical.

Bentonite beds are very useful as time correlation markers in some areas because they may have been formed by geologically instantaneous deposition from volcanic eruptions. They are evidenced by very low resistivities and SP deflections and are sometimes referred to as "super shales" because of their log response. Figure 7-11 shows the log response of thin bentonite beds within a shale section.

Diatomites and diatomaceous shales also exhibit very low resistivities. Figure 7-12 illustrates a long diatomite section of exceptionally low resistivity. Due to the very fine pores containing saline water and very little invasion of drilling fluid filtrate, both the deep- and shallow-investigation curves record a very low resistivity. Exceptionally high porosity (low density) is recorded by the density log opposite the diatomaceous shales in Fig.7-7. A similar high porosity is indicated by acoustic logs in diatomaceous intervals.

Conglomerates sometimes can be identified, once experience has been gained in a particular area, by their high resistivities relative to more porous sands (Fig.7-13). Poor sorting and large rock fragments (cobbles, etc.) may be responsible for the high observed resistivities.

In Fig.7-14, the SP and resistivity curves of a silty conglomerate, which is a distinctive formation over a widespread area, is compared with a fractured oil shale. There are apparent similarities; however, familiarity with logging patterns and close examination reveal several distinct differences. The fractured shale has high resistive peaks that do not always align themselves with the SP peaks and the envelope shapes of the resistivity curves do not show the vertical lithology and fluid changes indicated by the SP curve. This example illustrates the hazards of lithology determination from logs in an unfamiliar area and in the absence of rock and fluid sample data. The

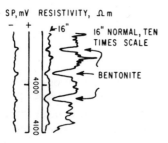

Fig.7-11. SP and resistivity log response to bentonite beds in shales. (Courtesy of Standard Oil Co. of California.)

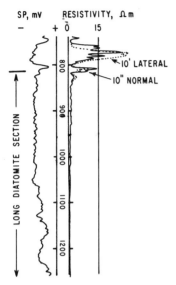

Fig.7-12. SP and resistivity logs in a diatomite section, showing exceptionally low resistivities. (Courtesy of Standard Oil Co. of California.)

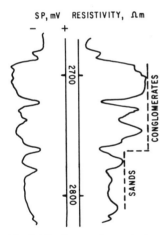

Fig.7-13. Comparison between resistivities of sands and adjacent conglomerates. (Courtesy of Standard Oil Co. of California.)

erratic resistivity values in the fractured shale are due to fracture porosity and cementation and not necessarily to fluid content. The siltstone matrix and thin interbedding of the conglomerate give it the appearance of a channel or lagoon deposit (see Fig.7-32, p.388).

Igneous rocks, if unaltered, exhibit unusually high resistivities and a

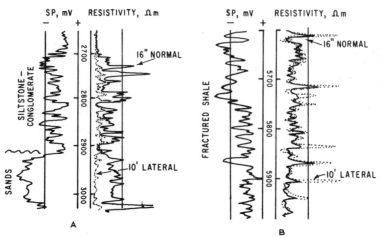

Fig.7-14. SP and resistivity log response of (A) siltstone-conglomerate, and (B) fractured oil shale. (Part B, courtesy of Standard Oil Co. of California.)

depressed or "sluggish" SP curve without distinctive characteristics. Their log appearance is often related to the amount of weathering to which they have been subjected. Granitic or metamorphic "basement" rocks are often overlain by other resistive rocks such as conglomerates, which exhibit a similar log appearance. Igneous sills usually can be identified by their high resistivity, when present in a sedimentary section. A knowledge of the area, or other log data, however, may be necessary for positive identification. For example, coal beds have the same SP–resistivity characteristics. Coal can be identified from the response of the acoustic, density and neutron logs. A coal bed will typically exhibit low acoustic velocities, low density and a high hydrogen content, whereas igneous rocks usually have high acoustic velocities, high bulk density, and low hydrogen content. Electric logs for three typically resistive rock types are presented in Fig.7-15.

Lithology determination from radioactivity logs

Gamma ray–neutron log response has been compared previously with electric logs in Fig.7-7 and 7-8. As a general rule, shales exhibit a relatively high counting rate on the gamma ray log and a low counting rate on the neutron log. Conversely, sands, conglomerates, and other coarse-grained rocks have a low gamma and a high neutron response. Radioactivity logs are not always good lithology identification tools in clastic sections, although they are very reliable in many areas.

There are some rock types, such as radioactive shales, that are easily identified by their radioactivity. Radioactive shales may not have any

IDENTIFICATION OF SEDIMENTS FROM WELL LOGS

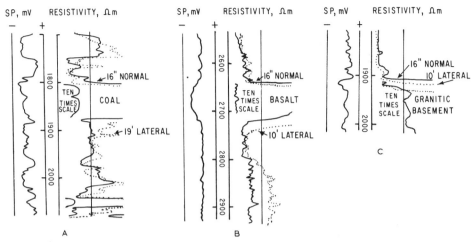

Fig.7-15. SP and resistivity log response to several types of resistive rocks. A. Coal. B. Basalt. C. Granite. (Courtesy of Standard Oil Co. of California.)

distinguishing characteristic on any other type of log. Diatomaceous shales that are distinctive on the porosity logs cannot be identified on the gamma log, because they are siliceous and are not radioactive (Fig.7-7). Also, in this illustration, no detectable difference can be seen between the oil and water saturated sands on either the gamma or neutron logs.

In sands containing glauconite, which often are called "green sands" due to the coloration added by this mineral, radioactivity logs can be very helpful. Owing to their radioactivity, glauconitic sands are very prominent on the gamma log (Fig.7-16), but cannot be distinguished on an electric log. Beds having distinctive radioactive characteristics sometimes can be correlated over wide distances, enabling stratigraphic ties to be made that might be impossible using any other method.

Lithology determination from "porosity logs": acoustic, density, and neutron

Although porosity logs were not designed to be lithology identification tools, they have proved to be very useful in this respect, particularly in carbonate sections. These tools are also excellent indicators of degree of compaction in sediments because most compaction takes place at the expense of porosity (see Chapter 2).

Inasmuch as each of the porosity tools measures a different rock property, porosities derived from them often vary according to rock type. If lithology is not known, it often can be determined by "cross plotting" the

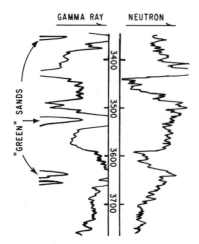

Fig.7-16. Gamma ray—neutron log of glauconitic (green) sands having a high level of radioactivity. (Courtesy of Standard Oil Co. of California.)

apparent porosities obtained from two or more porosity tools. The resultant data can then be used in selecting Δ_{ma} and ρ_m for use in the porosity equations. Figure 7-17 is a chart for determining lithology from density and neutron log data. The logging service companies supply similar charts for other tool combinations. This technique is very useful in carbonate sections.

The acoustic log is probably the best qualitative method for determining lithology in clastic sections; however, it is not always a good porosity tool, particularly when the sediments are uncompacted or uncemented. Low-porosity, cemented intervals and chert beds are easily identifiable by their very short transit time (high velocity), whereas shales and porous sands have much longer transit times. Shales may or may not have longer transit times than the sand intervals. Unconsolidated and poorly indurated shales typically have a higher transit time than do porous sands. Conversely, older, well-indurated shales in which porosities have been reduced to 10% or less, may have a shorter transit time than the nearby sands. Figure 7-18 compares electric and acoustic log responses in several rock types. As shown in this figure, the presence of gas in the pore space sharply increases the acoustic travel time in sands.

A density and neutron porosity log overlay in a low porosity, sand—shale—lime sequence is presented in Fig.7-19. In the shale-free sand interval, the density log records a higher apparent porosity than the neutron log, whereas in the shaly intervals the neutron log records an apparent higher porosity. This difference is due to the neutron log response to hydrogen. The shale-water content in the shaly sand is recorded as porosity by the neutron log. The separation between the porosities, as measured by the density and

IDENTIFICATION OF SEDIMENTS FROM WELL LOGS 371

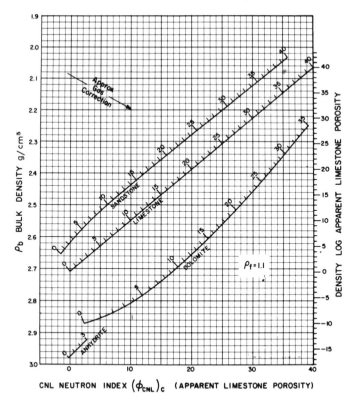

Fig.7-17. Density and neutron porosity log cross-plot for identification of lithology. (Courtesy of Schlumberger Well Services, Houston, Texas.)

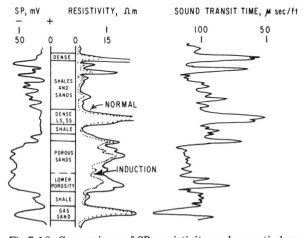

Fig.7-18. Comparison of SP—resistivity and acoustic log response in different rocks.

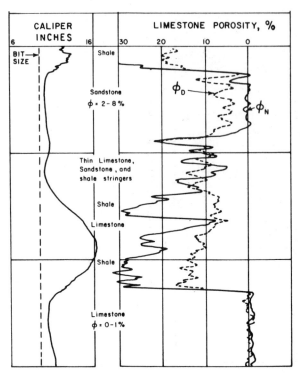

Fig.7-19. Comparison of borehole caliper, density and neutron porosity logs in various rocks. The porosities shown are recorded as though the entire section were limestone and corrections must be made for the clastic sections. (Courtesy of Schlumberger Well Services, Houston, Texas.)

Fig.7-20. Analog display of a computer-calculated sandstone analysis log. (Courtesy of Schlumberger Well Services, Houston, Texas.)
Integrated Porosity Feet: The open pips on the right edge of the depth track correspond to the integration of porosity. The distance between two consecutive pips represents 1 ft of 100% porosity or 7,758 barrels of pore volume per acre. Every tenth pip is enlarged.
Integrated Hydrocarbon Feet: The distance between two consecutive dark pips to the right of the depth track represents 1 ft of hydrocarbon in place, or 7,758 barrels of hydrocarbons per acre. Every tenth pip is enlarged.
V_{sh} = Bulk volume fraction of shale, both wet clay and silt. V_{sh} is computed from neutron—density data and from a special logic relating other shale indicators (SP, GR and Resistivity) to the volume of shale. This is an excellent correlation curve which should permit the differentiation between sands, shale and shaly sands.
k = Permeability index of the formation. The scale for this curve is a 5-cycle logarithmic scale.
S_w = Fraction of pore volume filled with formation water.
$\phi \cdot S_{hyr}$ = Residual hydrocarbons per bulk volume where S_{hyr} is residual hydrocarbon saturation.

IDENTIFICATION OF SEDIMENTS FROM WELL LOGS

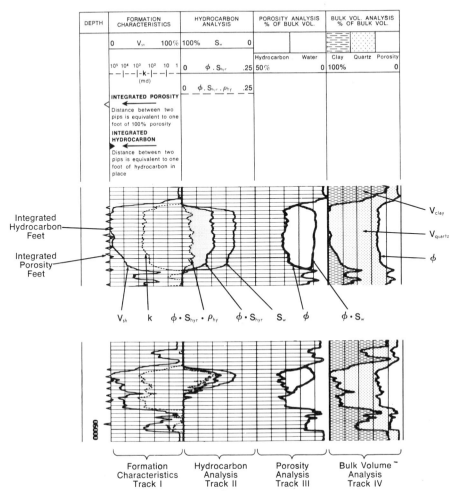

$\phi \cdot S_{hyr} \cdot \rho_{hy}$ = Weight of residual hydrocarbons per bulk volume where ρ_{hy} is the density of the hydrocarbons. The two curves, $\phi \cdot S_{hyr}$ and $\phi \cdot S_{hyr} \cdot \rho_{hy}$, converge in oil zones because the density of oil is close to unity. In light hydrocarbon zones, the two curves diverge.

The ratio of $\phi \cdot S_{hyr} \cdot \rho_{hy}$ to $\phi \cdot S_{hyr}$ is the hydrocarbon density.
The values of hydrocarbon density derived from the computation appear on the tabular listing.

V_{clay} = SARABAND logic assumes shale to consist of wet clay and silt. V_{clay} represents only the bulk volume fraction of wet clay, whereas V_{sh} of Track I represents the total shale bulk volume.

V_{quartz} = Bulk volume fraction of rock containing sand and silt.
ϕ = Formation porosity corrected for hydrocarbon and shale effect.
$\phi \cdot S_w$ = Water-filled porosity.
The area between the two curves corresponds to hydrocarbon-filled porosity.

neutron logs, is sometimes referred to as the difference between the effective and total porosity. The true porosity in the sand and shale intervals is higher than the apparent limestone equivalent porosity by several porosity percent. Corrections can be made by using various charts that can be obtained from the logging service companies. Limestone has been used as a porosity calibration standard for the radioactive logs because it is typically shale-free and both the density and neutron logs can be adjusted to give similar readings.

When a suite of logs is run, detailed calculations can be very time

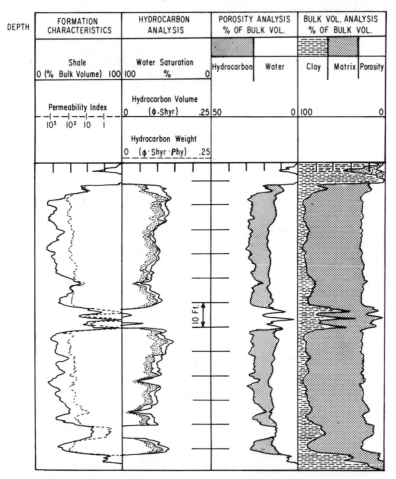

Fig.7-21. Analog display of a computer-calculated sandstone analysis log. Redrawn from original. ϕ = porosity, S_{hyr} = saturation of residual hydrocarbons, ρ_{hy} = density of hydrocarbons, matrix = skeletal structure. (Courtesy of Department of Oil Properties, City of Long Beach, California.)

consuming. By the use of computer programs, log data can be combined into a direct reading analog form. In Fig.7-20 and 7-21 a suite of logs has been combined into a continuous calculation of various rock and fluid properties. Similar analog presentations are available from the logging service companies for use in carbonate sections. Calculated logs such as these are like all other logging systems in that they require the judgement of an analyst who is familiar with the geologic section in the area.

THE IDENTIFICATION OF DEPOSITIONAL ENVIRONMENTS FROM WELL LOGS

The identification of clastic depositional environments from well logs basically relates to the identification of sand bodies, the texture, fabric, and geometry of which reflect their origin. In compaction studies, proper identification of environment and sand body type is important because differential compaction, which occurs both vertically and horizontally, is dependent upon the type and lineation of the sediment deposit. A knowledge of depositional environment is imperative in identification and forecasting sand body sequence and alignment when limited data are available. If a particular depositional pattern can be identified, better forecasting can be done as to the alignment of that body and where other sand bodies of interest might occur. This is of great importance to the geologist and well log analyst because these logs, which usually were run to evaluate a particular location, also may be used in exploring for nearby hydrocarbon reservoirs.

Shales furnish the reference (base) line on both electric and radioactivity logs in clastic sections. Exceptions to this general rule have been noted previously, i.e., bentonite beds on the electric log and radioactive shales on the gamma log. The nature of the SP log makes it particularly suitable for identification of depositional environments. Gross particle size changes are usually reflected by changes in the SP magnitude. Interspersed clay or silt in a sand body normally act to sharply reduce the amplitude of the SP curve. These changes also are reflected on the resistivity curves, particularly in sand bodies that contain hydrocarbons. All hypothetical log examples presented herein portray a resistivity curve as it might appear if the sand body contained hydrocarbons. A modified but similar log appearance is developed by the short normal resistivity curve if filtrate invasion occurs from an essentially fresh water drilling mud.

Formation dip logs have recently gained recognition as a source of information on depositional environment and sand body types. Development of new instrumentation and calculating techniques have supplied additional information that only a few years ago was either not recognized or was ignored. Detail dip data, which were recognized previously as representing stratigraphic dips within the clastic units, obscured the determination of structural dip, the value of which usually was being sought. Detailed information on formation dips, used in conjunction with the electric and/or other logs, can be considered as almost indispensable in stratigraphic and environmental determinations.

There is no depositional identification panacea to be found in well logs, because log character is based upon certain specific and limited petrophysical measurements made in a more-or-less vertical borehole and is not directly related to depositional conditions at all. Environment can be inferred by analyzing log patterns and relating them to known sand body or depositional sequence patterns. Well logs themselves will have no meaning without a prior knowledge of sand body types, depositional sequences, and granulometric and mineralogical compositions. It cannot be overemphasized that in determining sand body type and depositional environment from well logs, one must use all available data. Sand body origin often is vague and subject to disagreement even when an entire unit is exposed on the surface for direct examination.

Many different types of sand deposits appear to have similar log characteristics. Brief descriptions of what are considered to be typical sand body structures and geometries are given to enable the log patterns discussed to be related to rock properties.

Seven log patterns representing various types of sand bodies or depositional sequences are presented here. One of the most important depositional sequences, however, is not represented by an illustration, i.e., that of a shallow-water marine environment. Several log patterns represent sands that might be of shallow-water marine origin; however, each one represents differing depositional features rather than a definable sand body "type". No particular set of parameters is known to the writer that would permit meaningful construction of a generalized, theoretical log profile for shallow-water marine deposits.

In reviewing the log pattern types, it should be remembered that most sedimentary bodies are the result of a sequence of interrelated events rather than being the product of a single, isolated framework of environmental conditions.

Basic log patterns

Almost any log pattern, that might be characteristic of a particular environment, usually will share common elements with other patterns. Log patterns reflect measurements of a very few basic petrophysical properties such as granulometric composition, rock permeability, vertical layering

IDENTIFICATION OF SEDIMENTS FROM WELL LOGS 377

sequence, and, in some cases, mineral composition. Features such as crossbedding and slump structures are not readily identifiable from most well logs because of vertical restriction in their size. These features may be represented only by very small changes in curve deflection. These features are sometimes clearly recognizable when detailed dip data are available.

Five log patterns often observed are illustrated in Fig.7-22. They depict the following depositional conditions: (1) transitional, bottom or top; (2) abrupt, bottom or top; (3) oscillatory; (4) massive; and (5) interbedded. All of these patterns can be of varying magnitude and familiarity with an area may be required before the log analyst can properly evaluate what constitutes a significant change. These patterns are important in that each one represents either the end or beginning of a particular stage of deposition, or the continuity of deposition during a stage. The patterns are basic "clues" to changes in environmental conditions that can be related to depositional processes and to the types of sand bodies they produce. All of the patterns shown in Fig.7-22 reflect the changes in the grain size of material being deposited. Basically, clays, shales and silts are deposited from a low-energy transporting medium, whereas sands and coarser materials require a higher-energy environment.

The transitional patterns in Fig.7-22 indicate a gradual change in the size of the particles being deposited and a smooth transition from one

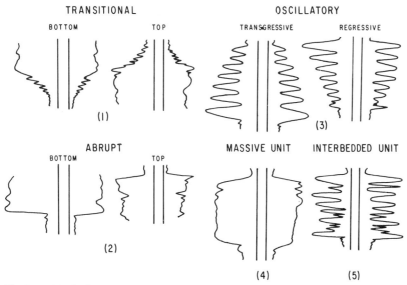

Fig.7-22. Basic log patterns representing either changes in deposition or depositional sequences. Hypothetical SP curve is to the left and hypothetical resistivity curve is to the right.

depositional environment to another. Three major environmental factors often are responsible for all 5 patterns: (1) progressive change in water depth, (2) gradational change in the energy of the transporting current, and (3) change in proximity to the sediment source. The physical conditions that might cause the above conditions include: (a) basin or shelf subsidence or uplift, (b) sea level changes, (c) climatic cycles, and (d) changes in the current path by obstructions.

The oscillatory, transgressive and regressive marine patterns are generally believed to represent long-term changes in water depth, caused by advancing or retreating seas and the consequent change in current type and sediment carrying power. The transgressive sea pattern represents a relative increase in water depth. Fons (1969) related the shape of the transgressive and regressive patterns to that of a Christmas tree and an upside-down Christmas tree, respectively. The shale interbeds are considered to be the result of cyclic changes in the sediment supply due to rhythmic sea level and water depth fluctuations. The above explanation is not universally applicable, particularly in the case of deep basin and similar environments.

The abrupt bottom and top patterns are indicative of sudden changes in the particle size being deposited. Abrupt contacts may represent unconformities or sharply differing environmental conditions. The abrupt bottom pattern is usually indicative of deposits such as those found in incised channel fills in littoral deposits, distributary channels cutting through a delta front, or at the bottom of turbidites.

Massive sand units, as indicated by a relatively uniform SP pattern, may represent periods of uniform conditions, a rapid deposition of material, or multistoried layers of shallow-water marine, alluvial or turbidite—tractionite sands.* A uniform or smooth SP curve usually indicates a homogeneous internal structure. Multistoried turbidite—tractionite sands, although having layers of varying grain size (graded bedding), may exhibit an SP curve shape that is usually indicative of uniform sand deposition, because the sorted layers are not sufficiently thick to cause a significant log response.

Interbedding of sands, silts, and shales of varying thicknesses indicate a series of changing conditions or environments that might exist in a shifting river channel or flood plain. Similar patterns are found where intermittently flooded marsh or lagoonal conditions persisted for a long period of time. A series of thin turbidites, interbedded with clays or shales, also may have this pattern.

Pirson (1970, chapter 2) also attributed recognizable secondary structural and sedimentary features to SP sedimentological patterns. The

---

*The term turbidite—tractionite is used here to define sand deposits that occur in basins distal to the sediment source. The nature of the carrying currents is not fully understood; however, they appear to be bottom currents that follow basin floor contours.

latter include: (1) convex, concave or linear shape of the slope of the SP pattern of transgressive and regressive marine sands, which indicate the relative speed of water depth change; and (2) converging, diverging, or parallel symmetry to bed envelopes on the SP curve (as related to the shale base line) (Fig.7-23). Pirson also suggested that in the case of an offshore bar sand, the location of a well with respect to sand bar edge can be determined by observing the amount of envelope divergence (Fig.7-23B). He further stated that in the case of channel fill, the convergence pattern will remain essentially constant regardless of well position with respect to bar edge.

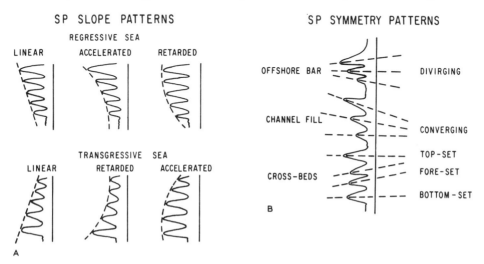

Fig.7-23. Theoretical SP slope and symmetry patterns identifying (A) depositional environments, (B) sand features.

Transgressive—regressive sea patterns are found quite often; however, because of the number of factors that can influence the appearance of the SP curve, it is not known how well they represent the speed of water depth change. The SP symmetry patterns are often either poorly defined or not present and, consequently, their use is limited. These same data often can be obtained from formation dip measurements as discussed later in this chapter.

Pirson (1970, chapter 1) discussed the redox log, which involves measurement of the oxidation—reduction potential of the formations penetrated by the wellbore. He considers the SP curve to be a subdued redox potential curve and the redox log to be superior to the SP in sedimentological and fluid identification studies. The redox log is not discussed here because it is still considered to be an experimental tool by most observers and the ultimate extent of its use has not been established. In addition, it is not readily available at the present time.

Classification of sand bodies

The classification of sand bodies by their depositional environment takes many forms and various classification systems are used. Potter (1967) gave the following definition for sedimentary environment: "A sedimentary environment is defined by a set of values of physical and chemical variables that correspond to a geomorphic unit of stated size and shape." He described the environment and characteristics of sand bodies of six major origins (see Table 7-II). Visher (1965) described six environmental sequences or processes which he called models (Table 7-II). Several types of sand bodies might be included in any one model as an inherent part of a continuing process. Natland (1967) offered a new classification for sand deposits based on their transport mode and energy (Table 7-II). The sand bodies and sequences illustrated in this chapter, and their environments as identifiable from well logs, do not fit into any one of these classifications alone. A comprehensive classification of depositional environments also was given by Crosby (1972).

Seven log patterns of either individual sand body types or depositional

TABLE 7-II

Comparison of various classifications of depositional environments, sand bodies and sequences

COMMON USAGE		AFTER VISHER (1965)		AFTER POTTER (1967)	AFTER NATLAND (1967)	
TERRESTIAL	CONTINENTAL	FLUVIAL		DESERT EOLIAN		
				ALLUVIAL		
		LACUSTRINE				
LITTORAL	SHELF	REGRESSIVE MARINE	TRANSGRESSIVE MARINE (MINOR PROCESS)	DELTAIC / CHANNEL–LAGOONAL*	TIDAL	TRACTIONITE	GRAVITITE
				BARRIER–ISLAND		
NERITIC	SHELF MARGIN			SHALLOW–WATER MARINE		
	BASIN			TURBIDITE	TURBIDITE	HEMIPELAGITE
		BATHYAL				
OCEANIC	OCEANIC	ABYSSAL PELAGIC				

* NOT IN VISHER OR POTTER CLASSIFICATION, ADDED BY THE WRITER.

sequences that might reasonably be identified from well logs were chosen as illustrations. They are: (1) alluvial—fluvial, (2) deltaic, (3) barrier-island, (4) tidal channel, (5) lagoonal, (6) turbidite, and (7) turbidite—tractionite. Other patterns may be identifiable by a log analyst familiar with a particular geographic area or depositional sequence.

Alluvial—Fluvial sands

It is almost impossible to ascertain the non-marine origin of a sediment from examination of one well log alone, although this may be possible when a group of logs is available. Alluvial sand bodies may closely resemble any one of several marine or deltaic deposits. The type illustration chosen (Fig.7-24) is similar in appearance to that of a transgressive marine environment. Due to their proximity to the source rock, continental clastics tend to be coarser than those of marine origin. The bottom layer of river channel deposits often is a conglomerate. Individual sand lenses within a channel deposit normally show a pronounced down-stream dip. Distribution of fluvial deposits commonly is erratic. When a coalescent sheet sand is formed, it is usually heterogeneous and poorly stratified. Grain size and the thickness of individual lenses tend to decrease in an upward direction. Silty intervals are common within the unit. The presence of carbonaceous wood and fragmental plant remains in alluvial deposits distinguish them from deltaic deposits with a similar log appearance, but having a marine and brackish-water fauna. A cross section of a channel fill oil reservoir is shown in Fig.7-25. Valley and channel fills may be detected by their characteristic abrupt lateral termination if log data from surrounding wells are available.

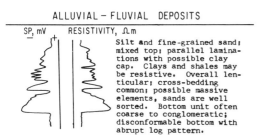

Fig.7-24. Hypothetical log pattern and characteristics of an alluvial—fluvial sand body.
Sand types: Valley fill, point-bar, fluvial channel.
Depositional environment: Non-marine, low to medium energy.
Geometry: Can be very long, perhaps 75—100 miles; width much narrower, perhaps up to 30 miles. Narrow channels are much more common with widths of 100 ft to ½ mile. Thickness is commonly a few tens of feet, may be coalesced into 100 ft+. Individual sands are lenticular; and the deposit usually has a concave-upward shape in cross section. Sand body is elongate downdip and at right angles to the paleo-shoreline, if present. Often arcate, sinuous, anastomosing.

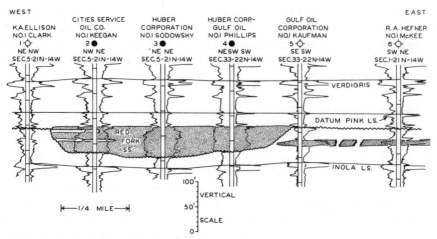

Fig.7-25. Cross section of the Red Fork Sandstone, which is a channel-fill deposit; Cheyenne Valley Field, Oklahoma. (After Winthrow, 1968, fig.12, p.1651. Courtesy of Bull. Am. Assoc. Pet. Geol.)

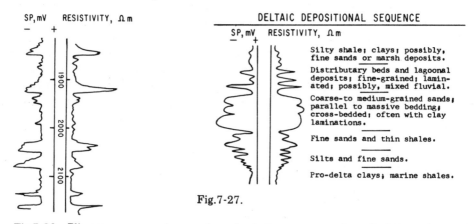

Fig.7-26. Pliocene, non-marine sand and shale sequence, deposited in a lacustrine environment.

Fig.7-27. Hypothetical log pattern and physical characteristics of a delta vertical profile.
Sand types: Bar-finger, channel, bar-crest, stream-mouth, delta-front, bar-slope.
Depositional environment: Shallow marine to brackish-lagoonal or fluvial. Combination of environments, any one of which may be predominant.
Geometry: Variable, elongate down-dip or crescent-fan shaped. Bar-fingers narrow upslope; bar-slope sands thin down dip. Sand sequence may be plano-concave or lens-shaped in cross section. Individual sands interfinger laterally; units usually contain more silt toward edge. Tens to hundreds of feet thick in coalesced sections. Individual sands up to a mile wide, a few miles long, systems may be many miles long or wide.

IDENTIFICATION OF SEDIMENTS FROM WELL LOGS 383

Sand bodies of alluvial fan origin may resemble turbidites quite closely. They are fan-shaped, generally become thinner down-dip, and may coalesce into very thick sections. Figure 7-26 shows a log of a Pliocene, non-marine deposit. Based on examination of the log pattern, the individual sands have all the characteristics of turbidites. These sand bodies probably were deposited as underwater fans in a lacustrine environment. The graded bed pattern of each sand is distinct on the log.

Deltaic sequence

Inasmuch as no one sand body or environment is characteristic of a delta vertical profile, no one log pattern can be truly representative of a deltaic sequence. Delta environments range from regressive, shallow-water marine through tidal—lagoonal and possibly to fluvial. The deltaic sequence or process is a result of the interrelation of a number of environments, each having its own characteristics. The log pattern in Fig.7-27 assumes a model that begins with deposition of pro-delta marine shales or clays, grades into silts and then sands of prograding delta-front, channel, or bar-finger deposits, and terminates with a lagoonal or marsh environment.

Multistoried sequences of distributary and delta-front sands are common due to channel migrations, prograding, and variations in river discharge. Identifiable log patterns of Pennsylvanian age delta units in Oklahoma are shown in Fig.7-28. Tidal environment units of deltas, such as bar-finger sands, may resemble fluvial deposits; and shallow-water marine sands may easily be confused with delta-front deposits. Quite often, incremental units of a deltaic sequence cannot be separately identified either by using well logs or other methods.

Two distinct sand body lineations may be present: (1) channel and bar-finger sands that are elongate downdip and generally cross the shore line and channel mouth, and (2) delta-front and bar-crest sands that are fan-shaped and oriented more or less parallel to the shore line. The deltaic sequence is characterized by a series of overlapping and intercalated sand units that do not have a very wide extent individually, but have an extensive genetic sequence. Figure 7-29 is a cross section of a deltaic sequence showing the various units representing different depositional environments.

Barrier-island sands

The term "barrier-island" is sometimes used to refer to a group of sand bodies such as offshore bars, cheniers, dune-capped bars, etc. All of these sand bodies are shallow-water, near-shore formations that are generally considered to contain more-or-less winnowed, mature sands. Sedimentologists might be able to differentiate between these sand bodies when sufficient data are available; however, their logs have similar appearance and

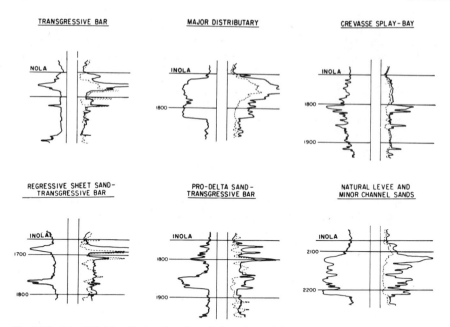

Fig.7-28. Identifiable Pennsylvanian delta depositional patterns in Oklahoma. (After Visher et al., 1971, fig.10, p.1223. Courtesy of *Bull. Am. Assoc. Pet. Geol.*)

represent comparable environments. According to the long-established definition, a barrier-island bar separates lagoonal and alluvial deposits from marine. Typically, they either were parallel to the paleo-shoreline or once formed the shoreline. They may be found laterally and at right angles to tidal-pass or delta-mouth sands. Davis et al. (1971) noted that some sand bodies referred to as bar sands do not exhibit a parallel alignment with the shoreline nor do they have the ocean—bar—lagoon configuration.

Cross-bedding in barrier-island sand bodies is common but is not necessarily a predominant feature. Stratigraphic dips often are bimodal. Dip components may represent either shore-face sands dipping seaward, or backshore deposits dipping landward. In cross-section, the shape may be that of a wedge thinning seaward.

The log pattern of a barrier-bar deposit is essentially that of an alluvial deposit but reversed (Fig.7-30). Figure 7-31 shows the relationship between a simple delta system and barrier-island complex, based on environmental relationships deduced from well logs.

Tidal channel sands

Tidal channel deposits (Fig.7-32) are formed in passes, estuaries and inlets that cross the beach and inter tidal zone. According to Potter (1967),

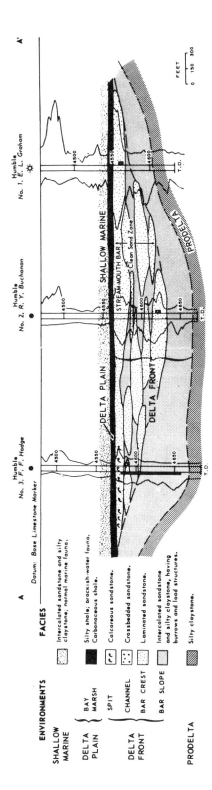

Fig.7-29. Cross-section of a delta sequence capped by a marine transgression, West Tuscola Field, Texas. (After Shannon and Dahl, 1971, fig.7, A–A¹, pp.1200–1201. Courtesy of *Bull. Am. Assoc. Pet. Geol.*)

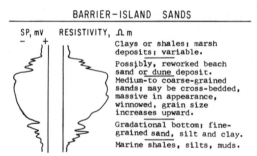

Fig.7-30. Hypothetical log pattern and physical characteristics of a barrier-island sand body.
Sand types: Barrier islands, offshore bars, cheniers, dunes.
Depositional environment: Shallow marine; wave front to tidal.
Geometry: Usually parallel to shore line; length many times its width. Low tens to hundreds of feet thick; hundreds of feet to a few miles wide; up to about 60 miles long. Multistory deposits common. Low angle bedding dipping seaward with individual and gross units thinning in that direction.

they resemble alluvial channel and shallow-water marine sands. Skeletal debris and glauconite, when cores are available, may serve to distinguish them from alluvial sands. As a general rule, tidal sand bodies exhibit a maturity not observed in alluvial sands. Lagoonal—tidal flat beds are the normal sequence found laterally.

Lagoonal sands

Although lagoon deposits do not represent a specific type of sand body, they can exhibit a very distinctive log pattern (Fig.7-33). The lagoonal environment is characterized by rapid changes in sedimentation and exposure to rapid oxidizing conditions. Water in a lagoon is not deep and strong currents are rarely present. Thin sands, silts, and carbonaceous deposits are interbedded in a rapidly alternating succession. Coarse-grained sands may be present in minor amounts. Inasmuch as barrier-island sands often constitute the seaward limit of a lagoonal sequence, log identification of the lagoonal environment can be important.

Examples of logs through channel and lagoonal deposits are compared in Fig.7-34. The thin beds and high—low resistivity contrasts of the lagoon deposits cause frequent reversals on the lateral resistivity curve. These reversals would not be present when using the induction log.

Turbidites and turbidite—tractionites

Turbidites are usually defined as deposits formed by differential settlement of material from a turbid suspension flow (often called a density current). It is believed that turbidites could give rise to thick masses of sands

IDENTIFICATION OF SEDIMENTS FROM WELL LOGS 387

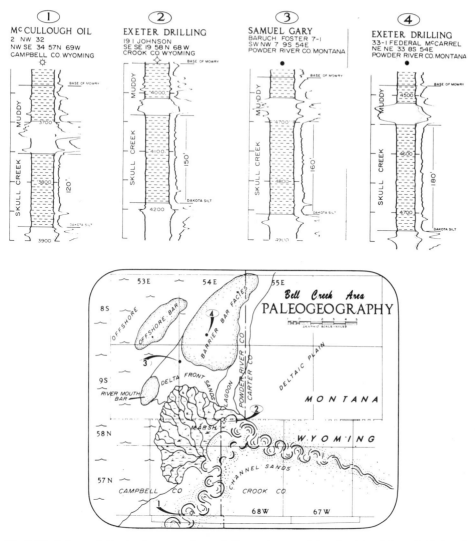

Fig. 7-31. Interpretation of a one-story delta system and barrier-island bar, Bell Creek Field, Montana. (After McGregor and Biggs, 1968, fig. 15—20, pp. 1879—1883. Courtesy of *Bull. Am. Assoc. Pet. Geol.*)

and shales in deep basin locales such as those of the Ventura and Los Angeles Basins in California. Criteria for recognizing turbidites include (1) their deposition from a suspension flow by differential settlement, as evidenced by a lower unit having graded bedding; (2) upper units having parallel and laminated bedding; and (3) widespread deposition of the overall sequence.

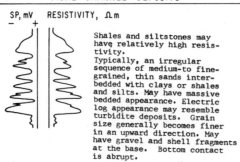

Fig.7-32. Hypothetical log pattern and physical characteristics of tidal-channel deposits.
Sand types: Distributary channel, estuary, inlet. Often similar to bar-finger and delta-front sands.
Depositional environment: Shallow-water marine, low energy. May exhibit either transgressive or regressive log character.
Geometry: Elongate downdip, long axis is at right angles to shore line. Width is narrow relative to length. About 50 ft + wide for inlets, etc., and up to 1,000 ft for major estuaries. Individual sands 10—25 ft thick, not likely to persist for any great distance. Moderately meandering, dendritic. Bimodal grain size distribution; cross-bedding typical.

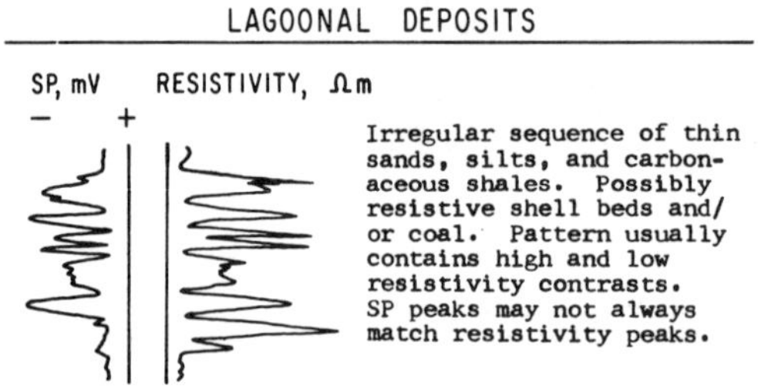

LAGOONAL DEPOSITS

SP, mV   RESISTIVITY, Ω m

Irregular sequence of thin sands, silts, and carbonaceous shales. Possibly resistive shell beds and/or coal. Pattern usually contains high and low resistivity contrasts. SP peaks may not always match resistivity peaks.

Fig.7-33. Hypothetical log pattern and physical characteristics of lagoonal deposits.
Sand types: Thin, laminated with marsh deposits. No distinct type.
Depositional environment: Brackish to marine; shallow-water, low energy; may mark the end of a regressive log sequence.
Geometry: Parallel to shore line or meandering. Individual sands of limited lateral extent. Units may be a few miles wide and many miles long: possibly, in the low tens to a few hundreds of feet thick.

Turbid suspension flows and differential settlement from them are logical occurrences in near-source, submarine canyons and to some extent near their mouths (Fig.7-35A). Slumping and gravitational sheet-sliding have also been proposed as possible depositional mechanisms for turbidites. The presence of a thick series of turbidity current sand deposits in a distal basin location is more difficult to comprehend.

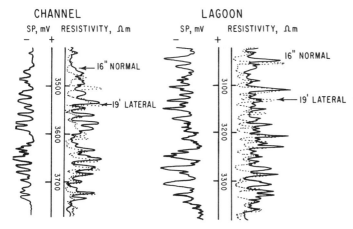

Fig.7-34. SP—resistivity log response of channel and lagoon deposits. (Courtesy of Standard Oil Co. of California.)

Conrey (1959) and Sullwold (1961) both attributed the widespread, thick (many thousands of feet) Pliocene and Miocene deposits in the Los Angeles Basin, California, to turbidity current deposition. Shepard et al. (1969) studied the present-day La Jolla, California, submarine canyon and environs. They noted that although all evidence points to sand and other material being transported down the existing canyon and into the fan-valley continuations, most recent investigations suggest depositional mechanisms of a more complex nature than those of turbidity currents alone. Much of the material being deposited there today does not show graded bedding.

Natland (1967) suggested a new classification for water-laid clastic sediments based on their mode of deposition and transport. His description of offshore-marine, bottom-current transported material (tractionites), coupled with the recent investigations of material being deposited today, suggests that the masses of sand present in the center of California basins might not be the result of turbidity current deposition alone. According to Natland's (1967) classification, tractionites are winnowed and do not contain appreciable amounts of silt and clay, whereas turbidites contain from 10 to 30% silt and clay.

Sullwold (1961) listed three turbidite depositional locations: (1) submarine channel, (2) submarine fan, and (3) basin floor. Walker (1966) separated turbidites into two divisions, proximal and distal, with reference to their distance from the sediment source. Stanley and Unrug (1972) concluded that deposition by turbidity currents represents only one of several important processes that are active on marine depositional slopes. In the opinion of the writer, there is a need for either a reclassification or a subdivision of turbidite deposits into classes that better reflect the nature of their depositional environment.

Figure 7-35B illustrates a log pattern that is typical for deep marine basin deposits. It is suggested that these sand—shale sequences might represent a combination of turbidite and tractionite deposits, or at the least, an intermediate depositional stage. Figure 7-36 illustrates two logs for deep-water turbidite—tractionite sand sequences. The indicated transgressive and regressive patterns are believed to have been caused by factors such as variation in sediment supply rather than changes in water depth.

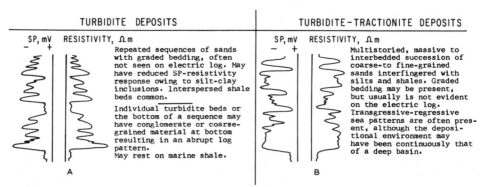

Fig.7-35. Hypothetical log patterns and physical characteristics.
A. *Turbidite deposits*
Turbidite sand types: Submarine channel fills and fans. Graded bedding is typical with coarse-grained to conglomeratic sands at the bottom with a grain size decrease upwards. Deposited from density flows. Contain dispersed silts and clays. Greywackes are considered by many authorities to be turbidites.
Depositional environment: Medium to deep marine current channels, submarine canyons, and fore-basins. Proximate to the sediment source.
Geometry: Long to fan-shaped, sheet-like. May be many miles wide and very thick. Individual sand units often cannot be correlated over short distances, whereas the sequence may be widely recognizable. Dips are parallel to the basin or channel floor. Channel fills may contain slump structures.
B. *Turbidite—tractionite deposits*
Turbidite—tractionite sand types: Marine basin, widespread sheets and fans. Contain silts and clays but in smaller amounts than turbidites. Carrying currents not well understood at present. Sometimes called "contourites". Graded bedding seen in well cores and outcrops but are not evident on well logs. May be winnowed to some degree.
Depositional environment: Medium to deep marine basins, distal to the sediment source, and where submarine canyons are not deeply incised.
Geometry: Sheet or blanket-like. Units may persist for many miles, whereas individual sands do not. Beds thin over old structures and thicken in basin troughs. Many thousands of feet thick. Dips are parallel to the basin or channel floor. Slump structures not seen.

IDENTIFICATION OF SEDIMENTS FROM WELL LOGS

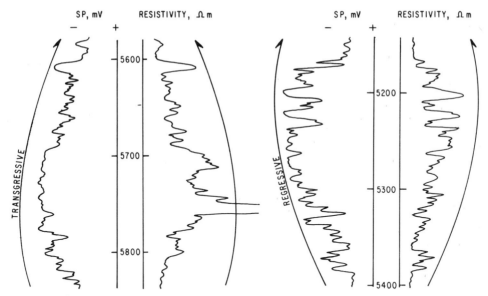

Fig.7-36. Transgressive and regressive log patterns in deep-marine basin turbidite—tractionite deposits.

USE OF FORMATION DIP MEASUREMENTS IN THE IDENTIFICATION OF SAND BODIES AND DEPOSITIONAL ENVIRONMENTS

Interpretation of stratigraphy and depositional structures are of prime importance in that sand body type and, consequently, environment can be inferred. Many formation dip patterns are similar and no stratigraphic interpretation is possible without first considering the structural implications. Recent advances, both in mechanical equipment and interpretive techniques, have made borehole dip measurements an important part of stratigraphic work. The tools used by all logging companies are simple in concept; however, calculations can be time consuming and difficult for the inexperienced person. Dip measurements are not read directly from the well log as recorded, but must be calculated. Detailed computations are made by the service companies upon request.

Principles of operation

Presently-used dip tools record 3 or 4 focused, microresistivity curves from electrodes spaced uniformly around the tool body. The recording pads are pressed against the borehole wall, either by springs or powered arms.

Dips are determined by correlating bed boundaries on each curve, assuming that these chosen points lie on a plane, and calculating the dip and strike of that plane. Older instruments, although closely resembling those in current use, were designed primarily for optical correlations and basically yielded structural information.

In order to calculate dips and strikes from the recorded bed boundaries, other data are needed. The electrodes must be oriented with respect to north, and borehole deviation angle and bearing must be known. A schematic diagram of a dip tool and its recording elements are presented in Fig.7-37. Dip data may be calculated: (1) by hand, (2) by semi-automatic hand-machine methods where log measurements are read visually and computations are done by machine, and (3) by a fully automatic method where the log is scanned and computed automatically. The automatic program requires that a correlation "search length" be specified; that is, how long will be the interval over which correlations are sought? This length is dependent upon the bed dip anticipated, the angle of the borehole, and whether stratigraphic data are desired. Steeply dipping beds require a longer correlation interval. A standard method of presentation of formation dip—strike data is shown in Fig.7-38. The placement of the dot indicates the degree of bed dip and the line (tail) indicates the direction. This presentation has been termed the "tadpole" plot for obvious reasons.

Gilreath and Maricelli (1964) identified three basic formation dip patterns that have stratigraphic and structural meaning: (1) consistent regional or structural dip, (2) dip increasing with depth, and (3) dip decreasing with depth. Structural dip was assumed to be represented by recurring measurements of a more-or-less uniform azimuth and magnitude, even though these may represent only a small part of the data. Campbell (1968) modified the above patterns by adding unconformities to pattern *A* of Gilreath and Maricelli (1964) and faults and current bedding to their pattern *B* (Fig.7-39). For ease of interpretation, he colored these patterns on the logs as follows: (1) green for structural dip, (2) red for the *A* pattern, and (3) blue for the *B* pattern. Recognition of sand bodies or events (folding, faulting, etc.) from dip patterns necessitates: (a) a thorough background in the geometry of the structural features that might be encountered and (b) a thorough knowledge of the internal and external features that might be seen in the sands themselves.

Inasmuch as certain log patterns might represent specific sedimentary features, the use of illustrations represents the best method to enable one to get a three-dimensional concept of the sand body that might be represented by a specific pattern (Fig.7-39—7-42). The number of patterns and various combinations that must be considered is large if folding, faulting and stratigraphy are all examined. The illustrations listed above are theoretical and

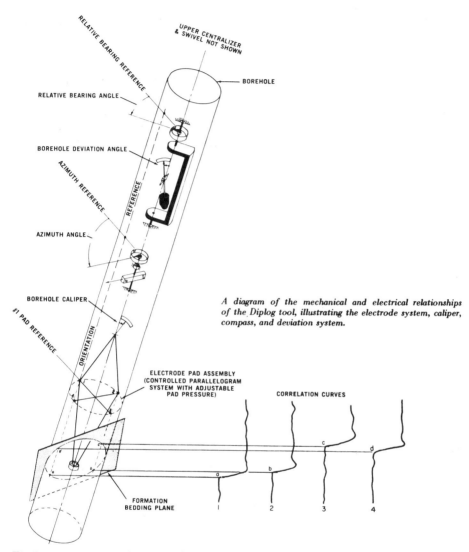

Fig.7-37. Schematic diagram of a formation dip measuring device and its recording pattern. (Courtesy of Dresser Atlas, Houston, Texas.)

simplified, but all are based on the work of many specialists. When interpreting sand body dip patterns, one has to consider not only the pattern itself, but whether the pattern appears above, below, or continues through the sand body in question. This, naturally, requires reference to other logs. Extensive information and literature on dip pattern interpretation are available from several logging companies.

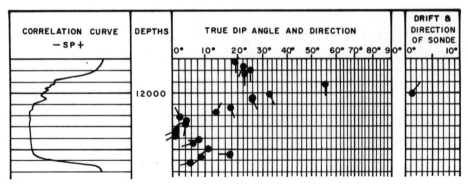

Fig.7-38. Presentation of formation dips and borehole inclination. (Courtesy of Schlumberger Well Services, Houston, Texas.)

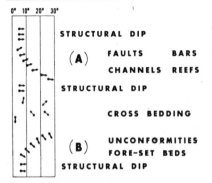

Fig.7-39. Typical dip patterns and possible interpretations. (After Campbell, 1968, fig.2, p.1702. Courtesy of *Bull. Am. Assoc. Pet. Geol.*)

Two types of sand bodies and one depositional feature are generally recognized as being readily identifiable from dip data: (1) barrier-bar sands, (2) channel- or trough-fill sands, and (3) current and cross-bedded features. Barrier-bar sands (Fig.7-40A) usually have a distinctive dip pattern when penetrated on either the shoreward or seaward slopeface. Due to slope deposition and differential shale compaction above the unit, dips above the sand body steepen in a down-slope direction. The direction of dips within the sand body is away from the paleo-shoreline or, generally, at right angles to the elongate extension of the sand body. Regional structure dips are normally seen below the sand unit. Azimuth frequency data are plotted in Fig.7-41A as they might appear on opposite sides of a barrier-bar sand. Cross-bedding may be recognizable within the sand body and a bimodal distribution of dip data is possible. In the presence of bimodal expression, the depositional dip slope is represented by a greater number of readings.

IDENTIFICATION OF SEDIMENTS FROM WELL LOGS 395

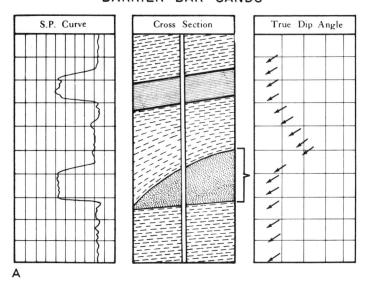

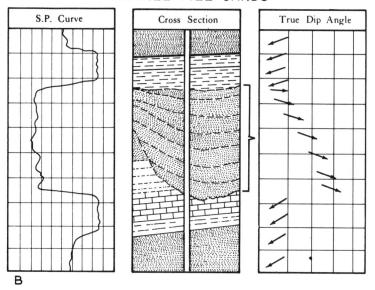

Fig.7-40. Hypothetical barrier-bar and channel-fill sand dip patterns. (Courtesy of Dresser Atlas, Houston, Texas.)

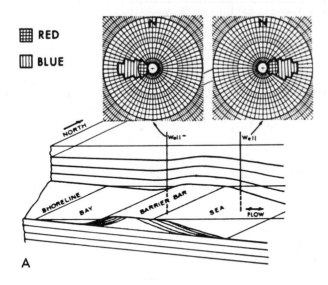

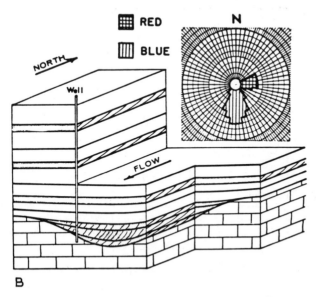

Fig.7-41. Azimuth frequency plots. A. Barrier-bar sands. B. Hypothetical channel-fill sands. (Courtesy of Schlumberger Well Services, Houston, Texas.)

Reefs may exhibit the same pattern; however, within the reef itself, random dips are likely to be recorded due to a lack of stratification.

Channel-fill sands (Fig.7-40B and 7-41B) are identified by an increasing dip towards the bottom of the unit. This pattern is pronounced towards the lateral extent of a channel-fill because of trough-bottom geometry. Direction of dip within the sand unit is towards the channel center. A bimodal pattern, when present, shows bed stratification in the direction of current flow and towards channel center. Dips towards the channel center (Campbell, 1968, p.1703) begin above the channel fill and continue through the sand body. The lower surface shows the channel-bottom slope. The axis of the channel body is oriented at approximately right angles to the direction of the sand body major dip pattern.

Current- and cross-bedded sand patterns are shown in Fig.7-42. The reliability of dip measurements for distinguishing current-bedding is somewhat controversial. Campbell (1968, pp.1706—1707) stated that present techniques can reliably detect current-bedding in units over two feet thick and that these thicknesses are common. Current-bedding, as used here, refers to the depositional slope of thin sand interbeds, also called tangential bottom foresets. The direction of current-bedding is the same as that of the depositing current flow. When this direction is known, it can be used to determine the deposit geometry. When detailed dip data are used for stratigraphic interpretations, structural dip data are vectorially removed. The appearance of random cross-beds is shown in Fig.7-42B. Figure 7-43 illustrates the *A* and *B* patterns as related to a fault and an unconformity, respectively. Drag along the downthrown fault block causes dips to increase in magnitude above the fault plane. The reverse is seen below the fault plane. Figure 7-43B depicts decreasing dips below an angular unconformity. The pattern most often observed in such cases is an abrupt change in dip magnitude and direction, but without the *B* pattern.

Other methods for plotting dip frequency data are used for various purposes and in different areas. To the trained observer, even very minor changes in dip patterns are important. Concentrated study in a particular geographic area is required to detect and interpret minor variations in dip patterns.

In conclusion one can say that various logging techniques hold great promise in deciphering the origin, geometry, and porosity of various sand bodies, all of which aid in determining the degree of compaction. These techniques are being refined continuously.

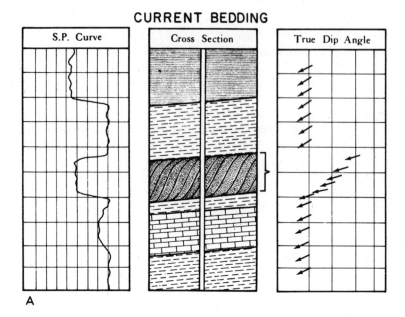

Fig.7-42. Hypothetical current and cross-bedded dip patterns. (Courtesy of Dresser Atlas, Houston, Texas.)

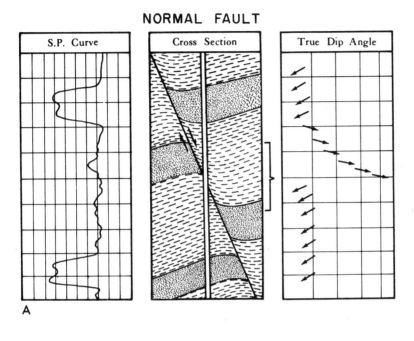

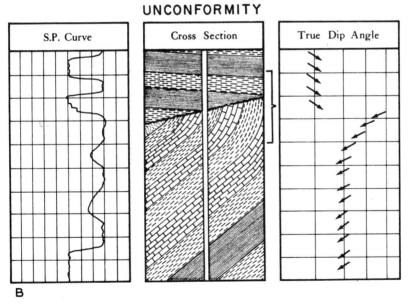

Fig.7-43. Hypothetical fault and angular unconformity dip log patterns. (Courtesy of Dresser Atlas, Houston, Texas.)

ACKNOWLEDGEMENT

The review of the manuscript and valuable suggestions by W. Lang, K. Wolf, and G. V. Chilingarian are greatly appreciated by the author.

REFERENCES

Berg, R. R. and Davis, D. K., 1968. Origin of Lower Cretaceous muddy sandstone at Bell Creek Field, Montana. *Bull. Am. Assoc. Pet. Geol.*, 52: 1888—1898.
Busch, D. A., 1971. Genetic units in delta prospecting. *Bull. Am. Assoc. Pet. Geol.*, 55: 1137—1154.
Campbell Jr., R. L., 1968. Stratigraphic applications of dip-meter data in Midcontinent. *Bull. Am. Assoc. Pet. Geol.*, 52: 1700—1719.
Carrigy, M. A., 1971. Deltaic sedimentation in Athabasca Tar Sands. *Bull. Am. Assoc. Pet. Geol.*, 55: 1155—1164.
Conrey, B. L., 1959. *Sedimentary History of the Early Pliocene in the Los Angeles Basin.* Thesis, Univ. South. Calif., Los Angeles, Calif., 273 pp.
Crosby, E. J., 1972. Classification of sedimentary environments. In: J. K. Rigby and W. K. Hamblin (Editors), *Recognition of Ancient Sedimentary Environments. Soc. Econ. Paleontol. Mineral., Spec. Publ.*, 16: 1—11.
Davis, D. K., Ethridge, G. G. and Berg, R. R., 1971. Recognition of barrier environments. *Bull. Am. Assoc. Pet. Geol.*, 55: 550—565.
Dresser Atlas, 1971. *Log Review 1 (Document).* Dresser Atlas Inc., Wireline Services, Houston, Texas, 117 pp.
Fons, L., 1969. Geological application of well logs. In: *Tenth Annual Logging Symposium Transactions.* Soc. Prof. Well Log Anal., Pap., AA: 44 pp.
Gilreath, J. A. and Maricelli, J. J., 1964. Detailed stratigraphic control through dip computations. *Bull. Am. Assoc. Pet. Geol.*, 48: 1902—1910.
Gorsline, D. S. and Emery, K. O., 1959. Turbidity current deposits in San Pedro and Santa Monica Basins off Southern California. *Bull. Geol. Soc. Am.*, 70: 279—290.
Guyod, H. and Shane, L. E., 1969. *Geophysical Well Logging, I.* Hubert Guyod, Houston, 256 pp.
Klein, G. de V., 1967. Paleocurrent analysis in relation to modern marine sediment dispersal patterns, *Bull. Am. Assoc. Pet. Geol.* 51: 366—382.
Krueger, W. C., 1968. Depositional environments of sandstones as interpreted from subsurface measurements — an introduction. *Bull. Am. Assoc. Pet. Geol.*, 52: 1825 (abstr.).
Lock, G. A. and Hoyer, W. A., 1971. Natural gamma-ray spectral logging. *The Log Analyst*, 12(5): 3—9.
McGregor, A. A. and Biggs, C. A., 1968. Bell Creek Field, Montana: A rich stratigraphic trap. *Bull. Am. Assoc. Pet. Geol.*, 52: 1869—1887.
Morgan, J. P. and Shaver, R. H. (Editors), 1970. *Deltaic Sedimentation, Modern and Ancient. Soc. Econ. Paleontol. Mineral., Spec. Publ.*, 15: 312 pp.
Natland, M. L., 1967. New classification of water-laid clastic sediments. *Bull. Am. Assoc. Pet. Geol.*, 51: 476 (abstr.).
Pirson, S. J., 1963. *Handbook of Well Log Analysis.* Prentice-Hall, Englewood Cliffs, N.J., 326 pp.
Pirson, S. J., 1970. *Geologic Well Log Analysis.* Gulf Publ., Houston, Texas, 370 pp.

Potter, E. P., 1967. Sand bodies and sedimentary environments: a review. *Bull. Am. Assoc. Pet. Geol.*, 51: 337—365.
Rich, J. L., 1951. Three critical environments of deposition and criteria for recognition of rocks deposited in each of them. *Bull. Geol. Soc. Am.*, 62: 1—20.
Rigby, J. K. and Hamblin, W. K. (Editors), 1972. *Recognition of Ancient Sedimentary Environments. Soc. Econ. Paleontol. Mineral., Spec. Publ.*, 16: 340 pp.
Schlumberger Well Surveying Corporation, 1958. *Introduction to Well Logging, Document 8.* Schlumberger Well Surveying Corporation, Houston, Texas, 4th ed., 176 pp.
Schlumberger Ltd., 1969. *Log Interpretation Principles (Document).* Schlumberger Ltd., New York, 110 pp.
Schlumberger Ltd., 1970. *Fundamentals of Dipmeter Interpretation (Document).* Schlumberger Ltd., New York, N.Y., 145 pp.
Schlumberger Ltd., 1972. *Log Interpretation Charts (Document).* Schlumberger Ltd., New York, N.Y., 92 pp.
Shannon Jr., J. P. and Dahl, A. R., 1971. Deltaic stratigraphic traps in West Tuscola Field, Taylor County, Texas. *Bull. Am. Assoc. Pet. Geol.*, 55: 1194—1205.
Shelton, J. W., 1967. Stratigraphic models and general criteria for recognition of alluvial, barrier bar, and turbidity current sand deposits. *Bull. Am. Assoc. Pet. Geol.*, 51: 2441—2461.
Shepard, F. P., Dill, R. F. and Van Rad, U., 1969. Physiography and sedimentary processes of La Jolla submarine fan and fan-valley, California. *Bull. Am. Assoc. Pet. Geol.*, 53: 390—420.
Silver, B. A. and Todd, R. G., 1969. Permian cyclic strata, northern Midland and Delaware Basins, west Texas and southeastern New Mexico. *Bull. Am. Assoc. Pet. Geol.*, 53: 2223—2251.
Stanley, D. T. and Unrug, R., 1972. Submarine channel deposits, fluxoturbidites and other indicators of slope and base-of-slope environments in modern and ancient marine basins. In: J. K. Rigby and W. K. Hamblin (Editors), *Recognition of Ancient Sedimentary Environments. Soc. Econ. Paleontol. Mineral., Spec. Publ.*, 16: 287—340.
Sullwold Jr., H. H., 1960. Tarzana Fan, deep submarine fan of late Miocene age, Los Angeles County, California. *Bull. Am. Assoc. Pet. Geol.*, 44: 433—457.
Sullwold Jr., H. H., 1961. Turbidites in oil exploration. In: *Geometry of Sandstone Bodies. Am. Assoc. Pet. Geol., Spec. Publ.*, 63—81.
Visher, G. S., 1965. Use of vertical profiles in environmental reconstruction. *Bull. Am. Assoc. Pet. Geol.*, 49: 41—61.
Visher, G. S., Saitta, B. S. and Phares, R. S., 1971. Pennsylvanian delta patterns and petroleum occurrences in eastern Oklahoma. *Bull. Am. Assoc. Pet. Geol.*, 55: 1206—1230.
Walker, R. G., 1966. Deep channels in turbidite-bearing formations. *Bull. Am. Assoc. Pet. Geol.*, 50: 1899—1917.
Winthrow, P. C., 1968. Depositional environments of Pennsylvanian Red Fork Sandstone in northeastern Anandarko Basin, Oklahoma. *Bull. Am. Assoc. Pet. Geol.*, 52: 1638—1654.

Chapter 8

MATHEMATICAL ANALYSIS OF SAND COMPACTION

R. RAGHAVAN and F. G. MILLER

INTRODUCTION

For over three decades, subsidence of the land surface in various areas of the world has attracted the serious attention of a large number of scientists in a wide variety of disciplines. More recently, there has been an added attraction mainly due to an "environmental awakening", because man's activities have begun inadvertently to produce some harmful effects of considerable magnitude and of possible alarming consequence. As pointed out by Allen (1969), land surface subsidence and rock compaction may be due to a wide variety of causes, either natural or resulting from man's activities. In the present chapter, mathematical models for a few important problems, which arise as a result of compaction of porous solids, are discussed. First, the classical work of Terzaghi (1923, 1943) is examined. One of the strongest incentives for the creation of the science of soil mechanics has been the remarkable success of Terzaghi's analysis in predicting the one-dimensional behavior of a wide variety of porous media. Accordingly, several modifications, such as the variation in permeability, compressibility, etc., have been incorporated. These have been summarized by many authors, for example, Taylor and Merchant (1940), Jaeger (1964), and Wu (1966). Substantial improvements, however, have taken place only in the last six years (Philip, 1968; Smiles and Rosenthal, 1968; Smith, 1971). These modern developments are examined in detail.

Terzaghi's solution of the consolidation problem involves the consideration of a one-dimensional shrinkage of a porous column due to a single-step function change in pressure at the base of the column (either finite or semi-infinite). The Terzaghi's solution neglects the movements of the solid particles of the porous column. A number of publications have extended Terzaghi's analysis to include the motion of the solid matrix. In many instances, however, Darcy's (1856) Law has been often stated incorrectly. This misstatement is due to the failure to distinguish between the rate of flow of the fluid relative to the moving grains of the matrix (which obeys Darcy's Law) and the rate of flow of fluid across the fixed boundaries of a volume element in which the matrix is deforming (which does not obey Darcy's Law). Mathematical models governing the movement of both the fluid and solid may be developed in either the Eulerian (stationary) or Lagrangian (moving or deforming) reference coordinates. The advantage of using a moving reference frame is that it does not explicitly involve the velocity of sand grains. A recent paper by Philip (1968) has examined this aspect in detail. Philip (1968) has shown that on neglecting the movement of the solid matrix the behavior of the more mobile fluid component is obscured. Philip (1968) has also shown that even though the problem is nonlinear, analytical solutions exist for a class of problems of great practical interest. The analytical solutions, though restricted in applicability, provide a basic understanding of the various physical processes which play a role in the consolidation or expansion rebound processes. In addition to solving a classic problem, the model as shown by Smith (1971) can also be used to predict the evolution of pore pressure in sediments. In considering the classical Terzaghi analysis, the approach presented by Philip (1968), Smiles and Rosenthal (1968), and Smith (1971) is followed in this chapter. Expressions for the instantaneous distribution of solids content, local mean velocities of solid and liquid, and displacement history of the solid particles are also developed. The effects of the overburden potential and gravity are also investigated. As a considerable amount of confusion exists regarding the choice of coordinate system, i.e., Eulerian or Lagrangian, the relationship between fixed and moving coordinate systems is examined and results are developed in both systems.

For water and oil reservoir rocks, which are "normally" compacted, changes in porosity (or pore volume) and permeability are fairly small, and these changes usually can be neglected. Laboratory data and field experience, however, indicate that in some instances these effects are important and should be considered. A number of studies have been made on the effect of internal and external pressure on various rock properties, and efforts have been made to incorporate variations of these properties in mathematical models. The variation of rock properties has been portrayed as a function of either pore pressure or effective stress. The effective stress is the difference between the external, confining or overburden pressure and the internal or fluid pressure within the pores of the rock. As the reservoir pressure decreases, the effective stress increases, causing changes in porosity, permeability and grain orientation, and some crushing.

The second section in this chapter considers a mathematical model which describes the idealized radial flow of a compressible liquid, with pressure-dependent properties, toward a well located at the center of a closed radial reservoir undergoing compaction. This single-well reservoir is assumed to be producing at a constant rate. Lateral components of deformation are assumed to be negligible. The change in thickness of the reservoir is accounted for through calculations using the compressibility of the reservoir rock. Flow is assumed to be strictly radial and fluid pressure gradients due to changes in thickness are neglected. It is shown that a second-order, nonlinear partial differential equation is obtained for a system in which reservoir porosity, permeability and compressibility are treated as functions of pressure. This equation has been solved by numerical means and solutions are presented. The main objective of this analysis is to demonstrate techniques of analyzing transient behavior in deformable systems.

Nonlinear equations described in this chapter generally have to be solved for each specific case of interest and solutions obtained are not of general utility. One of the principal objectives of this chapter is to provide solutions of general utility that can be obtained without resorting to a *digital computer*. This goal is achieved through transformation of nonlinear equations without incorporating any further assumptions, whereby the variable-property digital computer solutions to the nonlinear equations are correlated with constant-property analytical solutions. The resultant correlation allows solutions to be obtained rapidly for engineering purposes. Inasmuch as the transformation represents a scale change in the dependent variable (pressure or void ratio), it may be designated as a pseudo-void ratio or pseudo-pressure. It has a number of additional advantages which are discussed later.

In general, the nonlinear equations of the type considered here have to be solved by numerical methods through the use of a digital computer. Many numerical solution techniques are available whereby these equations can be solved. A few of these numerical methods are discussed in the third part of this chapter. The methods suggested are by no means exhaustive. The numerical methods comprise a sample and, in the opinion of the authors, have proved to be useful in terms of man- and machine-time.

From the viewpoint of mathematical physics, the major extension of Terzaghi's theory lies in extending the analysis to two- and three-dimensional systems. The extension to three-dimensional systems was first examined by Biot (1935) and since that time, multi-dimensional effects have received considerable attention. The basis for the three-dimensional theory was laid down essentially by Biot. In a series of papers Biot (1935, 1941a, b, 1955, and 1956) has shown that the deformation of a porous medium can be described theoretically as a generalization of the theory of elasticity. Many other workers in a wide variety of fields have also examined multi-dimensional effects. For example, in ground-water hydrology these problems have been examined by Jakob (1940, 1950), Verrujt (1965, 1969), and Cooper (1966).

The analogy between the thermal stresses (thermoelasticity) and the stresses in porous materials was first pointed out by Lubinski (1954) and later by Geertsma (1957b). On the basis of this analogy, Geertsma (1966) introduced the term "poroelasticity" to describe the bulk stress—strain relationships in deformable porous medium. The science of thermoelasticity is older and, hence, more developed than poroelasticity. Thus, many of the theories already developed in that field can be readily applied to the field of poroelasticity. Geertsma (1966, 1972) and Evangelisti and Poggi (1970) have used the solutions developed in the theory of elasticity by Mindlin and Cheng (1950) and Sen (1950), to describe land-surface subsidence due to the compaction of buried porous sediments. The importance of this problem, particularly to earth scientists, requires little emphasis. The main complication which has discouraged a theoretical analysis is the inclusion of the overlying soil masses. Mandel (1957, 1961) was the first to analyze such effects. The recent paper of Geertsma (1973) incorporates and demonstrates these effects in a very elegant way. This work should become a classic reference and serve as a basis for future work.

The fourth part of this chapter briefly examines the three-dimensional deformation of porous solids. The bulk stress—strain relationships governing such deformations are examined. General methods of solutions are discussed. The application of the ones developed in thermoelasticity and elasticity is also presented.

The writers would like to emphasize that the topics considered in this chapter are not intended to represent an entirely complete or balanced view of the subject of compaction. For example, the nonlinear aspects of consolidation (Van der Knaap, 1959; Davis and Raymond, 1965; Gibson et al., 1967) are not examined. The topics chosen are common, however, to many problems of subsidence in various fields of engineering and science. They deal with some of the basic aspects of compaction and, to a large extent, reflect the authors' interest. Inasmuch as the text has evolved through the incorporation of material from a wide variety of scientific disciplines, some overlap of nomenclature is unavoidable. A nomenclature list appears on p.514. It is hoped that the symbolism in places where there is an overlap is clear, and that no undue inconvenience would result to the readers.

Basic concepts of the compaction process

In order to establish an initial concept of the process of compaction, an analysis is made of the pressures in a tank filled with unconsolidated (loose) rock material and water, which support a load on a piston (Fig.8-1). The tank is equipped with a drainage valve, V. The total load, F_t, may be supported jointly by the fluid and the rock material:

$$F_t = Ap = Ap_g + Ap_p \tag{8-1}$$

where A is the cross-sectional area of the tank, p is the pressure exerted by the piston, p_g is the grain-to-grain pressure in the rock material, and p_p is the fluid pressure. If the valve, V, is closed, no fluid can escape from the tank,

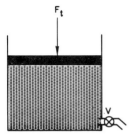

Fig.8-1. Schematic diagram of the tank compaction model. F_t = total load, V = valve.

and as the water would be much less compressible than the unconsolidated rock, the entire piston load would be borne by the fluid pressure:

$$F_t = Ap = Ap_p \qquad (8\text{-}2)$$

Thus:

$$p = p_p \qquad (8\text{-}3)$$

If the valve, V, is opened, some of the water drains out from the tank. As the fluid volume decreases, the piston descends, compacting the unconsolidated rock material and increasing grain-to-grain pressure. As the grain-to-grain pressure increases and the piston load is transferred to the rock material, the portion of the piston load borne by the fluid decreases until a point is reached where the fluid pressure is zero. All the piston load is then borne by the rock material, and the rock material will not compact any further unless the piston load is increased. In buried sediments, the load borne by the rock and fluid material is the load of the overburden rocks and their interstitial fluids, which is analogous to the piston load, F_t, in Fig.8-1. The corresponding pressure is equal to:

$$p_{ob} = (1 - \phi)\rho_g g z_s + \rho_w g z_w \phi \qquad (8\text{-}4)$$

where p_{ob} is the pressure due to the overburden load (also commonly known as overburden pressure), ϕ is the fractional porosity, ρ_g is the mass per unit volume (density) of rock grain material, ρ_w is the mass per unit volume (density) of water, g is the gravitational constant, z_s is the depth of the sediments, and z_w is the depth below the water table.

In many cases, z_s and z_w are approximately equal and eq.8-4 is written as:

$$p_{ob} = [(1 - \phi)\rho_g g + \phi \rho_w g] z_s \qquad (8\text{-}5)$$

The term in the square brackets is approximately equal to 144 lb/ft³ for common values of ρ_g and ϕ. When divided by 144 inch²/ft², a value of approximately 1 psi/ft is obtained, commonly known as the overburden "pressure gradient".

If the overburden load is analogous to the piston load in eq.8-1, then:

$$p_{ob} = [(1-\phi)\rho_g g + \phi\rho_w g]z = p_g + p_p \tag{8-6}$$

for $z = z_s = z_w$. Rearranging eq.8-6, the grain-to-grain pressure becomes equal to:

$$p_g = p_{ob} - p_p = [(1-\phi)\rho_g g + \phi\rho_w g]z - p_p \tag{8-7}$$

Equation 8-7 may be written as:

$$p_{ob} = [(1-\phi)\rho_g g + \phi\rho_w g)]z - p_p + p_p \tag{8-8}$$

If the fluid pressure is expressed as:

$$p_p = p_h + p_{ex} \tag{8-9}$$

where p_h is the hydrostatic fluid pressure due to a fluid head, z, that is equal to $\rho_w gz$, and p_{ex} is the excess fluid pressure or the pressure by which the total fluid pressure exceeds p_h, then eq.8-8 becomes:

$$p_{ob} = [(1-\phi)\rho_g g + \phi\rho_w g]z - p_h - p_{ex} + p_p \tag{8-10}$$

or:

$$p_{ob} = [(1-\phi)\rho_g g + \phi\rho_w g]z - \rho_w gz - p_{ex} + p_p \tag{8-11}$$

or:

$$p_{ob} = [(1-\phi)\rho_g g + \phi\rho_w g - \rho_w g]z - p_{ex} + p_p \tag{8-12}$$

Comparison of eq.8-12 and 8-6 shows that the sum of the first two terms on the right-hand side of eq.8-12 is equal to the grain-to-grain pressure. The grain-to-grain pressure is also known as the "net overburden pressure" and is the effective compaction pressure. Thus, the two expressions for grain-to-grain compaction pressure are as follows:

$$p_g = [(1-\phi)\rho_g g - (1-\phi)\rho_w g]z - p_{ex} \tag{8-13}$$

and:

$$p_g = [(1-\phi)\rho_g g + \phi\rho_w g]z - p_p \tag{8-14}$$

The two equations give the same answer and show how the density of the overburden is corrected for the buoyancy effect according to Archimedes Principle (the $-\rho_w g$ term in the brackets of eq.8-13).

Referring back to Fig.8-1, one can see that the rate at which compaction occurs and the piston load is transferred from the fluid to the rock material is determined by the rate at which the water is drained from the system. The rate of water drainage, in turn, is dependent on the permeability of the rock and the size of the valve. In the case of buried sediments, fluids from low-permeability layers drain into high-permeability layers. Excess fluid pressure, p_{ex}, in natural sediments is in some cases due to the fact that fluids did not have sufficient time to "bleed off". It may also be noted that the decrease in bulk volume of the rock and fluid in the tank is equal to the amount of fluid which is drained from the tank.

Terzaghi's (1923) compaction model

The Terzaghi formulation of the model governing the compaction process in a porous medium is presented here by considering the piston and cylinder assembly filled with a water-saturated sand (Fig.8-2). A constant force, F_t, is applied on the piston, causing an overburden pressure F_t/A at $z = H$. As a result, water in the sand is squeezed out and drains downward through the porous retainer at the bottom of the assembly. The cylinder walls are impermeable.

Two piezometers are fitted to the apparatus as shown in Fig.8-2. At any arbitrarily selected instant during compression, the head of water in the piezometer attached at level z is h, and the head of water in the piezometer attached at $(z + \delta z)$ is $(h + \frac{\partial h}{\partial z} \delta z)$. At the same instant, the weight rate of flow of water vertically downward at level, z, is W. At level $(z + \delta z)$, the corresponding instantaneous rate is $(W + \frac{\partial W}{\partial z} \delta z)$, assuming that δz is fixed and sufficiently small. At the face of the piston, the rate of water flow is zero, whereas at the bottom of the sand, where $z = 0$, it is a maximum. On this basis, the instantaneous rate at which the element of sand between z and $(z + \delta z)$ becomes depleted of water is $-(\partial W/\partial z)\delta z$. This rate of depletion is in lb/sec if W is expressed in lb/sec.

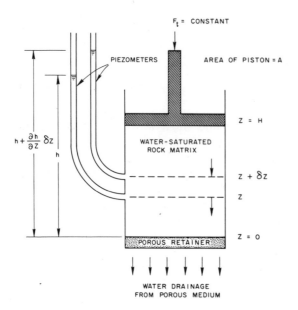

Fig.8-2. Consolidation model of a water-saturated sand.

The weight of water in the element of sand at the instant of observation is equal to $V_p \rho_w g$, where V_p is the instantaneous pore volume, ρ_w is the density of water, and g is the gravitational acceleration. If this weight is differentiated with respect to time, the instantaneous weight rate of depletion of water in the element is found to be $-\rho_w g(\partial V_p/\partial t)$, the minus sign indicating that the weight of water in the element, δz, is a decreasing function of time. The density of water, ρ_w, is assumed to be constant.

The continuity equation is then obtained by equating the weight rate of depletion in the following relation:

$$-\frac{\partial W}{\partial z} \delta z = -\rho_w g \frac{\partial V_p}{\partial t} \tag{8-15}$$

According to Darcy's Law:

$$W = \frac{Ak(\rho_w g)^2}{\mu_w} \frac{\partial h}{\partial z} \tag{8-16}$$

where k is the permeability of the medium and μ_w is the viscosity of the water. Substitution of the right hand side of eq.8-16 for W in eq.8-15 results in:

$$-\left\{\frac{\partial}{\partial z}\left[\frac{Ak(\rho_w g)}{\mu_w}\frac{\partial h}{\partial z}\right]\right\}_t \delta z = -\left(\frac{\partial V_p}{\partial t}\right)_z \tag{8-17}$$

To evaluate the right-hand side of eq.8-17, the chain rule of differentiation may be applied to show that:

$$\left(\frac{\partial V_p}{\partial t}\right)_z = \left(\frac{\partial V_p}{\partial p_p}\right)_z \left(\frac{\partial p_p}{\partial t}\right)_z \tag{8-18}$$

in which p_p is the pore water pressure at level z in the piston and cylinder assembly. The head of water in the piezometer corresponding to level z is h; hence,

$$p_p = p_a + (h - z)\rho_w g \tag{8-19}$$

in which p_a is atmospheric pressure. Differentiating eq.8-19 with respect to time, t, one obtains:

$$\left(\frac{\partial p_p}{\partial t}\right)_z = \rho_w g \left(\frac{\partial h}{\partial t}\right)_z \tag{8-20}$$

Substituting in eq.8-18 the value of $\left(\frac{\partial p_p}{\partial t}\right)_z$ given by eq.8-20, yields:

$$\left(\frac{\partial V_p}{\partial t}\right)_z = \left(\frac{\partial V_p}{\partial p_p}\right)_z \rho_w g \left(\frac{\partial h}{\partial t}\right)_z \tag{8-21}$$

The isothermal coefficient of compressibility of a porous medium, c_p, is defined by Ramey (1964):

$$c_p = +\frac{1}{V_p}\left(\frac{\partial V_p}{\partial p_p}\right)_T \tag{8-22}$$

where T is a constant temperature. Equation 8-22 can be presented as:

$$\left(\frac{\partial V_p}{\partial p_p}\right)_T = c_p V_p \tag{8-23}$$

A more detailed discussion of pore volume compressibility is given later.

Before proceeding further, the sign convention used in eq.8-22 requires clarification. As fluid pressure is increased within the pores of the porous medium, the rock grains are compressed and the bulk volume of the rock is increased due to the interaction of the internal fluid pressure and the external stress. The effect of both factors is to increase pore volume as fluid pressure increases. Thus, by the general definition of compressibility, the rock compressibility in terms of pore volume should be a negative quantity. The effect of an increase in pore volume with an increase in pressure, however, can also be viewed as an increase in the size of a container holding the fluid in the porous medium. If the container volume is held constant, the same effect can be obtained by an additional shrinking of fluids. Thus, the pore volume compressibility may be considered a positive quantity additive to the fluid compressibility. In this chapter, this interpretation of the definition of pore volume compressibility is used.

Substituting the value of $(\partial V_p / \partial p_p)$ from eq.8-23 in eq.8-21, the following relation is obtained:

$$\left(\frac{\partial V_p}{\partial t}\right)_z = c_p \, V_p \rho_w g \left(\frac{\partial h}{\partial t}\right)_z \tag{8-24}$$

Finally, substitution of the right-hand side of eq.8-24 for $(\partial V_p / \partial t)_z$ in eq.8-17, gives the basic differential equation of the compaction process:

$$\frac{\partial}{\partial z}\left(\frac{Ak}{\mu_w}\frac{\partial h}{\partial z}\right)\delta z = c_p \, V_p \frac{\partial h}{\partial t} \tag{8-25}$$

in which the subscripts have been dropped for convenience. Inasmuch as $A \cdot \delta z$ is the fixed bulk volume, V_b, of the element, and assuming that k and μ_w are both constant, eq.8-25 may be expressed as:

$$\frac{\partial^2 h}{\partial z^2} = \left(\frac{\mu_w}{k}\right) c_p \left(\frac{V_p}{V_b}\right) \frac{\partial h}{\partial t} \tag{8-26}$$

In soil mechanics the "coefficient of volume compressibility", m_v, is often used. It is defined as:

$$m_v = -\frac{1}{V_b}\left(\frac{\partial V_p}{\partial p_g}\right) \tag{8-27}$$

where p_g is the "grain-to-grain pressure" or net overburden pressure. To relate c_p in eq.8-26 to m_v in eq.8-27, it is necessary to note first that:

$$p_g = p_{ob} - p_p \tag{8-28}$$

where p_{ob} is the overburden pressure and p_p is the fluid pressure. At any level z, the overburden pressure is equal to $[(F_t/A) + (h-z)\rho_b g]$, where ρ_b is the bulk density of the water-saturated sediment. Grain-to-grain pressure, p_g, is equal to the net overburden pressure. Referring to eq.8-22:

$$\frac{\partial V_p}{\partial p_p} = \frac{\partial V_p}{\partial p_g} \frac{\partial p_g}{\partial p_p} \tag{8-29}$$

Differentiating eq.8-28 with respect to p_p:

$$\left(\frac{\partial p_g}{\partial p_p}\right)_z = -1 \tag{8-30}$$

Thus, c_p in eq.8-22 may be expressed as:

$$c_p = -\frac{1}{V_p}\left(\frac{\partial V_p}{\partial p_g}\right)_T \tag{8-31}$$

Division of eq.8-31 by eq.8-27 leads to:

$$c_p = m_v \frac{V_b}{V_p} \tag{8-32}$$

Substituting for c_p in eq.8-26, its value given by eq.8-32:

$$\frac{\partial^2 h}{\partial z^2} = \left(\frac{\mu_w}{k}\right) m_v \frac{\partial h}{\partial t} \tag{8-33}$$

In soil mechanics the "coefficient of consolidation", c_v, is defined as:

$$c_v = \frac{k}{\mu_w m_v} \tag{8-34}$$

Equation 8-33, therefore, may be expressed in a compact form:

$$\frac{\partial h}{\partial t} = c_v \frac{\partial^2 h}{\partial z^2} \tag{8-35}$$

Equation 8-35 is the differential equation which describes mathematically the process of compaction. If k is in ft^2, μ is in (lb sec)/ft^2, and m_v is in ft^2/lb, then c_v is in ft^2/sec (eq.8-34). In eq.8-35, $(\partial h/\partial t)$ is in ft/sec and $(\partial^2 h/\partial z^2)$ is in ft^{-1}.

Compressibility of porous media

As fluid is expelled from within the pores of the reservoir rock, a change in volume of the fluid and rock takes place. If, for example, the external or overburden pressure is fixed and if the fluid movement out of the pores is caused by a reduction in fluid pressure, as might be caused by a well penetrating the rock, then the fluid would expand, the rock matrix (solids only) would expand, and both the bulk volume and pore volume of the rock would shrink.

A number of instances exist wherein the volume change of the rock solid material is neglected. Changes in bulk rock properties, such as porosity and permeability, take place as a result of compaction. A considerable amount of experimental data on compressibility of rocks is available, exemplified by the works of Adams and Williamson (1923), Botset and Reed (1935), Hall (1953), Fatt (1958a, b), and Cleary (1959). Despite the large number of experimental studies, however, much confusion exists in the literature concerning compressibility.

Geertsma (1957a) recognized the following three types of compressibility of porous media:

(1) Rock matrix compressibility, c_r, which represents the change in the volume of the solid material (excluding the pores) per unit change in pressure, may be expressed as:

$$c_r = + \frac{1}{V_r} \frac{dV_r}{dp_p} \tag{8-36}$$

where V_r is the volume of the rock matrix and p_p is the fluid pressure.

(2) Rock bulk compressibility is the fractional change in bulk or total volume per unit change in uniform pressure.

(3) Pore compressibility is the fractional change in pore volume per unit change in pressure.

While considering the second and third definitions of compressibility, two types of stress variations have to be considered. The first is the internal, or pore stress, variation, wherein the fluid pressure in the pores is varied while the external stress is kept constant. The pore stress is due to the fluid pressure and is, therefore, always hydrostatic. The second stress variation is the external, or bulk stress, variation, where the external stresses are varied and the pore pressure is kept constant. The outside stress, however, may be a result of rock stresses (due to surrounding formations) and thus can have different values in various directions.

The stress tensor (Fig.8-3) for a homogeneous, isotropic matrix, which has a structure of continuous matter, can be presented in the conventional way:

$$S \equiv \begin{bmatrix} \sigma_x & \tau_{xy} & \tau_{xz} \\ \tau_{yx} & \sigma_y & \tau_{yz} \\ \tau_{zx} & \tau_{zy} & \sigma_z \end{bmatrix} \qquad (8\text{-}37)$$

where σ_x, σ_y, and σ_z represent the normal stresses in x, y, and z coordinate directions and τ's are the shear stresses acting on a cube of porous material (Biot, 1941a; Van der Knaap, 1959). As only isotropic situations are considered here, $\tau_{yx} = \tau_{xy}$, $\tau_{zy} = \tau_{yz}$, etc. This representation implicitly assumes that the elemental volume considered is large enough compared to the size of the pores, so that the porous medium may be treated as being homogeneous, and at the same time small enough, compared to the scale of the macroscopic phenomena of interest, so that it may be considered to be infinitesimal in the mathematical treatment. Geertsma (1957a) and Van der Knaap (1959) have shown that this tensor can be broken up into hydrostatic and deviatoric components:

$$P \equiv \begin{bmatrix} \overline{\sigma} & 0 & 0 \\ 0 & \overline{\sigma} & 0 \\ 0 & 0 & \overline{\sigma} \end{bmatrix} \qquad (8\text{-}38)$$

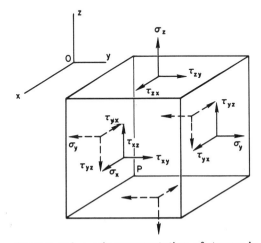

Fig.8-3. Schematic representation of stresses in a porous medium.

where:

$$\bar{\sigma} = \frac{\sigma_x + \sigma_y + \sigma_z}{3} \tag{8-39}$$

and:

$$D \equiv \begin{bmatrix} \sigma_x - \bar{\sigma} & \tau_{xy} & \tau_{xz} \\ \tau_{yx} & \sigma_y - \bar{\sigma} & \tau_{yz} \\ \tau_{zx} & \tau_{zy} & \sigma_z - \bar{\sigma} \end{bmatrix} \tag{8-40}$$

The hydrostatic stress causes only volume changes, whereas the deviatoric stress component gives rise to a change in shape only. As shown in eq.8-40, the sum of the stresses along the diagonal vanish. It is to be noted that $\bar{\sigma}$ represents the average of the normal forces on the matrix and fluid per unit bulk area (solid and pores) in the three principal directions. Geertsma (1957a) has shown that compressibility effects are completely determined by the value of the average stress, $\bar{\sigma}$. On considering a volume element representative of the porous medium, the pore volume, V_p, and the bulk volume, V_b, on the basis of the above discussion, would be a function of the average stress, $\bar{\sigma}$, and the pore pressure, p_p. Accordingly, a change in the pore volume, V_p, and the bulk volume, V_b, are given, respectively, by:

$$dV_p = \left(\frac{\partial V_p}{\partial p_p}\right)_{\bar{\sigma}} dp_p + \left(\frac{\partial V_p}{\partial \bar{\sigma}}\right)_{p_p} d\bar{\sigma} \tag{8-41}$$

and:

$$dV_b = \left(\frac{\partial V_b}{\partial p_p}\right)_{\bar{\sigma}} dp_p + \left(\frac{\partial V_b}{\partial \bar{\sigma}}\right)_{p_p} d\bar{\sigma} \tag{8-42}$$

where the partial derivatives are, in general, functions of pressure and stress. The average external stress is $\bar{\sigma}$ and the pore stress is p_p. According to Geertsma (1957a), when the homogeneous porous matrix is under a fluid pressure, p_p, and an external stress, $\bar{\sigma}$, when there is no shearing stress, and if body forces are absent, then:

$$\frac{dV_p}{V_p} = \frac{dV_b}{V_b} = \frac{dV_r}{V_r} = c_r dp = c_r d\bar{\sigma} \tag{8-43}$$

where V_r and c_r are the rock-matrix volume and compressibility, respectively. Implicitly assuming that the porosity is constant, and that as $\bar{\sigma}$ and p_p

are the only two forces acting, Geertsma (1957a) suggested that any change in one should counteract the other, i.e., $d\bar{\sigma} = dp_p$ or $(\bar{\sigma} - p_p)$ = constant. Substituting for dV_p and dV_b in eq.8-43 from eq.8-41 and 8-42, an expression for c_r may be written as:

$$c_r = \frac{1}{V_p}\left(\frac{\partial V_p}{\partial \bar{\sigma}}\right)_{p_p} + \frac{1}{V_p}\left(\frac{\partial V_p}{\partial p_p}\right)_{\bar{\sigma}} \tag{8-44}$$

or:

$$c_r = \frac{1}{V_b}\left(\frac{\partial V_b}{\partial \bar{\sigma}}\right)_{p_p} + \frac{1}{V_b}\left(\frac{\partial V_b}{\partial p_p}\right)_{\bar{\sigma}} \tag{8-45}$$

Inasmuch as two hydrostatic loadings are considered, the reciprocal theorem of Betti and Rayleigh (in: Love, 1934, pp.16, 44), which is based on energy considerations, would apply. The reciprocal theorem as stated by Geertsma (1957a) is as follows: "The whole work done by the forces of the first set (as a result of $d\bar{\sigma}$) acting over the displacements produced by the second set (as a result of dp_p) is equal to the whole work done by forces of the second set acting over the displacements produced by the first." Thus, using eq.8-41 and 8-42, this theorem leads to:

$$\left(\frac{\partial V_b}{\partial p_p}\right)_{\bar{\sigma}} = -\left(\frac{\partial V_p}{\partial \bar{\sigma}}\right)_{p_p} \tag{8-46}$$

One may define the bulk compressibility from eq.8-42 as (Geertsma, 1957a):

$$c_b = \frac{1}{V_b}\left(\frac{\partial V_b}{\partial \bar{\sigma}}\right)_{p_p} \tag{8-47}$$

This bulk compressibility can be determined through measurements of the bulk volume changes of a jacketed sample under variable hydrostatic pressure, keeping the pore pressure constant. The matrix-rock compressibility, c_r, can be determined by removing the jacket from the core and allowing the compressing liquid to enter the pores. Using the definitions discussed above, Geertsma (1957a) has shown that:

$$\frac{dV_p}{V_p} = -\frac{1}{\phi}[c_b - (1+\phi)c_r]\,dp_p + \frac{1}{\phi}(c_b - c_r)\,d\bar{\sigma} \tag{8-48}$$

and:

$$\frac{dV_b}{V_b} = -(c_b - c_r)\,dp_p + c_b\,d\bar{\sigma} \tag{8-49}$$

where ϕ is the porosity (ratio of the pore volume to the bulk volume). As pointed out by De Witte and Warren (1957), the above discussion (eq.8-41 and 8-47) implicitly assumes that the changes in porosity are negligible. This assumption in some instances may be incompatible with the theorem of Betti and Raleigh. This assumption, however, is only of minor importance and can be avoided, if necessary (Biot, 1941a). If $c_b \gg c_r$, these equations can be simplified:

$$\frac{dV_p}{V_p} = \frac{c_b}{\phi} (d\bar{\sigma} - dp_p) \qquad (8\text{-}50)$$

and:

$$\frac{dV_b}{V_b} = c_b (d\bar{\sigma} - dp_p) \qquad (8\text{-}51)$$

The above definitions of compressibility are by no means unique. For example, Hall (1953) and Fatt (1958a, b) used the following expression for the pore volume compressibility (these authors called it pore volume compaction):

$$c_p = \frac{1}{V_p} \left(\frac{\partial V_p}{\partial \bar{\sigma}} \right)_{p_p} \qquad (8\text{-}52)$$

whereas Chierici et al. (1967) used the equation:

$$c_p = -\frac{1}{V_p} \left(\frac{\partial V_p}{\partial \bar{\sigma}} \right)_{p_p} \qquad (8\text{-}53)$$

The above examples are given to demonstrate that the term "rock compressibility" used without qualification can be numerically almost meaningless or, at best, be wrongly interpreted. Thus, the measuring conditions need to be specified. Fatt (1958b) used the following definitions for the bulk compressibility and pseudo-bulk compressibility, respectively:

$$c_b = \frac{1}{V_b} \left(\frac{\partial V_b}{\partial \bar{\sigma}} \right)_{p_p} \qquad (8\text{-}54)$$

and:

$$c_{bt} = \frac{1}{V_b} \left(\frac{\partial V_b}{\partial p_p} \right)_{\bar{\sigma}} \qquad (8\text{-}55)$$

The pore compressibilities are defined by Fatt as:

$$c_p = \frac{1}{V_p}\left(\frac{\partial V_p}{\partial \bar{\sigma}}\right)_{p_p} \tag{8-56}$$

and:

$$c_{pt} = \frac{1}{V_p}\left(\frac{\partial V_p}{\partial p_p}\right)_{\bar{\sigma}} \tag{8-57}$$

and the matrix compressibility was expressed as:

$$c_r = \frac{1}{(V_b - V_p)}\left[\frac{\partial(V_b - V_p)}{\partial p_p}\right] \tag{8-58}$$

It is stressed that the above definitions (eq.8-54 and 8-58) are those of Fatt.

In addition to distinguishing between various measurements of compressibility, Fatt has also provided petrographic descriptions of the samples studied by him. This provides an additional advantage, particularly in the correlation of experimental results. Based on consideration of stresses and strains, Knutson and Bohor (1963) have introduced yet another definition of compressibility as follows:

$$c_p = \frac{1}{V_p}\left(\frac{\partial V_p}{\partial p_p}\right)_{\sigma_z, \epsilon_x, \epsilon_y} \tag{8-59}$$

where ϵ_x and ϵ_y are the strains in the x and y directions, respectively, and σ_z is the stress in the vertical direction. This definition of pore volume compressibility is most commonly used in practice (Ramey, 1964; Matthews and Russell, 1967, p.7). Normally isothermal conditions are encountered and, thus, the subscript T is also added. Unless otherwise mentioned, in this chapter eq.8-59 will be used as the definition of pore volume compressibility. An aspect which has not been discussed so far is the determination of the change in hydrostatic stress, $d\bar{\sigma}$, as a function of the pore pressure variation, dp_p. In order to determine this relationship, the boundary conditions have to be determined. The following possibilities need to be considered (Geertsma 1957a):

(a) Specification of principal stresses in the rock bulk material. For example, if the stress on the porous matrix is specified to be uniform, then $d\bar{\sigma} = 0$. This condition is most commonly encountered in laboratory experiments.

(b) Change in deformation of bulk volume. If the boundary deformations are given, then dV_b/V_b can be calculated and eq.8-49 can be used to determine the change in hydrostatic stress.

(c) Specification of a change in the value of principal stresses in a given number of principal stress directions, and the change in deformation in the remaining principal directions. This implies that the change in lengths have to be specified. This kind of situation is very likely to occur in nature. For example, in an oil reservoir the deformation in the horizontal directions is negligible and the stress in the vertical direction is constant.

COMPACTION AND REBOUND (EXPANSION) OF ONE-DIMENSIONAL POROUS COLUMNS

A review of the progress made in the last five years in the field of consolidation and rebound expansion of one-dimensional porous columns is presented, and modern methods of analysis are examined in this section. Darcy's Law and its implication as related to a mobile matrix are also examined. The specific problem considered is a one-dimensional shrinkage or expansion of a porous system to a single-step function change in pressure at the base of the column, which may be considered to be either finite or semi-infinite. The term, single-step function, implies that the pressure at the base of the column is changed instantaneously from an equilibrium or initial value to a lower (compaction) or higher (rebound expansion) value.

The theory presented here includes general forms of formation conductivity and pressure, and couples the movement of soil particles and the associated mass flow of liquid. It is shown that the latter effect is extremely important. Further, the differences between compaction and expansion processes are examined. Even though a non-linear equation results, analytical solutions exist for one class of problems. These analytical solutions are used to illustrate some of the basic aspects of a compaction or compaction process. The non-linear equation has also been solved numerically and the solutions obtained have been correlated with constant property solutions. As mentioned in the Introduction (p.405), the correlation is achieved through the use of a transformation. The transformation provides a powerful approach for developing rapidly a "correlation" solution to a rather difficult non-linear problem. These "correlation" solutions are estimates which should suffice for engineering or practical purposes. The effects of the overburden and gravity are examined separately. One aspect of compaction and rebound, which has resulted in a considerable confusion, is the choice of either a moving or stationary coordinate system to develop the mathematical model. This aspect is also examined in detail. The various elements of this problem are considered next.

Darcy's Law for the flow of liquids through deformable porous media

The use of Darcy's Law to describe the flow of liquids through non-deformable porous media is fairly well established. This law states that the macroscopic velocity of the fluid in the direction of flow at any particular point within the porous medium is proportional to the instantaneous gradient in fluid head or flow potential at that point. Symbolically, this may be expressed as:

$$\vec{U}_{ma} = -\frac{k\rho}{\mu}\nabla\Phi \quad (8\text{-}60)$$

where $\nabla \equiv \frac{\partial}{\partial x} + \frac{\partial}{\partial y} + \frac{\partial}{\partial z}$, $\vec{U}_{ma}$ is the macroscopic flow velocity, k is the permeability, μ is the viscosity of the liquid, and Φ is the flow potential. It should be noted that both k and μ may be considered as functions of the flow potential, which can be expressed as follows:

$$\Phi = g(z - z_o) + \int_{p_{ps}}^{p_p} \frac{dp_p}{\rho} + \frac{\overline{U}^2}{2} \quad (8\text{-}61)$$

where p_p is the fluid pressure, ρ is the liquid density, g is the acceleration due to gravity, $\overline{U}$ is the average microscopic velocity of the liquid and $\overline{U}^2/2$ represents the kinetic energy per unit mass of the liquid, and z is the elevation above the arbitrarily selected datum, z_o. The standard-state pressure at $z = z_o$ is p_{ps}. Equation 8-61 also assumes that $\overline{U}^2 = 0$ at z_o. Thus Φ is equal to zero at z_o, the datum. Because of the large surface area, viscous forces generally control the motion of fluids through porous media, and the flow is viscous. As a consequence, the kinetic energy term on the right-hand side of eq.8-61 may be dropped and the potential Φ can be expressed as:

$$\Phi = g(z - z_o) + \int_{p_{ps}}^{p_p} \frac{dp_p}{\rho} \quad (8\text{-}62)$$

$\vec{U}_{ma}$ is a macroscopic velocity and, as such, is not the true velocity of the liquid at any particular point. In fact, the calculation of true velocities within the porous medium is intractable, for the geometry of the porous channels is so complicated that it is impossible to specify them. An average microscopic velocity across a section can be determined if the porosity, ϕ, of the rock matrix is known:

$$\vec{U}_{mi} = -\frac{k\rho}{\mu\phi}\nabla\Phi \tag{8-63}$$

where $\vec{U}_{mi}$ is the microscopic velocity. As ϕ represents a statistical average value of the porosity, $\vec{U}_{mi}$ is essentially a statistically averaged microscopic velocity. A more detailed examination of the above concepts as related to non-deformable media may be found in extensive discussions of Darcy's Law by Hubbert (1940, 1956).

The above specification of Darcy's Law does not bring out one of the most important facets of this law, which is essential to the problem under consideration. In the above enunciation of Darcy's Law it is assumed that the velocity is measured with respect to a fixed coordinate system outside the porous medium and, thus, it is assumed that the porous medium does not move or distort. Only under this condition would the measured velocity correspond to that given by eq.8-60. This implies that Darcy's Law applies to fluid flow velocities relative to the solid matrix only and that for a deformable porous medium eq.8-60 can be generalized to:

$$\vec{U}_{ma,r} = -\frac{k\rho}{\mu}\nabla\Phi \tag{8-64}$$

where $\vec{U}_{ma,r}$ is the macroscopic velocity relative to the motion of the solid particles. If the flow velocity is not measured relative to the solid matrix, the value of the permeability calculated would be meaningless, as it would depend on the magnitude of the translation at various points of the grains of the porous medium or of its solid matrix. Extremely different calculated permeabilities result from fixed fluid-head gradients of different magnitude if the sand grains move or translate at different velocities. To the knowledge of the authors, the concept that Darcy's Law applies to fluid flow velocities relative to the solid matrix was first put forward by Gerzevanov in 1937 (see also Florin, 1948). In 1956, Biot mentioned that Darcy's Law applies relative to the movement of the sand grains. More recently this concept has been developed independently by Zaslavsky (1964), Cooper (1966), and Smith (1971).

Formulation and extension of the Terzaghi problem

The compaction system considered next is shown in Fig.8-4. The porous medium is imagined as a water-saturated skeleton of particles wherein both solid and liquid may move. First, a semi-infinite system is considered with the base ($z = 0$) of the skeleton being rigid. Attention is restricted to the one-dimensional movement of liquid and solid in the non-rigid porous

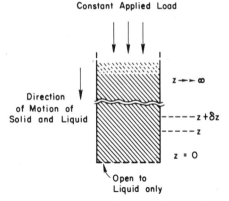

Fig.8-4. Schematic diagram of a model of water-saturated skeleton of particles wherein both the solid matrix and the liquid can move. The bottom of the system is rigid. The dashed lines represent infinity.

medium, which is free to expand or contract in the direction of fluid movement, but is laterally confined. The effects of gravity and overburden are considered later. The liquid is free to pass into or out of the system through the base; however, the solid particles cannot move through the base. The motion of the solid and liquid can be related by Darcy's Law. An elemental volume $A\delta z$ located between the planes z and $z + \delta z$ (in a fixed coordinate system) is considered. The area A of the elemental volume is assumed to be normal to the plane of the paper. As in all mathematical analyses, it is implicitly assumed that the elemental volume is large enough compared to the size of the pores so that the porous medium may be treated as being homogeneous and at the same time small enough, compared to the scale of the macroscopic phenomena of interest, so that it may be considered to be infinitesimal in the mathematical sense. The area A of the elemental volume may be thought of as a composite of rock and fluid where the rock occupies an area proportional to $1 - \phi$, whereas the liquid occupies an area proportional to ϕ, which is a function of z and t.

The equation of continuity for the rock in the volume $A\delta z$ may be written as:

$$\frac{\partial [U_r \rho_r (1-\phi)]}{\partial z} = - \frac{\partial [\rho_r (1-\phi)]}{\partial t} \tag{8-65}$$

where $U_r(z, t)$ is the absolute velocity of the rock relative to the fixed z axis, ρ_r is the density of the rock matrix, and t is the time. If the rock grains are assumed incompressible, eq.8-65 can be integrated with respect to z from a to b to yield:

$$[U_r(1-\phi)]_{z=b} - [U_r(1-\phi)]_{z=a} = \int_a^b \frac{\partial \phi}{\partial t} dz \qquad (8\text{-}66)$$

at an arbitrary time t. Noting that the model specifies no movement of solids through the base, the absolute velocity, U_r, of the rock must be zero at $z = 0$, at the bottom of the column of porous medium. Thus, if the lower limit $a = 0$ and if b is taken as an arbitrary level z, eq.8-66 reduces to:

$$U_r(z,t) = \frac{1}{[1-\phi(z,t)]} \int_0^z \frac{\partial \phi}{\partial t} dz \qquad (8\text{-}67)$$

It is to be noted that the absolute velocity of the rock is a function of the porosity, ϕ.

The movement of the liquid which is also assumed to be incompressible can be analyzed similarly for a range of z from a to b to find that:

$$(U_L\phi)_{z=a} - (U_L\phi)_{z=b} = \int_a^b \frac{\partial \phi}{\partial t} dz \qquad (8\text{-}68)$$

at an arbitrary time t. Again, if $a \to 0$, where $U_L \neq 0$, and if b is taken as an arbitrary level z, eq.8-68 reduces to:

$$(U_L\phi)_{z=0} - U_L\phi = \int_0^z \frac{\partial \phi}{\partial t} dz \qquad (8\text{-}69)$$

at an arbitrary time t. Equation 8-69 may be rearranged as:

$$U_L(z,t) = \frac{1}{\phi(z,t)} (U_L\phi)_{z=0} - \frac{1}{\phi(z,t)} \int_0^z \frac{\partial \phi}{\partial t} dz \qquad (8\text{-}70)$$

where $U_L(z,t)$ is the microscopic or absolute velocity of the liquid which is a function of z and t. The first term on the right-hand side of eq.8-70 would be equal to zero if the base were impermeable to the liquid. Darcy's Law for the system under consideration is written (macroscopic form) as:

$$u_{Lr}(z,t) = +\frac{k}{\mu}\frac{\partial p_p}{\partial z} = -\frac{k}{\mu}\frac{\partial \psi}{\partial z} \qquad (8\text{-}71)$$

where $u_{Lr}(z, t)$ is the macroscopic velocity of the liquid relative to the rock matrix, p_p is the fluid-pressure, and ψ is the moisture potential. Both p_p and ψ are dependent upon z and t. Equation 8-71 requires some clarification. It is identical to eq.8-64 except the fluid is assumed to be incompressible and gravity effects are neglected. The symbolism for the macroscopic velocity has been changed from $U_{ma,r}$ to u_{Lr}, mainly for purposes of convenience. The function, ψ, is defined to be identical to $-p_p$. The function ψ is commonly used in the soil physics literature where it is known as the swelling pressure or moisture potential. Later, it will be seen that experimental data are available mainly in terms of ψ, and that the use of this function is mathematically convenient.

The microscopic fluid velocity relative to the matrix is denoted by U_{Lr}, which is related to u_{Lr} through eq.8-63. If the permeability k is assumed to be a unique function of ϕ, then eq.8-71 may be rewritten as:

$$u_{Lr}(z, t) = -D \frac{\partial \phi}{\partial z} \tag{8-72}$$

where D is:

$$D(\phi) \equiv \frac{k(\phi)}{\mu} \frac{d\psi}{d\phi} \tag{8-73}$$

It is to be noted that D is a function of z and t and with respect to fluid flow in a deformable porous medium it is analogous to the diffusivity of a non-deformable porous medium.

The microscopic velocity of the flow of liquid relative to the solid can be calculated by subtracting eq.8-70 from eq.8-67. The corresponding relative macroscopic velocity can also be obtained by multiplying this difference by the porosity, ϕ. That is, the macroscopic velocity of the liquid relative to the solid can also be written as:

$$u_{Lr}(z, t) = \phi(z, t) U_{Lr}(z, t) = \phi(z, t) [U_L(z, t) - U_r(z, t)] \tag{8-74}$$

Using eq.8-67, 8-70, 8-72, and 8-73, eq.8-74 may be rewritten after simplification as:

$$(U_L \phi)_{z=0} - \left[\frac{1}{1 - \phi(z, t)} \right] \int_0^z \frac{\partial \phi}{\partial t} dz = -D \frac{\partial \phi}{\partial t} \tag{8-75}$$

Inasmuch as the right-hand members of eq.8-66 and 8-68 are the same, the left-hand members can be equated for any arbitrary time, t. Thus, the following is obtained:

$$(U_L \phi)_{z=a} + [U_r(1-\phi)]_{z=a} = (U_L \phi)_{z=b} + [U_r(1-\phi)]_{z=b} \quad (8\text{-}76)$$

The first term on the left-hand side of eq.8-76 is the volumetric rate of flow of liquid per unit of bulk area perpendicular to the direction of flow. Similarly, the second term is the volumetric rate of flow of rock per unit of bulk area transverse to the direction of flow. Thus, as a whole, the left-hand side of eq.8-76 is the volumetric rate of flow per unit of bulk area of liquid plus rock at level $z = a$, at time t. The right-hand side of the equation correspondingly is the volumetric rate of flow of liquid plus rock at the level $z = b$, at time t. Inasmuch as a and b were selected arbitrarily, eq.8-76 discloses that the total volumetric rate of flow of liquid plus rock is the same at all levels of the column of porous medium, at time t. If $a \to 0$ and if b represents an arbitrary z, eq.8-76 can be written as:

$$(U_L \phi)_{z=0} = U_L \phi + U_r(1-\phi) \quad (8\text{-}77)$$

In eq.8-77 and in the subsequent equations it shall be implicitly assumed that U_L, U_r, ϕ, etc., are functions of z and t. Applying eq.8-74 to the level where $z = 0$, it can be found that the left-hand member of eq.8-77 is equivalent to $(u_{Lr})_{z=0}$. Thus, at $z = 0$ and considering eq.8-72, it can be written as:

$$(u_{Lr})_{z=0} = [U_L \phi + U_r(1-\phi)]_{z=0} = -\left(D \frac{\partial \phi}{\partial z}\right)_{z=0} \quad (8\text{-}78)$$

Inasmuch as only liquid can flow out of the system, it follows from eq.8-78 that:

$$(u_{Lr})_{z=0} = (U_L \phi)_{z=0} = -\left(D \frac{\partial \phi}{\partial z}\right)_{z=0} \quad (8\text{-}79)$$

for all time t. Substituting the second expression on the right-hand side of eq.8-79 for $(U_L \phi)_{z=0}$ in eq.8-75, the following is obtained:

$$\int_{z=0}^{z} \frac{\partial \phi}{\partial t} \, dz = (1-\phi) \left[D \frac{\partial \phi}{\partial z} - \left(D \frac{\partial \phi}{\partial z}\right)_{z=0} \right] \quad (8\text{-}80)$$

Equation 8-80 is the fundamental equation describing the flow of liquids in a deformable porous medium. Differentiation of eq.8-80 with respect to z yields:

$$\frac{\partial \phi}{\partial t} = \frac{\partial}{\partial z}\left[(1-\phi)D\frac{\partial \phi}{\partial z}\right] + \left(D\frac{\partial \phi}{\partial z}\right)_{z=0} \frac{\partial \phi}{\partial z} \tag{8-81}$$

The second term on the right-hand side of eq.8-80 may be simplified further by differentiating with respect to z through the use of eq.8-78, and the following equation results:

$$\frac{\partial}{\partial z}\left[(1-\phi)^{-1}\int_0^z \frac{\partial \phi}{\partial t} dz\right] = \frac{\partial}{\partial z}\left(D\frac{\partial \phi}{\partial z}\right) \tag{8-82}$$

It should be emphasized that eq.8-80, 8-81, and 8-82 are all equivalent.

Comparison of the rigorous theory with Terzaghi's (1923) consolidation theory

It has been mentioned that the fluid pressure p_p in Terzaghi's analysis is identical to $-\psi$. Rearranging and carrying out an analysis similar to that in the previous section with p_p as the dependent variable, one obtains an expression equivalent to eq.8-80:

$$\frac{\partial p_p}{\partial t} = \left(\frac{\partial p_p}{\partial \phi}\right)\frac{\partial}{\partial z}\left[\frac{(1-\phi)k}{\mu}\frac{\partial p_p}{\partial z}\right] + \left(\frac{k}{\mu}\frac{\partial p_p}{\partial z}\right)_{z=0}\frac{\partial p_p}{\partial z} \tag{8-83}$$

For purposes of demonstration, the permeability of medium and viscosity of the liquid are treated as functions of pressure. As already seen, the isothermal pore volume compressibility can be defined as:

$$c_p = \frac{1}{V_p}\left(\frac{\partial V_p}{\partial p_p}\right)_T \tag{8-84}$$

Substituting $Az\phi$ for V_p, noting that A is a constant, simplifying, and substituting the resulting expression for $\partial p_p/\partial \phi$ in eq.8-83, the following equation is obtained:

$$\phi\left(c_p - \frac{1}{z}\frac{\partial z}{\partial p_p}\right)\frac{\partial p_p}{\partial t} = \frac{\partial}{\partial z}\left(\frac{k}{\mu}\frac{\partial p_p}{\partial z}\right) +$$

$$+ \phi \left(c_p - \frac{1}{z} \frac{\partial z}{\partial p_p} \right) \left(\frac{k}{\mu} \frac{\partial p_p}{\partial z} \right)_{z=0} \frac{\partial p_p}{\partial z} - \frac{\partial}{\partial p_p} \left(\frac{\phi k}{\mu} \frac{\partial p_p}{\partial z} \right) \qquad (8\text{-}85)$$

The equation developed by Terzaghi, assuming k and μ to be constant, is:

$$c_p \phi = \frac{\partial p_p}{\partial t} \frac{k}{\mu} \frac{\partial^2 p_p}{\partial z^2} \qquad (8\text{-}86)$$

Comparison of eq.8-85 and 8-86 shows the effect of including the motion of the solid matrix. The solutions obtained with both analyses are examined later. Equations 8-80 and 8-83 are identical, except that the dependent variable is different.

A general solution to the problem of flow through a semi-infinite deformable porous medium

In this section a fundamental understanding of the problem of flow through a semi-infinite deformable porous medium is sought. The simple problem of a step function change in pressure (or pore volume) at the base $z = 0$ is considered. Initially, the skeleton is at equilibrium, with initial uniform values of porosity ϕ_i and potential ψ_i. The new potential at $z = 0$ is ψ_0 and the corresponding porosity is ϕ_0. As stated previously, a semi-infinite system is assumed. Thus, if the analysis were to be applied to a finite column, it would be valid only for a period $0 \leqslant t \leqslant T^1$, where T^1 is the time when significant changes in ϕ and ψ occur at the surface (top of the finite column). The analysis presented in this section is similar to the classic work of Philip (1968).

In the present instance solutions may be obtained by solving either eq.8-80 or 8-83. Examination of eq.8-80 and 8-83 shows that eq.8-80 is of a simpler form than eq.8-83. Thus, for purpose of mathematical convenience, eq.8-80 is used. The relevant initial and boundary conditions, respectively, for the problem are:

$$\left. \begin{array}{ll} t = 0: & z \geqslant 0, \quad \phi = \phi_i \\ t > 0: & z = 0, \quad \phi = \phi_0 \\ t > 0: & z \to \infty, \quad \phi \to \phi_i \end{array} \right\} \qquad (8\text{-}87)$$

Equations 8-80 and 8-87 offer a solution through the use of a transformation similar to the Boltzman transformation (1894). One may reduce the partial differential equation to an ordinary differential equation through the use of the similarity relationship given by:

$$\lambda(\phi) = zt^{-1/2} \tag{8-88}$$

Formally, using the above substitution in eq.8-80, one obtains the ordinary differential equation:

$$-\frac{\lambda}{2}\frac{d\phi}{d\lambda} = \frac{d}{d\lambda}\left[(1-\phi)D\frac{d\phi}{d\lambda}\right] + D\left(\frac{d\phi}{d\lambda}\right)_{\lambda=0}\frac{d\phi}{d\lambda} \tag{8-89}$$

Equation 8-89 can be integrated between ϕ_0 and ϕ to obtain:

$$\int_{\phi_0}^{\phi} \lambda d\phi = 2(1-\phi)\left[\left(D\frac{d\phi}{d\lambda}\right)_{\lambda=0} - D\frac{d\phi}{d\lambda}\right] \tag{8-90}$$

The boundary conditions are transformed through the use of eq.8-88 as:

$$\lambda = 0 \text{ at } \phi = \phi_0 \text{ and } \lambda \to \infty \text{ for } \phi = \phi_i \tag{8-91}$$

which also implies that $\frac{d\phi}{d\lambda} \to 0$ as $\phi \to \phi_i$. Using eq.8-90 and 8-91, it can be shown that:

$$(1-\phi_i)^{-1}\int_{\phi_i}^{\phi_0} \lambda d\phi = -2\left(D\frac{d\phi}{d\lambda}\right)_{\lambda=0} \tag{8-92}$$

For the purpose of convenience, the parameter α is defined as:

$$\alpha = (1-\phi_i)^{-1}\int_{\phi_i}^{\phi_0} \lambda d\phi = -2\left(D\frac{d\phi}{d\lambda}\right)_{\lambda=0} \tag{8-93}$$

In the following analysis it is assumed that α and λ are known and the various quantities, such as the instantaneous matrix and liquid velocities, cumulative volume change, and displacement history of solid particles initially at various positions in the column, will be calculated. The instantaneous velocity of the solid matrix, $U_r(z, t)$, is then calculated from eq.8-67 as:

$$U_r(z, t) = \frac{1}{2}\left[1 - \phi(z, t)\right]^{-1} t^{-1/2} \int_{\phi}^{\phi_0} \lambda d\phi \tag{8-94}$$

It is also convenient to define a volume flux density or macroscopic velocity for the rock, $u_r(z, t)$, as:

$$u_r(z, t) = U_r(z, t) [1 - \phi(z, t)] \tag{8-95}$$

i.e.:

$$u_r(z, t) = \frac{1}{2} t^{-1/2} \int_\phi^{\phi_0} \lambda d\phi \tag{8-96}$$

The microscopic velocity of the liquid, $U_L(z, t)$, may also be calculated from eq.8-70 as:

$$U_L(z, t) = \frac{1}{2\phi(z, t)} \left(\alpha - \int_\phi^{\phi_0} \lambda d\phi \right) t^{-1/2} \tag{8-97}$$

If a volume flux density or macroscopic velocity, $u_L(z, t)$, for the liquid is defined in a manner similar to eq.8-95, one obtains:

$$u_L(z, t) = \frac{1}{2} \left(\alpha - \int_\phi^{\phi_0} \lambda d\phi \right) t^{-1/2} \tag{8-98}$$

The velocity of the liquid relative to solid is:

$$u_{Lr}(z, t) = \frac{1}{2} t^{-1/2} \left\{ \alpha - \frac{1}{[1 - \phi(z, t)]} \int_\phi^{\phi_0} \lambda d\phi \right\} \tag{8-99}$$

which may further be simplified as:

$$u_{Lr}(z, t) = \frac{1}{2} t^{-1/2} \alpha - U_r \tag{8-100}$$

The volume of liquid out of the system, V_L, may be easily calculated from eq.8-98 as:

$$V_L = (u_L \phi)_{\phi = \phi_0} = \frac{1}{2} \alpha t^{-1/2} \tag{8-101}$$

and the cumulative volume change over time t can be obtained by integrating eq.8-101 with respect to time. The cumulative outflow from the system is given by:

$$V_{cum} = \alpha t^{1/2} \tag{8-102}$$

Of prime interest in an analysis of this type is the displacement history of the rock particle, which is considered to be initially at z_i and at $z = \tilde{z}$ at any other instant. This can be easily calculated by assuming that the solid content between the plane $z = \tilde{z}$ and $z = 0$ is equal to that between $z = 0$ and $z = z_i$. Mathematically this can be expressed as:

$$\int_0^{\tilde{z}} (1-\phi)\,dz = z_i\,(1-\phi_i) \tag{8-103}$$

Integrating eq.8-103, the following relationship is obtained:

$$(1-\phi)\,\tilde{z} = z_i\,(1-\phi_i) + t^{1/2}\,[1-\phi(z,t)]\left[2D\frac{d\phi}{d\lambda} + \right.$$

$$\left. + (1-\phi_i)^{-1/2} \int_{\phi_i}^{\phi_0} \lambda\,d\phi \right] \tag{8-104}$$

Upon simplification, it can be shown that the following limits hold:

$$\tilde{z} = z_i + \alpha t^{1/2} \text{ if } t z_i^{-2} \text{ is small} \tag{8-105}$$

$$\tilde{z} \to \frac{z_i\,(1-\phi_i)}{(1-\phi_0)} \text{ as } t \to \infty \tag{8-106}$$

Analytical solution to flow through a semi-infinite deformable porous medium

So far the consolidation or the expansion of the porous medium was examined in an extremely general manner. It is evident that the solution requires that $D(\phi)$ or $k(\phi)$ and $\psi(\phi)$ should be known. To date, very little experimental information is available on the nature of $D(\phi)$ for unconsolidated sediments. The information available relates essentially to very fine sands, chalk or clay*. The experiments carried out by Vasquez (1961) (see Pirson, 1963, p.63) indicates that k is of the form:

$$k = \frac{k_c \phi^a}{(1-\phi)^b} \qquad (8\text{-}107)$$

where k_c, a and b are constants to be experimentally determined. The expression for k, given by eq.8-107, may also be justified on theoretical grounds. The well-known model for porous media proposed by Kozeny (1927) and Carman (1937) postulates that k should be directly proportional to the cross section available for flow, ϕ, and to the square of the characteristic separation length between the matrix particles, $[\phi/(1-\phi)]^2$. On the basis of the Kozeny–Carman model, one may assign values of 3 and 2 to a and b, respectively. Experimental information is also available on the relationship between ψ and ϕ for clay pastes. Experimental data indicate that gravimetric solution (ratio of mass of solution to mass of solid) content varies inversely with the moisture potential (see Philip, 1968; Darley, 1969). Thus, one can express ψ as:

$$\psi = -\psi_c \left(\frac{\phi}{1-\phi}\right)^c \qquad (8\text{-}108)$$

where ψ_c and c are arbitrary constants. Experiments indicate that the value of c varies between -2 and -5, with the value of -2 being appropriate for small values of ψ. For $c = -2$, the diffusivity, D, is given as:

$$D = D_c (1-\phi)^{-1} \qquad (8\text{-}109)$$

where D_c is a constant, provided one assumes the fluid to be Newtonian (i.e., μ is constant). Substituting the right-hand side of eq.8-109 in eq.8-73 for D, the following relationship is obtained:

$$\frac{\partial \phi}{\partial t} = D_c \frac{\partial^2 \phi}{\partial z^2} + D_c (1-\phi_0)^{-1} \left(\frac{\partial \phi}{\partial z}\right)_{z=0} \frac{\partial \phi}{\partial z} \qquad (8\text{-}110)$$

Introducing constants τ and β defined by:

---

*The analysis in this section, as in all others, is general. The fact that the example chosen is for clays or fine-grained sands should not be considered as a limitation. The analytical solution is presented mainly as an illustrative example so as to provide an insight into the various processes which are taking place. It should also be noted that the model proposed by Kozeny (1927) and Carman (1937) is valid for all porous solids.

$$\tau = D_c t \tag{8-111}$$

and:

$$\beta = \frac{-(D_c t)^{1/2}}{(1-\phi_0)} \left(\frac{\partial \phi}{\partial z}\right)_{z=0} \tag{8-112}$$

eq. 8-110 becomes:

$$\frac{\partial \phi}{\partial \tau} = \frac{\partial^2 \phi}{\partial z^2} - \beta \tau^{-1/2} \frac{\partial \phi}{\partial z} \tag{8-113}$$

It should be noted that β in eq.8-112 and α in eq.8-93 are related as follows:

$$\alpha = 2 D_c^{1/2} \beta \tag{8-114}$$

Before proceeding further, a few points deserve emphasis. First, it should be noted that β is a constant and second, that β is a function of ϕ and t. The fact that β is a constant follows from the solution of eq.8-113 and the similarity form of the solution, as given by eq.8-88. The solution of the problem consists of solving for both β and ϕ.

The similarity solution of eq.8-113, which is obtained next, is subject to the following initial and boundary conditions:

$$\left.\begin{array}{l} \tau = 0: \ z \geq 0, \ \phi = \phi_i \\ \tau > 0: \ z = 0, \ \phi = \phi_0 \\ \tau > 0: \ z \to \infty, \ \phi = \phi_i \end{array}\right\} \tag{8-115}$$

It is to be noted that eq.8-115 is a special case of eq.8-87. Through the similarity transformation:

$$\eta = z \tau^{-1/2} \tag{8-116}$$

eq. 8-113 and 8-115 become, respectively:

$$-\left(\frac{1}{2}\eta - \beta\right) \frac{d\phi}{d\eta} = \frac{d^2 \phi}{d\eta^2} \tag{8-117}$$

and:

$$\eta = 0, \ \phi = \phi_0; \ \eta \to \infty, \ \phi = \phi_i \tag{8-118}$$

The solution for ϕ can now be readily obtained from eq.8-117, subject to conditions given by eq.8-118 as:

$$\phi = \phi_i + (\phi_0 - \phi_i) \frac{\text{erfc}\left(\frac{1}{2}\eta - \beta\right)}{\text{erfc}(-\beta)} \qquad (8\text{-}119)^*$$

An expression for β may now be obtained from the transcendental equation as:

$$\pi^{1/2} \beta \exp(\beta^2) \text{erfc}(-\beta) = \frac{(\phi_0 - \phi_i)}{(1 - \phi_0)} \equiv \omega \qquad (8\text{-}120)$$

The limiting forms of β may be obtained from eq.8-120; when $\omega \to 0$, $\beta \to 0$ and for small $|\omega|$, β is given by:

$$\beta = \pi^{-1/2} \omega \qquad (8\text{-}121)$$

When ω is negative (consolidation), β is also negative and the following asymptotic relationship holds in the limit as $\omega \to -1$:

$$1 + \omega \simeq \frac{1}{2\beta^2} - \frac{3}{4\beta^4} + \frac{15}{8\beta^6} - \ldots \qquad (8\text{-}122)$$

On the other hand, if ω is positive (expansion process), β is also positive. and for large ω one obtains:

$$\beta \exp(\beta^2) = \frac{\omega \pi^{-1/2}}{2} \qquad (8\text{-}123)$$

The matrix velocity and fluid velocity can now be calculated. For the purposes of illustration, the microscopic velocity of the rock matrix is

---

*erfc(x) is the error function complement. The error function of x, erf(x) is defined (see Abramowitz and Stegun, 1969, p.297) as:

$$\text{erf}(x) = \frac{2}{\sqrt{\pi}} \int_0^x \exp(-\xi^2) \, d\xi$$

The error function complement of x, erfc(x), is given by $1 - \text{erf}(x)$.

calculated below. Substituting the appropriate expressions for ϕ, z and t in eq.8-94 by using eq.8-119, 8-116, and 8-111, respectively, one obtains the following expression for the rock matrix velocity:

$$U_r = \frac{D_c \tau^{-1/2} [1 - \phi(z,t)]^{-1} (\phi_0 - \phi_i)}{2\sqrt{\pi} \operatorname{erfc}(-\beta)}$$

$$\left\{ \int_0^\eta \chi \exp\left[-\left(\frac{\chi}{2} - \beta\right)^2\right] d\chi \right\} \tag{8-124}$$

In eq.8-124 and the equations which follow, it is implicitly assumed that the velocities are functions of z and t. Equation 8-124 can be further simplified by integrating by parts and using the relations provided by Abramowitz and Stegun (1964, p.304):

$$U_r = D_c \tau^{-1/2} (\phi_0 - \phi_i) \tag{8-125}$$

$$\left[\frac{\pi^{-1/2} \left\{ \exp(-\beta^2) - \exp\left[-\left(\frac{\eta}{2} - \beta\right)^2\right] \right\} + \beta \left[\operatorname{erf}\left(\frac{\eta}{2} - \beta\right) - \operatorname{erf}(-\beta) \right]}{(1 - \phi_i) \operatorname{erfc}(-\beta) + (\phi_0 - \phi_i) \operatorname{erfc}\left(\frac{1}{2}\eta - \beta\right)} \right]$$

Similar expressions for $U_L(z,t)$, $U_{Lr}(z,t)$, etc., may be determined. It has been shown that these expressions are dependent on the integral $(\tfrac{1}{2} \int_{\phi_i}^{\phi_0} \lambda \, d\phi)$.

For the present example, the value of this integral, defined by γ, is given as:

$$\gamma = \frac{1}{2} \int_\phi^{\phi_0} \lambda \, d\phi \tag{8-126}$$

Simplifying by substituting appropriate expression for λ and ϕ, the value of γ is obtained:

$$\gamma = \frac{D_c^{1/2} (\phi_0 - \phi_i)}{\operatorname{erfc}(-\beta)} \left[\pi^{-1/2} \left\{ \exp(-\beta^2) - \exp\left[-\left(\frac{\eta}{2} - \beta\right)^2\right] \right\} + \right.$$

$$\left. + \beta \left[\operatorname{erf}\left(\frac{\eta}{2} - \beta\right) - \operatorname{erf}(-\beta) \right] \right] \tag{8-127}$$

Equation 8-127 is also obtained using the relations provided by Abramowitz and Stegun (1964, p.304).

Having evaluated eq.8-127, one can obtain the microscopic velocity of the liquid as follows:

$$U_L(z,t) = \tilde{U}_L \, \tau^{-1/2} \tag{8-128}$$

where $\tilde{U}_L = D_c^{1/2}(\beta D_c^{1/2} - \gamma)/\phi$.

The microscopic velocity of liquid relative to the solid is given by:

$$U_{Lr} = \tilde{U}_{Lr} \, \tau^{-1/2} \tag{8-129}$$

where $\tilde{U}_{Lr} = D_c^{1/2}(\beta D_c^{1/2} - \frac{1}{1-\phi}\gamma)/\phi$.

The microscopic velocity of the matrix in eq.8-94 may be written as:

$$U_r = \tilde{U}_r \, \tau^{-1/2} \tag{8-130}$$

where $\tilde{U}_r = D_c^{1/2}\gamma(1-\phi)^{-1}$.

As mentioned earlier, the displacement history of the rock matrix is of prime interest in the analysis considered here. From eq.8-104 one may write the displacement history as:

$$\eta = \eta_i(1-\phi_i) + \int_0^\eta \left[\phi_i + \frac{(\phi_0-\phi_i)}{\text{erfc}(-\beta)} \text{erfc}\left(\frac{\eta'}{2}-\beta\right)\right] d\eta' \tag{8-131}$$

where η_i is the initial position of a particle and η' is the subsequent position of that particle at a subsequent time t. Integrating and again using the relations provided by Abramovitz and Stegun (1964, p.304), one obtains:

$$\eta = \eta_i + \frac{2(\phi_0-\phi_i)}{(1-\phi_i)\text{erfc}(-\beta)} \left\{ \left(\frac{\eta}{2}-\beta\right)\text{erfc}\left(\frac{\eta}{2}-\beta\right) - \pi^{-1/2}\exp\left[-\left(\frac{\eta}{2}-\beta\right)^2\right] + \beta\,\text{erfc}(-\beta) + \pi^{-1/2}\exp(-\beta^2) \right\} \tag{8-132}$$

Equation 8-132 can be further simplified to the form:

$$\eta = \eta_i + \frac{2(\phi_0 - \phi_i)}{(1-\phi_i)\operatorname{erfc}(-\beta)} \left\{ \left(\frac{\eta}{2} - \beta\right) \operatorname{erfc}\left(\frac{\eta}{2} - \beta\right) - \right.$$
$$\left. - \pi^{-1/2} \exp\left[-\left(\frac{\eta}{2} - \beta\right)^2\right] \right\} + 2\beta \tag{8-133}$$

Before discussing the solutions obtained, a number of important points should be considered. It is interesting to note that the solutions obtained in the present analysis are similar to Terzaghi's (1923) in the sense that the results are proportional to the square root of time. This comparison, however, only applies in a very gross sense. The analysis discussed above differs from the classical analysis in the following respects:

(1) The displacement of the rock and the associated mass flow (additional flow of liquid due to the movement of solid) are taken into account.
(2) The variation of permeability with pore volume (or pressure) is accounted for.
(3) The variation of the potential with pore volume is taken into account, including any nonlinearities.

The error, due to neglecting points 2 and 3, is not important. The first point, however, is very important, even in the limits of small strains. It has been shown by Philip (1968) that the Terzaghi's analysis of a small-strain compaction (or expansion rebound) experiment would indicate a value of k that is $(1 - \bar{\phi})^{-1}$ times the true value, where $\bar{\phi}$ is the mean value of ϕ during the experiment. For a mean value of $\bar{\phi} = 0.5$, the neglect of mass flow (additional flow of liquid due to movement of solids) by the Terzaghi's analysis leads to the over-estimation of k by a hundred percent. In this context, it should also be pointed out that many workers have examined consolidation and expansion by means of spring analogies, for example, Lambe (1951) and Christie (1964). These analogies, however, are very unconvincing for they fail to reproduce even the gross behavior of expansion or consolidation. The spring analogy indicates that an initial gross displacement is proportional to time, whereas Terzaghi's analysis and the present development show that it should be proportional to the square root of time.

Figure 8-5 shows the relationship between β and ω (also see Table 8-I). For the present time it is important to note that the portion of the curve representing the compaction process is steep, whereas that of the expansion process is rather flat.

Figure 8-6 shows the relationship between the volumetric liquid content or porosity and the dimensionless variable η for both the compaction and expansion processes. Comparison of the curves in Fig.8-6 brings out distinct differences between the two processes. It can be seen

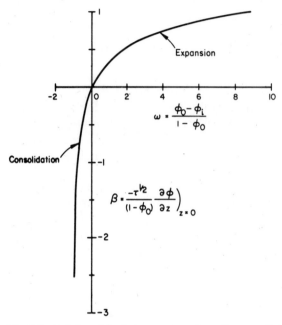

Fig.8-5. Relationship between β and ω for a soil for $a = 3$, $b = 2$ and $c = -2$. (After Raghavan, 1972, fig.1, p.16.)

TABLE 8-I

Relationship between ω and β

ϕ_0	ϕ_i	ω	β
0.50	0.95	−0.9	−1.940
0.50	0.90	−0.8	−1.175
0.50	0.80	−0.6	−0.595
0.50	0.70	−0.4	−0.311
0.70	0.60	0.33	0.155
0.70	0.50	0.66	0.269
0.70	0.40	1.0	0.359
0.70	0.30	1.33	0.429
0.90	0.50	4.00	0.751
0.95	0.60	9.00	1.0045

from this figure that the magnitude of the release of liquid during the consolidation process is twice the intake of liquid during the rebound expansion process. At the same time, it should be noted that the effect of the step function change (at the base of the column) for the expansion process has propagated three times as fast as for the consolidation process.

MATHEMATICAL ANALYSIS OF SAND COMPACTION

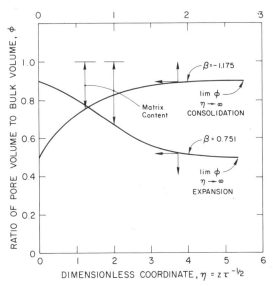

Fig.8-6. Relationship between porosity, φ and the dimensionless coordinate, η. (After Raghavan, 1972, fig.2, p.16.)

This is essentially due to the effect of mass flow of the solution, i.e., the additional flow of the liquid due to the movement of the solid, which is dependent on the initial value of ϕ_i. Undoubtedly ϕ_i is greater for the compaction process than for the expansion process.

Figure 8-7 demonstrates the behavior of the potential ratio ψ/ψ_i as a function of the dimensionless coordinate. Figures 8-8—8-11 show the behavior of the macroscopic velocities (volume flux densities) and microscopic velocities for both compaction and expansion processes, where the dominant effect of the mass flow of the solution can be seen. The relative velocity is zero at the base of the column and its magnitude is of the same order as the flow of the solution relative to solid particles where the compaction and rebound expansion processes are dominant. The classical analysis by Terzaghi (1923) does not bring out this aspect. In the upper part of the column the motion of the solution relative to the solid is zero. This implies that the predominant motion in this section is the bodily translation of the solid and liquid. Figures 8-12 and 8-13 demonstrate vividly the displacement of the solid matrix for both compaction and rebound expansion processes. Each curve represents the position of the particle initially at z_i relative to the base of $z = 0$ as a function of time. The parallel linear displacement histories essentially indicate bodily translation of solid and liquid with no relative motion and imply that the compaction or the rebound expansion process has not penetrated to the distance under consideration.

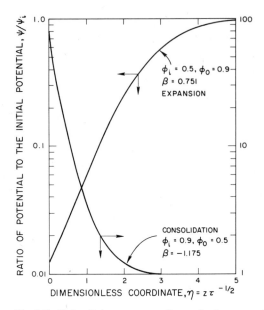

Fig.8-7. Potential, ψ, versus dimensionless coordinate, η, for consolidation and expansion processes. (After Raghavan, 1974, fig.2, p.1692.)

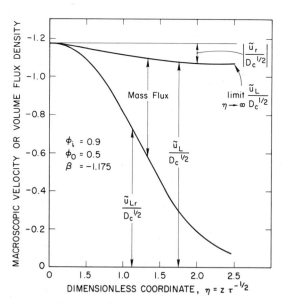

Fig.8-8. Macroscopic velocities of rock and liquid as a function of the dimensionless coordinate, η, for the compaction process. (After Raghavan, 1972, fig.5, p.17.)

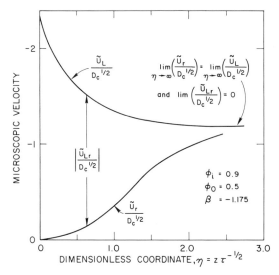

Fig. 8-9. Microscopic velocities of rock and liquid as a function of the dimensionless coordinate, η, for the compaction process. (After Raghavan, 1972, fig. 6, p. 17.)

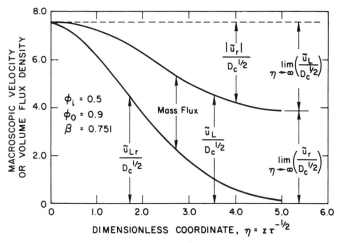

Fig. 8-10. Macroscopic velocities of rock and liquid as a function of the dimensionless coordinate, η, for the case of the expansion process. (After Raghavan, 1972, fig. 7, p. 17.)

Equations of motion in a moving coordinate system

In the present section, the Terzaghi problem is re-examined in a moving frame of reference. In the analysis which follows, a scale of length defined

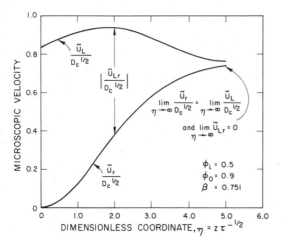

Fig.8-11. Microscopic velocities of rock and liquid as a function of the dimensionless coordinate, η, for the expansion process. (After Raghavan, 1972, fig.8, p.18.)

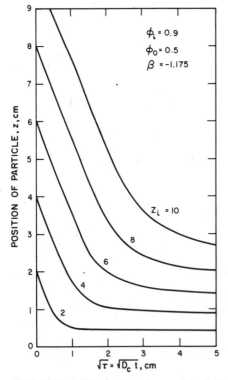

Fig.8-12. Displacement history of individual matrix particles as the semi-infinite column compacts. (After Raghavan, 1972, fig.9, p.18.)

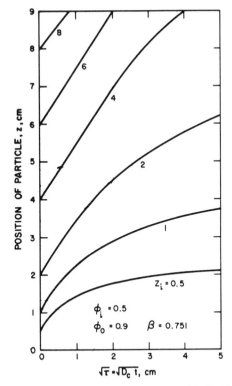

Fig.8-13. Displacement history of individual matrix particles as the semi-infinite column expands. (After Raghavan, 1972, fig.10, p.18.)

by the distribution of mass of the solid material is used. Accordingly, the position of the elemental volume is discussed in terms of a material coordinate system, m, defined as:

$$m(z, t) = \int_0^z [1 - \phi(z, t)] \, dz \qquad (8\text{-}134)$$

Physically m represents the cumulated volume of solids in cm^3 per cm^2 of cross section from the plane $z = 0$ to the plane z.

The choice regarding the scale is not a limited one. It would essentially have to depend on individual preference, but would probably be guided by experiments which will be performed to determine $D(\phi)$, $k(\phi)$, etc. Other appropriate choices for the variable length scale m would be:

$$m(z, t) = \int_0^z \frac{\rho_b}{\rho_r} \, dz, \; m(z, t) = \int_0^z [1 + e(z, t)]^{-1} \, dz, \text{ or:}$$

$$m = \int_0^z \left(1 + \frac{\rho_r \tilde{\Theta}}{\rho_w}\right)^{-1} dz \tag{8-135}$$

where ρ_b is the bulk density, e is the void ratio (ratio of the volume of voids to the volume of solid), and $\tilde{\Theta}$ is the ratio of the mass of water to the mass of rock. The void ratio, e, and the porosity, ϕ, are related as follows:

$$e(z, t) = \phi(z, t) [1 - \phi(z, t)]^{-1} \tag{8-136}$$

In this context it should be pointed out that in field and laboratory investigations, piezometers, pressure transducers, or other measuring instruments would not ordinarily remain fixed as the material deforms. In fact, they are likely to move with the sand grains. Hence, observed gradients are likely to be in terms of material coordinates, even though for some materials the difference between the coordinate systems could be negligible.

A further modification to the above system can also be made. The lower limit on the integral can be changed from 0 to $-\infty$. This has an inherent advantage in that the surface of the column is $m = 0$ regardless of the z coordinate. It should be noted that the relationships given by eq.8-134 and 8-135 are essentially realizable only in the laboratory and unlikely to occur in natural systems. In the following development it is again implicitly assumed that m, ϕ, etc., are functions of z and t and, further, the lower limit of the integral is changed from 0 to $-\infty$.

The requirement of continuity yields:

$$\frac{1}{(1-\phi)^2} \frac{\partial \phi}{\partial t} = -\frac{\partial}{\partial m}(u_{Lr}) \tag{8-137}$$

where the macroscopic velocity of the liquid relative to rock, u_{Lr}, is given by Darcy's Law. Substituting the right-hand side of eq.8-72 for u_{Lr} in eq.8-137, one obtains:

$$\frac{\partial \phi}{\partial t} = (1-\phi)^2 \frac{\partial}{\partial m}\left(D \frac{\partial \phi}{\partial z}\right) \tag{8-138}$$

A more compact form of eq.8-138 is obtained if one uses the void ratio, e, rather than ϕ. In this case eq.8-138 may be written as:

$$\frac{\partial e}{\partial t} = \frac{\partial}{\partial m}\left[\frac{D}{(1+e)^3} \frac{\partial e}{\partial m}\right] \tag{8-139}$$

In the following analysis it is shown that eq.8-138 and eq.8-82 are equivalent. Differentiating eq.8-134 with respect to z one obtains:

$$\frac{\partial m}{\partial z} = [1 - \phi(z, t)] \tag{8-140}$$

It is very important to note that $\frac{\partial \phi}{\partial t}$ in eq.8-82 and 8-138 implies $\left(\frac{\partial \phi}{\partial t}\right)_z$ and $\left(\frac{\partial \phi}{\partial t}\right) m$, respectively. Through the well-known chain rule of calculus, the following is valid:

$$\left(\frac{\partial \phi}{\partial t}\right)_m = \left(\frac{\partial \phi}{\partial t}\right)_z + \left(\frac{\partial z}{\partial t}\right)_m \left(\frac{\partial \phi}{\partial z}\right)_t \tag{8-141}$$

Noting that $U_r(z, t) = (\partial z/\partial t)_m$ and using eq.8-67, 8-70, and 8-71, the following is obtained:

$$\left[\left(\frac{\partial \phi}{\partial t}\right)_z + U_r \left(\frac{\partial \phi}{\partial z}\right)_t\right] (1 - \phi)^{-1} = \frac{\partial}{\partial z}\left(D \frac{\partial \phi}{\partial z}\right) \tag{8-142}$$

Substituting the value of $U_r(z, t)$, given by the right hand side of eq.8-67, it can be shown the left-hand sides of eq.8-142 and 8-82 are identical. (See also Philip, 1968 and Smiles and Rosenthal, 1968.)

Columns of finite length

A column of finite length is analyzed and the use of the moving coordinate system is demonstrated in this section. It should be evident that the solutions already presented can also be handled in a moving coordinate system. Recently, a new analytical technique has been proposed by Parlange (1971a, b, c; 1972a, b, c, d, e) to solve the differential equations used in the present section, provided the column is assumed to be semi-infinite.

The specific problem examined is that of a deformable porous solid, which is sealed at one end. (The effects of gravity and overburden are considered later.) An external load is applied at the other end and liquid is forced to move out of the system through the open end (Fig.8-14). The governing differential equation may be written as follows:

$$\frac{\partial e}{\partial t} = \frac{\partial}{\partial m}\left(D_m \frac{\partial e}{\partial m}\right) \tag{8-143}$$

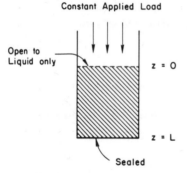

Fig.8-14. Schematic representation of the compaction or expansion of a porous column of finite length. (After Raghavan, 1972, fig.11, p.18.)

where $D_m(e)$, the diffusivity, is given by:

$$D_m = \frac{D}{(1+e)^3} \tag{8-144}$$

The associated initial and boundary conditions are:

$$\left.\begin{array}{ll} t = 0: & e = e_i \quad \text{for } 0 \leqslant m \leqslant M \\ t > 0: & e = e_0 \quad \text{at } m = 0 \\ t > 0: & \dfrac{de}{dM} = 0 \text{ at } m = M \end{array}\right\} \tag{8-145}$$

where M is the total volume of the solid matrix per unit area and is a constant. The quantity, M, can be determined by the expression:

$$M = \int_{-\infty}^{L_i} [1 + e(z,t)]^{-1} \, dz \tag{8-146}$$

where L_i is the initial length of the column.

Equation 8-143 is very similar to the diffusivity equation used to describe the flow of fluids through porous media, but in the present instance, however, the diffusivity is dependent upon the void ratio e. Thus, eq.8-143 is non-linear. Many have attempted to linearize eq.8-143 (for example, see Gibson et al., 1967). In the process of linearizing the equation, however, the movement of the solid and associated mass flux, etc., which play a dominant role in the problem under consideration, have been neglected.

MATHEMATICAL ANALYSIS OF SAND COMPACTION 447

Non-linear equations given by the system of eq.8-143 through 8-146 generally have to be solved for each specific case of interest and thus are not of general utility. Raghavan (1972), however, has pointed out that a general solution to eq.8-143 and 8-145 can be obtained. This general solution is obtained by means of a transformation which does not involve any additional assumptions. The transformation is obtained by another scale change defined by:

$$W(e) = \int_{e_b}^{e} D_m(e) de \qquad (8\text{-}147)$$

where e_b is a "base void ratio". The variable $W(e)$ has the dimensions of cm^2/sec. As D_m is a unique function of e (only two components — liquid and rock — are considered), the integral given by the right hand side of eq.8-147 is defined uniquely. It should be noted that for D_m to be unique, hysteresis effects must also be included. To present the principles involved, it is assumed that hysteresis is either negligible or that the hysteresis effect is accounted for. Equations 8-143 and 8-145 may be rewritten in terms of eq.8-147 as:

$$\frac{\partial^2 W}{\partial m^2} = \frac{1}{D_m} \frac{\partial W}{\partial t} \qquad (8\text{-}148)$$

with:

$$\begin{aligned}
t = 0: \quad & W(e) = W(e_i) \text{ for } 0 \leqslant m \leqslant M \\
t > 0: \quad & W(e) = W(e_0) \text{ at } m = 0 \\
t > 0: \quad & \frac{dW(e)}{dm} = 0 \text{ at } m = M
\end{aligned} \qquad (8\text{-}149)$$

Examination of eq.8-148 indicates that the equation is still non-linear. Raghavan (1972) has pointed out that eq.8-148 is similar to the pressure-dependent diffusivity equation, which describes the one-dimensional flow of ideal (Aronofsky and Jenkins, 1954) or real (Al-Hussainy et al., 1966) gases through porous media. The flow of liquids through porous media, where the properties of the liquid and rock are pressure dependent, can also be described by an equation similar to eq.8-148 (Raghavan et al., 1972). The transformation of the type used in eq.8-147 was first used in 1868 (see Carslaw and Jaeger, 1959) for analyzing steady-state heat conduction problems. More recently, this transformation was used by Al-Hussainy et al. (1966), in their study of the flow of real gases in porous media. The transformation used by Al-Hussainy et al., was designated as the real gas pseudo-pressure. Following Al-Hussainy et al., the function $W(e)$ was named the pseudo-void ratio by Raghavan (1972).

Equation 8-148 represents the fundamental partial differential equation which describes the one-dimensional flow through a deformable porous medium and was first presented by Raghavan (1972). Furthermore, eq.8-148 does not involve further assumptions. The real importance lies in the extreme utility of this equation to obtain rapid solutions for the flow of liquids through deformable porous media. This approach is indicated by the work of Aronofsky and Jenkins (1954), Al-Hussainy et al. (1966), Raghavan (1972), and Raghavan et al. (1972). Aronofsky and Jenkins found that for a constant rate of production of an ideal gas from a closed radial system, the pressures could be correlated with the solutions for transient liquid flow developed by Van Everdingen and Hurst (1949), provided the dimensionless time used was based on compressibility of the gas at the initial pressure in the system. On the basis of this empirical observation, Al-Hussainy et al. have shown that a similar correlation exists for the radial flow of real gases, provided the viscosity—compressibility product in the dimensionless time group is evaluated at the initial pressure. The basis for the correlation obtained by Al-Hussainy et al. is that the variation of the viscosity—compressibility product of the real gas with the pseudo-pressure is similar to variation of the compressibility of an ideal gas with pressure squared. Raghavan has shown that a similar basis exists in the present instance also. This is shown in Fig.8-15 where the diffusivity ratio, $\frac{D_{mi}}{D_m}$, in which D_{mi} is the initial value of diffusivity, is plotted as a function of e. For purposes of comparison, the variation of the viscosity—compressibility product with real gas pseudo-pressure is also shown. It can be seen that the variation is similar (convex to the origin). Thus it might be expected that the potential defined by eq.8-147 should correlate with the constant property solution as a function of dimensionless time group based on the initial properties of the system. Accordingly, a dimensionless time, t_D, is defined as:

$$t_D = \frac{D_{mi} t}{M^2} \qquad (8\text{-}150)$$

where M is defined by eq.8-146, t is the time, and D_{mi} is initial value of the diffusivity. For the purposes of this discussion, a dimensionless pseudo-void ratio, a dimensionless void ratio, a dimensionless distance and a dimensionless flow rate are defined by the following equations, respectively:

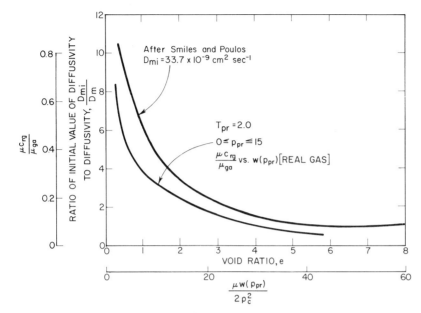

Fig.8-15. Relationship between diffusivity ratio, D_{mi}/D_m, and void ratio, e.

$$W_D(m_D, t_D) = \frac{W(m, t)}{W_i} \tag{8-151}$$

$$e_D(m_D, t_D) = \frac{e(m, t)}{e_i} \tag{8-152}$$

$$m_D = \frac{m}{M} \tag{8-153}$$

$$q_D(m_D) = \frac{Mu_{Lr}}{W_i} = -\frac{\partial W_D}{\partial m_D} \tag{8-154}$$

where W_i is the pseudo-void ratio at the initial condition, and u_{Lr} is the macroscopic relative velocity of the fluid. The subscript D signifies dimensionless variables. For the purposes of discussion, $\overline{W}_D(m_D, t_D)$ is defined as:

$$\overline{W}_D(m_D, t_D) = \frac{1}{q_D(m_D)}[1 - W_D(m_D, t_D)] \tag{8-155}$$

Physically $\overline{W}_D (m_D, t_D)$ represents the change in pseudo-void ratio with flux at any point in the system at any instant in time.

Solution to a deformable column of finite length (compaction)

The purpose of this section is to demonstrate the use of the pseudo-void ratio transformation discussed in the previous section. Using the data presented by Smiles and Poulos (1969), Raghavan (1972) solved the above system of equations (eq.8-148 and 8-149) by numerical means on a digital computer. He then correlated the variable property solutions with the constant property solutions through the use of eq.8-150—8-155. The correlations obtained by Raghavan (1972) are discussed next.

The $\overline{W}_D(m_D, t_D)$ computer solutions developed by Raghavan (1972) are compared to the constant property solutions in Fig.8-16. The constant property solution is given by (Miller, 1962):

$$e = e_i + (e_0 - e_i)\frac{4}{\pi} \sum_{n=1}^{\infty} \frac{1}{(2n-1)} \exp\left[-\frac{(2n-1)^2\pi^2 D_{mi}t}{4M^2}\right] \times$$

$$\sin\frac{(2n-1)\pi m}{2M} \qquad (8\text{-}156)$$

The triangular, square, and circular points represent the results obtained by solving the non-linear equations. The solid line represents the constant property solution. For purposes of illustration two points on the system have been chosen ($m_D = 0$ and $m_D = 0.5$). As shown in Fig.8-16, the correlation is good. The data presented are a severe test of the linearization process, for they represent a seven-fold change in the ratio of void ratio to the initial void ratio.

As shown in Fig.8-16, for large values of t_D, the constant property solutions and the variable property solutions diverge and the correlation becomes less useful. In addition, for large values of t_D, the correlations are a function of the ratio e_0/e_i. In other words, the solutions are "flux sensitive", particularly at large values of t_D. The variable property results approach the constant property solutions (1) as e_0 approaches e_i and (2) for small values of dimensionless time, t_D.

Figure 8-17 compares the dimensionless void ratio, e_D, for the constant property and variable property solutions. Values of e_D for the variable case is greater than that for the constant property case for all values of m_D considered for all times. This indicates that classical analysis would be conservative in estimating the length of the column during the compaction process.

MATHEMATICAL ANALYSIS OF SAND COMPACTION

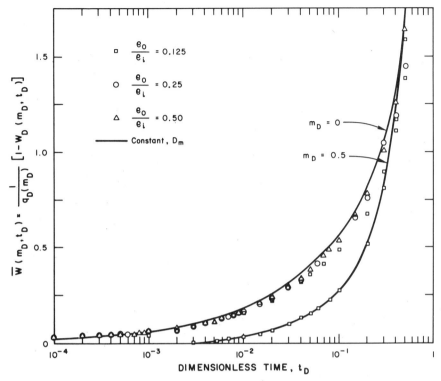

Fig.8-16. Computer solutions of $\overline{W}_D(m_D, t_D)$ superimposed on constant property solution (solid line) for a finite porous column due to a step function change in void ratio at $m_D = 0$ in the case of compaction. Curves for $m_D = 0$ and $m_D = 0.5$. (After Raghavan, 1972, fig.13, p.18.)

Figure 8-18 represents an attempt to correlate the dimensionless void ratio, e_D, with the dimensionless pseudo-void ratio $W_D(m_D, t_D)$. The solid line is the constant D_m solution and the results of the variable D_m solution are shown as circles, squares, triangles, etc. Figure 8-18 indicates that the results of two solutions correlate extremely well for small values of t_D. As the dimensionless time, t_D, increases, the two solutions diverge. This divergence, however, is a function of the dimensionless distance, m_D, for as m_D increases, the two solutions correlate for longer values of t_D. One reason for the divergence is that as $t_D \to \infty$, in the limit, $e_D \to 0.125$, whereas $W_D \to 0.0109$.

As mentioned earlier, the prime purpose of an analysis of this type is to trace the movement of the solid matrix and, thus, to calculate the length of

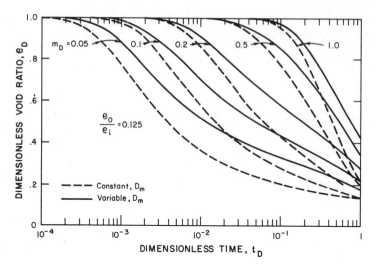

Fig.8-17. Comparison of void ratios as a function of dimensionless time, t_D, for constant and variable D_m. (After Raghavan, 1972, fig.14, p.19.)

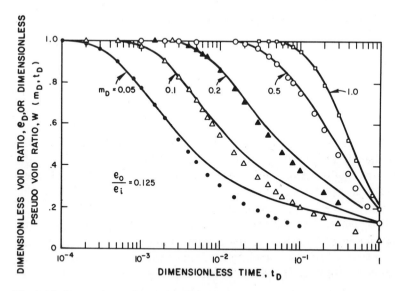

Fig.8-18. Comparison of pseudo-void ratio and void ratio (constant property) as a function of dimensionless time, t_D, in a porous column of finite length. (After Raghavan, 1972, fig.15, p.19.)

the porous column as a function of time. This is shown in Fig.8-19 where the dimensionless length of the porous column, L_D, defined by:

$$L_D = \frac{L_i - L(t)}{L_i - L_\infty} \qquad (8\text{-}157)$$

is graphed as a function of the dimensionless time, t_D. The instantaneous length, $L(t)$, of the column is obtained by using the relationship:

$$L(t) = \left(1 + e_i \int_0^1 e_D \, dm_D \right) M \qquad (8\text{-}158)$$

and L_∞ is the length of the column at infinite time. The dashed line in Fig.8-19 corresponds to the constant diffusivity, D_m, case. The solid line represents the variable D_m case. The figure shows that the constant property solution results in the overestimation of the rate of compaction because the change in D_m is neglected. Noting that D_m decreases very rapidly as e decreases, Smiles and Poulos (1969) have pointed out that the phenomenon of secondary compaction is probably a consequence of the neglect of the variation of D_m rather than a result of the anomalous behavior of compacting materials. This finding is extremely important considering the large number of publications which attribute the difference between experimental (or actual) results and theoretical predictions to this phenomenon.

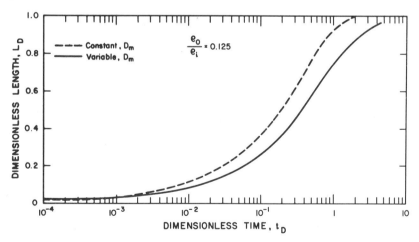

Fig.8-19. Relationship between dimensionless length, L_D, of a compacting porous column and dimensionless time, t_D, for constant and variable diffusivity, D_m. (After Raghavan, 1972, fig.17, p.19.)

Solution to a deformable column of finite length (rebound expansion)

On the basis of the discussion on the compaction process, it may appear that the results obtained by the solution of eq.8-143 and 8-145 for the expansion process would correlate equally well with the constant property solutions. Raghavan (1972) showed that the correlation of the variable property and constant property results is not a simple matter. Figure 8-20 shows the correlation obtained when $\overline{W}(m_D, t_D)$ is graphed as a function of the dimensionless time, t_D. It should be emphasized that the values of D_m used in calculating $\overline{W}(m_D, t_D)$ are the same as those presented in Fig.8-15 and are strictly valid only for compacting systems.

The solid curve in Fig.8-20 is the constant diffusivity case solution. The dashed curves show the correlation for the variable case. As shown in the figure, two definitions of dimensionless time have been used: one corresponding to D_m calculated at the initial void ratio, and the other based on D_m corresponding to the final void ratio. The correlation based on the initial (lower) values of the void ratio is no better or worse than that based

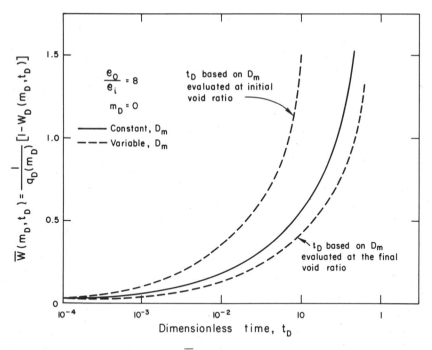

Fig.8-20. Computer solutions of $\overline{W}_D(m_D, t_D)$ term versus dimensionless time, t_D, superimposed on constant diffusivity solutions for an expanding porous column. (After Raghavan, 1972, fig.16, p.19.)

on the final (higher) value. As the constant-property case falls in between these two results, it may appear that a correlation using a dimensionless time based on physical properties half-way between the two extremes, i.e., "the theorem of the mean", might be better. Friedman (1958) has shown that the results must lie between the results evaluated at the two extremes. One cannot always assume, however, that evaluation at an average property will yield good results. Noting that the correlation corresponds to $e_0/e_i = 8$, based on the results obtained for the compaction case, one may say that better correlation can be obtained for smaller values of e_0/e_i.

Superposition of linearized solutions

The solutions presented in the previous sections assumed a single change in void ratio or porosity at one end of the porous medium. In many instances, however, a gradual change in pressure or void ratio (rather than an abrupt change) would represent conditions in nature more closely (see solid line on Fig.8-21). The concept of superposition is commonly used to solve situations of this kind. The basic idea behind this principle can be described by considering a gradual change in e from e_i to e_0 (Fig.8-21). The total time period of interest is to be divided into a number of equal-size increments: $(t_1 - 0)$, $(t_2 - t_1)$, $(t_3 - t_2)$, ... $(t_m - t_{m-1})$. Referring to Fig.8-21, the principle involves the replacement of the true, continuous, functional relationship between e and t by the matching step function shown.

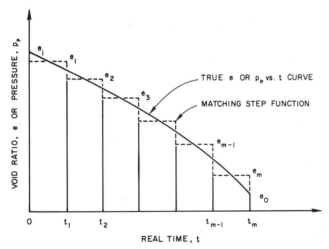

Fig.8-21. Continuous curve showing relationship between the void ratio, e, and real time, t, and the corresponding step function, which demonstrates the application of the superposition concept.

In applying the concept of superposition, the initial drop in the step function from e_i to e_1, imposed at time zero, is considered to be effective over the entire time interval $(t_m - 0)$. The drop $(e_1 - e_2)$ imposed at time t_1 is considered to be effective over the time interval $(t_m - t_1)$, and the succeeding drop $(e_2 - e_1)$ imposed at time t_2, is considered to apply over the time interval $(t_m - t_2)$. Continuing this process, the drop from e_{m-1} to e_m applies only over the last increment of time, $(t_m - t_{m-1})$. In this manner, the problem of gradually-decreasing void ratio or pressure is reduced to a number of sudden-drop-in-pressure problems of the kind already solved. The solution at any instant would then be represented as a sum of solutions obtained by considering the various step functions.

The principle of superposition is valid only for linear partial differential equations. As shown in the previous section, the correspondence between the variable-property and constant-property solutions is obtained by using the pseudo-void ratio function. Thus, superposition should be quite good for matching a variable void ratio change in any m_D. On the basis of the results obtained, it appears that superposition, which would require positive incremental values, would appear to be good. It should be noted, however, that the results obtained also depend slightly on the ratio e_0/e_i. When decreasing incremental values are considered, it is not evident that superposition can apply for negative incremental values. This is due to the fact that the expansion case does not correlate as well with the constant property solution as the compaction case. It should be noted, too, that as the ratio e_0/e_i is reduced, the correlations approach the constant-property case. This implies that superposition should hold for small increases of the e_0/e_i ratio.

As was the case for the pseudo-void ratio transformation, the validity of the superposition concept has to be demonstrated only through the comparison with finite difference solutions. This comparison can be easily made and is not shown principally due to the excellent comparison demonstrated by Al-Hussainy et al. (1966).

Effect of gravity and overburden potential on flow through deformable porous media

The concepts discussed so far may be extended to include the effect of gravity. To include gravitational effects, the potential, Φ, is defined as:

$$\Phi = \psi/\rho_L - zg \qquad (8\text{-}159)$$

where ψ is defined by eq.8-71, ρ_L is the density of fluid, z is the distance above a fixed datum ($z = 0$), and g is acceleration due to gravity. Equation 8-159 is similar to eq.8-62.

The inclusion of the overburden potential is slightly more complicated and requires some discussion. Philip (1969a, b) has shown that the overburden potential can be viewed as consisting of two components, solid and liquid, as follows (again it is assumed that the base ($z = 0$) of the column is constrained and the surface is unconstrained):

(1) Mechanical load, σ, on the column against which work has to be done if the column moves in an upward direction.

(2) Work has to be performed at every point where liquid is added, as the sediment is assumed to be capable of movement. Any addition of liquid to the sediment requires a local increase in the bulk volume so as to account for the extra liquid.

On the basis of the above considerations, Philip (1969a, b) has shown that the overburden potential, Ω, can be defined as:

$$\Omega = \sigma(z) + \int_{z=0}^{z} \gamma_b dz \tag{8-160}$$

where γ_b, the wet apparent specific weight, is given by:

$$\gamma_b = \frac{(\rho_L e + \rho_r)g}{\rho_L (1 + e)} \tag{8-161}$$

Thus the total potential including the effect of gravity and overburden is equal to:

$$\Phi_t = \psi/\rho_L - gz + \sigma(z) + \int_{z=0}^{z} \gamma_b \, dz \tag{8-162}$$

As presented above, it may be assumed that the overburden potential is limited to vertical systems only; however, this is not the case. For horizontal systems, if the unconstrained face is loaded, the overburden potential would exist. (The term overburden is somewhat a misnomer for this case, but the usage is consistent and systematic from a mathematical viewpoint.) In this instance the overburden potential is given by:

$$\Omega = \sigma(z) \tag{8-163}$$

The second component of eq.8-160 drops out as there are no external forces acting on the unconstrained face to consider and thus no work has to be performed. As pointed out by Philip (1969a), eq.8-160 may also be obtained on the basis of thermodynamic arguments.

Equilibrium solution to the porous column including gravity and the overburden potential

In this section the simple case of the vertical equilibrium of a porous column is examined. In the discussion which follows, the external load is assumed to act at $z = 0$. If Φ_{eq} is the equilibrium value of the potential, then one can equate this to the right hand side of eq.8-162 and the following equation would result:

$$\Phi_{eq} = \psi/\rho_L - gz + \sigma(0) + \int_{z=0}^{z} \gamma_b \, dz \tag{8-164}$$

Differentiating with respect to z and if $e = e_0$ at $z = 0$, the following relationship is obtained:

$$z(e) = \int_e^{e_0} \frac{1}{\rho_L(\gamma_b - g)} \frac{d\psi}{de} de \tag{8-165}$$

which may be simplified as:

$$z(e) = \frac{1}{(\rho_r - \rho_L)g} \int_e^{e_0} (1+e) \frac{d\psi}{de} de \tag{8-166}$$

As $\rho_r \neq \rho_L$, singular solutions are not possible for this case.

The above analysis applies only in the case of a two-component system, i.e., liquid and rock. The analysis in the presence of a third component (gas) is more complicated (see Philip, 1969a, b).

Steady and unsteady vertical flow including the effect of gravity and overburden potential

The solution for steady vertical flows may be written directly using eq.8-162. The following equation is obtained on substituting for Φ_t in Darcy's Law equation the right-hand side of eq.8-162:

$$u_{Lr} = -\frac{k}{\mu} \left[\frac{d\psi}{de} \frac{de}{dz} + \rho_L \gamma_b - \rho_L g \right] \tag{8-167}$$

Equation 8-167 can be rearranged as:

$$\frac{dz}{de} = \frac{-\frac{d\psi}{de}}{\frac{\mu u_{Lr}}{k} - \rho_L g + \rho_L \gamma_b} \qquad (8\text{-}168)$$

Integration of eq.8-168 would give the required values of $z(e)$. The right hand side of eq.8-168, however, is singular for:

$$u_{Lr} = \frac{k}{\mu} \frac{(\rho_L - \rho_r)g}{(1+e)} \qquad (8\text{-}169)$$

Appropriate singular and non-singular solutions can be obtained from eq.8-168 and 8-169. The equilibrium solution examined earlier (eq.8-166) can be used to delineate between steady upward and downward flow.

The case of unsteady vertical flows can also be easily examined. Using the continuity requirement given by eq.8-137 and the definition of the potential, Φ_t, from eq.8-162, the general equation describing deformable porous media may be written as:

$$\frac{\partial e}{\partial t} = \frac{\partial}{\partial m}\left[D_0 \frac{\partial e}{\partial m}\right] - \frac{\partial}{\partial m}\left[\frac{k}{\mu} \rho_L \left(g - \gamma_b\right)\right] \qquad (8\text{-}170)$$

where:

$$D_0 = \frac{k}{\mu(1+e)} \frac{d\psi}{de} \qquad (8\text{-}171)$$

The second term on the right hand side of eq.8-170 is of the order m^{-1} and for small values of time, t, can be neglected as compared to the first term. Thus, one reverts to the form of the equation already considered extensively (eq.8-143). Noting that D_0 and D_m are equivalent, short-time solutions to eq.8-170 can be obtained for various cases already examined. The pseudo-void ratio transformation can again be used. A detailed solution of eq.8-170 is yet to be presented. It may be solved by techniques suggested by Philip (1969c) or Parlange (1971a, b, c; 1972a, b, c, d, e).

Discussion of the modified Terzaghi problem

The main objective of the preceding discussion has been to examine anew the problem first considered by Terzaghi in 1923 in light of modern developments which have taken place over the past five years. In spite of the large number of papers which have been published, it has been shown that major elements have been neglected until recently. The perusal of proceedings of an international conference on land subsidence (1969), sponsored by the UNESCO, would give a good indication of the state of the art in the mid-sixties. This chapter, on the other hand, should provide a firm basis for future studies.

The analysis considered in this section has been restricted to the one-dimensional movement, principally because in nature most dimensional changes appear to be predominantly in, or limited to, the vertical direction. This appears to be due essentially to the horizontal constraints of the masses surrounding the compacting sediment. It should be noted, however, that individual solid particles may change in volume three-dimensionally (McNaab, 1960). In order to examine the three-dimensional bulk-volume changes, an idea of the stress distribution would also be essential. This aspect is examined in another section.

The pseudo-void ratio transformation presented herein is an important advance in the analysis of the problems examined. Transformations of this kind have been used since 1868 (see Carslaw and Jaeger, 1951). In the petroleum industry, the substitution was first used by Muskat (1949) in the study of potentiometric models. Fay and Prats (1951) have also suggested a similar procedure to solve nonlinear flow problems in porous media. Techniques for transforming partial differential equations, similar to the technique used in this chapter, have received marked attention in the petroleum industry due to the real gas—pseudo-pressure concept presented by Al-Hussainy et al. (1966). Instead of choosing a convenient value to correlate results, as suggested by Smiles and Poulos (1969), the transformation provides a sound basis for the analysis of the non-linear problem. On the basis of the results presented, the use of the transformation is justified. The proposed transformation should resolve the dilemma of Smiles and Poulos on the choice of the value of D_m to define a dimensionless time in presenting solutions obtained by numerical means.

The mathematical theory developed in this section, has been confirmed experimentally to a limited extent by Smiles and Rosenthal (1968) who considered the sediment columns to be semi-infinite. The experimental results agree very well with theoretical predictions. In the authors' opinion, a great need exists for determining diffusivities of various kinds of porous media for both compaction and expansion. It is hoped that more data will become available in the future so that $D(\phi)$, $k(\phi)$ and $\psi(\phi)$ can be described adequately.

On the basis of the above discussion, the principal conclusions may be summarized as follows:

(1) In an unstable porous medium, when there is a movement of the solid, Darcy's Law applies to flow relative to the solid particles.

(2) The magnitude of the mass flux, associated with the movement of sediment, can be of the same order of magnitude as the movement of liquid relative to the solid.

(3) The pseudo-void ratio transformation provides a basis for correlating non-linear solutions with constant-property solutions. Useful engineering and practical approximations can be obtained through the use of this transformation.

(4) The effects of gravity and overburden potential may be included in the mathematical analysis if required.

RADIAL FLOW TOWARDS A WELL LOCATED AT THE CENTER OF A CLOSED RADIAL COMPACTING RESERVOIR

The concept that the porous medium is absolutely rigid and nondeformable is a valid assumption for a wide range of problems of practical interest. It has been long realized that in some circumstances this assumption can lead to certain discrepancies, however, and that the use of "average" properties of the medium would reduce these errors. Considerable research effort has been made to study the effect of pressure-dependent rock properties (e.g., compressibility, porosity, and permeability) and fluid properties, using analytical and/or numerical techniques. As a result, numerous methods of solutions have been outlined in principle and a large number of particular problems have been solved by means of high-speed digital computers.

Rowan and Clegg (1962) presented a thorough review of the basic equations governing fluid flow in porous media. They demonstrated the effect of including parameters as functions of fluid pressure or space variables, and discussed the implications of linearizing the basic equations. The effect of discontinuities and non-homogeneities, etc., on flow through porous media has been discussed by a large number of workers (e.g., Mueller, 1962; Bixel et al., 1963; Carter, 1966; Bixel and Van Poollen, 1967; Closmann and Ratliff, 1967).

Similar problems are encountered in the case of heat conduction and a large variety of solutions is available for particular cases of transient and steady-state heat flow (e.g., Storm, 1951; Friedman, 1958; McMordie, 1962).

The flow of real gases through porous media is another similar, but more complex, problem. As mentioned earlier, the development of the theory and solutions of general utility have been considered in detail by

Al-Hussainy et al. (1966), both from a fundamental and an engineering point of view. Clegg (1968) has presented an approximate analytical solution for the same problem.

Even though the problem has been long recognized, general correlation solutions when rock and fluid properties are general functions of pressure have been developed only recently (Raghavan et al., 1972). The objective of this section is to demonstrate techniques of handling transients in a compacting porous medium where rock and fluid properties are general functions of fluid pressure. A single-well circular reservoir filled with a slightly compressible fluid and undergoing compaction is examined here. The well is located at the center of the reservoir, the outer boundary of which is assumed to be impermeable. The well is produced at a constant rate.

Mechanism of the flow of compressible liquids in compressible porous media

The mechanism of the flow of fluids through porous media is governed by the fluid pressure distribution, the physical properties of the rock matrix, the geometry of the media, and the $P-V-T$ relations of the fluids. The medium considered in the present section is assumed to be homogeneous and the compressibility, porosity, and permeability are arbitrary functions of fluid pressure.

The isothermal pore volume compressibility of the system is given by (see eq.8-59, p.419):

$$c_p = \frac{1}{V_p} \left(\frac{\partial V_p}{\partial p_p} \right)_T$$

As in the previous analysis, an elemental volume of the porous medium is considered. It is assumed that this volume is large enough compared to the size of the pores, so that it can be treated as homogeneous and, at the same time, small enough compared to the scale of the macroscopic phenomena, so that it can be considered to be infinitesimal from a mathematical viewpoint. If δx, δy and h represent the length, breadth and height of this elemental volume, then the pore volume associated with the elemental volume is $\delta x \cdot \delta y \cdot h\phi$, where ϕ is the fractional porosity. Then, if the compaction or expansion of the porous medium is limited to the vertical direction, the pore volume compressibility given by eq.8-59 becomes:

$$c_p = \frac{1}{h(p_p)\phi(p_p)} \left[\frac{\partial (h\phi)}{\partial p_p} \right]_T \qquad (8\text{-}172)$$

MATHEMATICAL ANALYSIS OF SAND COMPACTION

If the compressibility of the liquid, c_l, is assumed constant and if vertical components of the flow caused by changing thickness are neglected, the continuity equation for an elemental volume of thickness h yields:

$$\nabla \cdot [h(p_p)\rho(p_p)\vec{u}] = -\frac{\partial}{\partial t}[h(p_p)\rho(p_p)\phi(p_p)] \tag{8-173}$$

where $\rho(p_p)$ is the density of the fluid, $\vec{u}$ is the velocity of the fluid, and t is time.

Darcy's law for horizontal flow is:

$$\vec{u} = -\frac{k(p_p)}{\mu(p_p)} \nabla p_p \tag{8-174}$$

where $k(p_p)$ is the permeability of the medium and $\mu(p_p)$ is the viscosity of the fluid. Both the permeability and viscosity are assumed to be functions of pressure. It should be noted that in the present section the symbols $\nabla \cdot$ and ∇ relate only to the horizontal plane.

The following equation is obtained on substituting the value of $\vec{u}$ given by eq.8-174 in eq.8-173:

$$\nabla \cdot \left[\frac{h(p_p)\rho(p_p)k(p_p)}{\mu(p_p)} \nabla p_p \right] = \frac{\partial}{\partial t}[h(p_p)\rho(p_p)\phi(p_p)] \tag{8-175}$$

If the volume occupied by solid material in bulk volume $(\delta x \cdot \delta y \cdot h)$ is constant, then:

$$h(p_p)[1 - \phi(p_p)] = h_i(1 - \phi_i) \tag{8-176}$$

where the subscript i is used to denote initial conditions. Substituting for h its value as determined from eq.8-176 in eq.8-175 gives:

$$\nabla \cdot \left\{ \frac{h_i}{[1 - \phi(p_p)]} \frac{(1 - \phi_i)\rho(p_p)k(p_p)}{\mu(p_p)} \nabla p_p \right\} =$$

$$= \frac{\partial}{\partial t}[h(p_p)\rho(p_p)\phi(p_p)] \tag{8-177}$$

Application of the chain rule of differentiation to the right-hand member results in:

$$\frac{\partial}{\partial t}[h(p_p)\rho(p_p)\phi(p_p)] = \frac{\partial}{\partial p_p}[h(p_p)\rho(p_p)\phi(p_p)]\frac{\partial p_p}{\partial t} \qquad (8\text{-}178)$$

$$= h(p_p)\rho(p_p)\phi(p_p)\left[\frac{1}{\rho(p_p)}\frac{\partial \rho}{\partial p_p} + \right.$$

$$\left. + \frac{1}{h(p_p)\phi(p_p)}\frac{\partial (h\phi)}{\partial p_p}\right]\frac{\partial p_p}{\partial t} \qquad (8\text{-}179)$$

Referring to eq.8-172 and noting that the first term within the square brackets of the right-hand member of eq.8-178 is the liquid compressibility, c_l, eq.8-177 can be expressed in the following form:

$$\nabla \cdot \left\{\frac{h_i(1-\phi_i)\rho(p_p)k(p_p)}{[1-\phi(p_p)]\mu(p_p)}\nabla p_p\right\} =$$

$$= h(p_p)\rho(p_p)\phi(p_p)[c_l + c_p(p_p)]\frac{\partial p_p}{\partial t} \qquad (8\text{-}180)$$

Multiplying through by $(1-\phi_i)h_i$, noting eq.8-176 and that $(c_l + c_p)$ is the total compressibility c_t, for a system containing a liquid only, eq.8-180 reduces to:

$$\nabla \cdot \left\{\frac{\rho(p_p)k(p_p)}{[1-\phi(p_p)]\mu(p_p)}\nabla p_p\right\} = \frac{\phi(p_p)\rho(p_p)c_t(p_p)}{[1-\phi(p_p)]}\frac{\partial p_p}{\partial t} \qquad (8\text{-}181)$$

In a manner similar to that described in a previous section (p.447) a "pseudo-pressure" is defined by:

$$W(p_p) = \int_{p_{pm}}^{p_p} \frac{\rho(p_p)k(p_p)}{[1-\phi(p_p)]\mu(p_p)}\,dp_p \qquad (8\text{-}182)$$

where p_{pm} is the base pressure. On differentiating eq.8-182, the following relations are obtained:

$$\frac{\partial W(p_p)}{\partial t} = \frac{\rho(p_p)k(p_p)}{[1-\phi(p_p)]\mu(p_p)}\frac{\partial p_p}{\partial t} \qquad (8\text{-}183)$$

and:

$$\frac{\partial W(p_p)}{\partial x} = \frac{\rho(p_p) k(p_p)}{[1 - \phi(p_p)]\mu(p_p)} \frac{\partial p_p}{\partial x} \qquad (8\text{-}184)$$

A similar expression can be written for $\partial W(p_p)/\partial y$.

In light of eq.8-183 and 8-184, eq.8-181 can be written as:

$$\nabla^2 W(p_p) = \frac{\phi(p_p) \mu(p_p) c_t(p_p)}{k(p_p)} \frac{\partial W(p_p)}{\partial t} \qquad (8\text{-}185)$$

where $\nabla^2 \equiv \dfrac{\partial^2}{\partial x^2} + \dfrac{\partial^2}{\partial y^2}$

If ϕ, μ, c_t, and k are constants and the second degree terms are neglected, the conventional form of the diffusivity equation would result:

$$\nabla^2 p_p = \frac{\phi \mu c_t}{k} \frac{\partial p_p}{\partial t} \qquad (8\text{-}186)$$

Comparison of eq.8-185 and 8-186 indicates that the form of the diffusivity equation is preserved in terms of the pseudo-pressure, $W(p_p)$. However, important distinctions exist. First, eq.8-185 is nonlinear due to the fact that the diffusivity is pressure dependent. Second, in the development of eq.8-186, second-degree terms were dropped. A third distinction also exists: For a well producing at a constant rate in a finite, closed reservoir, it can be shown that every point in the reservoir will eventually experience constant rate of pressure decline with respect to time, provided that ϕ, μ, c_t, and k are each constant and that pressure gradients are everywhere small. Under these conditions the right-hand side of eq.8-186 is constant, reflecting a state of flow that has been called pseudosteady, quasisteady and, even, steady. Whatever the name used, this state represents a condition in which the outer boundary affects pressure distribution. In eq.8-185, the diffusivity is pressure dependent and, thus, the right-hand side of this equation cannot be a constant. This means that usage of the term "pseudosteady state" implies only an "apparent pseudosteady state", or a state of flow in which the reservoir boundary affects fluid production.

The specification of the problem is complete when the boundary and initial conditions are stated. The usual conditions used for reservoir study are constant pressure or constant mass rate of production at one of the boundaries of the system; for example, at the well wall. These would have to be expressed, however, in terms of the pseudo-pressure, $W(p_p)$. To make the conversion, it is necessary to note that the mass flow rate is given by:

$$\vec{q}\rho(p_p) = -\frac{A(p_p)k(p_p)\rho(p_p)}{\mu(p_p)} \nabla p_p \qquad (8\text{-}187)$$

$$= -\frac{A'(p_p)k(p_p)\rho(p_p)}{[1-\phi(p_p)]\mu(p_p)} \nabla p_p \qquad (8\text{-}188)$$

where $A'(p_p)/[1-\phi(p_p)] = A(p_p)$. Thus, in terms of the pseudo-pressure, the mass flow rate is given by:

$$\vec{q}\rho(p_p) = -A'(p_p) \nabla W(p_p) \qquad (8\text{-}189)$$

For the specification of constant pressure, eq.8-182 can be used directly for the definition of the pseudo-pressure. Thus, eq.8-182 and 8-189 can be used when pressure and/or rate are specified along the boundaries. In this context, it would be worthwhile to consider the definition of the pseudo-pressure used by Al-Hussainy et al. (1966):

$$W(p_p) = 2 \int_{p_{pm}}^{p_p} \frac{p_p}{\mu(p_p)z(p_p)} dp_p \qquad (8\text{-}190)$$

where $z(p_p)$ is the compressibility factor for the real gas and $\mu(p_p)$ is the viscosity of the real gas.

Although eq.8-182 and 8-190 are similar, significant differences do exist. The basic objectives of these two equations are the same, namely, to transform non-linear differential equations to a form similar to diffusion-type equations.

Equation 8-182 can also be used to calculate the pseudo-pressure function at the initial condition. Normally, the pressure in the reservoir is assumed to be uniform. Thus, $W(p_p)$ can be evaluated directly by eq.8-182.

Solution to transient radial flow

The specific problem considered in this section is the transient radial flow of variable-property liquids in a single-well radial reservoir, where the compressibility, porosity and permeability are arbitrary functions of pressure, when the outer boundary is impermeable (closed). A well producing at a constant rate is assumed to be located at the center of the reservoir. The initial pressure in the reservoir is assumed to be constant. Equation 8-181 for idealized radial flow becomes:

$$\frac{1}{r}\frac{\partial}{\partial r}\left\{r\frac{\rho(p_p)k(p_p)}{[1-\phi(p_p)]\mu(p_p)}\frac{\partial p_p}{\partial r}\right\} = \frac{\phi(p_p)\rho(p_p)c_t(p_p)}{[1-\phi(p_p)]}\frac{\partial p_p}{\partial t} \qquad (8\text{-}191)$$

and eq.8-185 can be expressed as:

$$\frac{1}{r}\frac{\partial}{\partial r}\left[r\frac{\partial W}{\partial r}\right] = \frac{\phi(p_p)\mu(p_p)c_t(p_p)}{k(p_p)}\frac{\partial W}{\partial t} \qquad (8\text{-}192)$$

Applying methods of numerical analysis, a number of authors have solved equations similar to eq.8-191 and 8-192 for certain special problems. The objective in this section, however, is to examine the general solution to these problems obtained by Raghavan et al. (1972). The general solution can be applied for engineering or practical purposes, provided the $W(p_p)$ function (eq.8-182) is available, without recourse to a digital computer. In order to examine the solution technique developed, however, it is necessary to first consider constant-property solutions developed by Van Everdingen and Hurst (1949).

The work of Van Everdingen and Hurst refers to the solution of eq.8-186, the constant-property equation, for idealized radial flow. The circular, closed reservoir is produced by a single well located at the center of the reservoir at a constant rate, q. The solution is given by:

$$p_D(r_D, t_D) = \frac{1}{2}[\ln(t_D/r_D^2) + 0.80907] \text{ for } t_D \leq \frac{1}{4}\left(\frac{r_e}{r_w}\right)^2 \qquad (8\text{-}193)$$

and:

$$p_D(r_D, t_D) = \ln r_D - 0.75 + 2t_D\left(\frac{r_w}{r_e}\right)^2 \text{ for } t_D > \frac{1}{4}\left(\frac{r_e}{r_w}\right)^2 \qquad (8\text{-}194)$$

where p_D is the dimensionless pressure drop, t_D is the dimensionless time, ln denotes the natural logarithm, and r_D is the dimensionless radial coordinate. The well radius is r_w and the reservoir radius is r_e. The dimensionless parameters of pressure, time and radial coordinate are defined, respectively, as:

$$p_D(r_D, t_D) = \frac{2\pi kh}{q\mu}[p_{pi} - p_p(r, t)] \qquad (8\text{-}195)$$

$$t_D = \frac{kt}{\phi\mu c r_w^2} \tag{8-196}$$

$$r_D = \frac{r}{r_w} \tag{8-197}$$

Equation 8-193 enables the solution for an infinite reservoir, whereas 8-194 represents the solution when boundary effects become dominant. It should be noted that in eq.8-194 the dimensionless pressure, p_D, is a linear function of dimensionless time, t_D. This implies that the rate of change of pressure with respect to time is constant. As mentioned earlier, this condition of flow is commonly called the pseudo-steady-state flow behavior.

Equation 8-192, together with the work of Aronofsky and Jenkins (1954) concerning the flow of ideal gases in porous media, and the work of Al-Hussainy et al. (1966) pertaining to real gas flow in porous media, provide a basis for the approach of Raghavan et al. (1972). The principal characteristics of the governing differential equations for the flow of ideal and real gases have been reviewed and compared by Al-Hussainy et al. (1966). The most important contribution by Aronofsky and Jenkins (1954) to the theory of flow of ideal gases through porous media was their conclusion that the production of gas at a constant rate from a closed radial flow system could be approximated very closely with the solutions of transient liquid flow developed by Van Everdingen and Hurst (1949), if the dimensionless time used is based on a compressibility corresponding to the initial pressure. Similarly, Al-Hussainy et al. (1966) showed that solutions for the flow of real gases could be correlated with the Van Everdingen and Hurst (1949) solutions, provided that the compressibility and viscosity factors in the dimensionless-time group were evaluated at the initial pressure. The basis for their correlation is that the variation of the viscosity—compressibility product with $W(p_p)$ for a real gas is similar to the variation of compressibility of an ideal gas ($1/p_p$) with pressure squared. The comparison is given by Al-Hussainy et al. (1966). Raghavan et al. (1972) pointed out a similar basis for the problem under discussion. In order to examine this basis, the following ratio $\tilde{\alpha}$, is defined as:

$$\tilde{\alpha} = \left[\frac{\phi\mu c}{k}\right]_{p_{pm}} \bigg/ \left[\frac{\phi\mu c}{k}\right]_{p_p} \tag{8-198}$$

The variation of $\tilde{\alpha}$ with the $W(p_p)/W(p_{pi})$ ratio is shown in Fig.8-22. Figure 8-22 also shows the variation of μc_g with $W(p_p)$ for a real gas flow in terms of reduced properties. Figure 8-22 indicates that the variation of $\tilde{\alpha}$ with respect to the ratio $W(p_p)/W(p_{pi})$ is similar to the variation of μc_g with

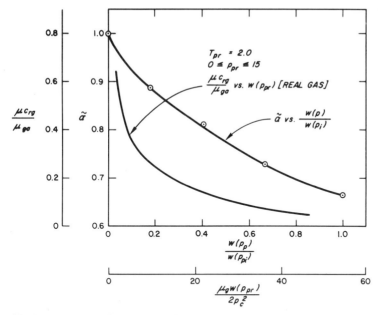

Fig.8-22. Pressure-dependent diffusivity term versus pseudo-pressure (After Raghavan et al., 1972, fig.1, p.270).

$W(p_p)$ for a real gas. One might, therefore, expect that the numerical solutions obtained by using eq.8-192 should correlate with the constant-property solutions, provided that the definition of a dimensionless time group, t_D, is based on initial values of viscosity, porosity, total system compressibility and permeability. Thus, the dimensionless time, t_D, is given by:

$$t_D = \frac{k(p_{pi})}{\phi(p_{pi})\mu(p_{pi})c_t(p_{pi})r_w^2} \, t \qquad (8\text{-}199)$$

A dimensionless potential drop, $W_D(r_D, t_D)$, and a dimensionless flow rate, $q_D(r_D, t_D)$, can be defined by the following equations, respectively:

$$W_D(r_D, t_D) = \frac{2\pi h(p_{pi})[1 - \phi(p_{pi})]}{q\rho(p_{pi})} [W(p_{pi}) - W(r, t)] \qquad (8\text{-}200)$$

and:

$$q_D(r_D, t_D) = \frac{\rho(p_{pi})}{2\pi h(p_{pi})[1 - \phi(p_{pi})][W(p_{pi}) - W(p_{pm})]} q(r, t) \qquad (8\text{-}201)$$

where $r_D = r/r_w$. The above dimensionless potential drop (eq.8-200) is similar to the dimensionless pressure drop (eq.8-195) of van Everdingen and Hurst (1949). It may be suggested that eq.8-191 and 8-192 can be written directly in terms of dimensionless quantities defined above; however, this leads to many errors. It should be noted that a basis for using dimensionless quantities must be established in advance (see Al-Hussainy et al., 1966). Particular care must be taken in evaluating W_D, t_D, etc., on the basis of initial values of system properties. If this is not done, neither the correlations developed in this section nor those developed by Al-Hussainy et al. (1966) are meaningful. Unfortunately, this aspect of correlation development has not been emphasized in earlier literature. Moreover, the part played by the pseudo-pressure (eq.8-182) in transforming the non-linear partial differential equation into a form similar to the diffusivity equation, but with a potential-dependent diffusivity, is not readily evident if the governing equation is written directly in terms of the dimensionless variables. In using the correlations described, these remarks should be borne in mind.

The $W_D(1, t_D)$ computer solutions developed by Raghavan et al. (1972) are compared to the van Everdingen—Hurst (1949) $p_D(1, t_D)$ theoretical solutions in Fig.8-23 ($r_D = 1$). The triangular-, square- and diamond-shaped data points represent the results of the Raghavan et al. (1972) study. The solid lines (which appear to be drawn through these points) are the Van Everdingen and Hurst (1949) curves. For each one of the finite ratios of radii presented (r_e/r_w = 251, 501, and 751), three dimensionless flow rates were considered (q_D = 0.00336, 0.00665, and 0.00933). The dimensionless flow rates refer to the flow rates at the well. Clearly, the correlation is excellent over the ranges of these variables. The variable property values used by Raghavan et al. in developing the data points are given in Table 8-II. However, any consistent set of similar data may be applied.

Departure from the Van Everdingen—Hurst (1949) solutions occurs for large values of t_D when the boundary effects are felt. This is seen more clearly in Fig.8-24 than in Fig.8-23. It is also seen in Fig.8-24 that this departure is a function of flow rate, indicating the solutions to be rate sensitive. Both Fig.8-23 and 8-24 disclose that at large times, reservoir pressures derived from the $W_D(1, t_D)$ solutions are lower than those from the $p_D(1, t_D)$ solutions. Thus, a terminal producing pressure is reached at a different time for the variable-property case than for the constant-property case. These observations emphasize the desirability of rigorously incorporating the rock and fluid properties that are pressure-dependent, rather than using average properties.

Raghavan et al. (1972) have shown that the concept of transient drainage radius, introduced by Aronofsky and Jenkins (1954), can be applied to the solution of liquid flow problems of interest here in a manner

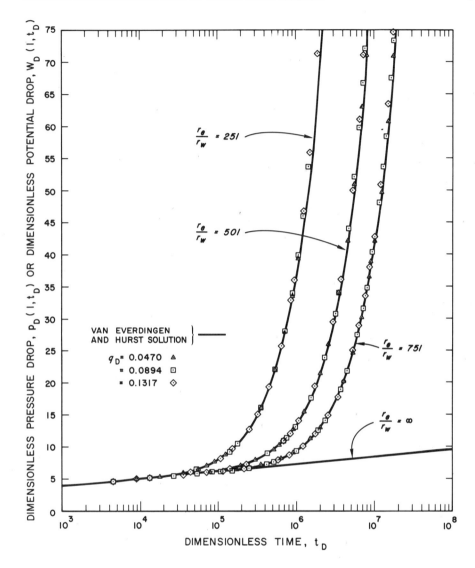

Fig.8-23. $W_D(1, t_D)$ computer solutions superimposed on the Van Everdingen and Hurst (1949) solutions, $p_D(1, t_D)$, for a closed reservoir producing at a constant rate. (After Raghavan et al., 1972, fig.2, p.271.)

TABLE 8-II

Data on oil reservoir rocks and liquid properties showing variation of properties with pore fluid pressure (After Raghavan et al., 1972, table 1, p.271.)

Pore fluid pressure, p_p (psi)	Porosity, $\phi(p_p)$	Permeability, $k(p_p)$ (md)	Compressibility, $c_p(p_p) \cdot 10^6$ (psi^{-1})	Viscosity[1], $\mu(p_p)$ (cp)	Density, $\rho(p_p)$ (lb/ft^3)
500	0.285	64.0	11.2	5.2	50.125
1,000	0.287	70.5	14.2	5.4	50.250
1,500	0.290	86.0	19.5	5.6	50.375
2,000	0.293	110.0	27.2	5.8	50.500
2,500	0.300	143.0	40.0	6.0	50.625

[1] After Amyx et al. (1960, p.441).

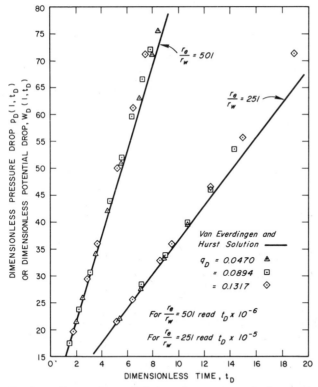

Fig.8-24. Comparison of $W_D(1, t_D)$ computer solutions with the Van Everdingen and Hurst (1949) solutions, $p_D(1, t_D)$, for idealized radial flow at large times. (After Raghavan et al., 1972, fig.3, p.271.)

similar to that used by Al-Hussainy et al. (1966) in their study of gas flow. The relevant equations in the case of liquid flow are (from Raghavan et al., 1972):

$$\ln \frac{r_d}{r_w} = W_D(p_{pw}) - W_D(\bar{p}_p) \tag{8-202}$$

where the transient drainage radius, r_d, can be correlated with t_D as:

$$\ln \frac{r_d}{r_w} = p_D(t_D) - 2t_d \left(\frac{r_w}{r_e}\right)^2 \tag{8-203}$$

In eq.8-202, $W_D(\bar{p}_p)$ is the dimensionless potential corresponding to the average pressure in the system, which can be evaluated using material balance calculations. The average pressure in the reservoir, $\bar{p}_p$, is the mean pressure and is of interest principally owing to two reasons: (1) the average pressure in the reservoir is a direct reflection of the quantity of fluid in place and is necessary to perform material balance calculations; and (2) if the well is shut in, the shut-in pressure of the well, p_{pws}, will approach the average pressure as the shut-in time, Δt, approaches infinity, i.e., $p_{pws} \to \bar{p}$ as $\Delta t \to \infty$.

The Aronofsky and Jenkins (1954) drainage radius correlation given by eq.8-202 and 8-203 is shown in Fig.8-25. The data points obtained by Raghavan et al. (1972) are only slightly lower than those of the constant-property solution. No theoretical justification exists for the use of eq.8-202 and 8-203; they are not solutions to eq.8-192. Figure 8-25 shows, however, that for $10^3 < t_D < 2 \cdot 10^7$, eq.8-202 and 8-203 yield a better correlation with the constant-property solutions than that reflected in Fig.8-23. This is true in spite of the fact that both eq.8-194 and 8-203 are solutions to the radial form of the constant-property differential equation given by eq.8-186. Al-Hussainy et al. (1966) also observed that the transient drainage concept yields better results with respect to their real-gas—pseudo-pressure correlations. The data presented in Fig.8-23 can be used to calculate rapidly the average pressure in the system.

Regarding the transient drainage radius concept, it is important not to confuse r_d with the radial coordinate, r_D. Furthermore, it is to be noted that Fig.8-25 indicates that for long production times, r_d becomes equal to $0.472 r_e$. For example, the curve of the figure representing $r_e/r_w = 501$ becomes a horizontal line for $t_D > 3.0 \cdot 10^4$. This horizontal line corresponds to $\ln(r_d/r_w) = 5.46$ and is a constant. Thus, $r_d/r_w = 236$ and $r_d/r_e(236/501) = 0.472$. Having r_d equivalent to $0.472 r_e$ does not mean that the physical drainage radius stabilizes about half way out in the reservoir. It merely means that the reservoir is behaving as though the flow were steady

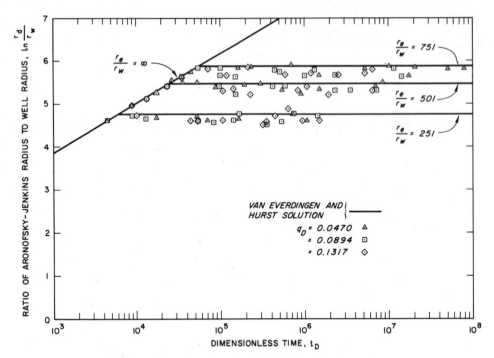

Fig.8-25. $W_D(1, t_D)$ computer solutions superimposed on $p_D(1, t_D)$ solutions using the Aronofsky and Jenkins drainage radius concept for a closed radial reservoir producing at a constant rate. (After Raghavan et al., 1972, fig.4, p.272.)

and the outer boundary is at a distance of 0.472 r_e instead of at r_e. Nonetheless, the whole reservoir from r_w to r_e is affected by the changing pressure distribution at all times.

Steady state and approximate pseudo-steady-state flow

The steady-state radial flow of a compressible liquid in a medium where compressibility, porosity, and permeability are arbitrary functions of pressure is considered in this section. For steady-state flow, Laplace's equation results. In axisymmetric coordinates, it is given by:

$$\frac{1}{r}\frac{d}{dr}\left[r\frac{dW(p_p)}{dr}\right] = 0 \qquad (8\text{-}204)$$

The boundary conditions are:

$$r = r_w: W(p_p) = W(p_{pw}) \qquad (8\text{-}205)$$

$r = r_e$: $W(p_p) = W(p_{pe})$ (8-206)

where p_{pw} is the fluid pressure at the well and p_{pe} is the fluid pressure at the external boundary. The pseudo-pressure functions $W(p_{pw})$ and $W(p_{pe})$ correspond to the pressures p_{pw} and p_{pe}, respectively. Integrating eq.8-204 and using the boundary conditions, the steady state potential distribution is given by:

$$W(p_{pr}) - W(p_{pw}) = \left[\frac{W(p_{pe}) - W(p_{pw})}{\ln \frac{r_e}{r_w}} \right] \ln \frac{r}{r_w} \qquad (8\text{-}207)$$

where $W(p_{pr})$ is the pseudo-pressure function corresponding to the pressure at distance r in the reservoir. All other symbols have been defined previously.

Equation 8-207 can be rearranged in the form analogous to the familiar radial flow equation as if the rate of production is constant:

$$q = \frac{2\pi h_i (1 - \phi_i)}{\rho(p_{pw})} \left[\frac{W(p_{pe}) - W(p_{pw})}{\ln \frac{r_e}{r_w}} \right] \qquad (8\text{-}208)$$

Equation 8-207 should also apply for the case where the pressure at the outer boundary is constant and when the dimensionless time is large. Monteiro (1970) has shown that for the constant-pressure outer boundary case, eq.8-207 can be written as:

$$W_D(1, t_D) = \ln \frac{r_e}{r_w} \qquad (8\text{-}209)$$

if t_D is large. Numerical computations on the other hand show that $W_D(1, t_D)$ values are significantly higher than the ones obtained by using eq.8-209. This corresponds to the higher values of $W_D(t_D)$ obtained when t_D is large for the closed outer boundary case. The reason for this is not evident and requires further study. It should be noted, however, that the real gas solutions are described very well by eq.8-209 (Wattenbarger, 1967).

It has already been shown that for long times, the flow equation for a closed outer boundary and constant production rate can be written as:

$$\ln \frac{r_d}{r_w} = \ln \frac{0.472 \, r_e}{r_w} = W_D(p_{pw}) - W_D(\bar{p}_p) \qquad (8\text{-}210)$$

It was also pointed out that eq.8-202 has no theoretical basis because true pseudo-steady-state does not exist for the model under consideration. Figure 8-25, however, suggests that for practical purposes such a condition does exist for the point $r_D = 1$. Figure 8-26 shows the relationship for values of $r_D = 1$, 26 and 501 and demonstrates flow behavior during the pseudo-steady-state period. The important feature of Fig.8-26 is that $W_D(r_D, t_D)$ versus t_D curves for various radial locations are essentially parallel. Thus the $W(p_p)$ profile is essentially independent of time and the $W(p_{pr})$ distribution with respect to radial distance can be obtained readily. For example, on the basis of Fig.8-26, the following equation should describe the $W(p_{pr})$ distribution at any instant in time adequately:

$$\ln \frac{0.606\, r_e}{r_w} = \frac{2\pi h_i (1 - \phi_i)}{q\rho(p_{pw})} [W(p_{pe}) - W(p_{pw})] \tag{8-211}$$

Discussion of radial flow in a compressible reservoir

The mathematical model described above (pp.462—476) offers a technique of general utility, whereby the behavior of a sand undergoing compaction due to withdrawal of a fluid can be predicted. This can be made in a general manner without further resort to a digital computer. A description of the procedure of obtaining fluid pressures for problems of interest has been described by Raghavan et al. (1972).

The method for transient radial flow presented here can also be applied to linear flow problems. Comparison of Fig.8-16 and 8-23, however, indicates that the pseudo-function correlations, which are to a large extent geometry-dependent for the radial flow solutions, appear to correlate better with their respective constant-property solutions rather than the linear flow numerical solutions and the corresponding constant-property equations. This has also been pointed out by Al-Hussainy (1966). Thus, care should be exercized in extending the transformation discussed here so far to other applications, particularly where the applicability has not been demonstrated. For example, the results presented in Fig.8-20 indicate that care should be taken in applying these results to injection (expansion) systems. It should be recalled that the pseudo-void ratio transformation is not adequate in obtaining an adequate correlation solution.

It should also be noted that even though the $W_D(t_D)$ and $p_D(t_D)$ solutions correlate very well, the actual wellbore pressures for the two cases are different. This is shown in Fig.8-27, where the ratio of wellbore pressure to initial pressure, p_{pi}, for the two cases is graphed as a function of dimensionless time, t_D. From Fig.8-27 it can be seen that the

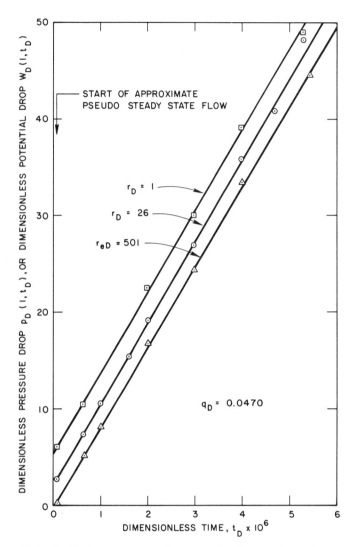

Fig.8-26. $W_D(r_D, t_D)$ distribution as a function of dimensionless time, t_D, for approximate pseudo-steady-state conditions.

variable-property solution is lower than the van Everdingen and Hurst solution. This indicates the need for including the variation in properties (rock and fluid) in a rigorous way.

It should be noted that the mathematical model developed here (pp.462–476) considers the properties of the porous matrix to be functions of the fluid pressure. This problem has also been solved by considering the

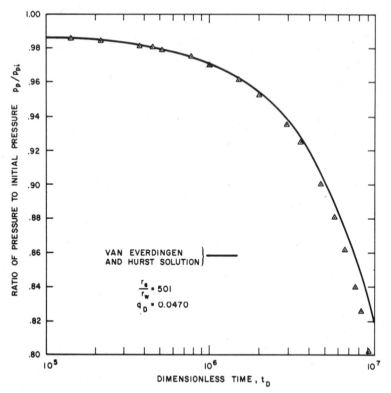

Fig.8-27. Comparison of wellbore pressures for constant property and variable property solutions. (After Raghavan et al., 1972, fig.5, p.273.)

various parameters (i.e., permeability, porosity, and compressibility) to be functions of effective stress. An elegant analysis by Cooper (1966) considers this aspect. Cooper has also assumed that the compaction of the sands occurs predominantly in the vertical direction; however, he does include the motion of the solid and liquid in the vertical direction. Cooper has also considered the problem in moving and stationary reference coordinates. Thus, he resolved an important controversy similar to the one involved in the Terzaghi (1923) model. In the model presented, this controversy is avoided because gradients in the vertical direction are ignored. Cooper's (1966) mathematical development of this problem is a special case of the three-dimensional theory developed by Biot (1935, 1941a). The three-dimensional effects are discussed in detail later.

The following specific conclusions can be drawn from the discussion of the model presented here (pp.462—476):

(1) Solutions of general utility can be obtained when reservoir and fluid properties are considered as arbitrary functions of pressure.

MATHEMATICAL ANALYSIS OF SAND COMPACTION 479

(2) When reservoir and fluid properties are considered as arbitrary functions of pressure, the solutions obtained can be correlated with the constant-property solution. At large values of time, there is a significant deviation from the van Everdingen and Hurst (1949) curves.

(3) The Aronofsky and Jenkins (1954) drainage radius concept represents an excellent engineering approximation to the solutions obtained to the nonlinear equation examined in this section, even though no theoretical basis exists whereby the validity of the concept can be demonstrated.

NUMERICAL SOLUTIONS

As shown in the previous sections, the equation governing the compaction or expansion (rebound) of a one-dimensional porous column is non-linear and, except for a class of problems already discussed, a digital computer is essential. In addition, the equation governing the flow to wells where the properties are pressure dependent is also a non-linear equation. The solution of the initial-value problems of the kind discussed so far has been studied by numerical analysts for a considerable length of time and, as a result, a number of numerical techniques are available. A few of the techniques are discussed in this section.

The numerical equation solution governing the one-dimensional consolidation (or rebound expansion) of a porous column is considered first. Techniques to solve both the semi-infinite and finite columns, in stationary and moving reference frames, are considered. Next, the equation governing the radial flow to the well at the center of a closed reservoir is examined. Even though the physical problems are different, the numerical techniques are similar. A brief introduction to finite difference approximations is also provided.

The methods presented herein have been used by the writers with success. They are presented here with a minimum of detail. The reader is referred to the following well-known texts for a more thorough discussion: Forsythe and Wasow (1960), Smith (1964), and Mitchell (1969).

A numerical solution technique to the one-dimensional, semi-infinite deformable porous column (eq.8-90)

The analysis discussed here is similar to the one suggested by Philip (1955, 1968). For the purpose of convenience the compaction problem is examined, but the analysis may be easily amended to the consideration of the expansion process. It involves taking λ (see eq.8-88, p.429) as the independent variable and employing a straightforward integral procedure, where one initial condition is determined approximately and then improved by the iteration. A modified version of eq.8-90 is the most convenient form for the application of the method discussed below.

The interval of interest $\phi = \phi_i$ to ϕ_0 is divided into i equal steps, where each step is given by $\delta\phi$. The $\lambda(\phi)$ curve is replaced by a histogram with λ_r and ϕ_r as coordinates of the midpoint of step r.

Using eq. 8-93, eq. 8-90 is written as:

$$\frac{(1-\phi)}{(1-\phi_i)} \int_{\phi_i}^{\phi_0} \lambda d\phi - \int_{\phi}^{\phi_0} \lambda d\phi = -2D(1-\phi)\frac{d\phi}{d\lambda} \tag{8-213}$$

where D is the diffusivity defined by eq.8-73.

Defining the following difference approximations:

$$I_{1/2} = \frac{1}{\delta\phi} \int_{\phi_i}^{\phi_0} \lambda d\phi \tag{8-214}$$

$$I_r = \frac{1}{\delta\phi} \int_{\phi_i}^{\phi_r} \lambda d\phi \tag{8-215}$$

$$\left[(1-\phi)D\frac{d\phi}{d\lambda}\right]_{r+1/2} = -\frac{\overline{\overline{D}}_{r+1/2}\,\delta\phi}{\lambda_{r+1} - \lambda_r} \tag{8-216}$$

where:

$$\overline{\overline{D}}_{r+1/2} = \frac{1}{\delta\phi} \int_{\phi_{r+1}}^{\phi_r} (1-\phi)D(\phi)d\phi \tag{8-217}$$

eq.8-213 can be written at the point $r + 1/2$ as follows:

$$\frac{1-\phi_{r+1/2}}{1-\phi_i} I_{1/2} - (I_{1/2} - I_{r+1/2}) = \frac{2\overline{\overline{D}}_{r+1/2}}{\lambda_{r+1/2} - \lambda_r} \tag{8-218}$$

or:

$$\lambda_{r+1} - \lambda_r = \frac{2\overline{\overline{D}}_{r+1/2}}{I_{r+1/2} - \left(\frac{\phi_{r+1/2} - \phi_i}{1-\phi_i}\right)I_{1/2}} \tag{8-219}$$

MATHEMATICAL ANALYSIS OF SAND COMPACTION 481

The iterative procedure developed by Philip (1968) is based on eq.8-219. Before the procedure can be formally applied, the following aspects have to be discussed:

(1) The determination of λ and $\int \lambda d\phi$ when $\phi = \phi_i$, i.e., the starting conditions. This is characteristic of the problem under consideration because as $\phi \to \phi_i$, $\lambda \to \infty$.

(2) The determination of $I_{1/2}$.

Philip (1955, 1968) has suggested that when ϕ is close to ϕ_i, an analytical solution can be obtained by assuming that $(1-\phi)D$ is represented by a constant value of $\overline{\overline{D}}_{i-1/2}$. If this assumption is made, then the governing differential equation reduces to the following form:

$$\frac{\partial \phi}{\partial t} = \overline{\overline{D}}_{i-1/2} \frac{\partial^2 \phi}{\partial z^2} - \frac{I_{1/2} \, \delta\phi}{2(1-\phi_i)} t^{-1/2} \frac{\partial \phi}{\partial z} \quad (8\text{-}220)$$

subject to the conditions:

$$\phi = \phi_{i-1}; \; z = \lambda_{i-1} t^{1/2}; \; \phi \to \phi_i; \; z \to \infty \quad (8\text{-}221)$$

The required solution of eq.8-220 and 8-221 is:

$$\lambda(\phi) = \frac{I_{1/2} \, \delta\phi}{1-\phi_i} + 2\sqrt{\overline{\overline{D}}_{i-1/2}} \, \text{inverfc} \left[\left(\frac{\phi - \phi_i}{\delta\phi}\right) \right.$$

$$\left. \text{erfc} \left(\frac{\lambda_{i-1} - (1-\phi_i)^{-1} I_{1/2} \, \delta\phi}{2\sqrt{\overline{\overline{D}}_{i-1/2}}}\right) \right] \quad (8\text{-}222)$$

where the notation "inverfc" (Philip, 1955, 1960) is the inverse of the error function complement. The error function complement, erfc(x) is defined by the equation:

$$\text{erfc}(x) = \frac{2}{\pi^{1/2}} \int_x^\infty \exp(-\zeta^2) \, d\zeta \quad (8\text{-}223)$$

Integrating eq.8-222 with respect to ϕ, and using the relation $\phi_r = \phi_0 - r\delta\phi$, the following equation is obtained:

$$I_{i-1/2} = \frac{\lambda_{i-1}}{2} + 2\sqrt{\overline{\overline{D}}_{i-1/2}} \left\{ \frac{\text{ierfc} \left[\dfrac{\lambda_{i-1} - (1-\phi_i)^{-1} I_{1/2} \, \delta\phi}{2\sqrt{\overline{\overline{D}}_{i-1/2}}}\right]}{\text{erfc} \left[\dfrac{\lambda_{i-1} - (1-\phi_i)^{-1} I_{1/2} \, \delta\phi}{2\sqrt{\overline{\overline{D}}_{i-1/2}}}\right]} \right\} \quad (8\text{-}224)$$

The function ierfc(x) is given by the following equation (Abramowitz and Stegun, 1964, p.299):

$$\text{ierfc}(x) = \int_x^\infty \text{erfc}\,\zeta\, d\zeta = \frac{1}{\sqrt{\pi}} \exp(-x^2) - x\,\text{erfc}(x) \qquad (8\text{-}225)$$

If the function $A(x)$ is denoted as:

$$A(x) = 2x\,\text{ierfc}(x)/\text{erfc}(x) \qquad (8\text{-}226)$$

then eq.8-224 may be rewritten as:

$$I_{i-1/2} = \frac{\lambda_{i-1}}{2} + \frac{2\sqrt{\overline{D}_{i-1/2}}}{[\lambda_{i-1} - (1-\phi_i)^{-1} I_{1/2}\,\delta\phi]} A$$

$$\left[\frac{\lambda_{i-1} - (1-\phi_i)^{-1} I_{1/2}\,\delta\phi}{2\sqrt{\overline{D}_i}_{1/2}} \right] \qquad (8\text{-}227)$$

Inspection of eq.8-227 indicates that $I_{i-1/2}$ can be determined if $I_{1/2}$ and λ_{i-1} are known.

Using eq.8-225, the function $A(x)$ can also be written as:

$$A(x) = \frac{2x \exp(-x^2)}{\sqrt{\pi}\,\text{erfc}(x)} - 2x^2 \qquad (8\text{-}228)$$

From eq.8-228 the following properties of $A(x)$ can be determined: $x = 0$, $\frac{dA}{dx} = 2\,\text{ierfc}(0) = 1.1284$, and $A(x) \to 1$ as $x \to \infty$.

Other values of $A(x)$ may be found as shown by Carslaw and Jaeger (1959, p.485).

The value of $I_{1/2}$ can be estimated by using a constant value of $(1-\phi)D$, which may be obtained from:

$$[(1-\phi)D]^1 = \frac{2}{(\phi_0 - \phi_i)^2} \int_{\phi_i}^{\phi_0} (\phi - \phi_i)(1-\phi)D(\phi)d\phi \qquad (8\text{-}229)$$

In eq.8-229, the superscript on the left-hand side represents the first guess. The integral $[I_{1/2} = \frac{1}{\delta\phi} \int_{\phi_i}^{\phi_0} \lambda d\phi]$ can be evaluated using the analytical solutions already developed.

The iterative procedure can be easily carried out as follows: Values of $\bar{\bar{D}}_{r+1/2}$ are tabulated from the known D function. A trial value $I_{1/2}$ obtained as described above and, assuming a value of λ_0 (in the case of compaction $\lambda_0 = 0$), λ_1 can be calculated using eq.8-219. $I_{3/2}$ can then be obtained from the relation:

$$I_{r+1/2} = I_{r-1/2} - \lambda_r \tag{8-230}$$

by using the following definitions:

$$\int_{\phi_r}^{\phi_{r+1/2}} \lambda d\phi = -\lambda_r \frac{\delta \phi}{2} \tag{8-231}$$

and:

$$\int_{\phi_{r-1/2}}^{\phi_r} \lambda d\phi = -\lambda_r \frac{\delta \phi}{2} \tag{8-232}$$

From eq.8-219 and 8-230, λ_r and $I_{r+1/2}$, respectively, can be evaluated in an alternate fashion from 1 to $[i-1]$. The value of $I_{i-1/2}$ obtained by using this procedure is then compared with the value obtained from eq.8-224. If the two values of $I_{i-1/2}$ are such that the relative difference between them is less than a predetermined tolerance, then the iteration is discontinued. If not, a new value of $I_{1/2}$ is assumed (on the basis of the computations) and the calculations are repeated. The above technique can also be applied when the problem is examined in the moving reference frame and the procedure is similar (see Philip, 1955, for details).

As in the case of non-linear problems, a formal analysis of the rate of convergence, stability, etc., is not possible. Undoubtedly, the rate of convergence would depend on the initial estimate of $I_{1/2}$, for it controls the entire sequence of calculations. On the basis of the results obtained by Philip (1955), it can be stated that the procedure converges rapidly. Consequently, with modern, high-speed digital computers no serious problems should arise.

Numerical solution to the one-dimensional finite deformable porous column by finite difference methods

The frequently used and readily applicable method in the numerical solution of non-linear partial differential equations is the method of finite differences. The one-dimensional problem is examined in the moving reference frame.

As usual, it is assumed that the function is single-valued, continuous, and finite. The goal of all finite difference schemes is to convert the continuous function (the dependent variable) defined on the space S (time and distance) by a set of discrete functions defined over a grid of discrete mesh points (see Fig.8-28). The basis for this is the Taylor Series, which for a function $u(x, y)$ can be expressed as follows (Abramowitz and Stegun, 1964, p.297):

$$u(x + h, y + k) = u(x, y) + h\, u_x(x, y) + k\, u_y(x, y)$$

$$+ \frac{1}{2} [h^2 u_{xx}(x, y) + 2hk\, u_{xy}(x, y) + k^2 u_{yy}(x, y)]$$

$$+ \frac{1}{6} \{h^3 u_{xxx}(x, y) + 3[h^2 k\, u_{xxy}(x, y) + hk^2 u_{xyy}(x, y)]$$

$$+ k^3 u_{yyy}(x, y)\} + 0(h^4 + k^4) \tag{8-233}$$

where h and k refer to mesh sizes in the x and y directions and the subscripts refer to partial derivatives. The last term on the right-hand side of eq.8-233 is the truncation error. The essence of the finite difference procedure is given by eq.8-233 and the required partial derivatives can be easily obtained using

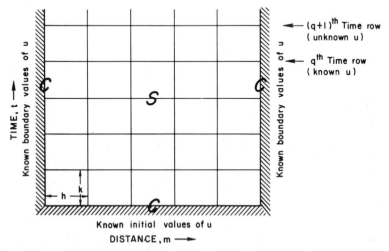

Fig.8-28. Schematic representation of finite difference technique.

this equation. For example, the partial derivative u_x at a point (x, y) represented by the mesh coordinates (p, q) can be expressed as:

$$\left(\frac{\partial u}{\partial x}\right)_{x,y} \equiv (u_x)_{p,q} = \frac{u_{p+1,q} - u_{p,q}}{h} + O(h) \tag{8-234}$$

Equation 8-234 is obtained by setting $k = 0$ in eq.8-233 and neglecting terms of the order h. As shown by eq.8-234, the truncation error is of the same magnitude as the mesh size. In order to increase the accuracy of the finite difference approximation, a central difference rather than a forward difference approximation is generally used. This approximation is given by the equation:

$$\left(\frac{\partial u}{\partial x}\right)_{x,y} \equiv (u_x)_{p,q} = \frac{u_{p+1,q} - u_{p-1,q}}{2h} + O(h^2) \tag{8-235}$$

Equation 8-235 can be obtained by setting $k = 0$ in eq.8-233 and writing out the Taylor Series for points $(x + h, y)$ and $(x - h, y)$. Simplification of the resulting equation yields eq.8-235. In a similar manner, the second derivative is given as:

$$\left(\frac{\partial^2 u}{\partial x^2}\right)_{x,y} \equiv (u_{xx})_{p,q} = \frac{u_{p+1,q} - 2u_{p,q} + u_{p-1,q}}{h^2} \tag{8-236}$$

The finite difference approximation to the partial system differential equation governing the one-dimensional compaction or rebound of a porous column on the moving reference frame is considered next. Equation 8-143 (p. 445) may be rewritten as:

$$\frac{\partial e}{\partial t_D} = \frac{\partial}{\partial m_D}\left(D_D \frac{\partial e}{\partial m_D}\right) \tag{8-237}$$

where $D_D = \frac{D_m}{D_{mi}}$, D_m is the diffusivity defined by eq.8-144, and e is the void ratio. The symbols m_D and t_D have been defined in eq.8-153 and 8-150, respectively. The initial condition is:

$$e = e_i; 0 \leqslant m_D \leqslant 1 \tag{8-238}$$

and the boundary conditions are:

$$t_D > 0: e = e_0 \text{ at } m_D = 0 \tag{8-239}$$

$$t_D > 0: \frac{de}{dm_D} = 0 \text{ at } m_D = 1 \tag{8-240}$$

In the following discussion, the subscript "D" is dropped for the purpose of convenience. On the basis of eq.8-236, the finite difference approximation on the right-hand side of eq.8-237 can be written as:

$$\frac{\partial}{\partial m}\left(D\frac{\partial e}{\partial m}\right)_{m,t} = \frac{1}{\Delta m^2}[D_{p+1/2,q+r}(e_{p+1,q+r} - e_{p,q+r}) -$$

$$- D_{p-1/2,q+r}(e_{p,q+r} - e_{p-1,q+r})] \tag{8-241}$$

In the above equation, p and q refer to the m and t coordinates, respectively, and Δm is the mesh spacing of the m coordinate. The subscript r can be set equal to 0, 1/2 and 1, which is discussed in detail later. The truncation error has been dropped in eq.8-241.

It should be noted that the values of D in eq.8-241 are to be evaluated at points halfway between the mesh points p and $p-1$ or p and $p+1$. As the value of D is dependent on e and as e is defined only at mesh points $p-1$, p, $p+1$, etc., eq.8-241 requires modification. This modification is achieved by approximating the value of D at $p+1/2$ by the average value of D at p and $p+1$. A similar approximation is used for the point $p-1/2$. Thus, if $D_{p+1/2, q+r}$ is replaced by $0.5 (D_{p, q+r} + D_{p+1, q+r})$ and $D_{p-1/2, q+r}$ is replaced by $0.5 (D_{p, q+r} + D_{p-1, q+r})$, then the right side of eq.8-241 may be written as:

$$\frac{1}{2\Delta m^2}[(D_{p,q+r} + D_{p+1,q+r})e_{p+1,q+r} -$$

$$- (D_{p+1,q+r} + 2D_{p,q+r} + D_{p-1,q+r})e_{p,q+r} +$$

$$+ (D_{p,q+r} + D_{p-1,q+r})e_{p-1,q+r}] \tag{8-242}$$

The left-hand side of eq.8-237 can be approximated as:

$$\left(\frac{\partial e}{\partial t}\right)_{m,t} \equiv (e_t)_{p,q} = \frac{e_{p,q+1} - e_{p,q}}{\Delta t} \tag{8-243}$$

where Δt is the time increment chosen. Thus, the finite difference formulation to eq.8-237 can be expressed as:

$$(D_{p,q+r} + D_{p+1,q+r})e_{p+1,q+r} - (D_{p+1,q+r} + 2D_{p,q+r} + D_{p-1,q+r})$$

$$e_{p,q+r} + (D_{p,q+r} + D_{p-1,q+r})e_{p-1,q+r} = \frac{2\Delta m^2}{\Delta t}(e_{p,q+1} - e_{p,q}),$$

(8-244)

where $1 \leqslant p \leqslant p_{max}$ and $1 \leqslant q \leqslant q_{max}$. The symbols p_{max} and q_{max} represent the maximum values of p and q. It should be noted that p_{max} is equal to $(1/\Delta m + 1)$, whereas q_{max} is governed by the total time of interest.

The boundary conditions can be approximated in a straightforward manner. Inasmuch as the condition at $m = 0$ is the first kind (Dirichlet, in: Mitchell, 1969, p.40),

$$e(0, t) = e_{1,q} = e_i \text{ for } 1 \leqslant q \leqslant q_{max} \tag{8-245}$$

The condition at $m = 1$ is of the second kind (Neumann, in: Mitchell, 1969, p.40). In this case, the approximation is as follows:

$$(e_m)_{1,t} = \frac{e_{p+1,q} - e_{p-1,q}}{2\Delta m} = 0, p = p_{max} \text{ and } 1 \leqslant q \leqslant q_{max} \tag{8-246}$$

which implies that:

$$e_{p+1,q} = e_{p-1,q}, p = p_{max} \text{ and } 1 \leqslant q \leqslant q_{max} \tag{8-247}$$

It should be noted that $(p_{max} + 1)$ is a fictitious point which is normally encountered with Neumann-type boundary conditions. The initial condition can also be approximated as:

$$e \equiv e_{p,1} = e_i \text{ for } 1 \leqslant p \leqslant p_{max} + 1 \tag{8-248}$$

Two important considerations, i.e., the choice of space and time increments, arise in the solution of the above finite difference equations. In the authors' experience, using about twenty increments along the m direction is adequate. The choice of the time step is more complicated. Undoubtedly, one would like to solve the problem with a minimum of computation effort. Additional factors, such as the facilities available and user expertise also play a role. As mentioned earlier, the subscript r can be set equal to 0, 0.5 or 1. The choice of the time step depends on the value of r. If zero is chosen for r, then the resulting difference is known as the explicit or forward difference scheme (Fig.8-29). This scheme is not

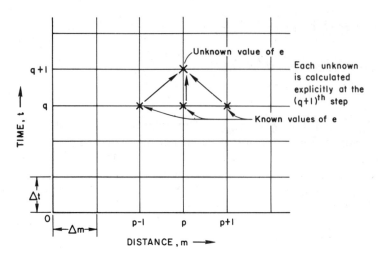

Fig.8-29. Schematic representation of the explicit finite difference scheme.

recommended, because Δt has to be extremely small. This leads to excessive computation time, even on a digital computer. This method is only recommended if the problem has to be solved using a desk calculator. Other schemes discussed below are difficult to handle on a desk calculator.

As the problem is nonlinear, the value of the time step Δt would have to be chosen empirically. Undoubtedly, the choice depends on D and Δm. The formula:

$$\Delta t \leqslant \frac{\Delta m^2}{2} \qquad (8\text{-}249)$$

which is only applicable if D is constant, can serve as a guide for the choice of the time step if the explicit scheme is to be used.

If the value of r is chosen to be unity, then the backward-difference, implicit technique results (Fig.8-30). This has two inherent advantages: (1) no stability conditions exist (i.e., there is no limit on the size of Δt) and (2) no oscillations in the solution result when Δt is chosen to be large. It should be noted, however, that the truncation error of the finite difference procedure is of the order of $(\Delta t + \Delta x^2)$. Thus, Δt should be chosen such that the truncation term does not dominate the solution.

If r is set equal to 0.5, the extremely popular Crank—Nicholson (1947) implicit procedure results. The functions at $q + 1/2$ are assumed to be equal to the average function at q and $q + 1$, i.e., $D_{p,\,q\,+\,1/2}$ is replaced by 0.5 $(D_{p,q} + D_{p,q\,+\,1})$. The approximation is of the order $(\Delta x^2 + \Delta t^2)$ and the scheme is unconditionally stable. The method has a serious defect, however,

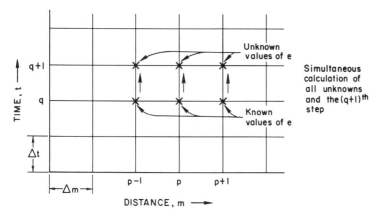

Fig.8-30. Schematic representation of the implicit finite difference scheme.

in that the solutions oscillate for the large values of Δt. For this reason, the backward-difference scheme is recommended as the best available method for the solution of this problem.

It should be noted that the use of the backward-difference or Crank—Nicholson scheme requires the value of D at $q + 1$ be known. This value is dependent on the value of e at the $q + 1^{th}$ step, which is unknown. The reason for describing these two techniques as implicit schemes should now be evident.

Both the Crank—Nicholson procedure and the backward-difference procedure may be represented as:

$$a_j e_{j-1} + b_j e_j + c_j e_{j+1} = d_j \qquad (8\text{-}250)$$

where $1 \leqslant j \leqslant R$. If the boundary conditions (eq.8-245 and 8-247) at $m_D = 0$ and 1 are used, then it can be shown that $a_1 = c_R = 0$. Values of d_j may be calculated using values of e_j at the current step. For the first time step, d_j would be given by the initial condition (eq.8-248). Thus, the right-hand side of eq.8-250 is known at the beginning of each time step. As already pointed out, a_j, b_j and c_j are functions of e, at the new time step as well as the current time step. An estimate of e_j at the new time step is made and values of D_j are calculated. Thus, values of a_j, b_j and c_j are obtained. As values of a_j, b_j and c_j are known, new values of e_j can be calculated by solving eq.8-250. These values of e_j are then used to compute D_j for the next iteration. The iteration for a particular time step is discontinued when values of e_j obtained on successive iterations are within a predetermined tolerance.

The algorithm to solve the system represented by eq. 8-250 was first suggested by Thomas (1949). The procedure described by algorithm can be summarized as follows. For each iteration, the first step involves the calculation of β_j and γ_j which are given by:

$$\beta_j = b_j - \frac{a_j \, c_{j-1}}{\beta_{j-1}} \quad \text{with } \beta_1 = b_1 \tag{8-251}$$

and:

$$\gamma_j = \frac{d_j - a_j \, \gamma_{j-1}}{\beta_j} \quad \text{with } \gamma_1 = \frac{d_1}{b_1} \tag{8-252}$$

The second step involves the calculation of e_j from the equations:

$$e_R = \gamma_R \tag{8-253}$$

and:

$$e_j = \gamma_j - \frac{c_j \, e_{j+1}}{\beta_j} \quad \text{for } R-1 \leqslant j \leqslant 1 \tag{8-254}$$

Numerical solution of the equation governing transient radial flow

Equation 8-192 (p.467), the partial differential equation to be solved by the method of finite difference, may be written as:

$$\frac{1}{r_D} \frac{\partial}{\partial r_D} \left(r_D \frac{\partial W_D}{\partial r_D} \right) = \widetilde{\alpha}_D \frac{\partial W_D}{\partial t_D} \tag{8-255}$$

where:

$$\widetilde{\alpha}_D = [\phi \mu c / k]_{p_p} / [\phi \mu c / k]_{p_{pi}} \tag{8-256}$$

The associated boundary conditions for $t_D \geqslant 0$ are:

$$\left(r_D \frac{\partial W_D}{\partial r_D} \right)_{r_D = 1} = - \left[\frac{\rho(p_p)}{\rho(p_{pi})} \right]_{r_D = 1} \tag{8-257}$$

and:

$$\left(\frac{\partial W_D}{\partial r_D}\right)_{r_{eD}} = 0 \qquad (8\text{-}258)$$

Initially, $W_D = 0$ everywhere in the reservoir.

The finite-difference procedures already discussed can be readily adopted. To facilitate computation, however, the coordinate system is modified by the transformation:

$$x_D = \ln r_D \qquad (8\text{-}259)$$

where ln represents natural logarithm. Using eq. 8-259, eq. 8-255 reduces to:

$$\frac{\partial}{\partial x_D}\left(\frac{\partial W_D}{\partial x_D}\right) = \tilde{\alpha}_D \exp(2x_D) \frac{\partial W_D}{\partial t_D} \quad \text{for } t_D > 0 \text{ and } 0 < x_D < x_{eD} \qquad (8\text{-}260)$$

On using eq. 8-259, the boundary conditions (eq. 8-257 and 8-258) are given by:

$$t_D > 0: \quad \left(\frac{\partial W_D}{\partial x_D}\right)_{x_D = 0} = -\frac{\rho(p_{pw})}{\rho(p_{pi})} \qquad (8\text{-}261)$$

and:

$$t_D > 0: \quad \left(\frac{\partial W_D}{\partial x_D}\right)_{x_D = x_{eD}} = 0 \qquad (8\text{-}262)$$

Initially, $W_D = 0$ for $0 \leqslant x_D \leqslant x_{eD}$.

Using the concepts already discussed, the finite-difference approximation can be written as:

$$\lambda(W_{p+1,q+1} - 2W_{p,q+1} + W_{p-1,q+1}) + (1-\lambda)(W_{p+1,q} - 2W_{p,q} + W_{p-1,q}) = \alpha_{p,q+\lambda} \exp(2x_p) \frac{\Delta x^2}{\Delta t}(W_{p,q+1} - W_{p,q}) \qquad (8\text{-}263)$$

where λ is a weighting factor. The explicit finite-difference scheme is obtained if λ is chosen to be zero in eq. 8-263. The backward-difference implicit procedure results if $\lambda = 1$ and the Crank and Nicholson formula is obtained for $\lambda = 1/2$. As discussed earlier, the backward-difference procedure is recommended. Again, for the purpose of convenience, the subscript "D" has been dropped.

The finite-difference approximations to the boundary conditions of eq.8-261 and 8-262 are:

$$4W_{2,q+1} - 3W_{1,q+1} - W_{3,q+1} = -2\Delta x \left[\frac{\rho(p_p)}{\rho(p_{pi})} \right]_{1,q+1} \quad (8\text{-}264)$$

and:

$$W_{p-1,q+1} = W_{p+1,q+1} \quad (8\text{-}265)$$

where $p = p_{max}$.

Equation 8-264 is a special difference form of $\frac{\partial W}{\partial x}$ and is obtained by considering two intervals, $(x + \Delta x)$ and $(x + 2\Delta x)$, in the forward direction. Its accuracy is of the order (Δx^2). Equation 8-263 can also be put in the form eq.8-250 and the Thomas algorithm can be used.

It should be mentioned that eq.8-244, the difference approximation, is essentially identical to eq.8-263, except that $x \equiv m$, $W_D \equiv e$ and $\exp(2x_p) \equiv 1$. Thus, even though the two problems are different, the finite-difference solution techniques are identical.

ASPECTS OF THREE-DIMENSIONAL DEFORMATION OF POROUS MEDIA

It is rather fortunate that in many instances the one-dimensional theory suffices to describe compaction of sediments, even though it is easy to point out many practical instances where purely geometric conditions can preclude the application of the one-dimensional analysis. Many limitations of the one-dimensional theory have been recognized for a long time. Undoubtedly, a general theory which includes the three-dimensional effects would describe practical problems more adequately. As a result, a large number of papers examining three-dimensional effects have appeared in the scientific and engineering literature based on the foundations laid down more than thirty years ago by Biot (1935). The progress in the application of the theory, however, has been slow for a number of reasons. The principal reason appears to be the inability to apply the theories developed to practical situations. There is a dearth of ready, practical solutions (Davis and Poulos, 1972). Another reason is that the three-dimensional bulk volume change requires assumptions regarding stress distributions in porous media. This aspect of the theory requires a more detailed and thorough examination (Philip, 1972).

In this section, the porous matrix is assumed to behave like an elastic solid. As shown here, the macroscopic stress—strain relation describing the three-dimensional deformation of a porous matrix is similar to that of the elastic non-porous solid bodies. In fact, the set of stress—strain relations are very similar to those used in the branch of elasticity concerned with thermal effects and which is called thermoelasticity. This analogy was first pointed out by Lubinski (1954) and later by Geertsma (1957b). In 1966, Geertsma again pointed out this comparison and coined the term "poroelasticity" to describe the macroscopic stress—strain relations encountered while examining the three-dimensional deformation of porous solids. As the theory of thermoelasticity is much more advanced than poroelasticity, the solutions developed in the former field may be easily applied to problems of interest in the latter field. A striking example of exploiting the analogy between thermoelasticity and poroelasticity is found in the work of Evangelisti and Poggi (1970) and by Geertsma (1966, 1973). These papers examine the movement of the land surface due to the compaction or rebound expansion of a porous body situated several hundreds or thousands of feet below the surface. As stated earlier, the solution to this problem has eluded many workers for a long period of time. Unfortunately, the work of Evangelisti and Poggi (1970) is in Italian and thus may not be readily accessible to many readers. Evangelisti and Poggi examined the movement of the land surface by both analytical and numerical techniques. The analytical technique is similar to that of Geertsma (1966), which is presented here. Readers interested in numerical solutions are referred to the original work of Evangelisti and Poggi (1970).

In the opinion of the writers, the work of Geertsma (1973) will become a standard reference and serve as a basis for future work. The results obtained by Geertsma are extremely general and have a wide range of applicability. The simplicity of the results obtained and the physical implications of the solution are striking.

In this section, the three-dimensional deformation of porous solids is examined in a simple way. The development of the theory of poroelasticity is along the classic lines presented by Biot (1935, 1941a, 1955, 1956). The analogy between thermoelasticity and poroelasticity is also demonstrated here. Finally, the results obtained by Geertsma are discussed and their implications analyzed.

Assumptions in the theory of poroelasticity

In the development of the theory of poroelasticity, the work of Biot (1935, 1941a, 1955, 1956) is closely followed. The following assumptions are made in the development of the theory presented here:

(1) The porous medium may be represented as an elastic medium.
(2) The porous material is isotropic.
(3) The stress—strain relations are reversible under final equilibrium conditions.
(4) The stress—strain relationships are linear.
(5) The strains are small.
(6) The fluid contained in the pores is incompressible; the fluid may, however, contain some gaseous bubbles.
(7) The fluid flows through the porous skeleton according to Darcy's Law.
(8) Body forces due to gravity are neglected.

As pointed out by Biot (1955), the assumption of isotropy may be easily taken into account. Of greatest concern to the consideration of coarse-grained sediments are assumptions 3, 4 and 5. Van der Knaap (1959) has pointed out methods by which assumptions 4 and 5 can be removed. The refinements which are necessary to eliminate these assumptions are not discussed here.

Development of stress—strain relationships in a porous medium

A small cubic element of the porous medium, with its sides being parallel to the coordinate axes (Fig.8-3, p.415) is considered. As usual, this element is taken to be large enough compared to the size of the pores so that it can be treated as homogeneous and at the same time small enough, compared to the scale of the macroscopic phenomena, so that it can be considered to be infinitesimal in the mathematical sense. It has already been shown (eq.8-37, p.415) that the stresses in a porous medium may be represented by:

$$S \equiv \begin{bmatrix} \sigma_x & \tau_{xy} & \tau_{xz} \\ \tau_{yx} & \sigma_y & \tau_{yz} \\ \tau_{zx} & \tau_{zy} & \sigma_z \end{bmatrix}$$

For the system under consideration, $\tau_{xy} = \tau_{yx}$, etc. (Biot, 1941a; Geertsma, 1966). That is, the stress tensor is symmetric with respect to its main diagonal. Thus, eq.8-37 can be written as:

$$S \equiv \begin{bmatrix} \sigma_x & \tau_z & \tau_y \\ \tau_z & \sigma_y & \tau_x \\ \tau_y & \tau_x & \sigma_z \end{bmatrix} \qquad (8\text{-}266)$$

where τ_{xy} has been replaced by τ_z, τ_{yz} by τ_x, etc.

Denoting u, v, and w as the displacement of the solid in the x, y, and z directions, respectively, the strain relations are given by:

$$\left. \begin{array}{ll} \epsilon_x = \dfrac{\partial u}{\partial x} & \epsilon_{xy} = \dfrac{\partial u}{\partial y} + \dfrac{\partial v}{\partial x} \\ \epsilon_y = \dfrac{\partial v}{\partial y} & \epsilon_{yz} = \dfrac{\partial v}{\partial z} + \dfrac{\partial w}{\partial y} \\ \epsilon_z = \dfrac{\partial w}{\partial z} & \epsilon_{zx} = \dfrac{\partial w}{\partial x} + \dfrac{\partial u}{\partial z} \end{array} \right\} \qquad (8\text{-}267)$$

The amount of liquid in the pores must also be considered in order to describe fully the macroscopic conditions of the porous medium. This can be represented by the ratio of the variation of the pore volume to the bulk volume, which is denoted by θ:

$$\theta = \frac{dV_p}{V_b} \qquad (8\text{-}268)$$

where V_p is the pore volume and V_b is the bulk volume.

As it has been assumed that changes in the sediment are reversible, the macroscopic condition of the sediment must be a definite function of the variables ϵ_x, ϵ_y, ϵ_z, ϵ_{xy}, ϵ_{yz}, ϵ_{zx}, and θ. These seven variables, in turn, must be definite functions of σ_x, σ_y, σ_z, τ_x, τ_y, τ_z, and p_p.

As only small strains are considered and the consequent change in pore volume is small, the relation between these variables may be considered to be linear. If the fluid pressure is taken to be equal to zero, the components of strain are functions only of the stresses. As the material is assumed to be isotropic and as the development of the theory is exclusively concerned with small strains, Hooke's Law for an isotropic, elastic, non-porous body is valid (Timoshenko and Goodier, 1970, p.8):

$$\epsilon_x = \frac{\sigma_x}{E} - \frac{\nu}{E}(\sigma_y + \sigma_z); \; \epsilon_y = \frac{\sigma_y}{E} - \frac{\nu}{E}(\sigma_z + \sigma_x); \; \epsilon_z = \frac{\sigma_z}{E} - \frac{\nu}{E}(\sigma_x + \sigma_y);$$

$$\epsilon_{xy} = \tau_z/G; \; \epsilon_{yz} = \tau_x/G; \; \epsilon_{zx} = \tau_y/G \qquad (8\text{-}269)$$

In eq.8-269, the constant E refers to Young's modulus or to the modulus of elasticity in tension, G is the shear modulus or modulus of rigidity, and ν is Poisson's ratio. A more detailed discussion regarding these elastic constants may be found in any standard textbook on elasticity (e.g., Sokolnikoff, 1956, chapter 1; Timoshenko and Goodier, 1970, chapter 1).

It is to be noted that E, G and ν are related by the expression:

$$G = \frac{E}{2(1 + \nu)} \tag{8-270}$$

The pore fluid pressure, p_p, may now be included by assuming that the sediment is isotropic. In this case, it cannot produce any shearing strain and, furthermore, it must have the same effect on all components of the strains ϵ_x, ϵ_y and ϵ_z. Thus, eq.8-269 may be expanded to include the effect of fluid pressure as follows:

$$\epsilon_x = \frac{\sigma_x}{E} - \frac{\nu}{E}(\sigma_y + \sigma_z) - \frac{p_p}{3H}, \; \epsilon_y = \frac{\sigma_y}{E} - \frac{\nu}{E}(\sigma_z + \sigma_x) - \frac{p_p}{3H},$$

$$\epsilon_z = \frac{\sigma_z}{E} - \frac{\nu}{E}(\sigma_x + \sigma_y) - \frac{p_p}{3H}$$

$$\epsilon_{xy} = \frac{\tau_z}{G}, \; \epsilon_{yz} = \frac{\tau_x}{G}, \; \epsilon_{zx} = \frac{\tau_y}{G} \tag{8-271}$$

where H is an arbitrary additional constant. The pore pressure does not affect the relationship between shearing stresses and strains, because the free expansion cannot produce angular distortion in an isotropic material. The effect of the ratio θ (eq.8-268) on these variables may be expressed as:

$$\theta = a_1 \sigma_x + a_2 \sigma_y + a_3 \sigma_z + a_4 \tau_x + a_5 \tau_y + a_6 \tau_z + a_7 p_p \tag{8-272}$$

where a_1, a_2, etc., are constants. As the material has been assumed to be isotropic, a change in the sign of τ_x, τ_y, and τ_z cannot affect the ratio θ. Thus, a_4, a_5 and a_6 must be equal to zero. Moreover, due to the requirement of isotropy a_1, a_2 and a_3 must all be identical. Thus, eq.8-272 can be written as:

$$\theta = \frac{1}{3H_1}(\sigma_x + \sigma_y + \sigma_z) - \frac{p_p}{\overline{R}}. \tag{8-273}$$

where H_1 and $\overline{R}$ are constants.

The relations given by eq.8-271 and 8-273 have five distinct constants. The constant H can be directly related to the bulk volume changes (Biot, 1941a):

$$\frac{1}{H} = -\frac{1}{V_b}\left(\frac{\partial V_b}{\partial p_p}\right)_{\overline{\sigma}} \tag{8-274}$$

where $\bar{\sigma}$ is a constant external stress. It has been shown by Biot that, based on the existence of potential energy considerations, H_1 and H are identical. According to Biot (1941a), this assumption implies that "... if the changes occur at an infinitely slow rate, the work done to bring the soil from the initial condition to its final state of strain and water content is independent of the way in which the final state is reached and is a definite function of the six strain components and the fluid content. This assumption follows quite naturally from that of the reversibility of the stress—strain relations introduced above, since the absence of a potential energy would then imply that an indefinite amount of energy can be drawn out of the soil by loading and unloading along a closed cycle." The constant $1/\bar{R}$ reflects the measure of the change in fluid content for a given change in fluid pressure. Thus $1/\bar{R}$ is equal to:

$$\frac{1}{\bar{R}} = \frac{1}{V_b} \left(\frac{\partial V_p}{\partial p_p}\right)_{\bar{\sigma}} \quad (8\text{-}275)$$

where $\bar{\sigma}$ is a constant external stress. If the changes in the porosity are assumed to be small and, thus, can be neglected, then the following relationship holds (Geertsma, 1957a, p.334):

$$\frac{1}{\bar{R}} = \{c_b - (1 + \phi)\, c_r\} \quad (8\text{-}276)$$

where c_b and c_r are the compressibilities given by eq.8-44 and 8-47. Equation 8-276 is presented merely to relate the work of Geertsma (1957a) and Biot (1941a). As mentioned above, one may use the definition given by eq.8-275 directly.

Equation 8-271 relates the strain and stress components. Often, the components of stress expressed as functions of the components of strain are needed. Adding the first three relations of eq.8-271 together and using the following notation:

$$\epsilon = \epsilon_x + \epsilon_y + \epsilon_z \quad (8\text{-}277)$$

the stress—strain relations become:

$$\left. \begin{array}{l} \sigma_x = 2G\left(\epsilon_x + \dfrac{\nu\epsilon}{1-2\nu}\right) + \bar{\alpha}p_p,\ \sigma_y = 2G\left(\epsilon_y + \dfrac{\nu\epsilon}{1-2\nu}\right) + \bar{\alpha}p_p, \\[2ex] \sigma_z = 2G\left(\epsilon_z + \dfrac{\nu\epsilon}{1-2\nu}\right) + \bar{\alpha}p_p \\[2ex] \tau_x = \epsilon_{yz}G,\ \tau_y = \epsilon_{zx}G,\ \tau_z = \epsilon_{xy}G \end{array} \right\} \quad (8\text{-}278)$$

where:

$$\bar{\alpha} = \frac{2(1+\nu)}{3(1-2\nu)} \frac{G}{H} \tag{8-279}$$

The quantity ϵ defined in eq.8-277 physically represents the expansion or dilation of the porous medium.

In an identical fashion, the variation of θ in eq.8-273 can be described as:

$$\theta = \bar{\alpha}\epsilon - \frac{p_p}{Q} \tag{8-280}$$

where:

$$\frac{1}{Q} = \frac{1}{R} - \frac{\bar{\alpha}}{H} \tag{8-281}$$

Comparison with thermoelasticity

It was mentioned earlier that the stress–strain relationship given by eq.8-278 is similar to the stress–strain relationships obtained in the field of thermoelasticity. As an example, the expression for σ_x is given by:

$$\sigma_x = 2G\epsilon_x + \frac{2G\nu}{(1-2\nu)}\epsilon - \frac{2G(1+\nu)}{(1-2\nu)}\bar{\bar{\alpha}}T \tag{8-282}$$

where $\bar{\bar{\alpha}}$ is the coefficient of thermal expansion and T is the temperature. All other symbols have the same meaning as before. Expressions for σ_y and σ_z can be obtained by substituting y or z for x in eq.8-282. The expressions for the τ's remain unchanged.

Comparison of eq.8-278 and 8-282 indicate that the stress–strain relationships in the case of thermal expansion (or contraction) and expansion (or contraction) due to variations in pore pressure are similar. The transformation between the two systems is given by:

$$\frac{p_p}{3H} \rightleftarrows -\bar{\bar{\alpha}}T \tag{8-283}$$

Thus, the reason for the introduction of the term poroelasticity by Geertsma (1966) becomes evident.

Equations of equilibrium for a porous elastic solid

In addition to the stress—strain relations discussed above, the equilibrium conditions which hold for the elastic, nonporous solids also apply for a porous solid. The equilibrium conditions, neglecting body forces due to gravity, are given by:

$$\left. \begin{array}{l} \dfrac{\partial \sigma_x}{\partial x} + \dfrac{\partial \tau_z}{\partial y} + \dfrac{\partial \tau_y}{\partial z} = 0 \\[2mm] \dfrac{\partial \sigma_y}{\partial y} + \dfrac{\partial \tau_z}{\partial x} + \dfrac{\partial \tau_x}{\partial z} = 0 \\[2mm] \dfrac{\partial \sigma_z}{\partial z} + \dfrac{\partial \tau_y}{\partial x} + \dfrac{\partial \tau_x}{\partial y} = 0 \end{array} \right\} \qquad (8\text{-}284)$$

Fundamentally, these equilibrium conditions describe the effect of variations of the stresses at any point and must be satisfied at all points throughout the body. At the surface they must be in equilibrium with the external forces on the surface of the body. For a detailed discussion of this concept the reader is referred to Timoshenko and Goodier (1970, chapter 8).

Darcy's Law applied to a poroelastic medium

As discussed earlier, Darcy's Law applies to the flow of fluids relative to the motion of the solids. A similar argument carries over to the three-dimensional case also. Thus, the equations of motion governing fluid flow in a porous medium are given as:

$$-\frac{k}{\mu}\frac{\partial p_p}{\partial x} = \dot{V}_x - \dot{U}_x; \quad -\frac{k}{\mu}\frac{\partial p_p}{\partial y} = \dot{V}_y - \dot{U}_y; \quad -\frac{k}{\mu}\frac{\partial p_p}{\partial y} = \dot{V}_z - \dot{U}_z \qquad (8\text{-}285)$$

where k is the permeability of the medium; μ is the viscosity of the fluid flowing through the porous matrix; $\dot{V}_x$, $\dot{V}_y$ and $\dot{V}_z$ are the averaged velocities of the liquid in the x, y, and z directions, respectively; $\dot{U}_x$, $\dot{U}_y$, and $\dot{U}_z$ are the averaged velocities of the solids in the x, y and z directions, respectively; and p_p is the pore fluid pressure.

Equations of equilibrium in terms of displacements

A method of solution generally followed in solving the elasticity problems is to eliminate the stress components in the equilibrium relations given by eq. 8-284. This is done by substituting the right-hand side of eq. 8-278 for the appropriate stresses. The procedure is demonstrated by considering the first relation in eq.8-284. Using the expressions of σ_x, τ_z, and τ_y provided by eq.8-278, the first relation of eq.8-284 becomes:

$$2G\frac{\partial \epsilon_x}{\partial x} + 2G\frac{\nu}{1-2\nu}\frac{\partial \epsilon}{\partial x} + \bar{\alpha}\frac{\partial p_p}{\partial x} + 2G\frac{\partial \epsilon_{xy}}{\partial y} + 2G\frac{\partial \epsilon_{zx}}{\partial z} = 0 \qquad (8\text{-}286)$$

Equation 8-286 can be simplified as:

$$G\nabla^2 u + \frac{G}{1-2\nu}\frac{\partial \epsilon}{\partial x} + \bar{\alpha}\frac{\partial p_p}{\partial x} = 0 \qquad (8\text{-}287)$$

where $\nabla^2 \equiv \frac{\partial^2}{\partial x^2} + \frac{\partial^2}{\partial y^2} + \frac{\partial^2}{\partial z^2}$. Similar expressions may be obtained for v and w. Thus, from the equilibrium conditions (eq.8-284) one obtains:

$$\left.\begin{array}{l} G\nabla^2 u + \dfrac{G}{1-2\nu}\dfrac{\partial \epsilon}{\partial x} + \bar{\alpha}\dfrac{\partial p_p}{\partial x} = 0 \\[2mm] G\nabla^2 v + \dfrac{G}{1-2\nu}\dfrac{\partial \epsilon}{\partial y} + \bar{\alpha}\dfrac{\partial p_p}{\partial y} = 0 \\[2mm] G\nabla^2 w + \dfrac{G}{1-2\nu}\dfrac{\partial \epsilon}{\partial z} + \bar{\alpha}\dfrac{\partial p_p}{\partial z} = 0 \end{array}\right\} \qquad (8\text{-}288)$$

In order to arrive at eq.8-288, the following expressions are required:

$$\frac{\partial^2 u}{\partial x^2} = \frac{\partial \epsilon_x}{\partial x}, \quad \frac{\partial^2 u}{\partial y^2} = \frac{\partial \epsilon_{xy}}{\partial y} - \frac{\partial \epsilon_y}{\partial x}, \quad \frac{\partial^2 u}{\partial z^2} = \frac{\partial \epsilon_{xz}}{\partial z} - \frac{\partial \epsilon_z}{\partial x},$$
$$\frac{\partial^2 u}{\partial x \partial y} = \frac{\partial \epsilon_x}{\partial y}, \quad \frac{\partial^2 u}{\partial x \partial z} = \frac{\partial \epsilon_x}{\partial z}, \quad \frac{\partial^2 u}{\partial y \partial z} = \frac{1}{2}\left(\frac{\partial \epsilon_{xz}}{\partial y} + \frac{\partial \epsilon_{xy}}{\partial z} - \frac{\partial \epsilon_{yz}}{\partial x}\right) \qquad (8\text{-}289)$$

The above six relationships are obtained by differentiating the relations given in eq.8-267 with respect to x, y and z, respectively. The second derivatives for the two other components of displacement v and w are obtained by cyclical interchange of the letters x, y and z and substitution of v and w for u. Thus, in all, eighteen relationships involving the second derivatives can be obtained.

It is to be noted that there are three equations with four unknowns u, v, w and p_p in eq.8-288. A complete system of four equations and four unknowns is obtained by incorporating Darcy's Law. In addition to Darcy's Law, the equation of continuity also holds. This is given by the expression:

$$\frac{\partial \theta}{\partial t} = \frac{\partial V_x}{\partial x} + \frac{\partial V_z}{\partial y} + \frac{\partial V_z}{\partial y} \qquad (8\text{-}290)$$

where V_x, V_y, V_z, etc., represent the average relative velocities of the fluid. Using eq.8-269, 8-285, and 8-290 one can obtain the following expression:

$$\frac{k}{\mu} \nabla^2 p_p = -\overline{\alpha} \frac{\partial \epsilon}{\partial t} + \frac{1}{Q} \frac{\partial p_p}{\partial t} \qquad (8\text{-}291)$$

Equation 8-291 is similar to the heat conduction equation except for the appearance of the term involving $\partial \epsilon / \partial t$. The system of relations given by eq.8-288 may be simplified further and combined into a single equation. Differentiating the first relation of eq.8-288 with respect to x, the second with respect to y, and the third with respect to z and adding, the following equation is obtained:

$$\frac{2G(1-\nu)}{(1-2\nu)} \nabla^2 \epsilon = -\overline{\alpha} \nabla^2 p_p \qquad (8\text{-}292)$$

Body forces generated by fluid movement

Lubinski (1954) has pointed out that the analogy between poroelasticity and thermoelasticity discussed above is not complete (identical in all respects) merely because the stress—strain relationships are similar. There are two fundamental differences. The first difference is the existence of body forces generated by fluid movement within the pores. The second difference involves the boundary conditions which are discussed later.

As the fluid flows within the porous medium, the viscous forces, which are generated, act on the walls of the pores and impart a body force in the direction of flow. By considering the forces acting within the pores of the porous medium, Lubinski (1954) has shown that the body forces are given by:

$$F_x = -\phi \frac{\partial p_p}{\partial x}, \; F_y = -\phi \frac{\partial p_p}{\partial y}, \; F_z = -\phi \frac{\partial p_p}{\partial z} \qquad (8\text{-}293)$$

where F_x, F_y and F_z are the body forces per unit bulk volume of the porous medium in the three principal directions x, y and z, respectively; ϕ is the porosity; and p_p is the fluid pressure.

Boundary conditions

The complete mathematical specification of any physical problem requires the consideration of the boundary conditions in addition to the governing equation. The specification of the boundary conditions in the present instance is complicated by the fact that the effects of void space and solid material have to be considered simultaneously. This represents the second distinct difference between poroelasticity and thermoelasticity. If the porous medium undergoing deformation is in contact with a permeable medium, then the microscopic force acting on the solid particles at the boundary is $-p_p$, where p_p is the fluid pressure. Thus the macroscopic boundary stress on the solid material is $-(1-\phi)p_p$. For this case the boundary forces per unit area, $\bar{p}_x$, $\bar{p}_y$, and $\bar{p}_z$, are given by the following relations:

$$\bar{p}_x = [-(1-\phi)p_p]\,l, \quad \bar{p}_y = [-(1-\phi)p_p]\,m, \quad \bar{p}_z = [-(1-\phi)p_p]\,n \quad (8\text{-}294)$$

in which l, m and n are the direction cosines of the external forces normal to the surface of the body at the point under consideration.

If the porous medium undergoing deformation is in contact with an impermeable boundary, then the boundary stresses $\bar{p}_x$, $\bar{p}_y$, and $\bar{p}_z$ in the three coordinate directions can be presented as follows (Lubinski, 1954):

$$\bar{p}_x = (\phi p_p - p'_{ex})\,l, \quad \bar{p}_y = (\phi p_p - p'_{ex})\,m, \quad \bar{p}_z = (\phi p_p - p'_{ex})\,n \quad (8\text{-}295)$$

where p'_{ex} is the pressure outside the impermeable boundary.

Solution of the general equations of poroelasticity

The general method of solving the poroelasticity problems is as follows:

(1) Displacements and stresses due to the pressures acting inside and on the boundaries of the porous medium are calculated. This is done by means of the stress—strain relations and equilibrium relations (eq. 8-271, 8-278, and 8-284). The body force and boundary conditions (eq. 8-293, 8-294, and 8-295) would have to be taken into account. In order to obtain the stress—strain relationships, however, the pressure field has to be determined. It will subsequently be shown that the pressure field can be calculated by using solutions already developed in the fields of heat conduction, fluid flow, and elasticity.

(2) The stresses and displacements due to surface and body forces, other than pressure, are then calculated using conventional methods of the classical theory of elasticity.

(3) The two effects (discussed in 1 and 2) are superimposed to obtain the total or resultant effects. Fundamentally, eq.8-291 and 8-292 have to be solved. The solution of eq.8-292, which is considered first, may be broken up into two parts, the complementary function and the particular solution. The complementary function is given by the equation:

$$\nabla^2 \epsilon = 0 \tag{8-296}$$

Methods of solution of eq.8-296 are well known (Timoshenko and Goodier, 1970). If a particular solution to eq.8-292 can be determined, then the solution of the poroelastic problem reduces to the solution of the general elastic problem.

One way of finding the particular solution is to define function $\bar{\psi}$ such that:

$$u = \frac{\partial \bar{\psi}}{\partial x}, \quad v = \frac{\partial \bar{\psi}}{\partial y}, \quad w = \frac{\partial \bar{\psi}}{\partial z} \tag{8-297}$$

In general, $\bar{\psi}$ is a function of x, y, z and t. In problems of thermoelasticity such a solution is called a thermoelastic displacement potential (Goodier, 1937; Timoshenko and Goodier, 1970, p.476). Accordingly, in the present instance, as pointed out by Geertsma (1966), it may be called the poroelastic displacement potential. Substituting for u, v, and w, the expressions given by eq.8-297 and noting that $\epsilon = \nabla^2 \bar{\psi}$ for $\epsilon = \frac{\partial u}{\partial x} + \frac{\partial v}{\partial y} + \frac{\partial w}{\partial z}$, eq.8-288 becomes:

$$\left. \begin{array}{l} \dfrac{\partial}{\partial x} \nabla^2 \bar{\psi} = -\dfrac{\bar{\alpha}}{2G} \dfrac{(1-2\nu)}{(1-\nu)} \dfrac{\partial p_p}{\partial x} \\[6pt] \dfrac{\partial}{\partial y} \nabla^2 \bar{\psi} = -\dfrac{\bar{\alpha}}{2G} \dfrac{(1-2\nu)}{(1-\nu)} \dfrac{\partial p_p}{\partial y} \\[6pt] \dfrac{\partial}{\partial z} \nabla^2 \bar{\psi} = -\dfrac{\bar{\alpha}}{2G} \dfrac{(1-2\nu)}{(1-\nu)} \dfrac{\partial p_p}{\partial z} \end{array} \right\} \tag{8-298}$$

All three expressions given in eq.8-298 are satisfied if the function $\bar{\psi}$ is taken as the solution of the equation

$$\nabla^2 \bar{\psi} = c_m p_p \tag{8-299}$$

where:

$$c_m = -\frac{\overline{\alpha}}{2G}\frac{(1-2\nu)}{(1-\nu)} \tag{8-300}$$

Solutions to equations of the type given by eq.8-299 are very common in the theory of potential (Timoshenko and Goodier, 1970, p.476). It is to be noted that eq.8-299 also implies that $\epsilon = c_m p_p$. The associated normal stress conditions given by eq.8-278 become:

$$\left. \begin{array}{l} \sigma_x = -2G(\epsilon - \epsilon_x) \\ \sigma_y = -2G(\epsilon - \epsilon_y) \\ \sigma_z = -2G(\epsilon - \epsilon_z) \end{array} \right\} \tag{8-301}$$

The shear stresses remain unchanged.

Following Timoshenko and Goodier (1970, p.476), the solution of eq.8-299 is:

$$\overline{\psi}(x,y,z,t) = -\frac{c_m}{4\pi}\iiint p_p(\xi,\eta,\zeta)\frac{1}{r'}d\xi d\eta d\zeta \tag{8-302}$$

where $p_p(\xi, \eta, \zeta)$ is the pressure at a typical point (ξ, η, ζ) at which there is an elemental volume $(d\xi d\eta d\zeta)$ and r' is the distance between the point (ξ, η, ζ) and the point (x, y, z). Equation 8-302 may be used to obtain the required solution.

The complete solution of eq.8-292 also requires that the complementary function, that is, the solution to eq.8-296, be determined. These involve solution of the general elastic problem. The general solution to eq.8-296 is given by Timoshenko and Goodier (1970, p.242):

$$\left. \begin{array}{l} u'' = \phi_1 - \overline{\beta}\dfrac{\partial}{\partial x}(\phi_0 + x\phi_1 + y\phi_2 + z\phi_3) \\[2pt] v'' = \phi_2 - \overline{\beta}\dfrac{\partial}{\partial y}(\phi_0 + x\phi_1 + y\phi_2 + z\phi_3) \\[2pt] w'' = \phi_3 - \overline{\beta}\dfrac{\partial}{\partial z}(\phi_0 + x\phi_1 + y\phi_2 + z\phi_3) \end{array} \right\} \tag{8-303}$$

where:

$$\overline{\beta} = \frac{1}{4(1-\nu)} \tag{8-304}$$

and u'', v'' and w'' are the displacements.

These solutions, which can be verified by formal substitution in eq.8-296, are called the Boussinesq (1885)—Papkovitch (1932) functions. The functions ϕ_0, ϕ_1, ϕ_2 and ϕ_3 are harmonic, that is, they satisfy the condition: $\nabla^2 \phi_0 = \nabla^2 \phi_1 = \nabla^2 \phi_2 = \nabla^2 \phi_3 = 0$.

It can be seen that ϕ_0 is identical to the solution given by $\nabla^2 \bar{\psi} = \epsilon = 0$, where $\bar{\psi}$ is defined in eq.8-302. Geertsma (1966) has shown that for practical problems of interest to earth scientists, ϕ_1, ϕ_2 and ϕ_3 are zero as rotation-free situations are being considered.

The formal solution is complete when the pressure distribution is determined by obtaining ϵ from eq.8-299 and 8-303 and substituting the value of ϵ in the pressure equation (eq.8-291). Substitution of the general solutions in eq.8-291 indicates two types of pressure fields (Biot, 1956; Geertsma, 1966). The first one is satisfied by the diffusion type equation:

$$\frac{k}{\mu} \nabla^2 p_p = \bar{\gamma} \frac{\partial p_p}{\partial t} \tag{8-305}$$

where:

$$\bar{\gamma} = \frac{1}{Q} - \bar{\alpha}\, c_m \tag{8-306}$$

Equation 8-305 governs the pressure distribution where there is no dilation and is similar to eq.8-186. Its solutions are well documented (Van Everdingen and Hurst, 1949; Carslaw and Jaeger, 1959; Miller, 1962; Rowan and Clegg, 1962; Matthews and Russell, 1967).

The second pressure field, p'_p, is the result of the dilation involved in that part of the displacement field that is described by ϕ_1, ϕ_2, and ϕ_3. It is given by the following equation:

$$p'_p = \frac{-2\bar{\alpha}\, G}{\left[\bar{\alpha}^2 + \dfrac{G}{Q(1-2\nu)}\right]} \left(\frac{\partial \phi_1}{\partial x} + \frac{\partial \phi_2}{\partial y} + \frac{\partial \phi_3}{\partial z}\right) \tag{8-307}$$

If ϕ_1, ϕ_2, and ϕ_3 are all zero and, consequently p'_p is zero, then the pressure distribution is governed entirely by eq.8-305.

From the above discussion it can be seen that in order to determine the displacements u, v, and w, the solution of the problem involves a solution of eq.8-299. Once the displacement and associated stress given by the particular solution are determined, then subtraction from the boundary values enables determination of boundary conditions which have to be satisfied by eq.8-303, 8-305 and 8-307. It has been shown that the solution of the latter equations is rather well documented in the fields of heat conduction, fluid flow through porous media, and in the theory of elasticity of nonporous solids. Thus, attention is focussed here on obtaining the particular solution of eq.8-299 or eq.8-292. A number of techniques are available (Timoshenko and Goodier, 1970; Sokolnikoff, 1956). Of particular interest is the "concept of strain nuclei" discussed by Mindlin and Cheng (1950) and Sen (1950).

The concept of strain nuclei

The concept of strain nuclei was first discussed by Lord Kelvin (Love, 1934, chapter 8). For an isotropic, non-porous, elastic solid of indefinite extent, the fundamental solution involving strain nuclei is that of a single force acting at a point in the solid body.

It has already been seen that in order to obtain a complete solution, the particular solution has to be obtained. The particular solution represents the effect of pressure. For the system under consideration, the Kelvin solution involves the case where the pressure is different from zero in a volume V and vanishes outside the volume V. The volume V may be the entire body or a part of the body. The fundamental solution involves considering the body to extend indefinitely in all directions and the volume V to be a part of it. The limit is obtained by diminishing V indefinitely. Thus, the basic solution involves the calculation of displacement due to a singularity (pressure) at a point in the system. The singularity represents a nucleus of strain. A more detailed discussion is provided by Love (1934). Of prime interest to the analysis considered here is the work of Mindlin and Cheng (1950) and Sen (1950). These authors have extended the results obtained by Lord Kelvin to the consideration of strain nuclei in semi-infinite solids with a traction-free surface.

The basic idea behind this concept may be understood more clearly by a closer examination of the particular solution already obtained (eq.8-302).

It has already been seen that the displacement can be represented by the poroelastic displacement potential defined in eq.8-302. From eq.8-297 and 8-302, the u component of the displacement can be written as:

$$u = -\frac{c_m}{4\pi} \frac{\partial}{\partial x} \iiint p_p(\xi, \eta, \zeta) \frac{1}{r'} \, d\xi \, d\eta \, d\zeta \qquad (8\text{-}308)$$

or:

$$u = \iiint p_p(\xi, \eta, \zeta) \, u^*(x, r) \, d\xi \, d\eta \, d\zeta \qquad (8\text{-}309)$$

where:

$$u^*(x, r) = -\frac{c_m}{4\pi} \frac{\partial}{\partial x}\left(\frac{1}{r'}\right) \tag{8-310}$$

Similar expressions may be obtained for v and w. The function $u^*(x, r)$ has a definite mechanical significance. It may be interpreted as the displacement of a point (x, y, z) in the x direction, due to the action of a center of a pressure situated at the point (ξ, η, ζ) of an elastic unbounded porous space (see Nowacki, 1962, p.13), and represents a nucleus of strain. Though the problem has been examined in an infinite space, Goodier (1937) has shown that this concept can also be applied to bounded systems. As mentioned earlier, of prime interest is the work of Mindlin and Chen (1950) and Sen (1950). These authors have developed solutions for problems of interest to compaction and subsidence.

Compaction and subsidence

The objective of this section is to analyze a reservoir filled with an incompressible fluid. For purposes of analysis, a cylindrical disc reservoir of constant thickness, h, is assumed. The cylindrical disc is assumed to be isolated from its surroundings by an impermeable barrier. (The effect and presence of sinks or sources are neglected, i.e., the details regarding the mechanism of fluid pressure change are unspecified). Thus, the entire porous body is assumed to behave like a tank with fluid withdrawal taking place uniformly throughout the system. If the pore fluid pressure is reduced uniformly, stresses and strains both inside and outside the body are changed. Both the reservoir and the surrounding rocks are assumed to have identical properties.

The nucleus of strain is assumed to be located at the point $(0, c)$ in Fig.8-31. Outside this nucleus, the displacement per unit volume and pressure change, $\Delta \bar{p}_p$, at a radial distance, r, from the nucleus is given by Geertsma (1966):

$$\vec{u}_e^* = \frac{c_m}{4\pi}\left\{\frac{\vec{R}_1}{R_1^3} + \frac{(3-4\nu)\vec{R}_2}{R_2^3} - \frac{6z(z+c)\vec{R}_2}{R_2^5}\right.$$

$$\left. -\frac{2\vec{k}}{R_2}\left[(3-4\nu)(z+c)-z\right]\right\} \tag{8-311}$$

where:

$R_1^2 = r^2 + (z-c)^2$, $R_2^2 = r^2 + (z+c)^2$, and $\vec{k}$ = unit vector in z-direction.

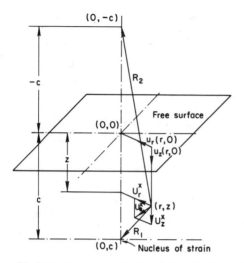

Fig.8-31. Geometry for the determination of the displacement field around the nucleus of strain in the half-space. (After Geertsma, 1966, fig.1, p.589.)

All these quantities are presented in Fig.8-31. The vertical displacement can be expressed as:

$$u_z^*(r, z) = \frac{c_m}{4\pi} \left[\frac{z-c}{R_1^3} + \frac{4\nu(z+c)-(z+3c)}{R_2^3} - \frac{6z(z+c)^2}{R_2^5} \right] \quad (8\text{-}312)$$

At the free surface ($z = 0$), eq.8-312 becomes:

$$u_z^*(r, 0) = -\frac{c_m(1-\nu)}{\pi} \frac{c}{(r^2+c^2)^{3/2}} \quad (8\text{-}313)$$

Equation 8-313, which shows the effect of subsidence at the free surface due to the nucleus of strain, corresponds to the case where there is no overburden or where the depth of burial is zero.

In the above discussion only the effect of the nucleus of strain has been examined. In order to consider the displacement for the disc-shaped reservoir having thickness h, eq.8-309 can be used. Substituting $\vec{u}_e^*$ for $\vec{u}^*(x, r)$ and $\vec{u}_e$ for u in eq.8-309, an expression for $\vec{u}_e$, the displacement at any point, can be obtained as follows (Geertsma, 1966):

$$\vec{u}_e = h\Delta\overline{p}_p \int_0^{r_e} \int_0^{2\pi} \vec{u}_e^*(r, z, \rho, \Theta)\overline{\rho}\, d\overline{\rho}\, dH \quad (8\text{-}314)$$

$$R_1^2 = \overline{\rho}^2 \sin^2\Theta + (r - \overline{\rho} \cos \Theta)^2 + (z-c)^2 \tag{8-315}$$

$$R_2^2 = \overline{\rho}^2 \sin^2\Theta + (r - \overline{\rho} \cos \Theta)^2 + (z+c)^2 \tag{8-316}$$

In eq.8-314, Δp_p is assumed to be constant throughout the disc-shaped reservoir. Equation 8-314 also implicitly assumes that $h/r_e \ll 1$. All quantities in eq.8-314, 8-315 and 8-316 are defined in Fig.8-32.

The evaluation of the integrals involves elliptic functions (Geertsma, 1966; Ter-Martirosyan and Ferronsky, 1969), which have been tabulated by Geertsma (1973). The integral, however, is easy to evaluate for the case $r = 0$, and eq.8-315 and 8-316, respectively, become:

$$R_1^2 = \overline{\rho}^2 + (z-c)^2 \tag{8-317}$$

and

$$R_2^2 = \overline{\rho}^2 + (z+c)^2 \tag{8-318}$$

An expression for the subsidence along the axis of the porous reservoir can now be calculated. This expression is given by:

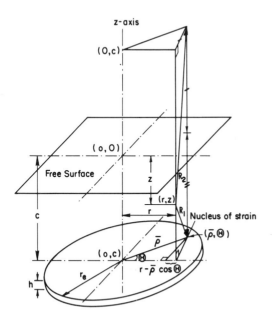

Fig.8-32. Geometry for the determination of the displacement field around a disc-shaped reservoir in the half-space. (After Geertsma, 1966, fig.2, p.590.)

$$\dot{u}_z(0,z) = -\frac{c_m h}{2} \left\{ \frac{C(\bar{Z}-1)}{[1+C^2(\bar{Z}-1)^2]^{1/2}} - \frac{(3-4\nu)C(\bar{Z}+1)}{[1+C^2(\bar{Z}+1)^2]^{1/2}} + \right.$$

$$\left. + \frac{2C\bar{Z}}{[1+C^2(\bar{Z}+1)^2]^{3/2}} + (3-4\nu+\epsilon') \right\} \Delta\bar{p}_p \qquad (8\text{-}319)$$

where $\bar{Z} = z/c$, $C = c/r_e$, $\epsilon' = -1$ for $z > c$, and $\epsilon' = +1$ for $z < c$. Thus, the subsidence at the free surface above the center of a disc-shaped depleted reservoir amounts to:

$$\dot{u}_z(0,0) = -2(1-\nu)\, c_m\, h\Delta\bar{p}_p \left(1 - \frac{C}{\sqrt{1+C^2}}\right) \qquad (8\text{-}320)$$

If the reservoir was situated at the free surface ($C = 0$), then:

$$\dot{u}_z(0,0) = -2(1-\nu)\, c_m\, h\Delta\bar{p}_p \qquad (8\text{-}321)$$

Equation 8-320 indicates that the subsidence is a function of the thickness, h, the drop in average reservoir pressure, $\Delta\bar{p}_p$, the coefficient c_m, and the ratio C. Normally, c_m would be large for loose or friable sands. These types of sands, therefore, should be of prime concern when subsidence effects are examined. Thus, for a given porous medium, noting that $\Delta\bar{p}_p$ would in general be a function of depth, the maximum subsidence is determined by the ratio C (ratio of the depth of burial to the lateral extent). If C is large ($C \gg 1$), then the term within the brackets on the right-hand side of eq.8-320 would become negligible. Consequently, even if $\Delta\bar{p}_p$ and c_m are large, $\dot{u}_z(0, 0)$ would be small. Land surface subsidence should not be an important factor in this case. On the other hand, if C is small, that is, the body is located at a shallow depth, then $\Delta\bar{p}_p$ would be small. This, in turn, implies that $\dot{u}_z(0, 0)$ is small. Thus, on the basis of eq.8-320, it can be concluded that the compaction of large, deeply-buried, loose or friable porous bodies should result in the subsidence of the land surface.

Geertsma (1966, 1973) has shown that lateral movements due to compaction can also be determined from eq.8-314. The analysis presented above is not necessarily restricted to disc-shaped reservoirs. Other arbitrary shapes can be easily included by evaluating numerically the integral on the right-hand side of eq.8-314. If required, the variation of h and c_m with space coordinates may also be included. In the latter event, however, expression for $\dot{u}_z(0, 0)$ is not rigorous. In this context, it should be noted that one of the basic assumptions of Geertsma's (1966, 1973) analysis is that it would be possible to determine a single value for c_m, which evidently will have to be determined experimentally in the laboratory. A discussion on the choice of an average value for c_m is not presented here. Another assumption regarding the theory presented by Geertsma (1966, 1973), which would require modification for application to practical problems, concerns the properties of the porous medium and of its surrounding rocks. This assumption that the properties are identical can be removed if numerical techniques are adopted (Evangelisti and Poggi, 1970).

It should be noted that in many practical cases it may be necessary to convert the results obtained by Geertsma (1966, 1973) to a time scale. In order to do so, the drop in average reservoir pressure $\Delta \bar{p}_p$ would have to be determined as a function of time. Thus, a model incorporating transients would be required. For example, a model similar to that described in the section on Transient Radial Flow may be used. A detailed discussion on the determination of pressures was provided in that section.

The importance and significance of the results presented by Geertsma (1966, 1973) requires little emphasis. Rather than using empirical techniques, for the first time a model has been developed which can be used to predict apriori the compaction of a porous medium and the resulting land surface subsidence in a rather simple fashion. In addition, a thorough understanding of the various parameters involved has been gained.

Discussion of three-dimensional deformation of porous solids

The objective of this section was to complement the discussion of one-dimensional theory considered earlier. This section provided a general background in handling mathematical techniques when three-dimensional effects are considered. In addition, it provided a starting point for further research efforts. The basic assumptions involved in the three-dimensional theory have been examined in detail. It has been shown that the stress—strain relationship governing the deformation of a compressible porous medium is similar to that of an elastic nonporous solid. The analogy and differences between poroelasticity and thermoelasticity have been demonstrated. As a result, the solutions already developed in the field of thermoelasticity can be used for a number of problems of practical interest in the field of poroelasticity.

At the beginning of this section it was pointed out that one of the main reasons for a lag in the application of three-dimensional theories developed over the past three decades has been a dearth of solutions which can be applied easily. It has also been shown that the works of Evangelisti and Poggi (1970) and Geertsma (1973) represent important contributions towards better understanding of compaction and subsidence processes.

SUMMARY

The primary objective of this chapter has been to provide an introduction to the various mathematical techniques which are commonly used to study rock compaction and land surface subsidence problems in different branches of science and engineering. As a result of the application of these mathematical models and techniques, it has been shown that both a quantitative information and better understanding can be obtained.

Starting from the simplest system that can be envisaged, that of a saturated porous matrix in a tank (Fig.8-1, p.407), some of the basic aspects of the compaction and rebound of porous materials, particularly in view of the developments that have taken place in the last few years, have been reviewed. Much progress has been made recently and has resulted in a better physical understanding of rock compaction and land surface subsidence.

The classical problem of Terzaghi (1923) has been re-examined and extended in the light of recent developments (Philip, 1968; Smiles and Rosenthal, 1968). Darcy's Law and its implication as related to a mobile matrix were also examined. The specific problem considered is a one-dimensional shrinkage or expansion of a porous system to a single step function change in pressure at the base of the column, which may be considered to be either finite or semi-infinite. The effects of the overburden and gravity were also included. One aspect of compaction or rebound (expansion), which has resulted in considerable confusion, is the choice of coordinate systems (namely, Eulerian or Lagrangian) to develop the mathematical model. This aspect was also examined in detail in this chapter.

The modified Terzaghi analysis presented herein (1) includes general forms of formation conductivity and pressure, and (2) couples the movement of soil particles and the associated mass flow of liquid. It has been shown that the mass flow effect is extremely important. Further, the differences between compaction and rebound (expansion) processes have been examined.

Even though a non-linear equation results, analytical solutions exist for a class of problems (Philip, 1968). These analytical solutions serve as illustrative examples and demonstrate the difference between the consolidation and expansion processes. As a result, increased physical understanding of the effect of the movement of the solid particles is achieved. The non-linear equation has also been solved numerically and the solutions obtained are presented.

Techniques of handling transients in a porous system have been demonstrated by considering the withdrawal of fluids from a well located at the center of a closed, radial reservoir. Changes in porosity, permeability, and compressibility of the rock, as well as liquid viscosity and density, due to fluid withdrawal are discussed. It has been shown that a second-order, non-linear, partial differential equation results when variations of the above parameters are considered. A solution to this equation when fluid is withdrawn at a constant rate is presented.

As mentioned earlier, one of the objectives of the writers was to enable research and/or field personnel to obtain answers within engineering accuracy while including the non-linearities discussed in this chapter. This goal is met through the use of the pseudo-void ratio (Raghavan, 1972) or pseudo-pressure transformations (Raghavan et al., 1972). Though the correlation solutions obtained relate only to specific instances, it has been shown that with care other correlation solutions may be obtained. To obtain more exact answers, numerical procedures are necessary. The section on numerical solutions should be of help in obtaining the desired results.

From the viewpoint of mathematical physics, the major extension of Terzaghi's analysis involves the consideration of two- and three-dimensional systems. The basic assumptions of the three-dimensional theory developed by Biot (1935; 1941a, b; 1956) have been examined in detail. It has been shown that compaction and rebound of sediments is one of the many divisions of the general theory of elasticity; and that the stresses, which result in a porous medium due to its deformation, are analogous to thermal stresses produced in non-porous elastic bodies. As a result of this analogy, the theories already developed in the field of thermoelasticity may be applied to problems of interest in compaction and subsidence. This analogy has been used by Geertsma (1966, 1973) and Evangelisti and Poggi (1970) to describe the subsidence of the land surface due to the compaction or rebound (expansion) of a porous layer located at depth. The work of Geertsma (1966, 1973) and Evangelisti and Poggi (1970) will serve as a basis for future research.

The models described in this chapter involve a radical simplification of real processes which occur in nature. The writers are acutely aware of the manifold limitations of mathematics to describe nature adequately. Their aim has been to provide a working knowledge of some useful methods of analysis. As in all other branches of science and engineering, their validity can be proven only on the basis of field experience. The models and theories suggested should provide a framework and serve as a guide to a better understanding of real processes. The writers hope that the information provided herein will serve as a stimulus for further analytical methods to be developed in the future. Sedimentologists involved in compaction work should definitely become familiar with mathematical techniques to further and improve the understanding of the subject.

ACKNOWLEDGEMENTS

With great pleasure the writers would like to express their appreciation to Dr. G. V. Chilingarian for inviting them to participate in this endeavor. They are also indebted to him for the critical review of the chapter. The authors would also like to express their gratitude to Dr. Henry J. Ramey Jr., of Stanford University, for reviewing portions of the manuscript and offering many constructive suggestions. Dr. Sullivan S. Marsden of Stanford University kindly allowed the writers to use his files on the subject.

The authors also take this opportunity to express their appreciation and grateful thanks to Mrs. Suzanne Y. Endow of Stanford University, Miss Lois L. Gray of System Science Incorporated, Palo Alto, and Miss Joyce Sowder of Cities Service Oil Company, Tulsa, for the expert typing and careful editing of this manuscript. Mrs. Maria Jedd of Stanford University was responsible for all the illustrations of this chapter.

The financial support of the Department of Petroleum Engineering, the School of Earth Sciences and Amoco Production Company of Tulsa are also gratefully acknowledged. Particular thanks are also extended to the Society of Petroleum Engineers of AIME, Dallas, for permission to reproduce some of the illustrations.

NOTATION

a = Arbitrary constant, defined by eq.8-107, dimensionless.
$a_1, a_2, a_3, \ldots a_7$ = Arbitrary constants, defined by eq.8-272, psi^{-1} or (dynes/cm^2)$^{-1}$.
A = Area, normal to the plane of the paper, ft^2 or cm^2.
$A(x)$ = Defined by eq.8-228.
b = Arbitrary constant, defined by eq.8-107, dimensionless.
c = Arbitrary constant, defined by eq.8-108, dimensionless.
c = Depth of burial, ft or cm.
c_b = Bulk volume compressibility, psi^{-1} or cm^2/kg.
c_g = Compressibility of real gas, psi^{-1} or cm^2/kg.
c_l = Compressibility of liquid, psi^{-1} or cm^2/kg.
c_m = Arbitrary constant, defined by eq.8-300, psi^{-1} or cm^2/kg.
c_p = Pore volume compressibility, psi^{-1} or cm^2/kg.
c_r = Rock compressibility, psi^{-1} or cm^2/kg.
c_{rg} = Reduced compressibility of real gas ($c_{rg} = c_g p_c$), psi^{-1} or cm^2/kg.
c_t = Total system compressibility, psi^{-1} or cm^2/kg.
c_v = Coefficient of consolidation, defined by eq.8-34, ft^2/sec or cm^2/sec.
C = Defined in eq.8-319, dimensionless.

D_c = Arbitrary constant, defined by eq.8-109, ft²/sec or cm²/sec.
D_D = Diffusivity, defined in eq.8-237, dimensionless.
D_m = Diffusivity, defined by eq.8-144, ft²/sec or cm²/sec.
D_o = Diffusivity, defined by eq.8-171, ft²/sec or cm²/sec.
$\underline{\underline{D}}(\phi)$ = Diffusivity, defined by eq.8-73, ft²/sec or cm²/sec.
$\bar{\bar{D}}_{r + 1/2}$ = Defined by eq.8-217, ft²/sec or cm²/sec.
e = Void ratio (ratio of volume of voids to volume of solids), defined by eq.8-136, dimensionless.
E = Modulus of elasticity in tension and compression, psi or dynes/cm².
F_x, F_y, F_z = Components of body force per unit volume, in the x, y, and z directions, respectively, lb/ft³ or kg/cm³.
g = Acceleration due to gravity, ft/sec² or cm/sec².
G = Modulus of elasticity of shear, psi or dynes/cm².
h = Thickness, ft or cm.
h = Fluid head, ft or cm.
h = Mesh size in the x direction, defined by eq.8-233.
H = Arbitrary constant governing three-dimensional deformation, defined by eq.8-271, psi or kg/cm².
H_1 = Arbitrary constant, defined by eq.8-273, psi or kg/cm².
I_r = Defined by eq.8-215, ft/sec$^{1/2}$ or cm/sec$^{1/2}$.
$I_{1/2}$ = Defined by eq.8-214, ft/sec$^{1/2}$ or cm/sec$^{1/2}$.
k = Mesh size in the y direction, defined by eq.8-233.
$\vec{k}$ = Unit vector in the z direction, defined in eq.8-311.
k_c = Arbitrary constant, defined by eq.8-107.
$k(p)$ = Permeability, md or cm².
$k(\psi)$ = Permeability, md or cm².
l, m, n = Direction cosines of the external normal to the surface of the porous body.
L = Length, ft or cm.
L_D = Dimensionless length.
m = Material coordinate, defined by eq.8-134, ft or cm.
m_D = Dimensionless material coordinate, defined by eq.8-153.
m_v = Coefficient of volume compressibility, defined by eq.8-27, psi^{-1} or cm²/kg.
M = Volume of solids per unit area, defined by eq.8-146, ft³/ft² or cm³/cm².
$\bar{p}$ = Average fluid pressure in the reservoir, psi or kg/cm².
p_c = Critical pressure of gas, psi or dynes/cm².
p_D = Dimensionless pressure drop, defined by eq.8-195.
p_{ex} = Excess pressure, defined by eq.8-9, psi or kg/cm².
p'_{ex} = External pressure on the boundary, defined by eq.8-295, psi or kg/cm².

p_g = Grain-to-grain pressure, psi or kg/cm².
p_{ob} = Overburden pressure, psi or kg/cm².
p_p = pore (fluid) pressure, psi or kg/cm².
p'_p = Pressure field due to dilation, defined by eq.8-307, psi or kg/cm².
p_{pi} = Initial fluid pressure, psi or kg/cm².
p_{pr} = Pseudo-reduced pressure, p_p/p_c, dimensionless.
p_{pr} = Fluid pressure at any radial location, psi or kg/cm².
p_{pw} = Fluid pressure at the well, psi or kg/cm².
p_t = Total pressure exerted by the piston (see Fig.8-1), psi or kg/cm².
$\bar{p}_x, \bar{p}_y, \bar{p}_z$ = Components of distributed surface force per unit area in the x, y, and z directions, respectively, psi or kg/cm².
q = Flow rate, ft³/day or cm³/sec.
q_D = Dimensionless flow rate, defined by eq.8-201.
Q = Arbitrary constant, defined by eq.8-281, psi or kg/cm².
r = Radius, ft or cm.
r' = Distance, defined in eq.8-308, ft or cm.
r_d = Aronofsky-Jenkins drainage radius, defined by eq.8-203, ft or cm.
r_D = Dimensionless radial distance, defined by eq.8-197.
r_e = Radius of circular reservoir (external radius) ft or cm.
$\bar{R}$ = Arbitrary constant, defined by eq.8-275, psi or kg/cm².
R_1, R_2 = Arbitrary constants, defined in eq.8-311 and by eq.8-315 and 8-316; also see Fig.8-30 and 8-31, ft or cm.
S = Stress tensor, defined in eq.8-37.
t = Time, sec.
T = Temperature, °R or °K.
T_c = Critical temperature of a real gas, °R or °K.
T_{pr} = Pseudo-reduced temperature of real gas, T/T_c.
u, v, w = Displacements in the x, y, and z directions, respectively.
$\vec{u}$ = Macroscopic velocity, ft/sec or cm/sec.
u^* = Strain nuclei due to displacement, defined by eq.8-310, psi/ft² or dynes/cm⁴.
$\vec{u}_e$ = Displacement due to pressure change in a disc-shaped reservoir, defined in eq.8-314, ft or cm².
$\vec{u}_e^*$ = Displacement per unit volume due to the nucleus of strain, defined by eq.8-311, ft/ft³ or cm/cm³.
u_L = Macroscopic velocity of liquid, ft/sec or cm/sec.
u_{Lr} = Macroscopic velocity of liquid relative to rock, ft/sec or cm/sec.
u_r = Macroscopic velocity or volume flux density of rock, ft/sec or cm/sec.
$u_{(x, y)}$ = Function, defined by eq.8-233.
$u_z^*(r, z)$ = Displacement in the vertical direction due to the nucleus of strain, defined by eq.8-312, ft or cm.

$\dot{u}_z(r, z)$ = Vertical displacement, defined by eq.8-319, ft or cm.
$\vec{U}$ = Average microscopic velocity, defined by eq.8-61, ft/sec or cm/sec.
U_L = Microscopic velocity of liquid, ft/sec or cm/sec.
U_{Lr} = Microscopic velocity of liquid relative to rock, ft/sec or cm/sec.
$\widetilde{U}_L$ = Defined by eq.8-128, ft²/sec or cm²/sec.
$\widetilde{U}_{Lr}$ = Defined by eq.8-129, ft²/sec or cm²/sec.
$\vec{U}_{ma}$ = Macroscopic velocity, defined by eq.8-60, ft/sec or cm/sec.
$\vec{U}_{ma,r}$ = Macroscopic velocity (relative), defined by eq.8-64, ft/sec or cm/sec.
$\vec{U}_{mi}$ = Microscopic velocity, defined by eq.8-63, ft/sec or cm/sec.
U_r = Microscopic velocity of rock, ft/sec or cm/sec.
$\widetilde{U}_r$ = Defined by eq.8-130, ft²/sec or cm²/sec.
U_x, U_y, U_z = Average velocities of the solid in the x, y, and z coordinate directions, respectively, defined by eq.8-285, ft/sec or cm/sec.
V = Volume, ft³ or cm³.
V_b = Bulk volume of porous material, ft³ or cm³.
V_{cum} = Cumulative outflow from system per unit area of the system, defined by eq.8-102, ft³/ft² or cm³/cm².
V_L = Volume of liquid out of the system, per unit area of system, defined by eq.8-101, ft³/ft²/sec or cm³/cm²/sec.
V_p = Pore volume, ft³ or cm³.
V_r = Volume of matrix (rock) material, ft³ or cm³.
V_x, V_y, V_z = Average relative velocities of the liquid in the x, y, and z directions, respectively, defined in eq.8-290, ft/sec or cm/sec.
$\dot{V}_x, \dot{V}_y, \dot{V}_z$ = Average velocities of the liquid in the x, y, and z directions respectively, defined in eq.8-285, ft/sec or cm/sec.
W = Weight rate of flow, lb/sec or kg/sec.
$W_D(m_D, t_D)$ = Dimensionless void ratio, defined by eq.8-151.
$W_D(r_D, t_D)$ = Dimensionless drop, defined by eq.8-200.
$\bar{W}_D(m_D, t_D)$ = Function, defined in eq.8-155.
$W(e)$ = Pseudo-void ratio, defined by eq.8-147, ft²/sec or cm²/sec.
$W(p_p)$ = Pseudo-pressure, defined by eq.8-182, lb/ft⁴/sec; also real gas pseudo-pressure.
x, y, z = Rectangular space coordinates, ft or cm.
x_D = Function, defined by eq.8-259.
$\tilde{z}$ = Position of particle at any instant, defined by eq.8-103, ft or cm.
z_o = Datum, defined by eq.8-61, ft or cm.
z_s = Depth of sediments, ft or cm.
z_w = Depth below water table, ft or cm.
Z = Defined in eq.8-319, dimensionless.
$Z(p_p)$ = Compressibility factor for a real gas, dimensionless.

Subscripts

b = bulk, base.
D = Dimensionless.
e = External boundary.
g = Gas.
i = Initial conditions.
l = Liquid.
Lr = Liquid relative to rock.
m = Moving reference frame.
p = Pore.
p, q, r = Mesh points in the finite difference procedure.
r = Reduced properties; rock; radial coordinate.
t = Total.
w = Inner boundary, the well.
x, y = Partial derivatives.

Mathematical symbols

∇ = Gradient, ft^{-1} or cm^{-1}.
$\nabla \cdot$ = Divergence, ft^{-1} or cm^{-1}.
∇^2 = Laplacian operator, ft^{-2} or cm^{-2}.
$\dfrac{d}{dx}$ = Ordinary derivative, differential.
$\dfrac{\partial}{\partial x}$ = Partial derivative.
δ = Variation (δx = the variation of x).

Greek symbols

α = Defined by eq.8-93, ft/sec$^{1/2}$ or cm/sec$^{1/2}$.
$\tilde{\alpha}$ = Constant defined by eq.8-198.
$\bar{\alpha}$ = Arbitrary constant, defined by eq.8-279, dimensionless.
$\bar{\bar{\alpha}}$ = Coefficient of thermal expansion, defined by eq.8-282, ($^\circ$F)$^{-1}$ or ($^\circ$C)$^{-1}$.
$\tilde{\alpha}_D$ = Dimensionless diffusivity, defined by eq.8-256.
β = Defined by eq.8-112, dimensionless.
$\bar{\beta}$ = Constant, defined by eq.8-304, dimensionless.
γ = Defined by eq.8-126, ft/sec$^{1/2}$ or cm/sec$^{1/2}$.
$\bar{\gamma}$ = Constant, defined by eq.8-306, dimensionless.
γ_b = Bulk specific weight, defined by eq.8-161, lb/ft^3 or kg/cm^3.

$\delta\phi$ = Increment in ϕ_r.
Δm_D = Increments in m_D for finite differences.
Δt_D = Increments in t_D for finite differences.
ϵ = Dilation, defined by eq.8-277, dimensionless.
ϵ_x, ϵ_y, ϵ_z = Unit elongation in the x, y, and z coordinate directions, respectively, dimensionless.
ϵ_{xy}, ϵ_{yz}, ϵ_{zx} = Shearing-strain components in rectangular coordinates, dimensionless.
η = Similarity transformation, defined by eq.8-116, ft/sec$^{1/2}$ or cm/sec$^{1/2}$.
θ = Variation of pore volume per unit bulk volume, defined by eq.8-272, dimensionless.
Θ = Angle defined in Fig.8-32.
λ = Similarity transformation, defined by eq.8-88, ft/sec$^{1/2}$ or cm/sec$^{1/2}$.
μ = Viscosity, lb sec/ft^2 or cp.
μ_g = Viscosity of a real gas, lb sec/ft^2 or cp.
μ_{ga} = Viscosity of a real gas at standard conditions, lb sec/ft^2 or cp.
μ_L = Viscosity of liquid, lb sec/ft^2 or cp.
μ_w = Viscosity of water, lb sec/ft^2 or cp.
ν = Poisson's ratio, dimensionless.
ξ, η, ζ = Location of elemental volume, eq.8-308.
ρ = Density, lb/ft^3 or gm/cm^3.
$\bar{\rho}$ = Radial distance, defined in Fig.8-32.
ρ_b = Bulk density of water-saturated sand, lb/ft^3 or g/cm^3.
ρ_g = Density of rock grain material, lb/ft^3 or g/cm^3.
ρ_L = Density of liquid, lb/ft^3 or g/cm^3.
ρ_r = Density of rock matrix, lb/ft^3 or g/cm^3.
ρ_w = Density of water, lb/ft^3 or g/cm^3.
$\sigma(z)$ = Mechanical load, component of overburden potential, defined by eq.8-160, psi/lb/ft^3 or kg/cm^2/g/cm^3.
$\bar{\sigma}$ = Average stress, defined by eq.8-39, psi or kg/cm^2.
σ_x, σ_y, σ_z = Normal stresses in the x, y, and z directions, respectively, in rectangular coordinates, psi or kg/cm^2.
τ = Defined by eq.8-111, ft^2 or cm^2.
τ_{xy}, τ_{yz}, τ_{zx} = Shearing-stress components in rectangular coordinates, psi or kg/cm^2.
ϕ = Porosity, ratio of pore volume to bulk volume, dimensionless.
ϕ_0, ϕ_1, ϕ_2, ϕ_3 = Harmonic functions, defined by eq.8-303.
Φ = Potential, various, defined by eq.8-61 and 8-62, psi or kg/cm^2.
Φ_t = Potential, defined by eq.8-162, psi/lb/ft^3 or kg/cm^2/g/cm^3.
ψ = Potential, swelling pressure defined by eq.8-71, psi or kg/cm^2.
ψ_c = Arbitrary constant, defined by eq.8-108, psi or dynes/cm^2.

$\bar{\psi}$ = Potential, defined by eq.8-297, ft² or cm² .
ω = Defined by eq.8-120, dimensionless.
Ω = Overburden potential, defined by eq.8-160, psi or kg/cm² .

REFERENCES

Abramowitz, M. and Stegun, I. A., 1964. *Handbook of Mathematical Functions with Formulas, Graphs and Mathematical Tables.* Applied Math. Series 55, U.S. Dep. of Commerce, National Bureau of Standards, Washington, D.C., 1046 pp.
Adams, L. H. and Williamson, E. D., 1923. The compressibility of minerals and rocks at high pressures. *Franklin Inst. J.*, 195: 475—529.
Al-Hussainy, R., 1966. *Transient Flow of Ideal and Real Gases Through Porous Media.* Thesis, Texas A & M Univ., College Station, Texas, 106 pp.
Al-Hussainy, R., Ramey Jr., H. J. and Crawford, P. B., 1966. The flow of real gases through porous media. *J. Pet. Tech.*, 20: 624—636.
Allen, A. S., 1969. Geologic settings of subsidence. *Geol. Soc. Am. Rev. Eng. Geol.*, 2: 305—342.
Amyx, J. W., Bass Jr., D. M. and Whiting, R. L., 1960. *Petroleum Reservoir Engineering.* McGraw-Hill, New York, N.Y., 610 pp.
Aronofsky, J. S. and Jenkins, R., 1954. A simplified analysis of unsteady radial gas flow. *Trans. AIME*, 201: 149—154.
Biot, M. A., 1935. Le probléme de la consolidation des matières argileuses sous une charge. *Ann. Soc. Sci. Brux.*, B55: 110—113.
Biot, M. A., 1941a. General theory of three-dimensional consolidation. *J. Appl. Phys.*, 12: 155—164.
Biot, M. A., 1941b. Consolidation settlement under a rectangular load distribution. *J. Appl. Phys.*, 12: 426—430.
Biot, M. A., 1955. Theory of elasticity and consolidation for a porous anisotropic solid. *J. Appl. Phys.*, 26: 182—185.
Biot, M. A., 1956. General solutions to the equations of elasticity and consolidation for a porous material. *J. Appl. Mech.*, 6: 91—96.
Bixel, H. C., Larkin, B. K. and Van Poollen, H. K., 1963. Effect of linear discontinuities on pressure buildup and drawdown behavior. *J. Pet. Tech.*, 17: 885—895.
Bixel, H. C. and Van Poollen, H. K., 1967. Pressure drawdown and buildup in the presence of radial discontinuities. *Soc. Pet. Eng. J.*, 7: 301—309.
Boltzman, L., 1894. Zur Integration der Diffusionsgleichung bei variablen Diffusionskoeffizienten. *Ann. Phys. (Leipz.)*, 53: 954—964.
Botset, H. G. and Reed, D. W., 1935. Experiment on compressibility of sand. *Bull. Am. Assoc. Pet. Geol.*, 19(7): 1053—1060.
Boussinesq, J., 1885. *Application des Potentiels à l'Étude de l'Équilibre et des Mouvements des Solides Élastiques.* Gauthier-Villars, Paris, France, pp.63—72.
Carman, P. C., 1937. Fluid flow through granular beds. *Trans. Inst. Chem. Eng. Lond.*, 30: 150—156.
Carslaw, H. S. and Jaeger, J. C., 1959. *Conduction of Heat in Solids.* Oxford Press, Oxford, 2nd ed., 510 pp.
Carter, R. D., 1966. Pressure behavior of a limited circular composite reservoir. *Soc. Pet. Eng. J.*, 6: 328—334.

Chierici, C. L., Ciucci, G. M., Eva, F. and Long, G., 1967. Effect of the overburden pressure on some petrophysical parameters of reservoir rocks. In: *Proc. World Pet. Congr., 7th, London.* Elsevier, Amsterdam, 2: 309—338.

Christie, I. F., 1964. A re-appraisal of Merchant's contribution to the theory of consolidation. *Geotechnique*, 14: 304—320.

Cleary, J. M., 1959. Hydraulic fracture theory, III. Elastic properties of sandstone. *Ill. State Geol. Surv., Circ.*, 281: 44 pp.

Clegg, M. W., 1968. *The Flow of Real Gases in Porous Media.* Annu. SPE Fall Meet., 43rd, Houston, Texas, Present. Pap., SPE 2091: 11 pp.

Closmann, P. J. and Ratliff, N. W., 1967. Calculation of transient oil production in a radial composite reservoir. *Soc. Pet. Eng. J.*, 7: 355—358.

Cooper Jr., H. H., 1966. The equation of ground water flow in fixed and deforming coordinates. *J. Geophys. Res.*, 71(20): 4785—4790.

Crank, J. and Nicholson, P., 1947. A practical method for the numerical evaluation of solutions of partial differential equations of heat conduction type. *Proc. Camb. Philos. Soc.*, 43: 50—67.

Darcy, H. J., 1856. Determination of the law of flow of water through sand. In: *Les Fontaines Publiques de la Ville de Dijon.* Libraire de Corps Impériaux des Pont et Chaussées et des Mines, Paris, pp.590—594.

Darley, H. C. H., 1969. A laboratory investigation of borehole stability. *J. Pet. Tech.*, 23: 883—892.

Davis, E. H. and Poulos, H. G., 1972. Rate of settlement under two- and three-dimensional conditions. *Geotechnique*, 22: 95—114.

Davis, E. H. and Raymond, G. P., 1965. A non-linear theory of consolidation. *Geotechnique*, 25: 161—173.

De Witte, A. J. and Warren, J. E., 1957. Discussion on the effect of fluid pressure decline on volumetric changes of porous rocks. *Trans. AIME*, 210: 331—340.

Evangelisti, G. and Poggi, B., 1970. Sopra i fenomeni di deformazione dei terreni da variazione della pressione di strato. *Atti. Accad. Sci. Inst. Bologna, Mem. Ser.*, II(6): 124 pp.

Fatt, I., 1958a. Pore volume compressibilities of sandstone reservoir rocks. *Trans. AIME*, 213: 326—364.

Fatt, I., 1958b. Compressibility of sandstone at low to moderate pressures. *Am. Assoc. Pet. Geol.*, 42(8): 1924—1957.

Fay, C. H. and Prats, M., 1951. The application of numerical methods to cycling and flooding problems. In: *Proc. World Pet. Congr., 3rd*, Brill, Leiden, II: 555—563.

Florin, V. A., 1948. *Teoriya Uplotneniya Zemlyanykh Pochv (Theory of Consolidation and Strengthening of Soils).* Moscow.

Forsythe, G. E. and Wasow, W. R., 1960. *Finite-Difference Methods for Partial Differential Equations.* Wiley, New York, N.Y., 444 pp.

Friedman, N. E., 1958. Quasilinear heat flows. *J. Heat Transfer, Trans. ASME*, 80(3): 635—644.

Geertsma, J., 1957a. The effect of fluid pressure decline on volumetric changes of porous rocks. *Trans. AIME*, 210: 331—338.

Geertsma, J., 1957b. A remark on the analogy between thermoelasticity and elasticity of saturated porous media. *J. Mech. Phys. Solids*, 6: 13—16.

Geertsma, J., 1966. Problems of rock mechanics in petroleum production engineering. *Proc. Congr. Int. Soc. Rock Mech., 1st, Lisbon*, 1: 585—594.

Geertsma, J., 1971. *Land Subsidence Above Compacting Oil and Gas Reservoirs.* Presented at the SPE Eur. Spring Meet., Amsterdam, Present. Pap., SPE 3730: 17 pp.

Geertsma, J., 1973. Land subsidence above compacting oil and gas reservoirs. *J. Pet. Tech.*, 25: 734—744.

Gerzevanov, N. M., 1937. *The Foundation of Dynamics of Soils.* Stroiizdat, Leningrad, 3rd ed.

Gibson, R. E., England, G. L. and Hussey, M. J. L., 1967. The theory of one-dimensional consolidation of saturated clays, I. *Geotechnique*, 17: 261—273.

Goodier, J. N., 1937. On the integration of the thermoelastic equations. *Philos. Mag.*, 7: 1017—1032.

Hall, H. N., 1953. Compressibility of reservoir rocks. *Trans. AIME*, 198: 309—311.

Hubbert, M. K., 1940. The theory of ground water motion. *J. Geol.*, 48: 785—944.

Hubbert, M. K., 1956. Darcy law and the field equations of the flow of underground fluids. *Trans. AIME*, 207: 222—239.

International Symposium on Land Subsidence, 1969. *IASH-UNESCO*, (Tokyo), Publ. 88 and 89: 324 and 661 pp.

Jaeger, J. C., 1964. *Elasticity, Fracture and Flow with Engineering and Geologic Applications.* Methuen, London/Wiley, New York, N.Y., 152 pp.

Jakob, C. E., 1940. The flow of water in an elastic artesian aquifer. *Trans. Am. Geophys. Union*, 2: 574—576.

Jakob, C. E., 1950. Flow of ground water. In: H. Rouse (Editor), *Engineering Hydraulics.* Wiley, New York, N.Y., pp.321—386.

Knutson, C. F. and Bohor, B. F., 1963. Reservoir rock behavior under moderate confining pressure. In: C. Fairhurst (Editor), *Rock Mechanics. Proc. Rock Mech. Symp., 5th, Univ. Minn.* Macmillan, New York, N.Y., pp.627—659.

Kozeny, J., 1927. Über kapillare Leitung des Wassers im Boden. *Sitzungber. Akad. Wiss. Wien, Math.-Naturw. Kl. Abt.*, 136(IIa): 271—306.

Lambe, T. W., 1951. *Soil Testing for Engineers.* Wiley, New York, N.Y. Chapman and Hall, London, 165 pp.

Love, A. E H., 1934. *A Treatise on the Mathematical Theory of Elasticity.* 4th ed. (1st ed., 1926), Cambridge, England, 643 pp.

Lubinski, A., 1954. The theory of elasticity for porous bodies displaying a strong porous structure. *Proc. U.S. Natl. Congr. Appl. Mech.*, 2nd, pp.247—256.

Mandel, J., 1957. Consolidation des couches d'argiles. *Proc. Int. Conf. Soil Mech., 4th, London*, 1: 360—367.

Mandel, J., 1961. Tassements produits par la consolidation d'une couche d'argile de grande epaisseur. *Proc. Int. Conf. Soil Mech., 5th, Paris*, 1: 733—736.

Matthews, C. S. and Russell, D. G., 1967. *Pressure Buildup and Flow Tests in Wells.* Soc. Pet. Eng. AIME, New York, N.Y., 172 pp.

McMordie, R. K., 1962. Steady state conduction with variable thermal conductivity. *J. Heat Transfer, Trans. ASME, Ser. C*, 84(1): 92—93.

McNabb, A., 1960. A mathematical treatment of one-dimensional soil consolidation. *Q.J. Appl. Math.*, 17: 337—347.

Miller, F. G., 1962. Theory of unsteady-state influx of water in linear reservoirs. *J. Inst. Pet.*, 48: 365—379.

Mindlin, R. D. and Cheng, D. H., 1950. Thermoelastic stress in the semi-infinite solid. *J. Appl. Phys.*, 21: 931—933.

Mitchell, A. R., 1969. *Computational Methods in Partial Differential Equations.* Wiley, New York, N.Y., 255 pp.

Monteiro, N., 1970. *Finite Difference Correlations of Pseudo-Pressure Functions with Oil-Well Flow Tests.* Dep. Pet. Eng., Stanford Univ., 33 pp. (unpubl. rep.).

Mueller, T. D., 1962. Transient response of nonhomogeneous aquifers. *Soc. Pet. Eng. J.*, 2: 33—43.

Muskat, M., 1949. The theory of potentiometric models. *Trans. AIME*, 179: 216—221.
Nowacki, W., 1962. *Thermoelasticity*. Addison Wesley, Mass./Pergamon Press, Oxford/ Polish Scientific Publishers, Warszawa, 628 pp.
Papkovitch, P. F., 1932. Solution génerale des equations fondamentales d'élasticité exprimée par trois fonctions harmoniques. *C. R. Acad. Sci.*, 195: 513—515, 754—756.
Parlange, J.-Y., 1971a. Theory of water movement in soils, 1. One-dimensional absorption. *Soil. Sci.*, 111: 134—137.
Parlange, J.-Y., 1971b. Theory of water movement in soils, 2. One-dimensional infiltration. *Soil Sci.*, 111: 170—174.
Parlange, J.-Y., 1971c. Theory of water movement in soils, 3. Two- and three-dimensional absorption. *Soil Sci.*, 112: 313—317.
Parlange, J.-Y., 1972a. Theory of water movement in soils, 4. Two- and three-dimensional steady infiltration. *Soil Sci.*, 113: 96—101.
Parlange, J.-Y., 1972b. Theory of water movement in soils, 5. Multi-dimensional unsteady infiltration. *Soil. Sci.*, 113: 156—161.
Parlange, J.-Y., 1972c. Theory of water movement in soils, 6. Effect of water depth over soil. *Soil Sci.*, 113: 308—312.
Parlange, J.-Y., 1972d. Theory of water movement in soils, 7. Multi-dimensional cavities under pressure. *Soil Sci.*, 113: 379—382.
Parlange, J.-Y., 1972e. Theory of water movement in soils: 8. One-dimensional infiltration with constant flux at the surface. *Soil Sci*, 114: 1—4.
Philip, J. R., 1955. Numerical solution of equations of the diffusion type with diffusivity concentration-dependent. *Trans. Farad. Soc.*, 61: 885—892.
Philip, J. R., 1960. The function inverc θ. *Aust. J. Phys.*, 13: 13—20.
Philip, J. R., 1968. Sorption and volume change in colloid pastes. *Aust. J. Soil Res.*, 6: 249—267.
Philip, J. R., 1969a. Moisture equilibrium in swelling soils, I. Basic theory. *Aust. J. Soil Res.*, 7: 99—120.
Philip, J. R., 1969b. Moisture equilibrium in swelling soils, II. Applications. *Aust. J. Soil Res.*, 7: 121—141.
Philip, J. R., 1969c. Hydrostatics and hydrodynamics in swelling soils. *Water Resour. Res.*, 5: 1070—1077.
Philip, J. R., 1972. Recent progress in the theory of irrigation and drainage of swelling soils. *Congr. Int. Comm. Irrig. Drain.*, 8th, 16 pp.
Pirson, S. J., 1963. *Handbook of Well Log Analysis*. Prentice-Hall, Englewood Cliffs, N. J., 326 pp.
Raghavan, R., 1972. *A Review of the Consolidation and Rebound Processes in One-dimensional Porous Columns*. Annu. SPE Fall Meet., San Antonio, Texas, Present Pap., SPE 4078: 20 pp.
Raghavan, R., 1974. Consolidation and rebound processes in one-dimensional porous columns. *J. Geophys. Res.*, 79(11): 1687—1698.
Raghavan, R., Scorer, J. D. T. and Miller, F. G., 1972. An investigation by numerical methods of the effect of pressure-dependent rock and fluid properties on well flow tests. *Soc. Pet. Eng. J.* 12: 267—275.
Ramey, H. J., Jr., 1964. Rapid methods for estimating reservoir compressibilities. *J. Pet. Tech.*, 18: 447—454.
Rowan, G. and Clegg, M. W., 1962. An approximate method for transient radial flow. *Soc. Pet. Eng. J.*, 2: 225—256.

Sen, B., 1950. Note on the stresses produced by nuclei of thermo-elastic strain in a semi-infinite elastic solid. *Q. J. Appl. Math.*, 8: 365—369.
Smiles, D. E. and Poulos, H. G., 1969. The one-dimensional consolidation of columns of soil of finite length. *Aust. J. Soil Res.*, 7: 285—291.
Smiles, D. E. and Rosenthal, M. J., 1968. The movement of water in swelling materials. *Aust. J. Soil Res.*, 6: 237—248.
Smith, G. D., 1964. *Numerical Solution of Partial Differential Equations.* Oxford Univ. Press, Oxford, 179 pp.
Smith, J. E., 1971. The dynamics of shale compaction and evolution of pore fluid pressures. *Math. Geol.*, 3: 239—263.
Sokolnikoff, I. S., 1956. *Mathematical Theory of Elasticity.* McGraw-Hill, New York, N.Y., 476 pp.
Storm, M. L., 1951. Heat conduction in simple metals. *J. Appl. Phys.*, 2: 940—951.
Taylor, D. W., 1948. *Fundamentals of Soil Mechanics.* Wiley, New York, N.Y., 700 pp.
Taylor, D. W. and Merchant, W., 1940. A theory of clay consolidation accounting for secondary compression. *J. Math. Phys.*, 19: 167—185.
Ter-Martirosyan, Z. G. and Ferronsky, V. I., 1969. Some problems of time-soil compaction in pumping liquid from a bed. In: *Int. Symp. Land Subsidence, 1969, Tokyo. IASH—UNESCO, Publ.* 88: 303—314.
Terzaghi, K., 1923. Die Berechnung der Durchlässigkeitsziffer des Tones aus dem Verlauf der Hydrodynamischen Spannungserscheinungen. *Sitzungsb. Akad. Wiss. Wien*, 132: 125—138.
Terzaghi, K., 1943. *Theoretical Soil Mechanics.* Wiley, New York, N. Y., 510 pp.
Thomas, L. H., 1949. *Elliptic Problems in Linear Difference Equations Over a Network.* Watson Sci. Comput. Lab. Rep., Columbia Univ., New York, N.Y.
Timoshenko, S. P. and Goodier, J. N., 1970. *Theory of Elasticity.* McGraw-Hill, New York, N.Y., 3rd ed., 567 pp.
Van der Knaap, W., 1959. Nonlinear behavior of elastic porous media. *Trans. AIME*, 216: 179—186.
Van Everdingen, A. F. and Hurst, W., 1949. The application of the Laplace transformation to flow problems in reservoirs. *Trans. AIME*, 186: 305—324.
Vasquez, H., 1961. *A Correlation between Permeability, Porosity and Grain-size for Consolidated Sandstones.* Thesis, Univ. of Texas, 39 pp.
Verrujt, A., 1965. Discussion, *Proc. Int. Conf. Soil Mech. Found. Eng.*, 6th, Montreal, 3: 401—402.
Verrujt, A., 1969. Elastic storage of aquifers. In: R. J. M. de Wiest (Editor), *Flow Through Porous Media*, pp. 331—376.
Wattenbarger, R. A., 1967. *Effects of Turbulence, Wellbore Damage, Wellbore Storage and Vertical Fractures on Gas Well Testing.* Thesis, Stanford Univ., Stanford, Calif., 139 pp.
Wu, T. H., 1966. *Soil Mechanics.* Allyn and Bacon, Boston, Mass., 431 pp.
Zaslavsky, D., 1964. Saturated and unsaturated flow equation in an unstable porous medium. *Soil Sci.*, 98: 317—321.

Appendix A

CONVERSION TABLE

Specific weight and pressure gradient of fluids and rocks

Specific weight			Pressure gradient	Specific weight			Pressure gradient
lb/gal	lb/ft³	g/cc	psi/ft	lb/gal	lb/ft³	g/cc	psi/ft
6.25	46.75	0.75	0.325	13.00	97.24	1.56	0.675
6.50	48.62	0.78	0.338	13.25	99.11	1.59	0.688
6.68	50.00	0.80	0.347	13.37	100.00	1.60	0.694
6.75	50.49	0.81	0.351	13.50	100.98	1.62	0.701
7.00	52.36	0.84	0.364	13.75	102.85	1.65	0.714
7.25	54.23	0.87	0.377	14.00	104.72	1.68	0.727
7.50	56.10	0.90	0.390	14.25	106.59	1.71	0.740
7.75	57.97	0.93	0.403	14.50	108.46	1.74	0.753
8.00	59.84	0.96	0.416	14.71	110.00	1.76	0.764
8.02	60.00	0.96	0.417	14.75	110.33	1.77	0.766
8.25	61.71	0.99	0.429	15.00	112.20	1.80	0.779
8.35	62.46	1.00	0.434	15.25	114.07	1.83	0.792
8.50	63.58	1.02	0.442	15.50	115.94	1.86	0.805
8.56	64.00	1.03	0.445	15.75	117.81	1.89	0.818
8.75	65.45	1.05	0.455	16.00	119.68	1.92	0.831
9.00	67.32	1.08	0.467	16.04	120.00	1.92	0.833
9.25	69.19	1.11	0.480	16.25	121.55	1.95	0.844
9.36	70.00	1.12	0.486	16.50	123.42	1.98	0.857
9.50	71.06	1.14	0.493	16.75	125.29	2.01	0.870
9.625	72.00	1.15	0.500	17.00	127.16	2.04	0.883
9.75	72.93	1.17	0.506	17.25	129.03	2.07	0.876
10.00	74.80	1.20	0.519	17.38	130.00	2.08	0.903
10.25	76.67	1.23	0.532	17.50	130.90	2.10	0.909
10.50	78.54	1.26	0.545	17.75	132.77	2.13	0.922
10.70	80.00	1.28	0.556	18.00	134.64	2.16	0.935
10.75	80.41	1.29	0.558	18.25	136.51	2.19	0.948
11.00	82.28	1.32	0.571	18.50	138.38	2.22	0.961
11.25	84.15	1.35	0.584	18.72	140.00	2.24	0.972
11.50	86.02	1.38	0.597	18.75	140.25	2.25	0.974
11.75	87.89	1.41	0.610	19.00	142.12	2.28	0.987
12.00	89.76	1.44	0.623	19.25	144.00	2.31	1.000
12.03	90.00	1.44	0.625	19.50	145.86	2.34	1.013
12.25	91.63	1.47	0.636	19.75	148.73	2.37	1.026
12.50	93.50	1.50	0.649	20.00	149.60	2.40	1.038
12.75	95.37	1.52	0.662	20.05	150.00	2.48	1.041

REFERENCES INDEX*

Numbers in italics refer to references lists

Abramowitz, M., 434, 435, 436, 482, 484, *520*
Ackroyd, L.W., 332, *343*
Adams, J.E., 89, *161*, 250, *289*
Adams, L.H., 414, *520*
Aitchison, G.D., 329, 331, 332, *345*
Alderman, A.R., 88, *161*
Al-Hussainy, R., 447, 448, 460, 462, 466, 468, 470, 473, 476, *520*
Allen, A.S., 403, *520*
Allen, D.R., 59, 60, 61, 63, 64, 66, 72, 74, 75, 77
Allen, J.R.L., 123, 124, *161*, 250, *289*
Allen, P., *289*
Amstutz, G.C., 152, *161*
Amyx, J.W., 472, *520*
Andreson, M.J., 262, 286, *289*
Andrews, J.H., 332, *344*
Andrews, P.B., 21, *40*
Anbah, S.A., 315, *343*
Applin, E.R., 205, 206, 209–212, *240*
Applin, P., 205, 206, 209–212, *240*
Armstrong, F.C., 221, 222, 224, 226, *240*
Armstrong, R.L., 220, 224, 226, 227, *240*
Arnal, R.E., 229, *240*
Aronofsky, J.S., 447, 448, 468, 470, 473, 474, 479, *520*
Arutyunova, N.M., *40*
Aschenbrenner, B.C., 314, *344*
Aslanyan, A.T., *240*
Athy, L.F., 75, 112, *161*, 276, *289*
Atwater, G.I., 93, *161*
Atwood, D.K., 88, 89, *161, 162*
Ayer, N.J., 89, 122, 130, 139, *161*

Baars, D.L., 218, *240*
Babaev, A.G., *240*
Bandy, O.L., 229, *240*
Bara, J.P., *240*
Barbat, W.F., 230, *240*
Bartlett, A.H., 340, *344*
Bass Jr., D.M., 472, *520*
Bathurst, R.G.C., 79, 87, 88, 91, 140, 143, 145, 146, 152, 158, *161*
Beal, L.H., *240*
Beall, A.O., *161*
Bebout, D.G., 113, *162*
Becker, G.F., 153, *161*
Beerbower, J.R., 250, *289*
Beeson, C.M., 315, *343*
Benson, L.V., 86, 88, *162*
Berg, R.R., *400*
Bergau, W., 307, *345*
Bergenback, R.E., *292*
Bernal, J.D., 83, 95, 102, *162*
Bernard, H.A., 250, *289*
Biggs, C.A., 387, *400*
Biot, M.A., 405, 415, 418, 422, 478, 492, 493, 494, 497, 513, *520*
Bird, J.M., 192, 193, 194, *240*
Bishop, W.F., 114, *162*
Bissell, H.J., *40*, 79, 89, *162*, 172, 186, 188, *240*
Bixel, H.C., 461, *520*
Bjerrum, L., 327, *344*·
Blatt, H., 2, *40*
Bloom, A.L., 230, 231, *240*
Blusson, S.L., 179, *242*
Bohor, B.F., 61, 76, 419, *522*

*Prepared by Dr. John O. Robertson, Jr.

Boltzman, L., 428, *520*
Bond, R.D., 340, *344*
Botset, H.G., 415, *520*
Boussinesq, J., 505, *520*
Bowen, V.T., 88, *166*
Bowin, C.O., 207, *245*
Bowyer, B., 227, *243*
Bubb, J.N., 88, 89, 161, *162*
Buchanan, F., 332, *344*
Bucher, W.H., 170, 171, 172, 220, *241*
Bullard, E., 193, *241*
Bunce, E.I., 132, *162*
Burst, J.F., 274, 276, *290*
Busch, D.A., 258, *290*, *400*
Busang, P.F., 300, *347*
Butler, G.P., 88, 89, *162*
Breeding, J.G., 87, *164*
Bricker, O.P., 87, 88, 144, 146, 162, *164*
Brill Jr., K.G., 218, *240*
Brink, A.B.A., 329, *344*
Brown Jr., L.F., 247—288, 289, *290*
Brown Jr., R.D., 228, *240*
Brown, R.L., 83, 95, 106, *162*
Brown, S.L., *290*

Cady, W.M., 199, *242*
Campbell Jr., R.L., 392, 394, 397, *400*
Cannon Jr., R.S., 179, *244*
Carman, P.C., 432, *520*
Carozzi, A.V., 114, *162*
Carpenter, C.B., 54, 62, *75*
Carrigy, M.A., *400*
Carter, R.D., 461, *520*
Carslow, H.S., 447, 460, 482, 505, *520*
Casagrande, A., *344*
Chamberlain, C.K., 213, 214, *241*
Chao, K.H., *344*
Chaplin, T.K., 322, *344*
Chapman, R.E., 249, 276, *290*
Chase, C.G., 234, *241*
Chave, K.E., 86, *162*
Cheney, M.G., 250, 266, *290*
Cheng, D.H., 406, 506, 507, *522*
Chepil, W.A., 335, *344*
Chierici, C.L., 418, *521*
Chilingar, G.V., 57, 59, 61, 63, 64, 66, *75*, 77, 79, 87, 89, 110, *162*, *164*, 206, 217, *241*, 306, 315, 318, 340, *343*, *344*, *346*

Chilingarian, G.V., 17, 28, 33, 34, 35, 37, 40, *41*, *42*, 133, *165*, 170, 315, *346*
Chiu, K.Y., *344*
Choquette, P.W., 146, 155, *162*, *163*
Christiansen, F.W., 224, *241*
Christie, I.F., 437, *521*
Cifelli, R., 88, *166*
Ciucci, G.M., 418, *521*
Clark, L.M., *241*
Clark, S.P., 340, *344*
Cleary, J.M., 414, *521*
Clegg, M.W., 461, 462, 505, *521*, *523*
Cline, L.M., 213, 214, *241*
Clipping, D.H., 229, *241*
Closmann, P.J., 461, *521*
Cohen de Lara, G., 317, *344*
Colazas, X.C., 59, 61, 64, 65, 66, *75*
Coleman, J.M., 250, *290*
Colle, J., 210, *241*
Coney, P.J., 234, *241*
Conrey, B.L., 389, *400*
Conybeare, C.E.B., 274, 276, 277, *290*
Coogan, A.H., 83, 95, 113, 114, 115, 116, 127, 132, 147, *162*
Cooke Jr., W.F., 201, *241*
Cooper Jr., H.H., 405, 422, 478, *521*
Correns, C.W., *344*
Craft, B.C., 32, *40*
Cramer, H.R., 186, *241*
Crank, J., 488, 489, 491, *521*
Crawford, P.B., 448, 460, 462, 466, 468, 470, 473, *520*
Crittenden Jr., M.D., 178, 231, 233, *241*
Crosby, E.J., 380, *400*
Crosby, G.W., 220, 221, 222, *241*
Cummins, W.F., 249, *290*

D'Appolonia, E., 309, 311, 329, *344*
Dahl, A.R., 385, *401*
Dallmus, K.F., 173, 174, 176, 190, 219, *241*
Dana, J.D., 171, 192, *241*
Darcy, H.J., 503, *521*
Darley, H.C.H., 432, *521*
Davis, D.K., 384, *400*
Davis, E.H., 406, 492, *521*
Davis, G.H., 44, 72, 77, 236, 238, 239, *244*
Davis, S.N., 239, *243*
Day, A.L., 153, *161*

REFERENCES INDEX

Deffeyes, K.S., 86, 88, 89, *162*
Demarest, D.F., 250, *292*
Denham, R.L., 201, *241*
De Sonza, J.M., *56*
Dewey, J.F., 192, 193, 194, 240, *241*
De Witt Jr., W., 250, *292*
De Witte, A.J., 418, *521*
Dickey, P.A., 202, 203, 204, *242*
Dickinson, G., 274, 276, 277, *290*
Dietz, R.S., 193, 195–197, 206, 207, 213–214, 228, *241*
Dill, R.F., 389, *401*
Donald, I.B., 316, *346*
Donath, F.A., 89, 128, 134, 135, 139, 141, *163*
Drake, C.L., 193, 197, *241*
Drake, N.F., 249, *290*
Dresser Atlas, 353, 355, 356, 393, 395, 398, 399, *400*
Duff, P.McL.D., 250, *290*
Dunham, R.J., 79, 83, 129, 137, *162*
Dunnington, H.V., 153, 154, 155, *162*

Eardley, A.J., 178, 190, 195, 208, 219, 220, 221, *242*
Eargle, D.H., 250, *290*
El-Nassir, A., 315, 318, *344*
El-Sohby, M.A., 323, *344*
Ellis, A.L., 103, *163*
Emerson, W.W., 340, *344*
Emery, K.O., 132, 162, 207, *242, 245, 400*
England, G.L., 406, 446, *522*
Eva, F., 418, *521*
Evangelisti, G., 406, 493, 511, 513, *521*
Everett, J.E., 193, *241*
Ewing, M., 193, 197, 201, *241, 245*

Fahhad, S., 77
Fatt, I., 39, 40, 56, 62, 71, *75, 76*, 133, *163*, 414, 418, 419, *521*
Fay, C.H., 460, *521*
Feather, J.N., 56, *76*
Fedorow, B.S., 336, *346*
Feofilova, A.P., 250, *290*
Ferguson, H.C., 210, *241*
Ferm, J.C., *292*
Ferronsky, V.I., 509, *524*
Finney, J.L., 83, 95, *162*
Fischer, A.G., 83, *161, 163*
Fisher, W.L., 269, *290*

Fisk, H.N., 250, 258, *290*
Florin, V.A., 329, *344*, 422, *521*
Folk, R.L., 2, 4, 5, 10, 12, 13, 14, 17, 18, 19, 21, 28, *40*, 79, *163*
Fons, L., 378, *400*
Foote, J.N., 340, *344*
Foote, P.D., 300, *347*
Forgotson Jr., J.M., *290*
Forsythe, G.E., 479, *521*
Fournier, R.O., 331, *346*
Fraser, H.J., 31, *41*, 83, 95, 96, 98, 101, 104, 105, 110, 112, 129, *163*, 300, 345
Frazier, D.E., 250, *290*
Frenzel, H.N., 250, *289*
Freund, J.E., *163*
Friedman, G.M., 86, 87, 140, *163*
Friedman, M., 56, *76*
Friedman, N.E., 455, 461, *521*
Friedman, S.A., 250, 258, *290*
Fruth Jr., L.S., 89, 128, 134, 135, 139, 141, *163*
Frydman, S., 300, *345*
Füchtbauer, H., *163*
Fujii, T., 319, *345*
Fukuzumi, R., 336, *347*
Fuller, M.L., 44, *76*
Fuller, W.B., 301, 302, *345*
Furer, L.C., 222, *242*

Gabrielse, H., 179, *242*
Gagliano, S.M., 250, *290*
Gaither, A., *40*, 106, 113, 114, 121, 122, *163*
Galley, J.E., 217, *242*
Galloway, W.E., 249, 250, 270, *290*
Garrels, R.M., 86, *162*
Garrison, R.E., 83, *163*
Gednetz, D.E., *292*
Geertsma, J., 55, *76*, 405, 406, 414, 415, 416, 417, 419, 493, 494, 497, 498, 503, 505, 508, 509, 510, 511, 513, *521, 522*
Gerard, R.D., 132, 162, *165*
Gerrard, C.M., 307, 322, *345*
Gerzevanov, N.M., 422, *522*
Ghose, S., 77
Gibson, R.E., 406, 446, *522*
Gilbert, C.M., 19, *42*
Gilbert, G.K., 168, 169, 170, 172, 192, 231, 232, *242*

Gilluly, J., 197, 198, 199, *242*
Gilman, Y.D., 336, *346*
Gilreath, J.A., 392, *400*
Ginsburg, R.N., 88, *165*
Glover, E.D., 91, *165*
Gomaa, E.M., 37, *40*
Goodier, J.N., 495, 499, 503, 504, 506, 507, *522, 524*
Goreau, T.F., 86, *164*
Gorsline, D.S., *242, 400*
Goss, L.F., 250, 266, *290*
Grabau, A.W., *41*
Grant, K., 331, 332, *345*
Grasly, R., *243*
Graton, L.C., 31, *41*, 83, 95, 96, 98, 101, 110, 129, *163*, 300, *345*
Gray, K.E., 62, *76*
Green, J.P., 238—239, *242*
Greenwood, D.A., 336, *345*
Gregory, A.R., 62, *76*
Griffiths, J.C., 5, 15, 16, *41*
Guber, A.L., *292*
Gupta, M.K., 328, *347*
Gustafsson, Y., 331, *346*
Guyod, H., 363, *400*
Guzman, E.J., 201—202, *242*

Hager, R., 55, *76*
Hall, H.N., 38, *41*, 55, *76*, 414, 418, *522*
Ham, W.E., 79, *163*
Hamblin, W.K., *401*
Handin, J., 55, 56, *76*
Hanna, W.F., 228, *240*
Harbaugh, J.W., 86, 88, *165*
Hard, H.N., 72, *76*
Hardin, B.O., 338, *345*
Harms, J.C., 155, *163*
Harris, C.C., 109, *163*
Harris, H.D., *242*
Harrison, E., *76*
Harrison, J.E., *242*
Hassey, M.J.L., 406, 446, *522*
Hathaway, J.C., 130, *163*
Hawkins, M.F., 32, *40*
Hawksley, P.G., 83, 95, 106, *162*
Hays, F.R., *163*
Hazzard, J.C., 179, *243*
Heald, M.T., 145, *163*
Hedberg, H.D., *76*, 276, *290*
Hess, H.H., 192, *242*

Holden, A., 329, 330, *345*
Holden, J.C., 193, 206—207, 228, *241*
Hough, B.K., *76*
Howell, J.B., 4, *41*
Hoyer, W.A., *400*
Huang, T., 86, *163*
Hubbert, M.K., 46, 49, *76*, 171, *242*
Hurst, W., 448, 467, 468, 470, 471, 472, 479, 505, *524*

Ingles, O.G., 300, 302, 314, 319, 322, 323, 329, 336, 340, *345, 346*
Ishihara, K., 321, 339, *345*
Itoh, N., 307, *345*
Itoh, T., 336, *347*
Ivanov, P.L., 329, *344*

Jackson, W.K., 250, *291*
Jaeger, J.C., 447, 460, 482, 503, 505, 520, *522*
Jakob, C.E., 405, *522*
Jam, L., 202, 203, 204, *242*
James, P.H., 274, *291*
Jenkins, R., 447, 448, 468, 470, 473, 474, 479, *520*
Jenkins Jr., W.A., 250, *292*
Jennings, J.E.B., 330, *345*
Jodry, R.L., 74, *76*
Johnson, D.P., 250, *289*
Johnson, D.W., 44, *77*
Johnson, R.H., 171, *242*

Kahle, C.F., 114, 116, *163*
Kahn, J.S., 114, 117, *163*
Kallstenius, T., 307, *345*
Kamen-Kaye, M., 174, 175, 176, 177, 192—193, 214, 218, 230, *242, 243*
Kantey, B., 329, *344*
Kawasumi, H., 327, *345*
Kay, M., 193, 220, *243*
Kelly, J., 53, *76*
Kelvin, Lord, 95, *163*
Kennedy, G.C., 91, *165*
Kersten, M.S., 340, *345*
Kiesnick Jr., W.F., *76*
King, F.H., 106, *164*
Kiselev, A.U., 300, *346*
Klein, G. de V., *400*
Knight, K., 57, 75, 330, *345*
Knox, S.T., 132, *162*

REFERENCES INDEX

Knudsen, F.P., 300, *346*
Knutson, C.F., 61, *76*, 419, *522*
Kohlhaas, C.A., 59, 61, 63, 64, 66, *76*
Kolb, C.R., 250, 290, *291*
Kolbuszewski, J., 303, *346*
Kozeny, J., 313, 346, 432, *522*
Kraft, J.C., 205, *243*
Kringstad, S., 327, *344*
Krivoy, H.L., 208, *243*
Krueger, W.C., *400*
Kruit, C., 250, *291*
Krumbein, W.C., 5, 7, 8, 15, 16, 17, *41*, 295, *346*
Kruyer, S., 301, *346*
Krynine, P.D., 4, *41*
Kuenen, Ph.H., 198, *243*
Kummeneje, O., 327, *344*
Kurzeme, M., 329, *346*

Lafeber, D., 293, 297, 298, 304, 319, 322, *345, 346*
Lambe, T.W., 437, *522*
Lamplugh, G.W., 330, *346*
Land, L.S., 86, *164*
Landes, K.K., 173, *243*
Lane, E.W., 6, *41*
Langnes, G.L., 35, *41*, 340, *346*
Larkin, B.K., 461, *520*
Larsen, G., 87, *164*, 306, 315, *346*
Larson, R.L., 234, *243*
Lawson, J.D., 316, *346*
Le Blanc, R.J., 87, *164*, 250, *289*
Lee, C.H., 103, *163*
Lee, I.K., 316, 323, *346*
Lee, W., 267, *291*
Lewis, D.W., 10, 12, 13, 17, 21, *40*
Lidz, L., 132, *162*
Lloyd, R.M., 88, *165*
Lock, G.A., *400*
Lofgren, B.E., 238, *243*
Lomise, G.M., 336, *346*
Long, G., 418, *521*
Longwell, C.R., 227, 231, *243*
Love, A.E.H., 417, 506, *522*
Lubinski, A., 405, 493, 501, 502, *522*
Lucia, F.J., 88, 89, *162*
Luyendyk, B.P., 207, *245*
Lye, B.R.X., 323, *347*

Mackenzie, F.T., 144, *164*
Maggio, C., 113, *162*
Maiklem, W.R., 83, 84, *164*
Main, R., 35, *40*
Maisano, M., 205, *243*
Mandel, J., 406, *522*
Mannon, R.W., 14, 37, *40*, 110, *162*, 206, 217, *241*
Maricelli, J.J., 392, *400*
Marsden Jr., S.S., 239, *243*
Martin, R., 274, *291*
Marvin, J.W., 95, 130, 131, 132, 141, *164*
Mason, J., 102, *162*
Masson, P.H., 116, 117, *164*
Matthews, C.S., 419, 505, *522*
Matthews, R.K., 86, 87, 88, 144, 162, *164*
Matthews, W.R., 53, *76*
Mattison, S., 331, *346*
Matzke, E.B., 95, 130, 132, *164*
Maxwell, J.C., 90, 91, 130, 132, *164*
Mayuga, M.M., 59, 60, 72, 74, *75*
McBride, E.F., 21, *41*
McCammon, R.B., 14, *41*
McCulloh, T.H., *76*
McCurdy, R.C., 54, *77*
McFarlan Jr., E., 250, *290*
McGeary, R.K., 301, *346*
McGill, G.E., *243*
McGowen, J.H., 248, 254, 269, 290, *291*
McGregor, A.A., 387, *400*
McGuire, W.J., *76*
McGuit, J.H., 201, *241*
McMahon, B., 293, *346*
McMordie, R.K., 461, *522*
McNabb, A., 460, *522*
Mecham, O.E., 54, *77*
Meinzer, O.E., *72*
Melson, W.G., 88, *166*
Menard, H.W., 234, *243*
Merchant, W., 503, *524*
Metcalf, J.B., 336, 340, *345*
Meyerhoff, A.A., 202, *243*
Middleton, G.V., *40*
Miller Jr., D.G., 130, *164*
Miller, E.E., 93, *161*
Miller, F.G., 59, 61, 63, 64, 66, *76*, 438, 439—443, 447, 450, 462, 467, 468, 469, 470, 471, 472, 473, 474, 476, 478, 505, 513, *522, 523*

Milliman, J.D., 86, 88, *164*, 207, *245*
Milton, C., *243*
Mindlin, R.D., 406, 506, 507, *522*
Minor, H.E., *76*
Misch, P., 179, *243*
Mitchell, A.R., 479, 487, *522*
Mollazal, Y., 4, *41*
Monteiro, N., *522*
Moore, D., 249, 250, *291*
Moore, R.C., 249, 251, *291*
Moore, R.T., 227, *243*
Morey, G.W., 331, *346*
Morgan, J.P., *400*
Morgan, J.R., 324, *346*
Morgenstern, M., 326, *347*
Morimoto, T., 336, *347*
Morron, N.R., 83, 85, 95, 96, 101, 106, 108, 109, *164*
Morrow, N.R., 109, *163*
Moshcheryakov, A.N., 336, *346*
Moss, A.J., 295, *347*
Mueller, J.C., 274, *291*
Mueller, T.D., 461, *522*
Murray, G.E., 172, 200–201, *243*
Murray, R.C., *40*, 87, 89, *164*
Muskat, M., 32, *41*, 460, *523*
Myers, D.A., 250, *292*

Natland, M.L., 380, 389, *400*
Neil, R.C., 323, *346*
Netterberg, F., 332, *347*
Newell, N.D., 130, *165*, 203, *243*
Newman, G.H., 66, 67, 70, *76*
Nicholson, P., 488, 489, 491, *521*
Nonogaki, K., 306, *348*
Norrish, K., 332, *347*
Nowacki, W., 507, *523*

Oda, M., 298, 319, *347*
Ojakangas, R.W., *243*
O'Keefe, J.A., 169, *244*
Oldershaw, A.E., 146, *164*
Onas, J., 307, 308, 322, *347*
Orchard, D.F., 323, *347*
Oriel, S.S., 221, 223, *240*
Orme, G.R., 89, 128, 134, 135, 139, 141, *163*

Page, B.M., 234, *243*
Pampeyan, E.H., 227, *243*

Papkovitch, P.F., 505, *523*
Park, Won-Choon (W.C.), 127, 152, *161*, *164*
Park Jr., C.F., 179, *244*
Parlange, J.Y., 445, 459, *523*
Peaker, K., 325, *347*
Peck, R.P., 44, 54, 77
Pellegrino, A., 309, 310, *347*
Pepper, J.F., 250, *292*
Peterson, J.A., *244*
Pettijohn, F.J., 2, 5, 7, 8, 11, 18, 19, *41*, 154, *164*, 250, *292*
Phares, R.S., 384, *401*
Philip, J.R., 403, 404, 432, 437, 445, 457, 458, 459, 479, 481, 483, 492, 512, *523*
Picard, L., 315, *347*
Pierce, J.W., 86, *163*
Pirson, S.J., 32, 41, 351, 378, 379, 400, 432, *523*
Pitman III, W.C., 234, *244*
Plummer, F.B., 249, 251, *292*
Podio, A.L., 62, *76*
Poggi, B., 406, 493, 511, 513, *521*
Poland, J.F., 44, 72, 77, 236, 238, 239, *244*
Potter, E.P., 380, 384, *401*
Potter, P.E., 2, 5, 19, *41*, 250, 258, 286, *292*
Poulos, H.G., 404, 453, 460, 492, *521*, *524*
Powell, J.W., *168*
Powers, M.C., 274, *292*
Prakash, S., 328, *347*
Prats, M., 460, *521*
Pratt, W.E., 44, 77
Pray, L.C., 79, 87, 163, *164*
Pressler, E.D., 205, *244*
Prokopovich, N., 330, *347*
Prozorovich, G.E., *166*
Pryor, A.J., 332, *347*
Pryor, W.A., 113, *165*, 250, *291*
Purdy, E.G., 88, 114, 134, *165*
Pyle, T.E., 208, *243*

Raghavan, R., 438, 439–443, 446, 447, 448, 450, 451, 452, 453, 462, 467, 468, 469, 470, 471, 472, 473, 474, 476, 478, 513, *523*
Rall, E.P., 250, *291*

Rall, R.W., 250, *291*
Ramey Jr., H.J., 411, 419, 448, 460, 462, 466, 468, 470, 473, *520, 523*
Ratliff, N.W., 461, *521*
Raupach, M., 331, *347*
Ray, J.R., 248, 254, *291*
Raymond, G.P., 406, *521*
Reed, D.W., 415, *520*
Reed Jr., J.C., 199, *242*
Reedy Jr., F., 201, *241*
Reeside Jr., J.B., 220, *244*
Renton, J.J., 145, *163*
Rhodes, M.L., 89, *161*
Rick, J.L., *401*
Richard, F.E., 338, *345*
Richards, A.F., 130, *164*
Ridgway, K., 83, 95, 105, *165*
Rieke III, H.H., 14, 34, 35, 37, *40, 41*, 59, 61, *75, 77*, 110, 133, *162, 165*, 170, 206, 217, *241*
Rigby, J.K., 203, *243, 401*
Rittenhouse, G., 16, *42*, 77, 83, 95, 113, 146, 147, 148, 154, 157, *165*
Roberts, J.E., 56, 57, 58, *77*
Roberts, R.J., 227, *243, 244*
Robertson, E.C., 130, *163, 165*
Robertson Jr., J.O., 35, *41*, 340, *346*
Robinson, R.B., 110, 127, *165*
Rochon, R.W., 274, *291*
Roddick, J.A., 179, *242*
Ronov, A.V., 236, *244*
Rosenthal, M.J., 403, 445, 450, 460, 512, *524*
Rothrock, H.E., 250, 267, *292*
Rowan, G., 461, 505, *523*
Rowe, J.J., 331, *346*
Rowe, P.W., 322, 323, 325, *347*
Rubey, W.W., 49, *76*
Rukhin, L.B., 23, 24, 25, *42*
Russel, D.G., 419, 505, *522*
Ryskewitch, E., 300, *347*

Said, R., *244*
Saito, T., 132, *162*
Sarkisyan, S.G., *40*
Sawabini, C.T., 59, 61, 63, 64, 66, *77*
Schaeffer, F.E., 178, *241*
Schat, E.H., 127, *164*
Scheidegger, A.E., 169, *244*
Schlanger, S.O., 90, 133, 153, *165*

Schlee, J., 132, *162*, 165
Scholle, P.A., *165*
Schuchert, C., 171, 192, 195, *244*
Schultze, E., 309, 313, 324, *347*
Scorer, J.D.T., 438, 439–443, 447, 462, 467, 468, 469, 470, 471, 472, 473, 474, 476, 478, 513, *523*
Scott, G.D., 102, *165*
Seals, M.J., 259, 261, 265, 272, *292*
Seger, W.R., 227, *244*
Sen, B., 406, 506, 507, *524*
Serruya, C., 315, *347*
Shane, L.E., 363, *400*
Shankle, J.D., 250, *292*
Shannon Jr., J.P., 385, *401*
Sharp, W.E., 91, *165*
Shaver, R.H., *400*
Shelton, J.W., *401*
Shepard, F.P., 389, *401*
Sheridan, R.E., 205, *243*
Shinn, E.A., 88, *165*
Shirkovskiy, A.I., 36, *42*
Siever, R., 2, 5, 19, *41*, 262, *292*
Silberling, N.J., 190, 191, *244*
Silver, B.A., *401*
Simonson, R.R., 229, *244*
Sinnokrot, A., 35, *40*
Sippel, R.F., 91, *165*
Skinner, H.C.W., 88, *161*
Slichter, C.S., 31, *42*
Sloss, L.L., 5, 16, *41*
Smiles, D.E., 403, 404, 445, 450, 453, 460, 512, *524*
Smith, A.G., 193, *241*
Smith, G.D., 479, *524*
Smith, J.E., 403, 404, 422, *524*
Smith, S.M., 234, *243*
Smith, W.O., 300, *347*
Sokolnikoff, I.S., 495, 506, *524*
Somerton, W.H., 71, *77*
Sommers, D.A., *243*
Spencer, G.B., 54, 62, *75*
Spencer, M., 205, *245*
Spieker, E.M., 220–221, 222–224, *245*
Sproll, W.P., 206–207, *241*
Stafford, P.T., 250, 267, *292*
Stanley, D.T., 389, *401*
Stearns, N.D., 106, 107, *165*
Stegun, L.A., 434, 435, 436, 482, 484, *520*
Stevens, R.G., 315, 318, *344*

Stewart, J.H., 180—182, *245*
Storm, M.L., 461, *524*
Sullwold Jr., H.H., 389, *481*
Sutton, J., 193, 197, *241*
Swann, D.H., 250, 258, *292*
Sykes, L.R., 130, *165*

Taft, W.H., 86, 88, *165*
Taher, S., 153, *165*
Talwani, M., 234, *244*
Tanimoto, K., 320, 321, *347*
Tarbuck, K.J., 83, 95, 105, *165*
Taylor, D.W., 503, *524*
Taylor, J.M., 73, 77, 121, 122, 125, 140, *165*
Teodorovich, G.I., 20, 22, *42*
Ter-Martirosyan, Z.G., 509, *524*
Termier, G., 334, *348*
Termier, H., 334, *348*
Terriere, R.T., 250, 267, *292*
Terry, R.D., 28, *42*
Terzaghi, K., 44, 54, 73, 77, 130, *166*, 311, *348*, 403, 409, 427—428, 437, 439, 459, 478, 512, 513, *524*
Terzaghi, R., 90, 92, *166*
Thomas, L.H., 490, *524*
Thompson, G., 88, *166*
Thompson, M.E., 86, *162*
Thompson, S.E., 301, 302, *345*
Tickell, F.G., 54, 77
Tiller, K.G., 332, *347*
Timoshenko, S.P., 495, 499, 503, 504, 506, *524*
Todd, R.G., *401*
Treadwell, R.C., 250, *292*
Trimble, D.E., 178, *241*
Trurnit, P., 152, *166*
Tryggvason, E., 202, 203, 204, *242*
Tsvetkova, M.A., 22, 34, 35, *42*
Tubb Jr., J.B., *292*
Turner, F.J., 19, *42*
Twenhofel, W.H., *42*

Uchupi, E., 207, *245*
Udden, J.A., 2, 6, *42*
Ueshita, K., 306, *348*
Unrug, R., 389, *401*
Uren, L.C., *42*

Van der Knaap, W., 56, 57, 58, 77, 406, 415, 494, *524*
Van der Vlis, A.C., 56, 57, 58, 77
Van Everdingen, A.F., 448, 467, 468, 470, 471, 472, 479, 505, *524*
Van Lopik, J.R., 250, *291*
Van Poollen, H.K., 461, *520*
Van Rad, U., 389, *401*
Van Siclen, D.C., 250, *292*
Vasquez, H., 432, *524*
Verrujt, A., 405, *524*
Visher, G.S., 24, 26, 27, *42*
Vlasic, Z.E., 332, *344*

Wadell, H., 17, *42*
Walker, R.G., 389, *401*
Waller, T.H., 248, 254, 285, *292*
Wang, C.S., 234, *245*
Wanless, H.R., 250, 274, 291, *292*
Ward, W.C., 14, *40*
Warren, J.E., 418, *521*
Wasow, W.R., 479, *521*
Wattenbarger, R.A., 475, *524*
Weaver, P., 201, *241*
Weeks, L.G., 221, *245*
Weiner, J.L., *292*
Welder, F.A., 250, *292*
Welder, J.M., 21, 22, 23, *42*, 79, 87, 90, 91, *166*, 274, 276, 277, *292*
Wengerd, S.A., 218, 219, *245*
Wentworth, C.K., 5, 6, *42*
Wermund, E.G., 250, *292*
Weyl, P.K., 86, 88, 89, 153, 162, *166*
Wilhelm, B., 71, 77
Wilhelm, O., 201, *245*
Wilkins, J.K., 317, *348*
Willis, D.G., 46, *76*
Willoughby, D.R., 297, 298, 304, 319, *346*
Wilson, M.D., 222, *245*
Winston, G.O., 206, 208, *245*
Winterkorn, H.F., 303, 304, 324, *348*
Winthraw, P.C., 382, *401*
Whalen, H.E., 77
White, R.J., *244*
Whiting, R.L., 472, *520*
Whitman, R.W., 338, 339, *348*
Williams, E.G., 250, *292*
Williams, F.J., *163*
Williams, H., 19, *42*
Williamson, E.D., 414, *520*

REFERENCES INDEX

Wolbert Jr., L.J., 250, *290*
Wolf, K.H., 17, 32, *42*
Woodward, L.A., 178, *241*
Wright, M.D., 250, *292*
Wu, T.H., 503, *524*

Yamanouchi, T., 309, 328, *348*

Yandell, W.O., 323, *347*
York, K.I., 314, 318, *348*
Young, J.C., *245*

Zaslavsky, D., 422, *524*
Ziegler, P.A., 179, *245*

SUBJECT INDEX

Acoustic log, 363, 364
Acoustic properties, sediments, 337—340
Aggregates, stabilization, 332
Alabama, 171
—, Appalachian geosyncline, 172
Alaska, 180, 190, 191
— earthquake, 327, 328
— sand, grading limits, 328
Alberta, Canada, 171
Algae, coralline, 84
—, plates, 84
Algal dolostones, 180
— sands, 84, 128, 129
Alluvial—fluvial deposits, logging patterns, 381—383
Almond Formation, Wyoming, 26
Amarillo Arch, Texas, 215
Anadarko Basin, Utah, 189, 214
— foredeep, 214
— downwarp, 214
Angle of internal friction, 312, 323, 327
— — repose, 324—325
Anisodimensional, definition, 341
Anisotropy, definition, 341
Anisotropic, stress field, 304
Anomia simplex Orbigny, 129
Antilles, 89
Antler Orogeny, 184
Appalachian Basin, 250
— geosyncline, 192—194, 197
— orogen, 193
Aragonite, 144
Arbuckle area, Oklahoma, 213
Archimedes Principle, 409
Arcturus Basin, 186
Ardmore—Anadarko geosyncline, 215
Arizona, 218, 227
Arkoses, 18, 19
Arkosic sands, 59, 64, 65, 70
— siltstones, 59
A.S.T.M., C131-69, 341

A.S.T.M. *(continued)*
—, D556-40T, 302
—, D557-40, 302
Atlantic Coastal Plain, 194, 199
Australia, 297, 329, 332, 333
Authigenic processes, effect upon permeability, 315
Avis Sandstone, Texas, 252

Bahamas, 87, 88, 89, 90, 114, 116, 124, 125, 127, 128, 130, 134, 136, 137, 138, 139, 148, 158, 159
—, cryptobasin, 207
—, oolites, 127
—, platforms, 203, 205, 206, 207, 216
Bar sandstones, 253, 255, 257
Bar-finger sandstones, 269, 274
Barrier-island, sand, logging patterns, 381, 383—384, 386
Basalt, SP and resistivity log, 368, 369
Baseline Formation, 227
Basement, log pattern, 368
—, subsidence, 175
Basin, 173, 174, 176, 208, 219
—, dynamic, 173, 174, 176, 208, 219, 230
—, sedimentary, 173, 174
—, subsidence, 176
Beach sand, 297
Bearing pressures, 324
— stress, 311
Beaverdam Mountains, Arizona, 227
Bell Creek Field, Montana, 387
Belt sandstones, 288
Beltian geosyncline, 178
— miogeocline, 182, 183, 189, 191, 197
— trough, 178
Benthic animals, effect upon compaction, 84
Bentonite clay beds, logging markers, 366
Berry Island, Bahamas, 128, 137
Big Cottonwood Series, Utah, 178

Bimodal, sorting, 13
Bimini Lagoon, Bahamas, 138
Binary mixtures, sorting, spheres, 103, 104, 105, 106, 107
Bird Spring Basin, Nevada, 218
Blach Ranch Limestone, Texas, 248, 252, 256, 263
Black Mesa Basin, Arizona, 218
Blake Plateau, Florida, 90, 132
Block-shaped grains, 84
Body force, 494, 501, 502, 515
Bolivar Coast, Venezuela, 57, 58
Boltzman transformation, 428
Bonaire, Antilles, 89
Bouguer gravity anomaly, 208
Boundary conditions, 421, 428, 431, 446, 465, 475, 491, 492, 502, 505, 506
—, Dirichlet, 487
—, Neumann, 487
Bowie elongate sandstone, 252, 256, 259, 260, 265, 268, 269
Brazos Island, Texas, 26, 269
Breckenridge Limestone, Texas, 252, 256, 259, 263, 282, 283
Bridging of particles, 9, 11, 72
British Columbia, Canada, 179
British Honduras, 87
British Standards Institution, 295, 296
Brittle, definition, 341
Browns Cay, Bahamas, 125, 127, 134, 158
Bucher's Law, 170, 171, 172
Buckling, 123, 135
Bulk densities, 52, 300, 301, 329, 330, 517
— —, collapse, 329—330
— flow, effect upon permeability, 315
— modulus, 311, 312, 320, 342
— specific weight, 30
— volume, 27, 29
Bunger Limestone, Texas, 248, 252, 256, 281, 282, 283, 284, 285
Buoyant force, 46, 47, 48

Caddo Chalk Formation, Texas, 248
Calcite, 89, 133, 144
Calcium/magnesium ratio, 87
Calclitharenites, 19—20
Calcrete, 330, 331
California, 44, 52, 59—62, 65, 66, 72, 132, 178, 180, 181, 190, 228, 230, 237, 238, 387

California bearing ratio, 321
Caliper log, borehole, 372
Callahan County, Texas, 252
Cambrian, 171, 172, 179, 180, 183
Camp Colorado Limestone, Texas, 248, 252, 254, 256
Camp Creek Shale, Texas, 248
Canada, 171, 178, 179, 190, 191, 202, 274
Canyon Group, Texas, 286
Capillarity, 36—37
—, pressure curves, 36, 106—110
Caprock, impermeable, 50, 51
Captiva Islands, Florida, 129
Carbonate sands, compaction, 134—140
— —, factors affecting compaction, 81—95
—, logging techniques, 350, 362, 369, 372, 374
Cementing material, source, 143, 144, 145
— —, —, pressure solution, 155—158
Cementation, 72, 82, 83, 87, 143—146, 323, 330—334
—, compaction, 73, 82, 83
—, composition, 146
—, grain size, 144
—, packing, 146—152
—, pH, 330—331
—, plasticity, 334
—, rate of, 145
—, source, 143—144
Cenozoic, 132
Chance packing, 101, 160
Channel fill, logging pattern, 378, 379, 381, 382, 394, 395, 396, 397
Chert arenite, 20
Cisco Group, Texas, 247—289
Classification, sands, coarse-grained, 293, 294
—, —, composition, 20, 21, 22
—, —, genetic, 24, 25
—, —, grain contacts, 72
—, —, maturity, 22, 23
—, —, particle shape, 296
—, —, — size, 2—4, 23—27
—, —, — symmetry, 299
—, —, porosity, 28
—, —, roundness, 15, 16
—, —, sphericity, 16
—, —, texture, 9—11, 12, 13
—, sand bodies, 380—391
—, water-laid clastics, 389

SUBJECT INDEX

Clastic sands, 133
— —, experimental compaction, 132—134
— —, logging patterns, 370, 375
— section, lithology from logs, 370
Clay, 56, 57, 60, 61, 65, 342
—, compaction, 277
—, illite, 57
Coal, SP and resistivity log, 368, 369
Coefficient, active earth pressure, 324, 325
—, compressibility, 39
—, passive earth pressure, 324, 325
—, volume compressibility, 412, 414, 515
Cohesion, 312
Coleman Junction Limestone, Texas, 248
Collapse of sands, 329—330
Colorado, 171, 215, 218, 221
— River Valley, Texas, 269
Compacted material, properties, 293
Compaction, apparatus, 63, 71
—, artificial, 336—337
—, clay, 276
— curve, sand, 307
—, formation, 38
— index, 127—130, 132, 158, 159, 160
—, lead shot, 111, 140
—, mud, 273, 276, 277, 279, 284—285
—, porosity, 308
—, process, 403, 406, 412, 414, 420, 437, 448, 452, 458, 483, 508, 509, 510, 511
—, rate, 56
—, secondary, 453
Comanche Series, Florida, 209—210
Complementary function, 502, 503, 504
Composition, sandstones, 18—19, 20, 21, 22
Compressibility, 37—40, 306—313, 420—462
—, bulk, 37, 38, 64, 416, 418, 419, 496—497, 514
—, effective, 55
—, elastic, 311
—, fluid, 414
—, formation, 37, 38
—, ideal gas, 466, 468, 514, 517
—, inelastic, 311
—, isothermal coefficient, 413
—, liquid, 463, 465, 514
—, measurement, 309
—, modulus, 309—311

Compressibility *(continued)*
—, pore volume, 37, 38, 64, 66, 413, 414, 416, 420, 421, 514
—, process, 49, 160
—, pseudo-bulk, 37, 39, 420
—, reduced, 514
—, rock, 37, 38, 414, 416, 418, 420, 421, 478, 497, 512
—, total, 469, 514
Concavo-convex contacts, 72, 122, 123, 139
Concho Platform, Texas, 266
Condensation Index, 123, 124, 160
Conductivity, electrical, 53
Cone penetrometer, 309
Confining pressure, effect on strength, 322
— —, effect upon velocity, 338, 339
Conglomerates, resistivity log, 366, 367, 368
Consolidated, sand, permeability, 34
— sandstone compressibilities, 67
Consolidation, 43, 44, 160
— process, 403, 406, 434, 437, 438, 479, 485
Contact angle, 340
Continental drift, 195, 198—199
— margin, 194
Continuity equation, 410, 420, 461, 498
Continuous subsidence, 195
Convergence, 483
Conversion table, 525
Coogan's packing density, 115
Coordination numbers, 98
Corcoran clay, 60
Cordilleran region, 172, 173, 175, 178, 181, 182—192, 197, 217, 219, 221, 222, 226, 227
— geosyncline, 172, 182, 216
— miogeosyncline, 186, 189
Coring, 350
Creep, 306, 307, 341
— compliance, 341
Cretaceous, 92, 93, 171, 172, 202
Crinoidal ossicle sands, 84
Critical void ratio, 325—326
Critical ratio of entrance, sphere packs, 104
Cross-anisotropy, 341
— —, packing, 297
Cross-bedding, dip data, 394, 397, 398
—, festoon, 255

Cross-bedding *(continued)*
—, logging patterns, 379
Crushing, effect upon porosity, 303
Crystal Falls Limestone, Texas, 248, 252, 256
Cubic packing, 83, 96, 97, 98, 99
Current bedding, dip-log patterns, 397, 398

Dakota sandstone, 171
Darcy's Law, 32, 317, 341, 403, 410, 420, 421, 422, 424, 425, 427, 444, 452, 461, 463, 499, 500
Death Valley, California, 178, 180, 181
Decompaction, process, 280, 283—285, 286
—, mudstone, 249, 262, 263, 273, 275, 276, 286
—, stratigraphic, 274, 276, 280
Definition, arkose, 19
—, anisodimensional, 341
—, anisotropy, 341
—, brittle, 341
—, bulk volume, 27
—, calclitharenites, 19—20
—, compaction, 43—45, 160
—, consolidation, 43—45
—, depocenter, 172
—, grain volume, 160
—, gravel, 342
—, greywacke, 19
—, haphazard packing, 160
—, heavy clays, 342
—, litharenite, 21
—, lithic sand, 19
—, orthoquartzite, 19
—, porosity, 27—28
—, pseudo-elastic, 343
—, rotund, 343
—, roundness, 11
—, sand, 2
—, silt, 343
—, sphericity, 17
—, texture, 3—5
—, till, 172
—, turbulent flow, 343
—, void ratio, 28
Delaware Basin, Texas, 216, 217
Deltaic, logging patterns, 381, 383, 387
— systems, 269, 270, 276, 288
Densification, 302, 314

Density, 27, 54, 293, 299—305, 314
—, apparent, 300, 301
—, bulk, 52, 300, 301, 329, 330, 517
—, liquid, 419, 422, 442, 510, 517
—, measurement of, 300
—, oil, 470
—, relative, 307
—, rock, 407, 421, 422, 517
—, sorting, 301
—, stiffness, 307
—, strength, 320, 321
—, water, 407, 410, 517
Density current, 381, 386—391
Density log, 362
Depocenter, 176, 178
—, subsidence of, 167, 169, 174
Depositional environment, 17—18, 23, 24, 26
— —, logging patterns, 375—390
— model, Texas, 266, 269
— patterns, sandstone, 258
— —, —, belt sandstone, 258
— —, —, distributary, 258
— systems, 267—269
Destin Beach, Florida, 26, 128
Deviator stress, 307, 322, 341
Devonian, 183, 184, 186
Diagenetic processes, effect upon permeability, 315
Diamictite, 178
Diastrophism, 168, 169
Diatomites, SP and resistivity log, 366, 367, 369
Difference approximation, backward, 487, 488, 489, 491
— —, central, 485
— —, Crank—Nicholson, 489
— —, foreward, 485, 489
Differential compaction, 264, 266, 274, 275, 283—286, 349
— equation, 412, 428, 445, 466, 468
— —, partial, 406, 460, 473, 485, 490, 513
— fluid densities, 74
— subsidence, 278
Diffusivity, 425, 446, 448, 453, 465, 480, 485, 515, 518
— equation, 446, 447, 465
Digital computer, 405, 467, 476, 486
Dilatancy, 341, 498, 505, 519
Dilation velocity, 339

SUBJECT INDEX

Dimensionless distance, 448, 451, 516
— flow rate, 448, 469, 470, 516
— length, 453, 515
— potential drop, 469, 470, 517
— pressure, 467, 468, 470, 515
— pseudo-void ratio, 448, 451
— radial coordinate, 467, 516
— time, 448, 450, 451, 453, 454, 460, 467, 468, 469, 470, 475
— void ratio, 448, 450, 451, 517
Dip measurements, calculations, 391—393
— log, 375—376, 391—399
— —, cross-bedding, 394, 397, 398
— —, current bedding, 394, 397, 398
— —, fault location, 392, 399
— —, patterns, 392, 393
— —, reefs, 397
— —, tool, 393
— —, unconformity, 399
Direct power compaction process, 336
Displacement potential, thermoelastic, 503, 520
— —, poroelastic, 503, 506, 520
Dolomite, 58, 89
Dolomitization, 87, 88—89
Drainage, effect upon compaction, 302
Drawdown test, permeability, 318
Drummond, Montana, 222
Dynamic basin, 173, 174, 176, 219, 230
— loading, 308, 327
— moduli, 323
— modulus, 342
— parameters, effect upon compaction, 81, 94—95

Earth, 167, 168, 170, 198, 207, 214, 219, 221
— core, 168
— crust, 168, 170, 198, 207, 214, 219, 221
— dimensions, 167—168
Earthquakes, subsidence, 238, 239, 327, 328
Edgewood Formation, Louisiana, 155, 156
Effective stress, 45, 46, 48
Effectiveness factor, pore pressure, 133
Egypt, 230
El Abra Formation, 159
Elastic, 342
— deformation, 55, 307

Elastic (continued)
— modulus, 342
— strain, 320, 321, 323
Elasticity, pore, 405, 406, 493, 499, 502, 506
Electric discharge compaction, 336
— log, clay effect, 364, 366, 375
— potential, natural, earth, 353
— well log, 350, 351, 354—356, 364—368
Electrical conductivity, formation, 357
Electrochemical, SP component, 354
— potential, 354
— treatment of clay, 315
Elko Basin, Nevada, 186
Elliptic functions, 507
Elongate sandstones, 253, 271, 272, 273, 288
— sandstone bodies, 258, 260, 261, 273, 288
Ely Basin, Nevada, 186
England, 327
Envelope shape, logs, 356
Eocene, 57, 58, 90
Epeirogen, 176
Epeirogenesis, 169, 192
Epeirogenic subsidence, 196
Epeirogeny, 169, 174, 175, 192
Equilibrium conditions, porous solids, 499, 500, 502
Equipment, compaction, 63, 71
Erosion, 334—336
—, air, 334—335
—, cementation effect upon, 334
—, factors controlling, 334
—, water, 334, 335
Error function, 434
— —, complement, 434, 481
Eskers, 342
Eugeosyncline, 183, 190, 193
Eureka, Nevada, 184
Excessive pore-fluid pressure, 74
Exogeosyncline 220
Expansion, artificially induced, 74
— process, 434, 437, 438, 439, 460, 479, 493
Explicit procedure, 488, 491

Fabric, 85—86
—, particle, 295, 297, 298
—, sediment, 293

Facies, sandstone, origin of, 252
Failure planes, beach sand, 297
Fault, location by logging, 392, 399
—, overburden pressure, 46
Feldspar, 58, 70
Feldspathic sands, compressibility, 61, 62
— sandstone, 329
Ferricrete, 330, 331, 335
Ferromagnesian minerals, 21, 23
Finis Shale, Texas, 248
Finite difference, approximation, 479, 483, 485, 486, 488, 491, 492
Fish Cay, Bahamas, 137
Fixed grain, 160
— —, Compaction Index, 124, 160
Flagstaff Limestone, Utah, 224
Floating grains, 122
— —, packing, 121, 123
Florida, 26, 86, 90, 127, 128, 129, 132, 133, 142, 199, 203, 205, 208, 209, 210
— Keys, 216
— Plateau, 203
— Platform, 206, 207
— Shelf, 208
Flow potential, 421, 422, 519
Fluid content, well logs, 349, 351, 358
— withdrawals, subsidence, 6, 236—238
Fluvial—deltaic, sandstones, 258, 259
— sandstones, 269
Flysch, 229
Foredeep Basin, 220, 222, 224
Forest Beach, North Carolina, 26, 27
Formation compaction, 55
— productivity, 420
— resistivity factor, 35
Formats, stratigraphic, 251—252
Fort Pierce Formation, Florida, 205
Fort Worth Basin, Texas, 247, 266, 267
Four Corners Area, 218
Fractured oil shale, SP and resistivity log, 368
Fractures, compaction, 135, 136, 137
—, patterns, sandstones, 274
Fracturing of sand grains, 56
Frequency curves, size distribution, 7, 8
Friable sandstones, compressibility, 68
Friction angle, porosity, 324
—, interparticle, 319
Fujiyama, Japan, 333

Fuller curves, 301—302
Furrows, 171, 172

Gaither Sandstone, Wyoming, 122
Gamma—gamma log, 361, 362, 366
Gamma radiation, natural, earth, 350, 358, 360
— ray log, 350, 351, 352, 358—360, 368—369
Gannett Group, 222
Gas, ideal, flow, 448, 468
—, logging response, 359, 362, 370
—, real, flow, 447, 448, 461, 466, 468
Gastropod, 84
— sands, 127
Geiger counter, logging, 358
Geometry, pore space, 99
—, sandstones, 287
Geostatic loading, 46, 72, 89—91, 127, 420, 423, 445, 457
— gradient, 50, 51, 52
— pressure, 45, 48, 49, 50, 51
Geosyncline, 171, 172, 173, 178, 192, 194, 197, 215
—, Beltian, 178
— couplet, Atlantic Coast, 196, 197
Geothermal gradient, 91, 92, 144—145
— reservoir, high-pressure, 74
Germany, 19
Glacial, 172
— sediments, compressibility, 310, 334
Glauconitic sands, logging, 369, 370
Golconda Thrust, 191
Gonzales Green Shale, Texas, 248
Goose Creek Oil Field, Texas, 71—72
Grading, 342
— curves, particle size, 302, 328, 336
—, particle, 301
Graham Formation, Texas, 248
Grain diameter, effective, 341
—, block-shaped, 84
— fabric, natural, 80
— floating, 121, 123
— orientation, 404
— shape, cementation, 145, 146
— —, compaction, 82—85
— size, distribution, 311, 319—320
— volume, 160
Grainstones, 79, 110, 113, 114, 115, 116, 117, 121, 122, 124, 127, 129, 130, 131, 136, 139, 140, 146, 147

Granulometric coefficients, 25
Grapestone, 136—137
— bars, 88, 128
Gravel, definition, 342
Gravity, in mathematical analysis, 404, 420, 421, 423, 446, 456, 457, 461, 494, 515
Great Basin, USA, 182, 226
Great Inagua Island, Bahamas, 89
Great Salt Lake, Utah, 175, 182
Green Mountains, 193
Green River Shale, 61, 62
Green sand, glauconite, 369
Greenbriar Formation, West Virginia, 143, 147, 150
Greywacke, 18, 19, 61—62
Ground water, hydrology, 405
Grouting, 318—319, 336—337
Gulf Coast Area, 72, 199, 200, 201, 217, 274
— —, beaches, 27, 48
— — geosyncline, 195, 201
— Coastal Plain, 201
Gulf of Mexico, 176, 195, 200, 201, 205, 214, 215, 216, 266
Gunsight Limestone, Texas, 248, 252, 256
Gypcrete, 335
Gypsum, 330

Hakodate Sand, 328
Haphazard packing, definition, 160
Harmonic functions, 505, 519
Harpersville Formation, Texas, 248
Hartz Mountains, Germany, 19
Heavy clays, definition, 342
Heterogeneous packing, spheres, 106, 107, 108
High-temperature gradient, 203, 204
Histogram, sorting, 6—7, 13
Hogup-terrace Mountains, Utah, 186
Home Creek Limestone, Texas, 248—252, 256, 272, 279, 281, 283, 284, 285
Hooke's Law, 48, 495
Horse Spring Formation, 227
Hour glass shape, logs, 352, 359
Houston, Texas, 201, 217, 236
Hydraulic permeability, 315—316
— properties, 83, 84
— radius, 304
Hydrocompaction, 238
Hydrogen log, see Neutron log

Hydrometric method, size analysis, 6
Hydrostatic gradient, 46, 49, 50, 51, 52
— loading, 59, 139
— stress, 45
Hypersubsidence, 176, 177, 178, 179, 186, 189, 190, 191, 199, 202, 205, 206, 207, 208, 216, 217, 219—220, 224, 228, 229, 230
Hysteresis, 444, 447

Ibex Limestone, Texas, 252
Idaho, 179, 184, 185, 191, 220, 221, 222
Illinois, 146, 258
Illite, 57, 61
—, compressibility, 61
Implicit procedure, 488, 489
Index properties, texture, 3, 5
India, 332
Indiana, 142, 143
Induced electrical response, log, see Induction log
Induction log, 350, 351, 352, 357—358, 360, 377, 379, 386
Inelastic strain, 320
Inherited parameters, compaction, 81, 82—87
Inhibitory parameters, compaction, 81, 87—94
Initial condition, mathematical analysis, 428, 433, 446, 463, 465, 466, 487, 491
Interbedded unit, log pattern, 377, 378
Interdeltaic system, 270
Intergranular loading, 46, 72, 89—91, 127, 420, 423, 445, 447
Interior Basin, 250
Internal friction angle, 312, 323, 324—325, 327
Invaded zone, mud filtrate, 353, 364, 375
Inverfc, 481
Irreducible fluid saturation, 109, 110, 160
— water saturation, 37, 160
Isopach maps, 180, 183, 184, 185, 197, 262, 273, 275, 278, 280, 281, 285
Isotropy, 494, 495, 496
Isostasy, 169, 198, 231
Isostatic subsidence, 230, 231
Isotopic dating, 168
Isotropic elastic behavior, 55
— rebound, 231, 232
— stress field, 304

Italy, 236
Iteration, 480, 481, 488
— procedure, 479—483
Ivan Limestone, Texas, 248, 252, 256

Japan, 236, 307, 333, 335, 336
Jingle Shell, 129
John's Valley Turbidite, Oklahoma, 213
Joulters Cay, Bahamas, 127, 148, 159
Jurassic, 93, 115, 116, 124, 125, 127, 142, 149, 150, 151, 158, 159

Kahn's Packing Index, 117
Kansas, 61, 62, 258
Kaolinite clay, 65
Kentucky, 143
Kingston Peak Formation, California, 181
Klamath Mountains, California, 228
Koipato Group, Nevada, 191
Kozeny equation, 313
Kulm Sandstone, Germany, 19

La Jolla, California, 389
La Panza Mountains, California, 228
Laboratory compaction tests, 54—71
Lake Bonneville, Utah, 168, 231, 232, 233
Lake Maracaibo, Venezuela, 44, 56, 60
Lake Mead, Nevada, 230, 231
Lake Okeechobee, 205
Laminar flow, 317, 318
Land surface, subsidence, 403, 493, 510, 512, 513
Lansing—Kansas City Limestone, 61, 62
Laplace's equation, 474
Las Vegas, Nevada, 180
— — Wasatch hinge line, 180, 190, 219—220, 224, 226
Lateral compressibility, sand, 307
Laterite, 332
Lime grainstones, 114, 129, 130, 131, 139, 140
Lime muds, 87
Limestones, 79
Liquefaction, 326—330, 334
Litharenite, 20, 21
Lithic arenite, 19
— sands, 19
Lithification, 87, 332
—, carbonate sands, 140—152

Lithology determination, logs, 349, 350—351, 361, 364—375
— —, porosity logs, 369—375
— —, radioactive logs, 368—369
Llano Uplift, 247
Load transfer, 49, 50, 51, 71
Loading, hydrostatic, 59, 139
—, triaxial, 46, 47, 54, 56, 59, 133, 134
—, uniaxial, 56, 59, 85
Loam, 342
Loess, 342
Log, acoustic, 363, 364
—, caliper, 372
—, density, 362
—, dip, 375—376, 391, 399
—, gamma—gamma, 361, 362, 366
—, induction, 350, 351, 352, 357—358, 360, 386
—, neutron, 350, 351, 352, 359, 361, 368—369
—, sonic, 363
—, SP, 350, 351, 353—354, 355, 356, 358, 360, 364, 366, 371, 377, 379
Logging, bar-finger sands, 382—383
—, bars, 384, 394, 395, 396
—, basalt, 368, 369
—, basement, 368
—, brackish water, 364
—, carbonates, 350, 359, 362, 369, 372, 374
—, channel-fill sands, 378, 379, 381, 382, 394, 395, 396, 397
—, clastics, 370, 375
—, coal, 368, 369
—, conglomerates, 366, 367, 368
—, current bedding, 397, 398
—, delta deposits, 381, 383—385, 387
—, depositional environment, identification, 375—390
—, diatomites, 366, 367, 369
—, fluid content, 349, 351, 358
—, fractured oil shale, 368
—, fresh water, 364
—, gas, effect upon, 359, 362, 370
—, glauconite, 369, 370
—, igneous rocks, 367—368, 369
—, lacustrine deposits, 382, 386
—, lagoonal deposits, 388, 389
—, lithology determination, 351, 376—379
—, offshore bar, 379

Logging *(continued)*
— patterns, alluvial—fluvial, 381—383
— —, barrier island, 381, 383—384, 386
— —, cross-bedding, 379, 394, 397, 398
— —, deltaic, 381, 383, 387
— —, lagoonal, 381, 386
— —, offshore bar, 379
— —, oscillatory, 377, 378
— —, secondary, 378—379
— —, SP slope, 379
— —, thin beds, 357
— —, tidal channel, 381, 384—386, 388
— —, turbidite—tractionite, 378, 381, 386—391
—, porosity, effective, 374
—, sands, 364, 366, 367
—, shale, 364, 368, 370, 371, 375
—, tadpole plot, 392, 394, 395
— techniques, 52—53
—, thin beds, 357, 386
— tool, configuration, 354, 355—356, 357, 359, 362, 391—393
—, unconformity, 399
Los Angeles Basin, California, 230, 387, 389
Louisiana, 48, 93, 126, 148, 149, 151, 155, 156, 158, 159, 202, 203, 204
Lower First Wall Creek, Wyoming, 122

Mapping, 174
—, correlation markers, 366
—, isopach, 262, 273, 275, 278
—, structural residuals, 272, 273
Maracaibo Lake, Venezuela, 44, 56, 60
Marathon geosyncline, 216
Marginal basin, 220
Masa Soil, Japan, 335
Mass flow of solution, 420, 437, 439
Massachusetts, 104
Massive unit logging, 377, 378
Masson's Packing Index, 177
Material balance, 47
Mature subsidence, 193
Maturity, concept, 21, 22, 23
—, mineralogical, 23
—, textural, 22
McCoy Creek Group, 180
Mean pore size, 304
Mega-atoll, 206
Mendota, California, 72

Mercury injection, 37
Mesa formations, 332, 333
Mesa Verde Sandstone, Wyo., 122
Mesh point, 484, 486
— —, coordinates, 485
— —, size, 485, 515
Mesocordilleran geanticline, 219
Mesozoic, geosyncline, 220
—, sediments, 173, 181, 182
Metaline District, Washington, 179
Mexican geosyncline, 220
Mexico, 150, 199, 201—202, 216, 220
Mica, 83, 129
Michigan Basin, 174
Microstylolitic contacts, 121, 122, 133
— solution, 153
Mid-Atlantic Ridge, 197
Midland Basin, Texas, 216, 217
Mineralogy of carbonates, 86, 87
Minus fraction, screen analysis, 5
Miogeocline, 182, 183, 189, 191, 197
Miogeosyncline, 183, 193, 220, 226
Mississippian, 142, 143, 158, 184, 185, 186
Missourian, 25, 26
Mode, sorting, 7, 9, 13, 14
Modulus, bulk, 311, 312, 320, 342
—, compressibility, 309—311, 312
—, elastic, 342, 495, 515
—, resilient, 322, 342
—, rigidity, 312, 320, 495
—, shear, 342, 495, 515
—, static, 342
—, Young's, 48, 309, 310, 312, 320, 340, 342, 495
Mohorovičić discontinuity, 168
Moisture barrier, 316
Moisture content, 319—320
Monocline, 343
Monosized, 343
Monotonic, 343
Montana, 178, 221, 222, 387
Moosejaw Synclinorium, North Dakota, 171
Moran Formation, Texas, 248
Morphology, particle, 293—295, 298
Morrison Formation, Wyoming, 122, 222
Mud compaction, 273, 277
Mudstone, 249, 262, 263, 273, 275, 276, 286

Multistory, sandstone pattern, 264—266, 271, 273, 288
Myers Beach, Florida, 129

Nahanni River Area, Canada, 179
Nelson Area, British Columbia, 179
Neutron log, 350, 351, 352, 359, 361, 368—369
— —, calibration, 361, 374
— —, formation gas, 359, 362
— —, porosity, 361, 362, 364, 370, 371
Nevada, 178, 179, 180, 181, 182, 186, 190, 191, 218
Nigeria, 332
Niigata earthquake, 327, 328
Niigata sand, 328
Noetia Ponderosa Say, 129, 130
Non-dilatant, 343
North Carolina, 26, 27
North Dakota, 171
North Leon Limestone, Texas, 252
North Queensland, Australia, 329, 332, 333
Norway, 327
Norwegian fiords, 327

Ocala Uplift, Florida, 211—212
Offshore bar, log patterns, 379
Oklahoma, 26, 171, 189, 213, 214, 215, 258, 266, 274
Olympic Peninsula, Washington, 228
One-dimensional flow, 403, 448, 492, 510
Ooliths, 156
Oolitic grainstones, 79, 110, 113, 115, 116, 117, 121, 124, 127, 128—130, 136, 139, 140, 147
— sands, 83, 86, 88, 110, 116, 127, 134
Oquirrh Basin, Utah, 186, 190
Ordovician, 171, 183
Oregon, 179, 190, 191
Orogen, 176, 192
Orogenesis, 169, 172
Orogeny, 169
Orthoquartzites, 18, 19
Orthorhombic packing, 96, 97, 98, 99, 113
Oscillatory, log pattern, depositional conditions, 377, 378
Ottawa Sand, 58, 61, 62, 65, 340

Ouachita geosyncline, 208—215
— Mountains, Texas, 208, 213, 260, 267, 288
— uplift, 267, 286
Outflow test, permeability, 318
Overburden loading, 46, 72, 89—91, 127, 420, 423, 445, 457
— potential, 457, 461, 520
Overpressured zone, 44, 51, 52, 53, 74

Pacific Coastal Plain, 227
— rim, 195
Packing, 85—86
—, beach sands, 113
—, chance, 101, 160
—, compaction, 83, 84
—, cross-anisotropic, 297
—, cubic, 83, 96, 97, 98, 99
—, definition, 160
—, density, 101, 114, 115, 117, 118—120, 122, 124, 127, 146, 158, 159, 300, 301
—, dune sand, 113
—, floating grains, 121, 123
—, grain size, 102—103
—, grain size mixture, 103—106
—, graphical extrapolation, 102
—, heterogeneities, 106—110, 160
—, Index, 116—121, 122, 124, 127, 131, 158, 159, 161
—, irregular grains, 110—113
—, Khan's Index, 117
—, orthorhombic, 96, 97, 98, 99, 113
—, particle shape, 297
—, porosity, 301
—, precompaction, 95—113
—, proximity, 117, 161
—, random, 100—103, 110—113, 161
—, regular, 96—100
—, rhombohedral, 298—299, 301
—, river sands, 113
—, sediments, 31
—, stability, 100, 101
—, unordered, 161
—, vibration effect on, 112—113
Pahrump Group, California, 180
Paleocene, 90
Paleogradients, sandstone, 262
Paleoslope, 258—260, 262, 269, 271, 279

SUBJECT INDEX

Paleosurface, evolution, 286—287
—, model, 280, 281, 283, 284, 285
Paleotopography, inherited, 278, 283
—, mapping, 262
Paleotransport routes, 260
Paleozoic, 143
— sediments, 172, 173, 180, 181, 189, 215—218, 276
Paradox Basin, Colorado, 218
Particle, mechanical interlock, 328
— orientation, 297—298
— size, 5
— —, classification, 2—4, 23—27
— —, effect of transportation, 24—27
Pelecypods, 84
Penetration test, 309
Pennsylvanian, 26, 172, 184, 186, 187
— sandstones, 247—289
Permeability, 32—35, 293, 300, 313—319, 404, 405, 410, 414, 421, 422, 425, 427, 437, 461, 462, 463, 466, 469, 472, 474, 478, 499, 512, 515
—, authigenic process, effect on, 315
—, collapsed structure, 315
—, consolidated sand, 34
—, effect of clay on, 34—35
—, function of moisture content, 314
—, function of porosity, 33
—, measurement of, 318
—, natural sediments, 313
—, unconsolidated sand, 34
Permian, 186, 188, 247
— Basin, Texas, 216, 217
— sandstones, 247—289
Persian Gulf, 89
Petrographic techniques, compaction study, 114—130
pH, effect upon cementation, 330—331
Phi scale, 5—6
Phyllarenite, 19, 20
Piezometer, 409, 442
Pigeon Cay, Bahamas, 138
Plasticity Index, 337
Plate bearing test, 311
Plate tectonics, 168, 169, 170, 181, 183, 191, 193—194, 195, 199, 207, 234
Pleistocene, 86, 90, 113, 127, 148, 159, 178, 230, 231
Plum Island Sand, 58
Pocattello, Idaho, 179, 186

Poisson's ratio, 48, 309, 311, 312, 321, 495
Polyaxial loading, 59
Prodelta mud, Texas, 260, 262, 269, 270, 286
Pore volume loss, pressure solution, 154, 155
— size distribution, permeability, 300
— structure, 293, 299—305
Poroelasticity, 405, 406, 493, 498, 499, 501, 502, 503, 511
Porosity, 27, 29, 31—32, 54, 56, 293, 297, 299—306, 338, 404, 405, 407, 414, 418, 421, 423, 424, 428, 437, 444, 461, 462, 466, 469, 472, 474, 478, 497, 501, 512, 519
—, angle of friction, 324
—, anisotropic, 307
—, applied pressure, 307
—, cementation, 306
—, compressibility, 306
—, effective, 32, 304
—, intraparticulate, 85
— logs, 361, 362, 363, 364, 369, 370, 371
—, measurement of, 303
—, surface area, 304
—, total, 32, 53
—, vibration, 304, 306, 308, 330, 336
Potential, overburden, 404, 457, 461, 520
—, flow, 421, 422, 519
—, moisture, 425
—, total, 457, 519
Potentiometric model, 460
Precambrian, 173, 178, 179, 180
Precompaction geometry, 95—113
Preconsolidation, 306, 309
Pressure, atmospheric, 411
—, average, 473, 515
—, bearing, 324
—, compaction, 408
—, critical, 518
—, effective, 306
—, excess fluid, 411, 515
—, excessive pore-fluid, 74
—, fluid, 404, 406, 407, 408, 413, 414, 421, 425, 427, 475, 496, 497
—, grain-to-grain, 404, 405, 406, 410, 411, 516
—, hydrostatic, 408, 414
—, initial, 448, 468, 476
—, moisture, 423
—, net-confining, 62
—, overburden, 408, 412, 413, 414, 516

Pressure *(continued)*
—, pore, 327, 404, 415, 416, 419, 472, 516
—, settlement, 311
—, standard state, 421
—, swelling, 425, 519
Primary consolidation, 44
Pseudo-elastic, definition, 343
Pseudo-pressure, 405, 448, 460, 464, 466, 473, 480, 512, 517
— —, real gas, 448, 460, 473, 480, 517
— —, reduced, 516
Pseudo-reduced temperature, 516
Pseudo-steady state, flow, 465, 468, 476
Pseudo-void ratio, 405, 448, 449, 450, 456, 459, 460, 512, 517
Pueblo Formation, Texas, 248
Putnam Formation, Texas, 248
Pyroclastic material, 20

Quantitative grain size, genetic classification, 24
Quartzite, 61, 62
Quartzose sands, 85, 93, 121
Quasisteady state of flow, 465
Quicksand, 49
Quinn Clay, Texas, 248

Radioactive logs, *see* gamma and neutron logs
Radius, external, 475, 516
—, dimensionless, 476, 516
—, well, 475
Rate of loading, effect on compaction, 92
Rebound process, 435, 479, 493, 512, 513
Reciprocal theorem, 417
Red Creek Metaquartzite, Utah, 178
Red Fork Sandstone, Oklahoma, 382
Red River Uplift, Texas, 247
Redox log, 379
Regressive sea, logging pattern, 379
Residual, structure mapping, 272, 273, 281, 282—283
Resilient modulus, 322, 342
Resistivity curve, conventional, 354—356
— —, induction, 357—358
—, drilling fluid filtrate, 354
—, lateral, 354, 355, 356
—, log, *see* Electric log
—, normal, 354, 355, 366
—, sediments, 340

Resistivity *(continued)*
—, short normal, 356, 357, 364
—, water, 354
Reynolds number, 317—318
Rhode Island, 58
Rhode Island Sand, 58
Rhodesia, 329
Rhombohedral packing, 83, 96, 97, 98, 99, 106
Rincon-Dos Cuadros Reservoir, California, 52
Rocky Mountains, 171
— — geosyncline, 220, 221, 223, 224, 225
Rolling, transportation, 24, 26, 27
Rotund, definition, 343
Roundness of particles, 11—12, 15, 17, 69, 295

Sacramento Valley, California, 228
Saddle Creek Limestone Texas, 252, 254, 256
Safe slopes, 324
Salinity, fluid, 53
Salt Creek Bend Shale, Texas, 248
Salt Lake, Utah, 178, 186
Saltation, transportation, 24, 26, 27
San Joaquin Basin, California, 229—230, 236
— — —, Colorado—New Mexico, 218—219
— — Valley, California, 72, 229
San Luis Potosi, Mexico, 150
San Pedro Basin, California, 229
Sand, beach, 298
— boils, 327
— definition, 2, 343
— dunes, 335
— grain fracturing, 56
—, oolitic, 83, 86, 88, 110, 116, 127, 134
—, Ottawa, 58, 61, 62, 65, 340
—, replacement test, 300
—, shear modulus, 342, 495
—, unconsolidated, 56, 68—69
Sandstones, bar, 253, 255
—, bar-finger, 269, 274
—, belt, 288
—, composition of, 18—22
—, cross-bedding, 255
—, delta, 255—257, 258, 259, 264, 269, 270, 276, 288
—, elongate, 252, 256, 259, 260, 265, 268, 269, 271, 272, 273, 288

Sand *(continued)*
—, fluvial, 253—255, 258, 269
—, fracture patterns, 274
—, geometry, 287
—, interdeltaic, 270
—, multistory pattern, 264—266, 271, 273, 288
—, offsetting, 264
—, sheet, 257—258
—, shelf, 270, 271
—, structures, 271—273
Sandy Cay, Bahamas, 136
Sanibel Island, Florida, 129
Santa Ana Branch Shale, Texas, 248
Santa Monica Basin, California, 229
Saskatchewan, Canada, 171
Saturation, effect on compaction, 302, 303, 308
—, effect on strength, 319—320, 321
Schell Creek Mountains, Nevada, 180
Scintillometer, logging, 358
Screen analysis, 3, 5, 13
Secant modulus, 322
Secondary compaction, 453
Sedimentary volumes, 199
Sedwick Limestone, Texas, 248, 252
Self potential, logging, 350, 351, 353—354, 356, 358, 360, 364, 366, 371, 377, 379
— —, — —, clay effect, 354
Sevier Arch, 226
Sevier orogenic belt, 220, 221, 226, 227
Shackelford County, Texas, 248, 252
Shale, 56, 72—73
—, logging, 364, 368, 370, 371, 375
Shannon, Wyoming, 122
Shape factor, 293
—, particle, 293—295, 296
Shattering point pressure, 57
Shear modulus, 342, 495, 515
— strength, 320
— —, beach sands, 298
— —, cohesionless sands, 326—327
Shelf systems, 270, 271
Shock treatment, 320
Siberia, 93
Sierra Madre Range, California, 229
Sierra Nevada, California, 228, 229
Sieving techniques, 3, 5
Silcrete, 330, 331, 335

Silt, definition, 343
Siltstone, 59, 60
Silurian, 155, 156, 183
Similarity relationships, 428—431, 433
Singular solution, mathematical analysis, 458, 459
Size frequency distribution, 6—9
Skeletal loading, 45, 49—51, 71
— sand, 128, 129, 137
Smackover Oolite, 93, 115, 116, 124, 125, 142, 146, 149, 150, 151, 158, 159
Soil mechanics, 44, 403, 412, 413
Soil physics, 425
Solubility, calcium carbonate, 91, 132, 143, 144, 331
—, gypsum, 330
—, iron, 331
—, silica, 331
Solution, pressure, 148, 149, 150, 152—159
—, —, effect on pore volume, 154, 155
Sonic log, 361, 363, 364, 366, 370, 371
— modulus, 323
— transit time, 53
— velocities, 337, 338
Sonoma orogeny, 191
Sorting, 302, 313
—, bimodal histogram, 13
—, binary mixtures, 103—107
— coefficient, 23, 24
—, effect upon compaction, 85
—, mode, 7, 9, 13, 14
—, plus fraction, screen, 5
—, skewness, 7, 9, 13, 14, 85
—, unimodal, 13
Sound propagation log, 361, 363—364, 366, 370, 371
— velocity, *see* Sonic transit time
SP log, *see* Self potential log
Spalling, 123, 135, 143
Spanish Fork Canyon, Utah, 224
Specific gravity, 30—31
— storage, 317
— weight, 30
Spheres, compaction of, 131—132
Sphericity, 12, 16, 17, 69
Ste. Genevieve Formation, 126, 142, 143, 146, 151, 158
Stability, 481

Static loading, 306, 308, 327
—, modulus, 323, 342
Stephens County, Texas, 248, 254, 256
Storage coefficient, 317
Stockwether Limestone, Texas, 248, 252, 254, 256
Stoke's equation, 6
Strength, cementation, 332
—, fabric, 319
—, sediments, 319—326
—, unconfined, 312
Stress, average, 416, 519
—, bulk, 412, 497
—, compaction, 45—53
—, effective, 45, 46, 341, 404, 408, 409, 412, 413, 478
—, external, 414, 416, 497
—, hydrostatic, 419, 420
—, internal, 416
—, neutral, 48—49
—, normal, 416, 504, 519
—, pore, 416
— ratio, 323
— -strain relationship, 493, 494, 497, 498, 499, 501, 502
— tensor, 415, 494, 516
Strain, nuclei, 504
—, nucleus, 506, 507, 508, 516
Structure, residual mapping, 272, 273
—, well log mapping, 349
Stylolites, 153, 155
Stylolitization, 152—159
Subarkose, 21
Sublett Basin, Idaho, 186, 189, 190
Sublitharenite, 20, 21
Submarine canyon, 389
Sub-plastic clays, 332
Subsidence, 169, 176, 278, 280, 281, 403, 406, 460, 508, 509, 510, 511, 512, 513
—, cause, 238—239
—, compaction stages, 234, 235
—, compactional, 266
—, depocenter, 172
—, fluid withdrawal, 6, 236—238
—, geologic, 167—168
—, measurement of, 239—240
—, miscellaneous causes, 238
—, plate tectonics, 234
—, rate, 169

Subsidence *(continued)*
—, surface, 44, 71—72
—, trends, 281—283
Subcrustal flow, 198
Superposition, mathematical analysis, 455, 456
Surface area, determination, 35, 36
— —, effect on compaction, 82
— —, porosity, 304, 305, 306, 307
— —, sediments, 35—36
— texture, particle, 295, 299, 308
Suspension, transportation of particles, 24, 26, 27
Sutured contacts, 72
Synclinoria, 192
Synorogenic, 186

Tadpole plot, dip log, 392, 394, 395
Taiwan, 307
Tangent modulus, 322
Tangential contacts, 122
Tangential-long contact, 72
Taphrogeny, 169
Taylor series, 484
—, Wyoming, 122
Tectonics, 173, 174, 175
—, controlling deposition, 249
—, earthquake, 238
—, forces, effect on fabric, 295—297, 306, 311, 327, 332
—, plate, 168, 169, 170, 181, 183, 191, 193—194, 195, 199, 207, 234
Temperature, distribution, 204
—, effect on compaction, 91
Tennessee, 126, 127, 171
Tertiary, 90, 133, 202, 274
Terzaghi's compaction model, 409—414
Texas, 26, 48, 71—72, 89, 93, 115, 116, 124, 125, 127, 146, 201, 214—218, 247—289
Textural classes, 12, 13
Thermal properties, sediments, 340
Thermoelasticity, 407, 408, 493, 498, 499, 502, 511, 513
Thin beds, logging pattern, 357
Thomas algorithm, 490
Thumb Formation, 227
Tidal channel deposits, logging patterns, 388
Till, definition, 343

SUBJECT INDEX

Tillites, 172
Time of burial, compaction, 92–93
Tortuosity, 34, 35
Traction, transportation by, 24, 26
Transgressive sea, log pattern, 379
Transmissivity coefficients, 317
Transportation process, grain-size
  distribution of sediments, 24—27
Transient drainage, radius, 466, 473, 479
Transit time, sonic log, 363
Triassic, 173, 183, 189, 190, 197
Triaxial loading, 46, 47, 54, 56, 59, 65,
  133, 134
Trinity Valley, Texas, 255
Trough, Beltian, 178–182
—, ocean, 169
Truncation error, 484, 485, 486, 488
Tuff, 70
Turbidites, 213, 228, 229, 230, 381, 386–
  391
— -tractionite deposits, 390
Turbulent flow, 316, 317, 318, 343
— —, definition, 343
Twin Creek Limestone, Idaho, 222
Tyler sieve sizes, 3

Uinta Mountains, Utah, 178
Uncompahgre Highland, Colorado, 218
Unconformity, dip log, 399
Unconsolidated sand, permeability, 34
— —, void ratio, 54
UNESCO, 460
Uniaxial loading, 56, 59, 65
Uniformitarianism, 168
Unimodal sorting, 13
Unordered packing, 161
USSR, 93, 336
Ust Balyk Field, USSR, 93
Utah, 168, 174, 175, 178, 179, 180, 182,
  186, 189, 190, 214, 220, 221, 222,
  224

Vadose zone, 88
Velocity, absolute, 423
—, liquid, 429
—, macroscopic, 421, 422, 425, 429, 444,
  516, 517
—, matrix, 427, 434, 435
—, microscopic, 422, 429, 436, 517
—, solid, 499
Venezuela, 44, 56, 57, 58, 60, 236

Ventura Basin, California, 132, 387
Vibration, effect on porosity, 304, 306,
  308, 330, 336
—, unconsolidated sand, 54
Vibratory roller, 304, 336
Vibroflotation, 329, 336
Virgil Series, Texas, 247, 248, 250—252,
  267
Virgin compaction range, 57
— Mountains, Arizona, 227
Virginia, 143, 171
Viscosity, fluid, 499
—, ideal gas, 468
—, oil, 472
—, real gas, 466, 468
—, water, 412
Viscous deformation, 55
Visual estimation of percentage, 28
Void ratio, 27, 28, 54, 57, 58, 60, 65, 300,
  338, 444, 447, 450, 451, 454, 456,
  515, 517
— —, critical, 325—326
— —, dimensionless, 448, 450, 451, 517
— —, effect of pressure, 326
— — -porosity relationship, 29—30
— —, pseudo, 448, 451
— —, unconsolidated sands, 34, 54
Volcanics, 179, 181
Volume, bulk, 412, 414, 416, 420, 492,
  496, 501, 517
—, matrix, 416, 517
—, pore, 410, 412, 414, 416, 428, 463,
  493, 517

Waldrip Limestone, Texas, 256
— Shale, Texas, 248
Wales, England, 327
Wasatch Mountains, Utah, 224
Washington, 179, 190, 227, 228
Wave agitation, 24, 26, 27
— transmission, 337—340
Wayland Shale, Texas, 248
Welts, 171, 172
Wentworth's scale, 2, 3, 5
West Surgut Field, USSR, 93
West Texas Basin, Texas, 247, 267, 271
West Tuscola Field, Texas, 385
West Virginia, 143, 147, 150
Western Canada sedimentary basin, 230
Wet Mountains, Colorado, 215
Williston Basin, Montana, 174

Willow Tank Formation, 227
Wilmington Oil Field, California, 44, 59—62, 65, 230, 236, 237
Windermere Group, Nevada, 181
Wolfcamp Series, Texas, 247, 248, 250—252, 267
Woodbine Sandstone, 61—62

Wyoming, 26, 122, 220, 221, 222
— overthrust belt, 220

Young's Modulus, 48, 309, 310, 312, 320, 340, 342, 495
Yucatan Peninsula, Mexico, 201—202, 216

Zeugogeosyncline, 216, 218